VOLUME 2

Surfactants in Solution

VOLUME 2

Surfactants in Solution

Edited by

K. L. Mittal

IBM Corporation, Hopewell Junction, New York

and

B. Lindman

University of Lund, Lund, Sweden

PLENUM PRESS • NEW YORK AND LONDON

7299-2803

CHEMISTRY

Library of Congress Cataloging in Publication Data

Main entry under title:

Surfactants in solution.

"Proceedings of an international symposium on surfactants in solution, held June 27–July 2, 1982, in Lund, Sweden"—T.p. verso.
Includes bibliographical references and indexes.
1. Surface active agents—Congresses. 2. Solution (Chemistry)—Congresses. 3. Micelles—Congresses. I. Mittal, K. L., 1945–   . II. Lindman, Björn, 1943–
TP994.S88   1983   668'.1   83-19170
ISBN 0-306-41483-X (v. 1)
ISBN 0-306-41484-8 (v. 2)
ISBN 0-306-41485-6 (v. 3)

Proceedings of an international symposium on Surfactants in Solution,
held June 27–July 2, 1982, in Lund, Sweden

© 1984 Plenum Press, New York
A Division of Plenum Publishing Corporation
233 Spring Street, New York, N.Y. 10013

Printed in the United States of America

PREFACE

This and its companion Volumes 1 and 3 document the proceedings of the 4th International Symposium on Surfactants in Solution
held in Lund, Sweden, June 27-July 2, 1982. This biennial event was
christened as the 4th Symposium as this was a continuation of earlier conferences dealing with surfactants held in 1976 (Albany)
under the title "Micellization, Solubilization, and Microemulsions";
in 1978 (Knoxville) under the title "Solution Chemistry of Surfactants"; and in 1980 (Potsdam) where it was dubbed as "Solution Behavior of Surfactants: Theoretical and Applied Aspects."The proceedings of all these symposia have been properly chronicled.[1,2,3]
The Lund Symposium was billed as "Surfactants in Solution" as both
the aggregation and adsorption aspects of surfactants were covered,
and furthermore we were interested in a general title which could
be used for future conferences in this series. As these biennial
events have become a well recognized forum for bringing together
researchers with varied interests in the arena of surfactants, so
it is amply vindicated to continue these, and the next meeting is
planned for July 9-13, 1984 in Bordeaux, France under the cochair-
manship of K. L. Mittal and P. Bothorel. The venue for 1986 is
still open, although India, inter alia, is a good possibility.
Apropos, we would be delighted to entertain suggestions regarding
where and when these biennial symposia should be held in the future
and you may direct your response to KLM.

The response to these biennial events has been growing and as
a matter of fact we had to limit the number of presentations in
Lund. Even with this restriction, the Lund Symposium program had
140 papers from 31 countries by more than 300 authors. So it is
quite patent that this meeting was a veritable international
symposium both in spirit and contents. It should be added that the
program contained a number of overviews by prominent researchers,
as it is imperative to include some overviews to cover the state-
of-knowledge of the topic under discussion.

As for these proceedings [containing 126 papers (2156 pages)
by 324 authors from 29 countries], these are arranged in nine
parts. Parts I and II constitute Volume 1; Parts III-VI comprise
Volume 2; and Parts VII-IX are the subject of Volume III. Apropos,

the papers in the proceedings have been rearranged (from the order they were presented) in a more logical manner. Among the topics covered include: Phase behavior and phase equilibria in surfactants in solution; structure, dynamics and characterization of micelles; thermodynamic and kinetic aspects of micellization; mixed micelles; solubilization; micellar catalysis and reactions in micelles; reverse micelles; microemulsions and reactions in microemulsions; application of surfactants in analytical chemistry; adsorption and binding of surfactants; HLB; polymerization of organized surfactant assemblies; light scattering by liquid surfaces; and vesicles.

A few salient aspects of these proceedings should be recorded for posterity. All papers were reviewed by qualified reviewers so as to maintain the highest standard. As a result of this, most papers were returned to respective authors for major/minor revisions and some did not pass the review. In other words, these proceedings are not simply a collection of unreviewed papers, rather the peer review was an integral part of the total editing process. It should be added that we had earnestly hoped to include discussions at the end of each paper or group of allied papers, but in spite of constant exhortation, the number of written questions received did not warrant undertaking such endeavor. However, it must be recorded that there were many spontaneous and brisk discussions both formally in the auditorium and informally in other more suitable (more relaxed) places. Most often the discussions were enlightening, but on occasions one could feel some enthalpy as these tended to be exothermic.

Also a general concern was expressed about the possibility of correlating research done in different laboratories. In particular in the microemulsion field it was felt that a few selected, stable and well-defined systems should be chosen for collaborative work between a number of active groups using a variety of techniques. The response to such discussion (initiated by Prof. M. Kahlweit, Göttingen) was very heartening and culminated in the so-called Lund Project (coordinator Prof. P. Stenius, Stockholm), which is a coordinated collaboration between a number of research groups in different countries. A report meeting was hosted by M. Kahlweit in Göttingen in the spring 1983 and further results will be presented in Bordeaux in 1984. In addition, throughout the meeting, small groups of people were seen to be leisurely discussing more specific topics of mutual interest. In other words, there were ample and lively discussions in various forms during the span of this symposium.

Coming back to the proceedings, even a cursory look at the Table of Contents will convince even the most skeptic that the field of surfactants in solutions has come a long way, and all signals indicate that the accelerated tempo of interest and research in this area is going to continue. Also it is quite clear that as we learn more about the amphiphilic molecules, more excit-

ing research areas and pleasant applications will emerge. It should
be added that these proceedings cover a wide spectrum of topics by
a legion of prominent researchers and provide an up-to-date coverage of the field. The coverage is inter- and multidisciplinary and
both overviews and original unpublished research reports are included. Also it should be pointed out that both the aggregation and
adsorption of surfactants are accorded due coverage. These proceedings volumes along with the earlier ones in this vein (total $\sim$ 5000
pages) should serve as a repository of current thinking and research dealing with the exciting field of surfactants in solution.
Also these volumes should appeal to both veteran and neophyte researchers. The seasoned researchers should find these as the source
for latest research results, and these should be a fountainhead of
new research ideas to the tyro.

Acknowledgements: One of us (KLM) is thankful to the appropriate management of IBM Corporation for permitting him to participate in this symposium and to edit these proceedings. His special
thanks are due to Steve Milkovich for his cooperation and understanding during the tenure of editing. Also KLM would like to
acknowledge the assistance and cooperation of his wife, Usha, in
more ways than one, and his darling children (Anita, Rajesh, Nisha
and Seema) for creating only low decibel noise so that Daddy could
concentrate without frequent shoutings. The time and effort of the
reviewers is sincerely appreciated, as the comments from the peers
are a desideratum to maintain standard of publications. We are
appreciative of Phil Alvarez, Plenum Publishing Corp., for his
continued interest in this project. Also we would like to express
our appreciation to Barbara Mutino for providing excellent and
prompt typing service. Our thanks are due to the members of the
local Organizing Committee (Thomas Ahlnäs, Thomas Andersson, Gunnar
Karlström, Ali Khan, Mary Molund, Gerd Olofsson, Nancy Simonsson
and Marianne Swärd) who unflinchingly carried out the various
chores demanded by a symposium of this magnitude. The financial
support of the Swedish Board for Technical Development, the Swedish
National Science Research Council, and the University of Lund is
gratefully acknowledged.

K. L. Mittal   B. Lindman
IBM Corporation   University of Lund
Hopewell Junction, NY 12533   Lund, Sweden

1.   K. L. Mittal, Editor, Micellization, Solubilization and
     Microemulsions, Vols. 1 & 2, Plenum Press, New York, 1977.
2.   K. L. Mittal, Editor, Solution Chemistry of Surfactants,
     Vols. 1 & 2, Plenum Press, New York, 1979.
3.   K. L. Mittal and E. J. Fendler, Editors, Solution Behavior
     of Surfactants: Theoretical and Applied Aspects, Vols. 1 & 2,
     Plenum Press, New York 1982.

CONTENTS

PART III. THERMODYNAMIC AND KINETIC ASPECTS
OF MICELLIZATION

Micellization as a Nucleation Phenomenon with Variable
Surface Tension and Cut-off at Zero Surface Tension
    T. S. Sørensen 709

Computation of the Micelle-Size-Distribution from
Experimental Measurements
    A. Ben-Naim 731

Thermodynamics of Micelle Formation in Aqueous Media =
"Second Virial Coefficient"
    K. S. Birdi, E. Stenby and D. K. Chattoraj 745

Thermodynamics of Micellization and Solubilization of
Pentanol in the System Water + Sodium N-Octanoate +
N-Pentanol at 25°C - Part 2. Solubilization of Pentanol
    J.B. Rosenholm and P. Stenius 755

Dissolution and Micellization of Long-Chain Alkylsulfonic
Acids and Their Sodium Salts in Water
    M. Saito, Y. Moroi and R. Matuura 771

The Effect of Cosolvents on the Formation of Micelles of
Cetyltrimethylammonium Bromide in Aqueous Solutions
    L. G. Ionescu, L. S. Romanesco and F. Nome 789

The Sphere to Rod Transition of CPX and CTAX Micelles in
High Ionic Strength Aqueous Solutions: The Specificity of
Counterions
    G. Porte and J. Appell 805

Salt-Induced Sphere-Rod Transition of Ionic Micelles
    S. Ikeda 825

Quasielastic Light Scattering Studies of the Micelle to
Vesicle Transition in Aqueous Solutions of Bile Salt and
Lecithin
    P. Schurtenberger, N. A. Mazer, W. Känzig and R.
    Preisig 841

The Applicability of Micellar Models to the Activity
Coefficients of Sodium Carboxylates in Aqueous Solution
    G. Douhéret and A. Viallard 857

Kinetic Applications of Bile Salt Amphiphiles
    C. J. O'Connor, R. G. Wallace and B. T. Ch'ng 875

Electrostatics of Micellar Systems
    J. Frahm and S. Diekmann 897

PART IV.   SOLUBILIZATION

Solubilization Equilibria Studied by the FT-PGSE-NMR
Multicomponent Self-Diffusion Technique
    P. Stilbs 917

Selective Solubilization in Aqueous Surfactant Solutions
    R. Nagarajan and E. Ruckenstein 923

Thermodynamics and Mechanisms of Solubilization of Alcohols
in Aqueous Ionic Surfactant Systems
    H. Høiland, O. Kvammen, S. Backlund and K. Rundt 949

Solubilization of Phenothiazine and its N-Alkyl Derivatives
into Anionic Surfactant Micelles
    Y. Moroi, K. Satō, H. Noma and R. Matuura 963

Cholesterol Solubilization and Supersaturation in Bile:
Dependence on Total Lipid Concentration and Formation of
Metastable Dispersions
    D. Lichtenberg, I. Tamir, R. Cohen and Y. Peled 981

Further Investigations on the Micellar Solubilization of
Biopolymers in Apolar Solvents
    P. Meier, V. E. Imre, M. Fleschar and P.L. Luisi 999

PART V. MICELLAR CATALYSIS AND REACTIONS
IN MICELLES

Micellar Effects on Reaction Rates and Equilibria
    L. S. Romsted 1015

Reversed Micellar Enzymology
    A. V. Levashov, Yu.L. Khmelnitsky, N. L. Klyachko and
    K. Martinek 1069

Comparison of Rate Enhancements in Micellar and Nonmicellar
Aggregates
    C. A. Bunton 1093

On the Validity of the Pseudo-Phase Model for Micellar
Catalysis
    L. G. Ionescu and F. Nome 1107

Analysis of the Effect of Micelles and Vesicles on the
Reactivity of Nucleophiles Derived from the Dissociation of
Weak Acids
    H. Chaimovich, J. B.S. Bonilha, D. Zanette and I. M.
    Cuccovia 1121

Micelle-Mediated Luminescence and Chromatography
    L. J. Cline Love, R. Weinberger and P. Yarmchuk 1139

Micellar Effects on Kinetics and Equilibria of Electron
Transfer Reactions
    E. Pelizzetti, E. Pramauro, D. Meisel and E.
    Borgarello 1159

Catalysis of Ester Hydrolysis by Functionalized Counterion
Surfactants
    M. Gobbo, R. Fornasier and U. Tonellato 1169

Specific Micellar Effects in the Temporal Behaviour of
Exctied Benzophenone: Consequences upon the Polymerization
Kinetics
    P. Jacques, D. J. Lougnot and J. P. Fouassier 1177

Quantitative Treatment for Salt Effects and Equilibria
Shifts in Micellar Solutions
    R. Gaboriaud, G. Charbit and F. Dorion 1191

The Nickel(II)-PADA Reaction as a Solubilization Probe in
Anionic Micellar Solutions
    J. R. Hicks and V. C. Reinsborough 1207

The Application of Surfactants in Spectrophotometric Determination of Metal Ions: The Interaction Between Cationic
Surfactants and Some Organic Dyes
    L. Čermáková 1217

        PART VI. ADSORPTION AND BINDING OF SURFACTANTS

Study of the Hydrophobic and the Electrostatic Interactions
in Microphases Concentrated in Surfactant via Adsorption at
Charged Interfaces
    D. Schuhmann, P. Vanel, E. Tronel-Peyroz and H. Raous ... 1233

Thermodynamics of Binding Cationic and Anionic Surfactants
to Binary and Ternary Mixtures of Proteins
    B. K. Sadhukhan and D. K. Chattoraj 1249

An Internal Reflection Infrared Spectroscopic Study of AOT
Adsorption onto Solid Surfaces
    K. McKeigue and E. Gulari 1271

Relation between Adsorption on a Metal Surface and Monolayer
Formation at the Air/Water Interface from Amphiphilic
Solutions
    T. Arnebrant, T. Nylander, P. A. Cuypers, P.-O. Hegg
    and K. Larsson 1291

Quantitative Ellipsometry of Protein Adsorption at
Solid-Liquid Interfaces
    P. A. Cuypers, J. W. Corsel, M. P. Janssen, J. M. M.
    Kop, H. C. Hemker and W. Th. Hermens 1301

Association of Surfactants in Dilute Aqueous Solutions:
Effect on Their Surface Properties
    D. Exerowa and A. Nikolov 1313

Polydisperse Non-Ionic Surfactants: Their Solution Chemistry
and Effect on the Wettability of Solid Surfaces
    G. G. Warr, P. Scales, F. Grieser, J. R. Aston,
    D. R. Furlong and T. W. Healy 1329

Conditions of Phase Separation, Both at the Interface and in
Solution. The Adsorption Isotherm and the Consequence of
Critical Phenomena on the Behaviour of the System
    M. Privat and R. Bennes 1339

Carotenoid Films at the Air/Water Interface
    E. Chifu and M. Tomoaia-Cotisel 1349

Comparison of Interfacial Active Properties of Glycolipids
from Microorganisms
    S. Lang, A. Gilbon, C. Syldatk and F. Wagner 1365

Study of the Interaction Between Surfactants and
Polyacrylamide of Various Hydrolysis Degree
    J. Sabbadin, J. Le Moigne and J. François 1377

Studies on the Interaction of Sodium Saccharin with
Alkyltrimethylammonium Bromides
    S. S. Davis, P. E. Bruce, P. Daniels and L. Feely 1391

Thermodynamic Studies on the Interaction Between Lysozyme
and Sodium n-Dodecyl Sulphate in Aqueous Solutions
    M. N. Jones and P. Manley 1403

About the Contributors 1417

Index xxiii

CONTENTS OF VOLUME 1

PART I: PHASE BEHAVIOR AND PHASE EQUILIBRIA IN SURFACTANT
SOLUTIONS

Principles of Phase Equilibria in Surfactant - Water Systems
    B. Jönsson, P.-G. Nilsson, B. Lindman, L. Guldbrand and
    H. Wennerström   3

On the Phase Behavior of Systems of the Type H_2O - Oil -
Nonionic Surfactant - Electrolyte
    M. Kahlweit and E. Lessner   23

Phase Equilibria in and Lattice Models for Nonionic
Surfactant-Water Mixtures
    J. C. Lang   35

Amphiphilic Aggregates in a Lyotropic Nematic Phase
    J. Charvolin, Y. Hendrikx and M. Rawiso   59

Liquid Crystalline Structures Occurring in Aqueous Systems
of a Totally Fluorinated Fatty Acid and Some of its Salts
    K. Fontell   69

Water ^{2}H and ^{17}O NMR in Dodecylammonium Chloride/D_2O
Lyotropic Mesophases
    B. Robin-Lherbier, D. Canet, J. P. Marchal and
    J. Brondeau   79

The Interaction Between Water and Ethylene Oxide Groups in
Oligo (Ethylene Glycol) Dodecyl Ethers as Studied by ^{2}H NMR
in Liquid Crystalline Phases
    T. Klason and U. Henriksson   93

Surfactant Alkyl Chain Mobility and Order in Micelles and
Microemulsions
    T. Ahlnäs, O. Söderman, H. Walderhaug and B. Lindman 107

Thermodynamics of Partially Miscible Micelles and Liquid
Crystals
    R. F. Kamrath and E. I. Franses 129

^{31}P and ^{2}H NMR Studies of Phase Equilibria in the Three
Component System: Monoolein-Dioleoylphosphatidylcholine-
Water
     H. Gutman, G. Arvidson, K. Fontell and G. Lindblom 143

Micelle Formation and Phase Equilibria of Surface Active
Components of Wood
     P. Stenius, H. Palonen, G. Ström and L. Ödberg 153

Fluid Microstructures of Sodium 4-(1'-Heptylnonyl)
Benzenesulfonate Mixtures
     W. G. Miller, F. D. Blum, H. T. Davis, E. I. Franses,
     E. W. Kaler, P. K. Kilpatrick, K. E. Nietering, J. E.
     Puig and L. E. Scriven 175

Phase Structures and Phase Diagrams of Some Surfactant
Systems with Divalent Counterions. Effect of Ca^{2+} and Mg^{2+}
Counterions on the Stability of Liquid Crystalline Phases
     A. Khan, K. Fontell and B. Lindman 193

A New Optically Isotropic Phase in the Dilute Region of the
Sodium Octanoate – Decanol – Water System
     W. J. Benton and C. A. Miller 205

NMR and Polarized Emission Studies of Cubic Phases and
Model Membranes
     P.-O. Eriksson, L. B.-Å. Johansson and G. Lindblom 219

The Use of Freeze-Fracture and Freeze-Etching Electron
Microscopy for Phase Analysis and Structure Determination of
Lipid Systems
     T. Gulik-Krzywicki, L. P. Aggerbeck and K. Larsson 237

The Structure of Plasma Lipoproteins: Evaluation by X-Ray
and Neutron Small-Angle Scattering
     P. Laggner 259

          PART II. STRUCTURE, DYNAMICS AND CHARACTERIZATION
                           OF MICELLES

The Packing of Amphiphile Chains in Micelles and Bilayers
     D. W. R. Gruen and E. H. B. de Lacey 279

Molecular Organization in Amphiphilic Aggregates
     K. A. Dill 307

The Nature of the Surfactant-Block Model of Micelle
Structure
    P. Fromherz 321

Structure in Micellar Solutions: A Monte Carlo Study
    P. Linse and B. Jönsson 337

Multi-Method Characterization of Micelles
    F. M. Menger 347

Tracer Self-Diffusion Studies of Surfactant Association
    N. Kamenka, M. Puyal, B. Brun, G. Haouche and
    B. Lindman 359

An Introduction to Neutron Scattering on Surfactant Micelles
in Water
    B. Cabane, R. Duplessix and T. Zemb 373

Light Scattering and Small-Angle Neutron Scattering
Investigations of Double-Tailed Surfactants in Aqueous
Solutions
    L. J. Magid, R. Triolo, E. Gulari and B. Bedwell 405

Viscoelastic Detergent Solutions
    H. Hoffmann, H. Rehage, W. Schorr and H. Thurn 425

The Effect of Intermicellar Interactions on Interpretations
of Micellar Diffusivities by Dynamic Light Scattering
    D. F. Nicoli, R. B. Dorshow and C. A. Bunton 455

Laser-Light Scattering Study of Nonionic Micelles in Aqueous
Solution
    V. Degiorgio and M. Corti 471

Light Scattering from Concentrated Solutions of Sodium
Octanoate Micelles
    M. Drifford, T. Zemb, M. Hayoun and A. Jehanno 487

NMR and ESR Studies of Dibutylphosphate Micellar Aggregates
    S. Belaïd and C. Chachaty 501

Conformational Change of Surfactants Due to Association;
Raman Scattering and Carbon-13 NMR Studies
    H. Okabayashi and K. Matsushita 517

An NMR Study of Paramagnetic Relaxation Induced in Octanoate
Micelles by Divalent Ions
    T. Zemb and C. Chachaty 527

ESR Study of Spin Labels in Surfactant Micelles
    P. Baglioni, M. F. Ottaviani, G. Martini and
    E. Ferroni 541

Spin Label Study of Molecular Aggregates
    M. Schara and M. Nemec 559

Micellar Structure and Water Penetration Studied by NMR and
Optical Spectroscopy
    K. A. Zachariasse, B. Kozankiewicz and W. Kühnle 565

Solubilization and Water Penetration into Micelles and Other
Organized Assemblies as Indicated by Photochemical Studies
    D. G. Whitten, J. B. S. Bonilha, K. S. Schanze and
    J. R. Winkle 585

Critique of Water Penetration Studies in Micelles using
Extrinsic Probes
    K. N. Ganesh, P. Mitra and D. Balasubramanian 599

The Size of Sodium Dodecyl Sulfate Micelles with Various
Additives: A Fluorescence Quenching Study
    M. Almgren and S. Swarup 613

Fluorescence Quenching Aggregation Numbers in a Non-Ionic
Micelle Solution
    J.-E. Löfroth and M. Almgren 627

Fluorescence Quenching Equilibria Studies in Ionic Micelles
in Aqueous Media
    K. S. Birdi, M. Meyle and E. Stenby 645

Fluorescence Quenching in Micellar Systems
    F. C. De Schryver, Y. Croonen, E. Geladé, M. Van der
    Auweraer, J. C. Dederen, E. Roelants, and N. Boens 663

FT-IR Studies of Aqueous Surfactants: The Temperature
Induced Micelle Formation
    H. H. Mantsch, V. B. Kartha and D. G. Cameron 673

About the Contributors 691

Index xxiii

CONTENTS OF VOLUME 3

PART VII. REVERSE MICELLES

Kinetic Consequences of the Self Association Model in
Reversed Micelles
    C. J. O'Connor and T. D. Lomax 1435

Dynamics of Reversed Micelles
    Z. A. Schelly 1453

Reverse Structures in a p-Nonylphenolpolyethyleneglycol
(9.6 Mole Ethylene Oxide) - Water System
    A. Derzhanski and A. Zheliaskova 1463

Reactivity Studies in A.O.T. Reverse Micelles
    M. P. Pileni, J. M. Furois and B. Hickel 1471

Preparation of Colloidal Iron Boride Particles in the
CTAB-n-Hexanol-Water Reversed Micellar System
    N. Lufimpadio, J. B. Nagy and E. G. Derouane 1483

PART VIII. MICROEMULSIONS AND REACTIONS IN MICROEMULSIONS

Microemulsions - An Overview
    Th. F. Tadros 1501

The Water-in-Oil Microemulsion Phenomenon: Its Understanding
and Predictability from Basic Concepts
    H.-F. Eicke, R. Kubik, R. Hasse and I. Zschokke 1533

Phase Behavior of Microemulsions: The Origin of the Middle
Phase, of Its Chaotic Structure and of the Low Interfacial
Tension
    E. Ruckenstein 1551

Influence of Cosurfactant Chemical Structure upon the Phase
Diagram Features and Electrical Conductive Behavior of
Winsor IV Type Media (So-Called Microemulsions)
        M. Clausse, J. Peyrelasse, C. Boned, J. Heil, L.
        Nicolas-Morgantini and A. Zradba 1583

Fluorescence Probe Study of Oil in Water Microemulsions
        R. Zana, J. Lang and P. Lianos 1627

Characterization of Microemulsion Structure Using Multi-
Component Self-Diffusion Data
        B. Lindman and P. Stilbs 1651

Photon Correlation Techniques in the Investigation of Water-
in-Oil Microemulsions
        J. D. Nicholson and J. H. R. Clarke 1663

Water/Oil Microemulsion Systems Studied by Positron
Annihilation Techniques
        S. Millán, R. Reynoso, J. Serrano, R. López and L. A.
        Fucugauchi 1675

Zeta Potential and Charge Density of Microemulsion Drops
from Electrophoretic Laser Light Scattering - Some
Preliminary Results
        S. Qutubuddin, C. A. Miller, G. C. Berry, T. Fort, Jr.
        and A. Hussam 1693

Low Temperature Dielectric Properties of W/O Microemulsions
and of their Highly Viscous Mesophase
        D. Senatra and C. M. C. Gambi 1709

Mutual and Self Diffusion Coefficients of Microemulsions
from Spontaneous and Forced Light Scattering Techniques
        A. M. Cazabat, D. Chatenay, D. Langevin, J. Meunier and
        L. Leger 1729

Percolation and Critical Points in Microemulsions
        A. M. Cazabat, D. Chatenay, P. Guering, D. Langevin,
        J. Meunier, O. Sorba, J. Lang, R. Zana and M. Paillette 1737

Structural and Dynamic Aspects of Microemulsions
        P. D. I. Fletcher, A. M. Howe, N. M. Perrins, B. H.
        Robinson, C. Toprakcioglu and J. C. Dore 1745

Structure of Nonionic Microemulsions by Small Angle Neutron
Scattering
        J. C. Ravey and M. Buzier 1759

Fluctuations and Stability of Microemulsions
    S. A. Safran 1781

Influence of Salinity on the Composition of Microemulsion
Pseudophases: Correlation Between Salinity and Stability
    J. Biais, B. Clin, P. Lalanne and M. Barthe 1789

Phase Studies and Conductivity Measurements in Micro-
emulsion-Forming Systems Containing a Nonionic Surfactant
    T. A. Bostock, M. P. McDonald and G. J. T. Tiddy 1805

Existence of Transparent Unstable Solutions in Three and
Four Components Surfactant Systems
    T. Assih, P. Delord and F. C. Larché 1821

Theory of Phase Continuity and Drop Size in Microemulsions
II. Improved Method for Determining Inversion Conditions
    J. Jeng and C. A. Miller 1829

Effect of the Molecular Structure of Components on Micellar
Interactions in Microemulsions
    D. Roux, A. M. Bellocq and P. Bothorel 1843

The Importance of the Alcohol Chain Length and the Nature of
the Hydrocarbon for the Properties of Ionic Microemulsion
Systems
    E. Sjöblom and U. Henriksson 1867

Transport of Solubilized Substances by Microemulsion
Droplets
    C. Tondre and A. Xenakis 1881

Light Scattering and Viscometric Investigations of Inverse
Latices Formed by Polymerization of Acrylamide in Water-
Swollen Micelles
    Y. S. Leong, S. J. Candau and F. Candau 1897

Application of the Ion-Exchange Model to O/W
Microemulsions
    R. A. Mackay 1911

    PART IX. GENERAL OVERVIEWS AND OTHER PAPERS

HLB - A Survey
    P. Becher 1925

Polymerization of Organized Surfactant Assemblies
    J. H. Fendler 1947

Light Scattering by Liquid Surfaces
      D. Langevin, J. Meunier, D. Chatenay 1991

Surface Charge Density Evaluation in Model Membranes
      C. Stil, J. Caspers, J. Ferreira, E. Goormaghtigh and
      J-M. Ruysschaert 2015

Breakdown of the Poisson-Boltzmann Approximation in Polyelectrolyte Systems: A Monte Carlo Simulation Study
      B. Jönsson, P. Linse, T. Åkesson and H. Wennerström 2023

Colloidal Stability of Liposomes
      L. Rydhag, K. Rosenquist, P. Stenius and L. Ödberg 2039

Fast Dynamic Phenomena in Vesicles of Phospholipids During
Phase Transitions
      V. Eck and J. F. Holzwarth 2059

Dynamic Light Scattering Study of DMPC Vesicles Coagulation
Around the Phase Transition of the Aliphatic Chains
      D. Sornette and N. Ostrowsky 2081

The Effect of Ginseng Saponins on Biochemical Reactions
      C. N. Joo 2093

Determination of Very Low Liquid-Liquid Interfacial Tensions
from the Shapes of Axisymmetric Menisci
      Y. Rotenberg, S. Schürch, J. F. Boyce and A. W.
      Neumann 2113

Mechanism of Using Oxyethylated Anionic Surfactant to
Increase Electrolyte Tolerance of Petroleum Sulfonate
      Y-C. Chiu 2121

Local Anesthetic-Membrane Interaction: A Spin Label Study of
Phenomena that Depend on Anesthetic's Charge
      S. Schreier, W. A. Frezzatti, Jr., P. S. Araujo and
      I. M. Cuccovia 2145

About the Contributors 2157

Index xxiii

Part III
Thermodynamic and Kinetic Aspects
of Micellization

MICELLIZATION AS A NUCLEATION PHENOMENON WITH VARIABLE

SURFACE TENSION AND CUT-OFF AT ZERO SURFACE TENSION

Torben Smith Sørensen

Fysisk-Kemisk Institut
Technical University of Denmark
DK-2800 Lyngby, Denmark

It is shown that micellar aggregation numbers can
be determined by a zero surface tension principle for
ionic as well as nonionic micelles. Above the cmc the
micelle nucleates and grows until the increase in surface density of polar or ionic heads leads to a sum of
short range and electric surface tension which is zero.
The micelles cannot be larger, since the surface tension
then becomes negative and the micelles are dissolved by
the Brownian motions. For ionic micelles a transition
range of ionic strengths between low aggregation numbers in the Coulombic limit and high aggregation numbers in the nonionic limit is predicted in accordance
with experimental facts.

The cmc behaviour of micelles can be understood
by generalizing classical nucleation theory to take into account a variable surface tension. The major part
of the difference between cmc of nonionic micelles and
the equilibrium concentration of the corresponding alkane is simply the surface tension term in the nucleation theory. For a precise evaluation of the surface
tension of the critical micelle, the interfacial tension of a plane and pure oil-water interface has to be
corrected for surface pressure and for curvature. For
ionic micelles, the influence of the surface potential
and the electrical contribution to surface tension has
to be accounted for. Experimental data for cmc of ionic
micelles seem to indicate that the real surface potentials are 17 % higher than calculated by the sperical,
non-linear Poisson-Boltzmann equation in the case of
cationic micelles and 40 % higher in the case of an-

ionic micelles. Anionic micelles have large, hydrated
ionorganic cations as counterions which are more efficient in depressing the dielectric constant near the
micellar surface than small, unhydrated ionorganic anions. This explains the larger deviation for the anionic micelles.

INTRODUCTION

The clusters of amphiphilic molecules called micelles have
intriguing properties from a physico-chemical point of view, since
they are positioned somewhere in the "no mans land" between macroscopic and microscopic systems. The radial extension of the globular
micelles is typically of the order of 20 Å with aggregation numbers of the order of 100 molecules, though much higher aggregation
numbers may be found for nonionic micelles and ionic micelles at
high salt concentrations.

It might be that a micelle is quite unstable and it has certainly a rapid exchange of material with the surrounding solution.
Nevertheless, it is impossible to think of an instantaneous "snap-
shot" of a micellar solution without thinking also about some instantaneous surfaces separating what is basically inside the micelle from what is basically outside. As with macroscopic interfaces
there is a certain degree of arbitrariness in the location of the
Gibbs dividing surface, since the density profiles in the transition zones can easily extend over as much as 10 Å.[1] For micelles
this means that the major part of the micelle (in volume) is actually an interfacial transition zone between hydrocarbon and aqueous solution with the polar head groups distributed approximately
"Gaussian" around some mean distance from the center. Because of
the small number of molecules, fluctuations of the "macroscopic"
parameters will be quite large, for example the surface tension
will experience vivid fluctuations. But (statistical) thermodynamics is still valid for the time averages, and if one denies the
micelle the right to have a surface tension, one is actually say-
ing that the mean value of the surface tension is zero, which is
certainly a most interesting postulate!

If a micelle had a surface tension close to that for pure oil-
water interfaces (50 mN/m at 25 °C) the (mean) Laplace overpressure
in a sperical micelle of radius 20 Å would be as much as 493 atm.
Such a Laplace overpressure has been invoked by some authors[2,3] in
order to explain for example the two fold decrease in O_2-solubility
in SDS-micelles compared to pure nonane or the fact that the ratio
of solubilized n-alkane decreases with the increase in volume of
the solubilizate. However, Menger[4] has given an alternative explanation of these facts and several others[5] in terms of a water penetration model of the micelle.

Obviously, both models cannot be correct at the same time, since they both predict the same facts. The viewpoint of Menger and others is that the "fluid drop" model is not a good model for micelles. I would like here to turn the arguments in a different direction, however: The assumption of a fluid droplet not much bigger than an interfacial transition region is in perfect agreement with the water penetration model of Menger, but then a mean surface tension must exist. Its value, however, has then to be quite close to zero in order to avoid the excessive Laplace overpressure. In comparison, it should be mentioned that Nielsen and Sarig[6] have studied nucleation of tribromomethane in an immiscibility gap in the tribromoethane-water-methanol system. Calculating the radii of critical nuclei from the Volmer-Becker-Döring theory of homogeneous nucleation, they found values of order of 20 Å. The calculated surface tension of the droplets was in perfect agreement with the macroscopic surface tension for a plane interface corresponding to the tie line considered, when it was corrected for curvature with a Tolman formula[7] with a reasonable correction distance ($\delta = 2.4$ Å).

The idea put forward by the present author in a previous publication[8] is the following: The micelle has to be treated as a nucleating oil droplet with a surface tension, which varies with the aggregation number. If the mean shape of the micelle is assumed sperical, with the head groups roughly in the interfacial region, the concentration of polar head groups on the surface will increase roughly proportional to the cubic root of the aggregation number. Considering the usual results found for adsorbed monolayers, the surface pressure will increase due to ideal gas effects and repulsive steric and electrostatic forces. Thus, the surface tension of the micelle decreases with increasing aggregation number. When the surface tension becomes close to zero, the micelle becomes very unstable due to the fluctuating impacts of the solvent molecules (Brownian motions) and it is easily disrupted. When the surface tension becomes negative, dissolution of the micelle occurs spontaneously. Negative surface tensions are only possible in transitory states. As a classical example from electrocapillary phenomena, Ilkovic[28] showed that when a mercury-water interface with adsorbed quaternary ammonium ions was brought to regions with negative surface tension (potential of the Hg-drop less than $\sim$ -2 V), a brown cloud of colloidal mercury is spontaneously formed in the aqueous solution adjacent to the interface. Therefore, phase separation will be interrupted at zero surface tension (ZST) and due to the principle of microscopic reversibility, the micelles will be in a very dynamic state dividing and coalescing with equal rates and taking all kinds of strange shapes with the "fluctuating peanut" shape as the predominant. Still, however, the mean shape will be spherical, so this shape is appropriate for calculation of mean values of thermodynamic parameters. The purpose of the present publication is to show how these ideas work for aggregation numbers as well as for cmc-values for nonionic and ionic micelles. A generali-

zation of classical nucleation theory is found to be necessary for
this aim.

1. FACTORS INFLUENCING THE SURFACE TENSION OF MICELLES

The interfacial tension of a pure and plane oil-water interface at 25 °C will be set equal to $\gamma_o = 50$ mN/m. In comparison, the
interfacial tension of clusters of amphiphilic molecules will be
lowered because of the presence of head groups and because of the
curvature. Another subdivision of the interfacial tension is to assume that γ is the sum of a contribution from short range forces
and a contribution from electrostatic forces:

$$\gamma = \gamma_{sr} + \gamma_E \quad (1)$$

Neglecting first the electric surface tension, the simplest possible
model is to consider the head groups at the micellar interface in
analogy with an adsorbed monolayer at an oil-water interface, more
specifically as a two dimensional gas with excluded area (a_{ex}) as
the sole interaction[8,9]. Furthermore, curvature is dealt with by
means of Tolman's approximate formula[7]. The result is:

$$\gamma_{sr}^{micelle} \approx \{\gamma_o + \frac{kT}{a_{ex}} \ln[1-\theta]\} \frac{R_{st}}{R_{st}+2\delta} \quad (2)$$

The fraction of the area covered by head groups is $\theta = a_{ex}/a$ where a
is the actual area/head group. R_{st} is the radius of the surface of
tension. Tolman considered small droplets in equilibrium with their
vapor. In that case δ is the distance between the surface of tension
and the surface of no superficial density. It is positive when the
surface of tension is inside the surface of no superficial density.
This is the case for droplets whereas for bubbles the opposite is
true, and the surface tension is increased by curvature. The interpretation of δ for droplets in a surrounding solution is more subt-
le[10,11]. A simplified interpretation of δ for a (nonionic) micelle
is shown on Fig.1.

For ionic micelles, there will be a negative contribution to
the interfacial tension from the electrostatic forces as is known
from the theory of electrocapillarity[12], the theory of colloid stability[13] and the thermodynamic and electrostatic theory of the electric double layer[14]. The simplest way to show this is to start with
the generalized Gibbs-Duhem equation for the interface (with the
electrochemical potential of the adsorbed species), assume an activity corresponding to a Langmuir-isotherm ($kT\ln(\theta/1-\theta)$) and inte-

grate from γ_0 to the final γ in two steps (reversible adsorption of neutral species to the final θ and reversible charging of the species with fixed θ to the final surface charge and potential regulating the external amphiphile concentration correspondingly in both steps, see ref.11 for details). The final result is for a plane interface (σ=charge density on surface, ψ_s=surface potential):

$$\gamma = \gamma_0 + \frac{kT}{a_{ex}} \ln(1-\theta) - \int_0^{\psi_s} \sigma \, d\psi_s \quad (3)$$

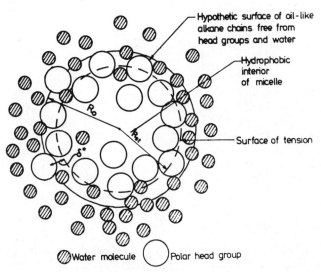

Figure 1. Schematic drawing of the mean configurations of polar heads and water molecules, and the location of the surface of tension in the present model. Hydrocarbon backbones have been left out for clarity. For ionic micelles charges have to be placed on head groups and ionic clouds to be distributed in the external solution.

The last term is γ_E and is clearly negative. For spherical interfaces we correct only the short range part for curvature (like in equation (2)), since the charge-potential integration automatically incorporates curvature effects.

However, the detailed theory of how to position the surface of tension and how to correct for curvature at charged interfaces has not yet been worked out. We are therefore bound to use reasonable approximations. An alternative to the charging procedure was devised by Sanfeld[14]. He suggested the use of the components of the Maxwell stress tensor in combination with the Bakker formula[15] in order to calculate γ_E. For plane interfaces the Sanfeld method gives identical values of γ_E as the charging method, but for spherical interfaces it is necessary to use some tricks in order to have equivalent results (see ref.(11)).

2. AGGREGATION NUMBERS OF NONIONIC MICELLES FROM ZST-PRINCIPLE

It is simplest to begin with aggregation numbers of nonionic micelles. Following the ideas outlined in the Introduction and in ref.(8) we shall assume the aggregation numbers to be governed by a zero surface tension (ZST) principle. In that case, the Tolman correction for curvature has clearly no importance. From equation (2) we obtain the condition for having zero surface tension in terms of the aggregation number (m):

$$m_o = \frac{36 \pi V_{HC}^2}{a_{ex}^3}[1-\exp(-\frac{\gamma_o a_{ex}}{kT})]^3 \quad (4)$$

V_{HC} is the mean volume of one hydrocarbon chain in the liquid hydrocarbon phase. In deriving equation (4), we have assumed that the radius appropriate for the calculation of the area per head group and θ is the surface of tension radius R_{st} and that this radius can be approximated by the hypothetical radius R_0 of a droplet of hydrocarbon of the same chain length as in the amphiphile and containing the same number of hydrocarbon molecules as the micelle. For n-alkanes at 25 °C we have approximately:

$$V_{HC}/Å^3 = 58 + 26.9 \cdot n_c \quad (5)$$

n_c is the number of carbon atoms in the alkane or in the hydrophobic part of the amphiphile. In general R_{st} will not coincide with R_0 and we may write:

$$R_{st} = R_o - \delta^* \quad (6)$$

If water penetrates the micelle to a high degree, we would expect δ^* to be positive and to have a magnitude of several Å. However, since R_0 is a quite arbitrary radius (it is not precisely the radius of no superficial density of oil) and since the Tolman correction distance δ has a complex meaning in solutions, we cannot simply identify δ^* and δ. The area per head group or θ in equation (2) should be calculated with reference to R_{st} and not R_0. The ZST-principle leads to:

$$m_0(\delta^*) = m_0(\delta^*=0)\left[1 - \frac{\delta^*}{\sqrt[3]{3V_{HC}m_0(\delta^*)/4\pi}}\right]^6 \quad (7)$$

$m_0(\delta^*=0)$ is the aggregation number calculated using equation (4). The correction factor in equation (7) is just a correction for the difference in total area between $4\pi R_0^2$ and $4\pi(R_0-\delta^*)^2$.

Table I.   Calculated Aggregation Numbers m_0 of Nonionic Micelles.

n_C	$a_{ex}=15\text{Å}^2$			$a_{ex}=20\text{Å}^2$			$a_{ex}=30\text{Å}^2$		
$\delta^*=$	0	2.5Å	5Å	0	2.5Å	5Å	0	2.5Å	5Å
8	1473	1003	549	800	489	180	289	133	–
10	2111	1546	1001	1146	772	410	414	226	–
12	2863	2203	1565	1555	1117	695	561	341	123
14	3729	2974	2242	2025	1524	1041	731	479	235
16	4709	3859	3033	2557	1993	1447	923	639	365

The simple calculation of m_0 in Table I has some pedagogic merit, since the aggregation numbers are of the correct order of magnitude for nonionic micelles. Furthermore, the Table indicates that the aggregation numbers should grow with increasing chain length of the hydrophobic part of the amphiphile, and that the aggregation numbers should diminish with increasing bulkiness of the head group. Both effects are observed with oligo oxyethylene alkohol ether surfactants ($C_{n_C}EO_{n_{EO}}$) where m increases with n_C and decreases with n_{EO} (see ref.16 and ref.8 Fig.5).

There is no absolute agreement between the data and any of the columns in Table I, however. Compare with the column with heading m(exp) in Table II. Most probably the reason is that the coverage fraction (θ) at zero surface tension is too large in the case of nonionic micelles for the Langmuir adsorption isotherm to be a valid approximation. We shall therefore proceed in a somewhat different way. Zero surface tension is reached in general, when the surface pressure of amphiphiles adsorbed on an oil-water interface (Π) is equal to the interfacial tension of a pure and plane oil-water interface γ_0. Curvature corrections are immaterial at ZST.

$$\Pi(a_o) = \gamma_o \quad (8)$$

Since Π is only a function of the area per head group (a), this area has to be constant for all nonionic micelles (equal to a_o). Remembering that a_o is measured on the surface R_{st} we have:

$$\frac{a_o}{4\pi} = \frac{\beta^2}{m_o^{1/3}}(1 - \frac{\delta^*}{\beta\sqrt[3]{m_o}})^2 \equiv F(m_o, \delta^*) \quad (9)$$

We have defined:

$$\beta \equiv \sqrt[3]{\frac{3V_{HC}}{4\pi}} \quad (10)$$

The hydrophobic radius is therefore given by:

$$R_o = \beta \cdot \sqrt[3]{m_o} \quad (11)$$

Assuming different values of δ^*, $F(m(exp), \delta^*)$ can now be calculated from experimental data. This has been done in Table II for $\delta^*=0$, 2.5 Å and 5 Å. For $\delta^* = 5$ Å, F is reasonably independent of m(exp). In comparison with data it should be remembered that measurements of aggregation numbers of nonionic micelles are quite uncertain due to easy aggregation of micelles to form "supermicelles" (sphere to rod transition). The value 10500 reported for $n_C=16$ and $n_{EO}=6$ is most certainly the aggregation number of such aggregated micelles. Nevertheless, from the mean value of F we calculate a value of $a_o=23\pm6$ Å^2. This area per head group cannot be far from the excluded area, and this explains why the Langmuir formula did not work for the nonionic micelles. Also, the distance $\delta^*=5$ Å does not seem unrealistic if water is assumed to penetrate deeply into the micelle. The relative constancy of the F-values for $\delta^*=5$ Å therefore lends credit to the hypothesis that the area per head group and the surface pressure are constant for all nonionic micelles with identical head groups.

3. AGGREGATION NUMBERS OF "PRIMITIVE" IONIC MICELLES

A "primitive" ionic micelle is a micelle which does not know that the Debye-Hückel theory is quantitatively wrong except at very low surface potentials. Again, the main raison d'etre for such a species is its pedagogic value. When the linearized Poisson-Boltzmann equation is solved in spherical coordinates, a linear relation

Table II. Radii and F-values Calculated from Experimental Aggregation Numbers[16] for Hexaoxyethylene alkanol ethers (n_{EO}=6) as a Function of n_C and δ^*.

n_C	m(exp)	R_o Å	F(m,0) Å^2	F(m,2.5Å) Å^2	(F(m,5Å) Å^2
8	32	12.8	5.10	3.30	1.89
10	73	17.9	4.37	3.23	2.27
12	400	33.1	2.74	2.35	1.98
14	3100	68.5	1.51	1.41	1.30
16	2430	65.7	1.78	1.64	1.52
16	(10500)	(107.0)	(1.09)	(1.04)	(0.99)

Mean : 1.8 ± 0.5 Å^2

between the surface potential and the surface charge density is obtained. Using the charging method described in Section 1, we obtain
for the electric surface tension:

$$\gamma_E = -\frac{1}{2\varepsilon a_E^3}(1+\kappa a_E)^{-1} \cdot (\frac{m e_o^z A}{4\pi})^2 \quad (12)$$

The aggregation number is m, the charge of the amphiphile is z_A
(units of elementary charge e_o), ε is the absolute permittivity of
pure water, a_E the radial position of the surface charge (we shall
put $a_E=R_o$) and κ is the inverse Debye-Hückel screening length. For
γ_{sr} we assume the Langmuir expression for simplicity. From the ZST-
principle we then obtain the following relation between the aggregation number (m_o) and κ:

$$\kappa \beta = \frac{1}{\sqrt[3]{m_o}}\left[(\frac{a_{ex}}{\lambda^2}) \frac{m_o}{8\pi[\frac{\gamma_o a_{ex}}{kT}+\ln\{1-\frac{a_{ex}\sqrt[3]{m_o}}{4\pi\beta^2}\}]} - 1\right] \quad (13)$$

The length β was defined in equation (10) and the charging length
of the micelle is defined as:

$$\lambda \equiv \sqrt{\frac{3kT\epsilon V_{HC}}{e_o^2}} \quad (14)$$

From equation (13) κ can be calculated assuming a given aggregation number. When m_o increases, we reach a point where the expression in the denominator becomes zero. This leads to $\kappa \to +\infty$ and to equation (4). We shall call this the nonionic limit, where the double layer is fully contracted towards the micelle. On the other hand, when m_o decreases we reach a point where κ becomes zero and then negative. The m_o value corresponding to $\kappa=0$ is then the minimum aggregation number found at very small ionic strengths. This limit will be called the Coulombic limit. This limit can be found by iteration on:

$$m_o(\kappa=0) = 8\pi(\frac{\lambda^2}{a_{ex}})[\frac{\gamma_o a_{ex}}{kT} + \ln\{1 - \frac{a_{ex}\sqrt[3]{m_o(\kappa=0)}}{4\pi\beta^2}\}] \quad (15)$$

Since the argument of the logarithm is usually close to unity, we have the following successive approximations:

$$m_o(\kappa=0) \approx 8\pi[\frac{\gamma_o\lambda^2}{kT} - (\frac{\lambda}{\beta})^2 \frac{\sqrt[3]{m_o(\kappa=0)}}{4\pi}] \approx 8\pi\frac{\gamma_o\lambda^2}{kT} \quad (16)$$

Iteration on the first approximation usually yields very precise results. The last approximation is within 10 % error. Thus, the aggregation number in the Coulombic limit is largely independent of the excluded area.

Fig.2 shows some calculations of m_o as a function of the inverse Debye-Hückel length (square root of ionic strength) for ionic micelles with $|z_A|=1$ and $n_C=8$, 10, 12, 14, 16. It is seen that the aggregation numbers increase sharply from the Coulombic limit at low ionic strengths to the nonionic limit at high ionic strengths. Such increases of aggregation numbers have indeed been observed by Mazer et al[17] (SDS, light scattering) and by Dalsager et al[18,19] (SDS, TTAB, membrane osmometry) and the aggregation numbers found are in the range from 100 (0.1 mol/dm³ added salt) to 500-700 (0.8 mol/dm³).

However, the calculation of concentration of added salt from the κ-values in Fig.2 is not worthwhile, since the simple Debye-Hückel equations cannot give any quantitative agreement at this point.

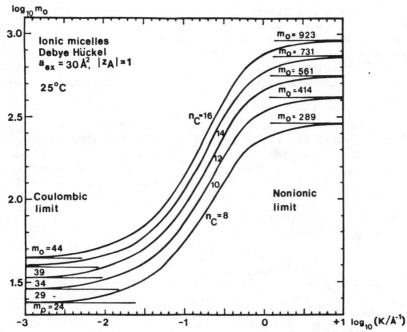

Figure 2.   Calculated aggregation numbers of Debye-Hückel ionic micelles by the ZST-principle. Abscissa: Inverse Debye-Hückel screening length.

4. CMC AS A CRITICAL NUCLEATION CONCENTRATION

Turning next to cmc, it has often been said that the cmc-
transition bears resemblance to a (not very sharp) phase transition.
It seems worthwhile then to treat micellization as a nucleation phenomenon with variable surface tension during nucleation.[8,11] We
shall here follow the outline of classical nucleation theory given
in ref.(10) with modifications to our present purpose.

The point of departure is the Helmholtz' free energy of formation of one "droplet" with aggregation number m from a large
amount of solution:

$$\Delta F_m = m \cdot \{ -A_m^{HC} + \Delta \mu_m + P_m + z_A e_o \psi_s (m) \} + \frac{1}{3} \gamma_m \cdot A_m \quad (17)$$

In equation (17), $\mathcal{A}_m^{HC}$ is the driving affinity for the <u>hydrocarbon</u> moiety of the amphiphile to go from the solution into the droplet with radius of the surface of tension $R_{st}=R_m$ with a corresponding Laplace overpressure Δp_m. Further, $\Delta\mu_m$ is the excess chemical potential in the m-cluster due to small phase effects. This might include a positive contribution from van der Waals forces like in thin films (ref.(9)) or positive contributions from trans-gauche statistical mechanics of the HC-chains with space restrictions as considered by Dill[20] or Gruen and De Lacey[21]. P_m is the contribution to the free energy of transfer from the polar head group arising from the nonelectrostatic short range forces. The electrostatic forces contribute with the term $z_A \cdot e_o \cdot \psi_s(m)$. The interfacial term $\frac{1}{3} \cdot \gamma_m \cdot A_m$ involves the area of the surface of tension and the interfacial tension γ_m. The affinity $\mathcal{A}_m^{HC}$ is derived from the usual "hydrophobic affinity" $\mathcal{A}_\infty^{HC}$ for a macroscopic oil phase ($R=\infty$) by means of a Laplace-Kelvin correction:

$$\mathcal{A}_m^{HC} = \mathcal{A}_\infty^{HC} - \frac{2\gamma_m \cdot v_{HC}}{R_m} \quad (18)$$

For the bulk affinity for n-alkanes we assume[22]:

$$\frac{\mathcal{A}_\infty^{HC}}{kT} = \ln \frac{c_A}{c_A^*} = 4.11 + 1.492 \cdot n_C + \ln \frac{c_A}{55.5} \quad (19)$$

The external concentration of amphiphile is c_A, and c_A^* is the solubility of the alkane corresponding to the amphiphile.

As is well known from nucleation theory, the <u>critical nucleus</u> is the nucleus which is in unstable equilibrium with the surrounding solution. Thus, the total driving affinity is zero:

$$+\mathcal{A}_{cr}^{HC} - \Delta\mu_{cr} - P_{cr} - z_A \cdot e_o \cdot \psi_s(cr) = 0 \quad (20)$$

The free energy of formation of the critical droplet is therefore given by the usual expression:

$$\Delta F_{cr} = \frac{4\pi}{3} \gamma_{cr} R_{cr}^2 \quad (21)$$

Combining equations (18) and (20) we obtain for the critical radius:

$$R_{cr} = \frac{2\gamma_{cr} v_{HC}}{\mathcal{A}_\infty^{HC} - \Delta\mu_{cr} - P_{cr} - z_A \cdot e_o \cdot \psi_s(cr)} \quad (22)$$

This expression may be introduced into the free energy of formation given by equation (21), and the concentration of critical nuclei may then be calculated from the law of mass action:

$$\ln \frac{c_{cr}}{55.5} = - \frac{\frac{16\pi}{3}(\frac{\gamma_{cr}}{kT})^3 v_{HC}^2}{[\ln \frac{c_A}{c_A^*} - \frac{\Delta\mu_{cr}+P_{cr}}{kT} - \frac{z_A \cdot e_o \cdot \psi_s(cr)}{kT}]^2} \quad (23)$$

From the theory of homogeneous nucleation it is known that the concentration of critical nuclei which is able to induce nucleation at a measurable rate is very low indeed. For nucleation of ice in water c_{cr} is about 10^{-18} mol/dm^3 when freezing is initialized (-40°C). For nucleation of water droplets from humid air $c_{cr}=10^{-22}$ mol/dm^3 (see ref.(10)). Not knowing the number for nucleating oil droplets in water we assume $c_{cr}=10^{-20}$ mol/dm^3. The exact number is not very important, since $c_{cr}/55.5$ is a sharply rising function of $\ln(c_A/c_A^*)$ in equation (23). Our assumption now is that insertion of $c_{cr}=10^{-20}$ mol/dm^3 in equation (23) leads to c_A=cmc. With a final reformulation we have:

$$\ln \frac{cmc}{55.5} = -4.11-1.492n_C+2.45\cdot10^{-2}(\frac{\gamma_{cr}}{\gamma_o})^{3/2}\cdot v_{HC}$$

$$+ \frac{\Delta\mu_{cr}+P_{cr}+z_A \cdot e_o \cdot \psi_s(cr)}{kT} \quad (24)$$

(γ_o=50 mN/m, v_{HC} in Å^3/HC-chain, T=298,15 K)

The critical nucleus corresponding to c_A=cmc will be called the critical micelle. The radius of the critical micelle is found from equation (22) with c_A=cmc as calculated from equation (23) with $c_{cr}=10^{-20}$ mol/dm^3. The result is:

$$R_{cr}(cmc) = \frac{2\gamma_o}{2.45\cdot10^{28}(m^{-3})kT}\sqrt{\frac{\gamma_o}{\gamma_{cr}}}(m)$$

$$=9.91\cdot\sqrt{\frac{\gamma_o}{\gamma_{cr}}}(Å) \quad (25)$$

The remarkably simple result suggests that the radius of the critical micelle is only dependent on the surface tension of the critical micelle.

5. COMPARISON WITH CMC–DATA FOR NONIONIC AND IONIC MICELLES

Figure 3 shows clearly that it is not at all a bad approximation to consider cmc of nonionic micelles to be a reflection of the
"supersaturation" exhibited by nucleating oil droplets in water. We
have simply put $\gamma_{cr}=\gamma_0$ and $(\Delta\mu_{cr}+P_{cr})/kT=0$. If one would adopt the
usual pseudophase model for the micelle, one is forced to assume
excess chemical potentials in the micelle as high as 5–8 kT. This
is quite unrealistic. Figure 2 shows, that the influence of the EO-
heads cannot be greater than the vertical bars (less than 1 kT). Also, the calculations of Gruen and De Lacey[21] have shown that the
free energy cost of forcing alkane chains with $n_c=12$ into a spherical micelle with aggregation number 53 is only 0.8 kT per chain.
Therefore, the interfacial term of the nucleation theory seems to
be the only rescue in order to explain the large difference between
cmc and solubility data.

A perfect fit to the experimental data is obtained assuming
$\gamma_{cr}/\gamma_0=0.725$ and $(\Delta\mu_{cr}+P_{cr})/kT=3.13$ as mean values for all n_c. Thus,
the surface tension of the critical micelles is lower than the in-
terfacial tension of a pure and plane interface as expected from
the considerations made in Section 1. The critical micelle is less
than the equilibrium micelle, where the accumulation of head groups
has reached a level leading to $\gamma_m=0$ (see Section 2).

The idea of an average value of γ_{cr}/γ_0 is basically wrong,
however. From equation (25) we then derive that all critical micelles have the same size irrespective of n_c. The aggregation numbers
will then be higher for small n_c than for high n_c. Therefore, the
area per head group is smaller for small n_c than for high n_c in contradiction to the assumption of a constant γ_{cr}. We shall instead
find consistent values by iteration on the equations given. Assuming
a Tolman correction distance $=2.5$ Å as found for nucleating tribromomethane droplets in water–methanol mixtures[6], we can use equation
(2) for the micellar surface tension. The best fit to the experimen-
tal data is found for $a_{ex}=15$ Å^2 and $(\Delta\mu_{cr}+P_{cr})/kT=4$. For details of
the calculation, see ref.(11).

The excluded area is quite low. Usually, excluded areas found
by monolayer studies using the Langmuir isotherm are around 30 Å^2 [9].
However, if a fully extended poly–oxyethylene chain is rotated
around the longitudinal axis, a "repulsion cylinder" with cross sec-
tional area 16.4 Å^2 is created (ref.(8)). We shall not forget to
mention that the estimations in Section 2 indicated that the Langmuir approach is not quantitative at zero surface tension for EO-
micelles. Nevertheless, it might well be better for the critical
micelles, where the coverage fraction θ is not so close to unity.

The above iterative calculations have lead us to the values
of γ_{cr} as a function of n_c. The radii of critical micelles then

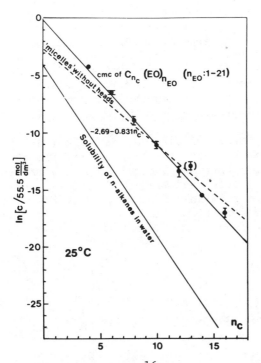

Figure 3.   Cmc of nonionic micelles[16] compared to solubility of corresponding n-alkanes and to the results of simple nucleation theory
with $\gamma_{cr}=\gamma_0$ and $(\Delta\mu_{cr}+P_{cr})/kT=0$ (dashed line). Vertical bars incorporate a small systematic variation with n_{EO} in the experimental da-
ta.

follow from equation (25). We find R_{cr} decreasing from 13.8 Å for
$n_C=4$ to 12.5 Å for $n_C=16$. Comparing with the R_0-values in Table II
we see that R_{cr} are indeed less than the radii of the equilibrium
micelles. The length of a fully extended HC-chain for $n_C=4$ is only
5-6 Å. This causes no problems in an interfacial model of the micelle, since the radial position of the head groups is distributed
statistically. In a sphere of radius 13.5 Å with a "central cavity"
of radius 5.5 Å, we need only to pull a volume fraction of HC-
chains equal to 7.2 % towards the center in order to fill up the
cavity.

Ionic micelles may be treated similarly, but we now have the additional complexities of calculating the surface potentials and the electric surface tension. Space does not permit details of calculations here. They will be given in ref.(11). So much can be said, however, that we utilize an approximate analytical solution to the spherical, nonlinear Poisson-Boltzmann equation for the calculation of ψ_S and a compromise between the charging method and the "Maxwell-Bakker-Sanfeld" approach for the calculation of γ_E (see Section 1).

The measurements of Dalsager[19] (from our laboratory) of cmc for the three anionic surfactants SDeS, SUS and SDS (Sodium Decyl Sulphate, Undecyl Sulphate and Dodecyl Sulphate) and the three cationic surfactants DTAB, TrTAB and TTAB (Dodecyl, Tridecyl and Tetradecyl Trimethylammonium Bromide) as a function of added NaCl or KBr were selected for investigation. The following half-theoretical and half-experimental function of cmc is defined:

$$G_{cmc} \equiv \ln\left[\frac{cmc}{55.5}\right] + 4.11 + 1.492 \cdot n_C - \frac{\Delta\mu_{cr} + P_{cr}}{kT}$$

$$-\left[\frac{\gamma_{cr}^{sr}}{\gamma_o} + \frac{\gamma_{cr}^E}{\gamma_o}\right]^{3/2} \cdot 2.45 \cdot 10^{-2} v_{HC}'(\text{Å}^3) \quad (26)$$

If the parameters are chosen correctly and the surface potentials calculated by the spherical, nonlinear Poisson-Boltzmann equation are correct, then by our equation (24) we should obtain a straight line for G_{cmc} through the origin and with a slope equal to z_A (+1 for cationic micelles and -1 for anionic), when plotted against $e_o\psi_S(cr)/kT$.

In order to obtain G_{cmc}-values through the origin we have to choose $(\Delta\mu_{cr} + P_{cr})/kT = 3$ and the following average values for the short range part of the surface tension:

$$\frac{\gamma_{cr}^{sr}}{\gamma_o} = 0.60 \pm 0.03 \, (\text{anionic}); \frac{\gamma_{cr}^{sr}}{\gamma_o} = 0.80 \pm 0.05 \, (\text{cationic}) \quad (27)$$

The electric surface tension for the critical ionic micelles is −5 mN/m or numerically smaller values, so some error in ψ_S and γ_E will not seriously affect the G_{cmc}-values. The radii of the critical micelles vary slightly with n_C and salt concentration. The radii of the critical cationic micelles lie within 11.6±0.2 Å. For the anionic micelles the figure is 14.2±0.4 Å. The figures in equation (27) correspond to $a_{ex} = 30$ Å^2 for the $-SO_4^-$ group as well as for the $-N(CH_3)_3$ group which is the normal value (ref.(9), p.116). However,

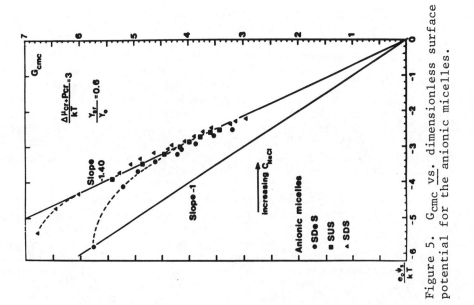

Figure 5. G_{cmc} vs. dimensionless surface potential for the anionic micelles.

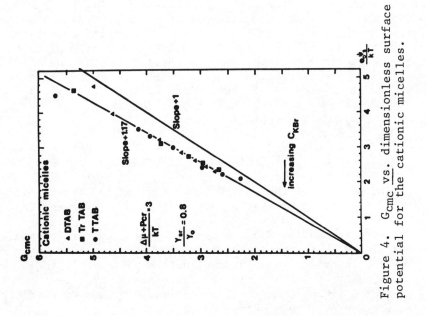

Figure 4. G_{cmc} vs. dimensionless surface potential for the cationic micelles.

the Tolman distances found were different. For the anionic micelles $\delta=2.5$ Å, for the cationic micelles $\delta=0.74$ Å.

The G_{cmc} vs. $e_0\psi_s(cr)/kT$ plot for the cationic micelles is given as Figure 4. Figure 5 shows the similar plot for the anionic micelles. Despite large differences in the original ln(cmc/55.5) data, the treatment here really reduces all the data for a given charge type to one single line through the origin (apart from some problems for small ionic strengths and high numerical surface potentials for the anionic micelles). The slopes are deviating from +1 and −1, however. We take this as an indication that the surface potentials calculated from the nonlinear Poisson-Boltzmann equation are not quite correct. If the real surface potentials were 17 % higher for cationic micelles and 40 % higher (numerically) for anionic micelles everything would be perfect.

Actually, calculations of Sanfeld[14] have indicated deviations from the nonlinear Poisson-Boltzmann equation in precisely that direction. The surface potentials are higher at the same surface charge densities because of the decrease of permittivity induced by the presence of the counterions and because of the volume effects on the ions in connection with the great Kelvin and Helmholtz pressures near the double layer. The deviations are greater for negatively charged micelles since those have positive counterions (Na^+) in their presence. Small, positive inorganic ions have an appreciable hydration shell, whereas anions like Br^- are approximately unhydrated[23,24]. Therefore, Na^+ occupies more space and is more efficient in lowering the dielectric constant of the solution in the immediate neighbourhood of the anionic micelles than Br^- in the case of cationic micelles. The calculations of Sanfeld predict a surface potential which is 6 % higher than the nonlinear Poisson-Boltzmann result in the case of a 0.1 mol/dm^3 NaCl solution at a plane double layer with a surface charge density equal to −0.067 Coulomb/m^2. Howeverm the charge densities found here for the anionic micelles are about double of that value. Extensions of Sanfeld's calculations to higher charge densities, higher salt concentrations and higher surface curvature are now being carried out.

CONCLUSIONS AND DISCUSSION

The success of the interfacial model of the micelle seems remarkable at this point. Let us sum up the most important features:

1) The theory predicts increasing aggregation numbers of nonionic micelles with n_c and decreasing aggregation numbers for more bulky head groups in accordance with experimental facts.
2) From the ZST-principle one can calculate an area per hexaoxyethylene head groups equal to 23±6 Å^2 from micelles with widely differing aggregation numbers and volumes.

3) Debye-Hückel micelles show a transition from low aggregation
   numbers in the Coulombic limit at low ionic strengths to high
   numbers in the nonionic limit at high strengths in accordance
   with experimental numbers for ionic micelles.
4) The interfacial term in the theory of nucleating oil droplets is
   able to account for most of the difference between cmc-values of
   nonionic micelles and solubility of oil in water.
5) A comparison between generalized nucleation theory and data for
   cmc of nonionic and ionic micelles leads to quite realistic values of excluded areas of the head groups and of Tolman correction distances.
6) The surface potentials are found to be numerically higher than
   predicted by the spherical, nonlinear Poisson-Boltzmann equation. The deviation is greater for anionic micelles with large,
   hydrated cations as counterions as compared with cationic micelles with small, unhydrated anions as counterions as predicted
   in earlier self-consistent calculations by Sanfeld.

Another fact that might find its explanation in a ZST model
for micelles is the rapid increase in aggregation numbers of nonionic micelles when the temperature is raised towards the cloud
point. Such micelles might well be "supermicelles" formed by aggregation of spherical micelles as indicated, for example, by NMR-stu-
dies[25]. Due to the low surface tension, the activation energy for
coalescence is very low and a Schmoluchowski type of coagulation
theory would be appropriate. When two micelles coalesce they cannot
form a shperical supermicelle, since the surface tension would then
be negative. Also, so many head groups would have to be buried in
the hydrophobic core in order to avoid a central cavity, that it
would be energetically unfavourable. Long cylinders or tortuous
"snakes" would be the result. Zero surface tension should also be a
guiding principle here, since any significant surface tension causes a fluid cylinder to break into droplets by the Rayleigh instability. Contrary to nonionic micelles, the ionic ones do not easily
aggregate at low or moderate ionic strengths due to the repulsion
between the double layers.

Finally, it should be possible to explain some of the findings of relaxation studies of micellization by Aniansson et al[26]
in terms of the surface tension theory of micelles. Certainly the
whole size distribution curve can be found by the method presented
here, and by generalizing the classical theory of nucleation rates,
it should be possible to understand the relaxation times. According
to Hoffmann et al[27] interesting information about the micellar critical nucleus can be extracted from the slow relaxation time (τ_2)
whereas the fast time (τ_1) reflects the rapid exchange of amphiphiles between the solution and the micelle. The question is, however, if the low surface tension of micelles does not lead to a
significant contribution to relaxation kinetics from micelle division into submicelles and the reverse coalescence process. This
will be the topic of a future publication.

Anyway, the ZST-principle must have dramatic consequences for our mental picture of a micellar solution. The very low surface tension makes the micelles easily deformable, and due to the impact of the Brownian collision in water molecules they will look more like "monsters" or "amebes" than nice spherical droplets. The spherical shape in our model is therefore to be understood only as a mean configuration in time or over an ensemble of micelles.

ACKNOWLEDGEMENTS

Valuable discussions with Professor A.Sanfeld, Bruxelles, and Professor E.Ruckenstein, Buffalo, are gratefully acknowledged.

REFERENCES

1.   S. Ono and S. Kondo, in "Handbuch der Physik", S. Flügge, Editor, Vol. X, p. 134. Springer Verlag, Berlin, 1960
2.   I. B. C. Matheson and A. D. King, Jr., J. Colloid Interface Sci., 66, 464 (1978)
3.   P. Mukerjee, Kolloid Z. Z. Polym., 236, 76 (1970)
4.   F. M. Menger, J. Phys. Chem., 83, 893 (1979)
5.   F. M. Menger, these proceedings
6.   A. E. Nielsen and S. Sarig, J. Crystal Growth, 8, 1 (1971)
7.   R. C. Tolman, J. Chem. Phys., 17, 333 (1949)
8.   T. S. Sørensen, Acta Chem. Scand., A36, 141 (1982)
9.   A. Sheludko, "Colloid Chemistry", pp. 114 - 119, Elsevier, Amsterdam, 1966
10.   R. Defay, I. Prigogine, A. Bellemans and D. H. Everett, "Surface Tension and Adsorption", Longman, Greens & Co., London, 1966
11.   T. S. Sørensen, Chem. Eng. Comm., No. 1-2 (1983)
12.   J. Newman, "Electrochemical Systems", Chap. 7, Prentice-Hall, Englewood Cliffs, N.J., 1973
13.   E. J. W. Verwey & J. Th. G. Overbeek, "Theory of the Stability of Lyophobic Colloids", Elsevier, Amsterdam, 1948
14.   A. Sanfeld, "Introduction to the Thermodynamics of Charged and Polarized Layers", Wiley-Interscience, New York, 1968
15.   G. B. Bakker, "Theory de la couche capillaire plane des corps purs", p. 14, Gauthiers-Villars, Paris, 1911-1912
16.   P. Becher, in "Nonionic Surfactants", M. J. Schick, Editor, Chapter 15, Marcel Dekker, New York, 1967
17.   N. A. Mazer, G. B. Benedek and M. C. Carey, J. Phys. Chem., 80, 1075 (1976)
18.   K. S. Birdi, S. U. Dalsager and S. Backlund, J. Chem. Soc. Faraday Trans. 1, 76, 2035 (1980)

19. S. U. Dalsager, "Teoretisk og Eksperimental Analyse af Micel-
     ledannelse", Ph. D. Thesis. Technical University of Denmark,
     Lyngby, 1980
20. K. A. Dill, these proceedings
21. D. W. R. Gruen and E. H. B. De Lacey, these proceedings
22. C. Tanford, "The Hydrophobic Effect. Formation of Micelles
     and Biological Membranes", 2nd edition, Wiley-Interscience,
     N. Y., 1980
23. T. S. Sørensen, Acta Chem. Scand., A32, 571 (1978)
24. T. S. Sørensen, Acta Chem. Scanc., A33, 583 (1979)
25. E. J. Staples and G. J. T. Tiddy, J. Chem. Soc. Faraday Trans.
     1, 74, 2530 (1978)

26. E. A. G. Aniansson, S. N. Wall, M. Almgren, H. Hoffmann, J.
     Kielmann, W. Ulbricht, R. Zana, J. Lang and C. Tondre, J. Phys.
     Chem. 80, 905 (1976)
27. H. Hoffmann, H. Nüsslein and W. Ulbricht in "Mizellization,
     Solubilization and Microemulsions", K. L. Mittal, Editor,
     Vol. 1 p. 280, Plenum Press, N. Y. 1977
28. D. Ilkovic, Coll. Trav. Chim. Tchecosl. 4, 480 (1932)

COMPUTATION OF THE MICELLE-SIZE-DISTRIBUTION FROM EXPERIMENTAL MEASUREMENTS

A. Ben-Naim

Department of Physical Chemistry
The Hebrew University of Jerusalem
Jerusalem 91904, Israel

A new method of extracting information on the micelle size
distribution (MSD) from experimental measurements is outlined. The method consists of analyzing the functional dependence of the osmotic pressure of a surfactant solution
on the total concentration of the surfactant. It is shown
that moments of the MSD are related to the derivatives of
these functions. Using a limited number of such moments,
and the mathematical procedure of information theory, one
can compute the "best guess" of the MSD. Some preliminary
numerical examples are presented.

INTRODUCTION

Recently there have been a few attempts to compute the micelle
size distribution (MSD) by using a statistical mechanical approach[1-9].
These calculations were all based on some very simplifying assumptions concerning the nature of the solute-solute and solute-solvent
interactions. With these assumptions, supplemented by some further
mathematical approximations, one can compute the MSD as well as
some thermodynamic quantities, such as the osmotic pressure of the
surfactant solution. Of course, since there is no experimental way
of measuring the MSD, the only way of assessing the success of a
particular model is the comparison between the computed and the
experimental values of the thermodynamic quantities only. Such a
comparison is not a stringent test of either the physical model or
the computational procedure. Many different models may lead to the
same thermodynamic quantities. In this paper we suggest that information on the MSD is inherently contained in and may be derived
from experimental data on the osmotic pressure of a surfactant
solution. Of course, the reliability of the MSD that may be obtained

731

by this method depends on the details and accuracy of the experimental data.

In what follows we outline a procedure for calculating what may be referred to as the "best" MSD, based on the available experimental data. In particular we use data on the osmotic pressure, but in principle other thermodynamic data, such as activity coefficients, vapor pressures etc. could be used as well.

The method consists of analyzing the experimental data in essentially three steps: 1) Computation of the monomer concentration; 2) Computation of the moments of the MSD; 3) Estimation of the "best" MSD that is consistent with the available experimental data. In the latter, we use the information-theoretical idea of maximizing the entropy defined by the MSD.[10,11]

Throughout the following analysis we use the assumption that the system, water-surfactant, is an associated-ideal solution, i.e. all interactions among the aggregates are neglected. This assumption appears to be valid for dilute solutions of non-ionic surfactants, but may not be valid for high concentrations, or for ionic surfactant solutions, where nonideality effects should be taken into account.

The two fundamental equations, for a system of a surfactant in a solvent are:

$$\rho_T = \sum_{i=1}^{\infty} i\rho_i \tag{1}$$

$$\beta\pi = \rho_c = \sum_{i=1}^{\infty} \rho_i \tag{2}$$

where ρ_i is the number (or molar) density of aggregates consisting of i monomers, ρ_T is the total concentration of the surfactant in the solution. π is the osmotic pressure and $\beta = (kT)^{-1}$, with k the Boltzmann constant (or the gas constant, if we use molarities concentrations throughout). ρ_c is the total density of the aggregates in the system, sometimes referred to as the apparent molarity of the surfactant. Using the assumption of ideality of the solution, we may relate each of the ρ_i to the monomer concentration by

$$\rho_i = K_i \rho_1^i \tag{3}$$

where K_i is the equilibrium constant. In terms of ρ_1, Equations (1) and (2) may be rewritten as

$$\rho_T = \sum_{i=1}^{\infty} i K_i \rho_1^i \tag{4}$$

$$\rho_c = \sum_{i=1}^{\infty} K_i \rho_1^i . \tag{5}$$

Let k_n be the equilibrium constant for the addition reaction

$$A_{n-1} + M \to A_n \tag{6}$$

$$k_n = \frac{\rho_n}{\rho_1 \, \rho_{n-1}} \tag{7}$$

where A_n is an aggregate built up of n monomers. In the simplest association process the addition of one monomer to an aggregate involves the same amount of work, independently of n, i.e. we take

$$k_n = k_2 \text{ for all n} . \tag{8}$$

The general relation between K_n and k_i is simply

$$K_n = \mathop{\pi}_{i=2}^{n} k_i \tag{9}$$

which reduces to $K_n = (k_2)^{n-1}$ for the simplest model in Equation (8). In principle k_2 may be obtained from experimental data. Having a plot of ρ_1 as a function of ρ_T (see next section) one finds

$$\lim_{\rho_T \to 0} \left(\frac{\partial^2 \rho_1}{\partial \rho_T^2} \right) = -4 \, k_2 . \tag{10}$$

In the most general case we define the correction function G_n as

$$K_n = k_2^{n-1} G_n . \tag{11}$$

Assuming that k_2 is known from experimental sources we may rewrite Equations (4) and (5) in a reduced form, namely

$$y = \sum_i i G_i x^i \tag{12}$$

$$f = \sum_i G_i x^i \tag{13}$$

where

$$x = k_2 \rho_1, \quad y = k_2 \rho_T \text{ and } f = \beta \pi k_2 \tag{14}$$

COMPUTATION OF THE MONOMER CONCENTRATION

We assume that experimental data on the osmotic pressure (or any equivalent thermodynamic quantity) as a function of the total concentration of the surfactant are available, i.e. we have the curve of ρ_c as a function of ρ_T. Both ρ_T and ρ_c are related through

ρ_1 in Equations (13) and (14). From these two equations we can easily obtain

$$\frac{\partial \rho_c}{\partial \rho_T} = \frac{\partial \ln \rho_1}{\partial \ln \rho_T} \quad . \tag{15}$$

We integrate this identity between $\rho_T(1)$ and $\rho_T(2)$

$$\ln \rho_1(2) - \ln \rho_1(1) = \int_{\rho_T(1)}^{\rho_T(2)} \frac{\partial \rho_c}{\partial \rho_T} d\ln \rho_T \tag{16}$$

If we choose $\rho_T(1)$ very small so that at this concentration we have $\rho_T(1) \approx \rho_1(1)$ (this follows from Equation (14) in the limit $\rho_1 \to 0$), we may rewrite Equation (16) as

$$\ln \rho_1(2) = \ln \rho_T(1) + \ln \rho_T(2) - \ln \rho_T(2) + \int_{\rho_T(1)}^{\rho_T(2)} (\frac{\partial \rho_c}{\partial \rho_T}) d\ln \rho_T$$

$$= \ln \rho_T(2) + \int_{\rho_T(1)}^{\rho_T(2)} [\frac{\partial \rho_c}{\partial \rho_T} - 1] d\ln \rho_T \quad . \tag{17}$$

Since the integrand is zero when $\rho_T \to 0$ we may replace the lower limit of the integral by $\rho_T(1) = 0$ to obtain the final result, for any concentration ρ_T

$$\rho_1 = \rho_T \exp\{\int_0^{\rho_T} [\frac{1}{\rho_T} \frac{\partial \rho_c}{\partial \rho_T} - \frac{1}{\rho_T}] d\rho_T\} \quad . \tag{18}$$

Clearly the entire right hand side of Equation (18) may be computed from experimental data, hence ρ_1 may be calculated at each ρ_T.

CALCULATION OF THE MOMENTS OF THE MICELLE-SIZE-DISTRIBUTION

The micelle-size-distribution MSD is defined by

$$P_i = \frac{\rho_i}{\sum_{j=1}^{\infty} \rho_j} \tag{19}$$

and the ℓ'th moment of the distribution is defined by

$$M_\ell = \sum_{i=1}^{\infty} i^\ell P_i = \frac{\sum_{i=1}^{\infty} i^\ell K_i \rho_1^i}{\sum_{i=1}^{\infty} \rho_i} = <i^\ell> \quad . \tag{20}$$

It is also convenient to define the sum

$$N_\ell = \sum_{i=1} i^\ell \rho_i = \sum_{i=1} i^\ell K_i \rho_1^i \tag{21}$$

which gives

$$\langle i^\ell \rangle = \frac{N_\ell}{N_o} \tag{22}$$

In particular we have

$$\langle i^o \rangle = 1 \tag{23}$$

$$\langle i^1 \rangle = \frac{N_1}{N_o} = \frac{\rho_T}{\rho_c} \tag{24}$$

Next we take the derivative

$$\frac{\partial \rho_c}{\partial \rho_T} = \frac{\partial \rho_c}{\partial \rho_1} \frac{\partial \rho_1}{\partial \rho_T} = \frac{\Sigma i K_i \rho^{i-1}}{\Sigma i^2 K_i \rho_1^{i-1}} = \frac{N_1}{N_2} = \frac{\langle i \rangle}{\langle i^2 \rangle} \tag{25}$$

Having computed $\langle i \rangle$ from Equation (24) we may obtain $\langle i^2 \rangle$ from knowledge of the first derivative of ρ_c as a function of ρ_T. This procedure may be continued to higher derivatives, e.g. the second derivative gives

$$\frac{\partial^2 \rho_c}{\partial \rho_T^2} = \frac{\langle i^2 \rangle^2 - \langle i \rangle \langle i^3 \rangle}{\rho_c \langle i^2 \rangle^3} \, . \tag{26}$$

Hence having computed $\langle i \rangle$ and $\langle i^2 \rangle$ we may obtain $\langle i^3 \rangle$ from Equation (26). Of course this procedure becomes more and more complicated, and less useful as we proceed to higher moments. A simpler procedure for a numerical computation is to use the sums N_ℓ defined in Equation (21). Thus

$$\frac{\partial \rho_c}{\partial \rho_T} = \frac{N_1}{N_2} \tag{27}$$

from which we compute N_2 (since $N_1 = \rho_T$) for any ρ_T. Having N_2 as a function of ρ_T we take its first derivative

$$\frac{\partial N_2}{\partial \rho_T} = \frac{N_3}{N_2} \tag{28}$$

From which N_3 may be determined at each ρ_T. Next we take

$$\frac{\partial N_3}{\partial \rho_T} = \frac{N_4}{N_2} \tag{29}$$

to obtain N_4 at each ρ_T. In general, having computed N_k at the k-th step we may proceed to obtain

$$\frac{\partial N_k}{\partial \rho_T} = \frac{N_{k+1}}{N_2} \tag{30}$$

wherefrom N_{k+1} may be computed. Clearly at each step the moments may be computed from the sums N_k, through

$$M_k = \frac{N_k}{N_o} \tag{31}$$

The procedure outlined above shows that, in principle, if we have a very accurate and dense set of measurements we could obtain as many moments as we wish. In practice, however, the procedure may be useful to obtain only the first few moments only. This allows one to estimate the best guess of the MSD which is consistent with the available experimental information.

COMPUTATION OF THE "BEST" MSD

We assume that we have already analyzed the experimental curves of ρ_c as a function of ρ_T, and obtained a few moments of the distribution, say

$$M_\ell = \sum_{i=1}^{\infty} i^\ell P_i \qquad \ell = 0, 1, 2, 3 \cdots L \tag{32}$$

Clearly from the knowledge of the distribution one can compute all the moments. The converse is also true, if we have <u>all</u> the moments of a distribution one can reconstruct the distribution itself. Here, however, we may expect to have only the first few moments, and therefore we can satisfy ourselves by asking, what is the "best" guess that we could obtain given the experimental data at hand. This kind of question may be answered by the well known procedures of information theory. One first defines the entropy, of any distribution, by

$$S = -\sum_{i=1}^{n} P_i \ln P_i \tag{33}$$

where we have assumed that the aggregate of maximum size consists of n monomers. The "best" guess of a distribution is obtained by maximizing S subject to the constraint that L moments are known.

The mathematical problem is thus to find an extremum to the function

$$S + \Sigma \lambda_\ell M_\ell \tag{34}$$

where λ_ℓ are the Langrange multipliers. Thus

$$\frac{\partial}{\partial P_j}[-\Sigma P_i \ln P_i + \sum_{\ell=0}^{L} \lambda_\ell \Sigma i^\ell P_i] = 0 \tag{35}$$

for each j = 1, 2, $\cdots$ n. Performing the differentiation with respect to P_j we obtain the condition

$$P_j = \exp[\lambda_0 - 1] \exp[\sum_{\ell=0}^{L} \lambda_\ell j^\ell] \tag{36}$$

Thus if we know all the λ_ℓ's we may compute the required distribution P_j. The λ_ℓ's may be determined from the L+1 equations defining the moments, namely,

$$M_k = \sum_{j=1}^{n} j^k P_j = \sum_{j=1}^{n} j^k \exp[\lambda_0 - 1] \exp[\sum_{\ell=1}^{L} \lambda_\ell j^\ell] \tag{37}$$

For k = 0, 1, $\cdots$ L. These are L+1 equations for the L+1 unknowns λ_ℓ. We can use the fact that $M_0 = 1$ to eliminate λ_0, i.e.

$$1 = \exp[\lambda_0 - 1] \sum_{j=1}^{n} \exp[\sum_{\ell=1}^{L} \lambda_\ell j^\ell] \tag{38}$$

Thus dividing each M_k in Equation (37) by Equation (38) we obtain

$$M_k = \frac{\sum\limits_{j=1}^{n} j^k \exp[\sum\limits_{\ell=1} \lambda_\ell j^\ell]}{\sum\limits_{j=1}^{n} \exp[\sum\limits_{\ell=1} \lambda_\ell j^\ell]} \tag{39}$$

where we now have L equations with the unknowns $\lambda_1 \ldots \lambda_L$. Once these are determined, one can use Equation (38) to compute λ_0 also, and thereby all of the P_j's in Equation (36) can be calculated.

SOME NUMERICAL EXAMPLES

There are very few reports on the measurements of the osmotic pressure as a function of the total surfactant concentration. For example Corkill et al.[12] have made vapor pressure measurements of aqueous solutions of n-octylhexaoxyethylene-glycol monoether. From these data they obtained a plot of the apparent molarity (ρ_c) as a function of the total molarity of the surfactant. A similar study on aqueous solutions of potassium octanoate was reported by Moule et al.[13]. However both of these studies have reported only a few measurements (about 10-12 points) and therefore may not be useful as an input to the procedure we have described in this paper.

The illustrations presented in the following use a theoretical model as an input, i.e. we start from a choice of a series of K_i. Then we choose values of ρ_T at close intervals. Next we invert Equation (4) to obtain the corresponding values of ρ_1 as a function of ρ_T. Once ρ_1 is known we compute all the ρ_i's through Equation (3) and hence the input-distribution P_i in Equation (19). Also we can use Equation (5) to compute ρ_c at each ρ_T and obtain the (theoretical) curve of ρ_c as a function of ρ_T for this particular model.

At this point we treat the curve of $\rho_c = \rho_c(\rho_T)$ as if it were a (hypothetical) experimental curve. We use it to follow the procedure as outlined in the previous sections. First we compute ρ_1 as a function of ρ_T from Equation (18) and compare the results with the values computed from the model. Secondly, we compute the moments of the distribution by the method outlined above and again compare with the moments computed from the model. Finally, we use this procedure to compute the "best guess" of a distribution and compare that with our input distribution. This procedure has one clear-cut advantage over a straightforward application of the method on real experimental data. In the latter case we may obtain a distribution but there is no way of assessing its relevance to the real distribution of the experimental solution. On the other hand, by using a model-input-distribution and following the above outlined procedure we obtain a computed output distribution which may be compared with the input distribution. In this way the effectiveness of the procedure is tested, and a detailed examination of its dependence on various parameters (number of points, their density, accuracy, etc.) may be studied).

In the following we use dimensionless quantities throughout, so that we could use either Equations (12) and (13) or the pair of Equations (4) and (5) (with $k_2 = 1$). The simplest model to start with is the monodispersed case, i.e.

$$
K_i = \begin{cases} 1 & \text{for } i = 1 \\ 1 & \text{for } i = n_{max} \\ 10^{-10} & \text{for } 1 < i < n_{max} \end{cases} \tag{40}
$$

where n_{max} is the largest aggregate that is assumed to exist in the solution. In our first "experiment" we take $n_{max} = 10$. We use 50 points for ρ_T, at each of which we compute ρ_1 from Equation (4) and substitute into Equation (5) to compute ρ_c. The two curves obtained are shown in (Figure 1) [and similarly Figure 2 for $n_{max} = 20$].

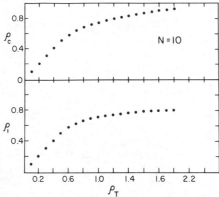

Figure 1. ρ_1 and ρ_c as a function of ρ_T for the monodispersed case with $n_{max} = 10$.

Using the values of ρ_c as a function of ρ_T we now follow the procedure of converting these data into MSD. First we compute ρ_1 from Equation (18) the results obtained agree with the original values and are almost indistinguishable from the results plotted in (Figure 1). Next we computed the moments $M_0 \cdots M_5$ (note $M_0 = 1$, $M_1 = \rho_T/\rho_c$). We note that the agreement between the computed moments and the original ones is satisfactory. There seems to be a better agreement for the lower moments, but even for M_5 the agreement is within less than one percent of their values. Finally, we proceed to compute the MSD. Here we start with a low value of ρ_T where we could guess initial values of λ_i's use, these as input parameter in an iterative procedure to compute λ_i's which solve the set of Equations (39). When increasing the total concentration, we use the values of the λ_i's from a previous concentration as an initial guess

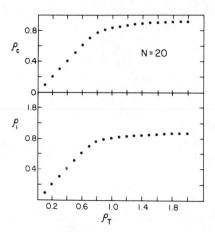

Figure 2. ρ_1 and ρ_c as a function of ρ_T for the monodispersed case with $n_{max} = 20$.

for the next higher concentration. The initial values of λ_i's for the first ρ_T are chosen as follows. We rewrite the distribution as

$$P_j = \frac{\rho_i}{\sum\limits_{i=1} \rho_i} = \frac{K_j \rho_1^j}{\rho_c} \tag{41}$$

$$\ln P_j = j \ln \rho_1 - \ln \rho_c + \ln K_j \tag{42}$$

Clearly, in the limit of $\rho_T \to 0$ both $\rho_1 \sim \rho_T$ and $\rho_c \sim \rho_T$, and only monomers exist in the solution. Thus for $j = 1$ we must have $P_j = 1$ and all other P_j's= 0. For $\rho_T \sim 0$ we must still assume that the first two terms on the right hand side of Equation (42) will be the leading terms, i.e.

$$\ln P_j = j \ln \rho_1 - \ln \rho_1 \qquad (\rho_T \sim 0) \tag{43}$$

Comparing with Equation (36)

$$\ln P_j = (\lambda_0 - 1) + \sum_{\ell=0}^{L} \lambda_\ell j^\ell \tag{44}$$

we arrive at the initial guess for the parameters

$$\lambda_0 = 1 - \ln\rho_1$$

$$\lambda_1 = \ln\rho_1 \qquad\qquad\qquad (45)$$

$$\lambda_2 = \lambda_3 = \ldots \lambda_2 = 0 .$$

These parameters are used in the iterative procedure for computation of λ_i's at the first ρ_T and so forth. Because of the inaccuracy of the procedure we expect that the MSD as determined at each concentration will be very inaccurate. Therefore we used an averaging procedure to determine K_i (since ρ_1 and ρ_c are known, the knowledge of either P_i or K_i are equivalent). The advantage of this procedure is that whereas the MSD is dependent on the concentration ρ_T, the series of values of K_i are independent of the concentration of the surfactant (which follows from the ideality assumption adopted at the beginning of our treatment). Thus at each concentration we first determine the distribution P_j, using Equation (41) then we determine the corresponding values of K_j. In an actual computation the latter do depend on the total concentration ρ_T. To take a concentration independent values of K_i we simple average

$$\langle K_i \rangle = \frac{\displaystyle\sum_{\text{Overall concentrations}} K_i (\rho_T)}{\text{Number of concentrations}}$$

In fact, because of the large inaccuracies of the computation at the very low and very high concentrations the average in Equation (46) was taken only for about 40 points excluding the concentrations at both edges. The values of $\langle K_i \rangle$ may be compared with the original input of K_i to test the reliability of the procedure. Of course one can recover the MSD at each concentration by resorting to Equation (41) where the average $\langle K_i \rangle$ are used. Tables I and II show values of K_i and $\langle K_i \rangle$ for the case of $n_{max} = 10$ and $n_{max} = 20$ respectively. It is seen that the agreement between the input-model and the output calculation is quite satisfactory, i.e. the values of $\langle K_i \rangle$ are almost equal to K_i for $K_i = 1$, and are very small when K_i equals 10^{-10}.

The success of this procedure to compute the MSD seems sufficiently promising for application on real experimental data. We hope that experimentalists in this field will be encouraged to undertake the measurement of osmostic pressures (or any equivalent quantity) as a function of the total surfactant concentration at closely spaced intervals so that a meaningful analysis of the data could be achieved.

Table I.   Input Values of K_i and Computed Values of the Average K_i for a System with $n_{max} = 10$ and Five Moments

i	Input K_i	Output $\langle K_i \rangle$
1	1.0	0.99
2	10^{-10}	0.54×10^{-4}
3	10^{-10}	0.12×10^{-6}
4	10^{-10}	0.29×10^{-8}
5	10^{-10}	0.10×10^{-8}
6	10^{-10}	0.97×10^{-8}
7	10^{-10}	0.56×10^{-6}
8	10^{-10}	0.54×10^{-4}
9	10^{-10}	0.38×10^{-2}
10	1.0	1.01

Table II.   Input Values of K_i and Computed Values of the Average K_i for a System with $n_{max} = 20$ and Five Moments

i	Input K_i	Output $\langle K_i \rangle$
1	1.0	1.0
2	10^{-10}	0.18×10^{-9}
3	10^{-10}	0.32×10^{-18}
4	10^{-10}	0.58×10^{-26}
5	10^{-10}	0.99×10^{-33}
6	10^{-10}	0.16×10^{-38}
7	10^{-10}	0.25×10^{-43}
8	10^{-10}	0.36×10^{-47}
9	10^{-10}	0.47×10^{-50}
10	10^{-10}	0.57×10^{-52}
11	10^{-10}	0.63×10^{-53}
12	10^{-10}	0.62×10^{-53}
13	10^{-10}	0.57×10^{-52}
14	10^{-10}	0.47×10^{-50}
15	10^{-10}	0.36×10^{-47}
16	10^{-10}	0.25×10^{-43}
17	10^{-10}	0.16×10^{-38}
18	10^{-10}	0.99×10^{-33}
19	10^{-10}	0.57×10^{-26}
20	10^{-10}	0.18×10^{-18}
21	10^{-10}	0.18×10^{-9}
22	1.0	1.00

REFERENCES

1. G. S. Kresheck, in "Water, A Comprehensive Treatise", F. Franks, Editor, Plenum press, New York 1975.
2. C. Tanford, "The Hydrophobic Effect", 2nd edition, J. Wiley, New York, 1980.
3. C. A. Hoeve and G. C. Benson, J. Phys. Chem., $\underline{61}$, 1149 (1957).
4. D. C. Poland and H. A. Scheraga, J. Phys. Chem., $\underline{69}$, 2431 (1965).
5. C. Tanford, J. Phys. Chem., $\underline{78}$, 2469 (1974).
6. E. Ruckenstein and R. Nagarajan, J. Phys. Chem., $\underline{79}$, 2622 (1975).
7. R. Nagarajan and E. Ruckenstein, J. Colloid Interface Sci., $\underline{60}$, 221 (1977).
8. A. Ben-Naim and F. H. Stillinger, J. Phys. Chem., $\underline{84}$, 2872 (1980).
9. E. Ruckenstein and R. Nagarajan, J. Phys. Chem., $\underline{85}$, 3010 (1981).
10. E. T. Jaynes, Phys. Rev., 106, 620 (1957).
11. E. T. Jaynes, in "Statistical Physics", Brandeis Summer Institute, 1962, W.A. Benjamin, Inc., New York, 1963.
12. J. M. Corkill, J. F. Goodman and J. R. Tata in "Hydrogen-Bonded Solvent Systems", A.K. Covington and P. Jones, Editors. Taylor and Francis, London, 1968.
13. D. Moule, P. White and G. C. Benson, Canadian J. Chem. $\underline{37}$, 2086 (1959).

THERMODYNAMICS OF MICELLE FORMATION IN AQUEOUS MEDIA =

"SECOND VIRAL COEFFICIENT"

K. S. Birdi[1], E. Stenby[1] and D. K. Chattoraj[2]
Fysisk-Kemisk Institut[1]
The Technical University of Denmark
Building 206, DK-2800 Lyngby, Denmark
Jadvapur University, Calcutta, India[2]

Certain amphiphiles, such as detergents or surfactants, form small or large aggregates (micelles) in aqueous media when the concentration is greater than the critical micelle concentration (c.m.c.). Although a great many studies about different physical properties of such micellar systems have been reported in the current literature, the number of studies related to micellar molecular weights (weight average, N_w; number average, Nn) and the second virial coefficient (B_2: as determined from light-scattering or osmotic pressure) have been very few. As is known from the theory of the behavior of macromolecules in solvents, if the solute (such as micelles) is many times larger than the solvent (water) then the knowledge of B_2 is extremely useful if a complete thermodynamic treatment is desired. In other words, micelles would exhibit B_2 as one would expect for other macromolecules, which is given as:

$$B_2 = B_{Don} + B_{MM} - B_{MS}$$

where subscripts denote Donnan (Don), excluded volume (MM) and micelle (M) - salt (S) interactions. This expression has been applied to different macroions, and in a previous study it was shown that important information is also obtained in the case of aqueous micellar solutions. In this study the B_2 data as obtained by osmotic pressure measurements for different nonionic and ionic micellar systems under varying experimental conditions (temperature, electrolyte concentration) are

analyzed. These data show that analysis of B_2 of macro-
ions, i.e., micelles, provides with a much useful information as regards the solution behavior of these systems.

INTRODUCTION

Amphiphiles, such as detergents or surfactants, form aggregates, i.e, micelles in aqueous media, when their concentration is
greater than the critical micelle concentration, c.m.c. Since the
size of micelles is many orders of magnitude larger than the solvent
(water), it is obvious that the solution properties of micelles
would necessarily be the same as is known for other macromolecules
(such as proteins or synthetic polymers).[1] In other words, below
the c.m.c. the solute in these systems would behave as in ordinary
solutions, while above the c.m.c. the solute would behave as macromolecules (micelles). This means that below the c.m.c. ordinary solute-solvent thermodynamics is applicable; above the c.m.c. these
systems need be analyzed with the help of macromolecule thermodynamics. Accordingly, the knowledge of micellar molecular weight and
the second virial coefficient, B_2, becomes extremely important in
order to provide a satisfactory thermodynamic analysis.

In a previous study in this series[1] of reports on the second
virial coefficient, B_2, of micelles, we described in detail the various forces which give rise to non-ideal behavior (i.e., $B_2 \neq 0$) of
such systems. In continuation of earlier reports we shall present
here an analysis of additional data so as to be able to provide a
better understanding of micellar systems.

Theory of Osmotic Pressure

Before describing the second virial coefficient, B_2, we will
briefly develop the necessary thermodynamic relations which lead
to this analysis.

From elementary solution theory[2,3,4,5] we find that the magnitude of osmotic pressure, Π, as measured for two solutions where
on the solvent side we have monomers ($\approx$ c.m.c.), and on the solution side there is a micellar solution (i.e., monomers$\leftrightarrow$micelle, and
if we assume that monomers $\approx$ c.m.c.) is given by:

$$\Pi = RTw_M[\frac{1}{M_n} + B_2w_M + B_3w_M^2 + \ldots] \qquad (1)$$

where w_M is the concentration of micelles (i.e., total surfactant
concentration-c.m.c.) in g/ℓ, B_2 and B_3 are the second and third
virial coefficients, and M_n is the number average micellar molecular weight. If the virial coefficients are zero, then Equation (1)

reduces to the van't Hoff equation. However, the van't Hoff equation is also the limiting form of Equation (1) as $w_M \to 0$ for non-ideal solutions:

$$[\frac{\Pi}{w_M}]_{w_M \to 0} = \frac{RT}{M_n} \quad (2)$$

In most systems the plots of (Π/w_M) versus (w_M) have been found to be linear, which leads us to conclude that the value of B_3 is negligible (within the experimental reproducibility),[2-5] as found for other macromolecular solutions also.

The above requirement that M_n or the aggregation number (N_n) determination <u>can only</u> be estimated from the condition for the validity of relation in Equation (2) is in many cases misinterpreted where values of N are reported from spectroscopic techniques (such as fluorescence: see Reference (6) in these proceedings). Similar thermodynamic arguments are applicable to the light-scattering relations:[7]

$$H(\frac{w_M}{\tau}) = RT[\frac{1}{M_w} + 2B_2 w_M + 3B_3 w_M^2 + \ldots] \quad (3)$$

where H is related to the wave-length of scattering light and difference in refractive indices of solvent and solution, τ is the turbidity, B_2 and B_3 are the second and third virial coefficients. Analogous to Equation (2) a plot of $H(\frac{w_M}{\tau})$ versus (w_M) yields the value of $(\frac{1}{M_w})$ as $(w_M) \to 0$, where M_w is the weight average micellar molecular weight. The slopes of plots of Equations (2) and (3) differ by a factor of two, from which the value of B_2 can be estimated. The values of B_2 as determined from light-scattering and osmometry methods are compared in the following, i.e., additional data than previously reported. All the values found in the literature indicate, convincingly, that the ratio, $M_w/M_n \sim 1$, i.e., the micelles are very narrowly dispersed.[4]

Analysis of Second Virial Coefficient (B_2)

As described elsewhere,[1,7] from the thermodynamic analysis of the second virial coefficient of macromolecules it is found that:

$$B_2 = B_{Don} + B_{MM} - B_{MS} \quad (4)$$

$$= (\frac{1000v_1}{M_m^2})[\frac{Z^2}{4m_s} + \frac{\beta_{MM}}{2} - \frac{\beta_{MS}^2 \, m_s}{(4 + 2\beta_{SS} m_s)}] \quad (5)$$

where B_{Don} arises from the Donnan effect in the case of macro-ions with charge Z, B_{MM} arises from the fact that macromolecules being many orders of magnitude larger than the solvent molecule exhibit excluded volume effect, and B_{MS} arises from the interactions between the charges of macro-ion and the (inorganic) ions accumulated in the vicinity of the former. The various terms are described as follows:

Donnan effect in charged micelles:

As given in Equation (5):

$$B_{Don} = (\frac{1000v_1}{M_m^2})(\frac{Z^2}{4m_s}) \quad (6)$$

where v_1 is the partial specific volume of the solvent (water)~1, m_s is the molal concentration ($\approx$ molar concentration) of the added electrolyte, in ionic micelles $Z=N_n=$aggregation number.

Excluded volume effect:

If the inner core of the micelle is assumed as rigid, then it leads us to define the excluded volume, B_{MM}, term as:[4]

$$B_{MM} = (-\frac{16\Pi N_A R^3}{3M_m^2}), \text{ spherical micelles} \quad (7)$$

$$= (\frac{16\Pi N_A R^3}{3M_m^2})f_{IS}, \text{ non-spherical micelles} \quad (8)$$

where R is the Stokes radius of the micelle, f_{IS} is the correction factor derived for the non-spherical micelle:

$$f_{IS}=1, \text{ spherical micelle} \quad (9)$$

$$\approx 1 + \frac{1}{15} e^4 + \frac{37}{60} e^6 + \ldots, \quad e \ll 1 \quad (10)$$

where e is the eccentricity:[1]

$$e^2 = \frac{a^2 - b^2}{a^2}, \text{ oblate} \quad (11)$$

$$= \frac{b^2 - a^2}{b^2}, \text{ prolate} \quad (12)$$

where a and b are the major and minor axes of an ellipsoidal macro-molecule. It remains to be determined whether a given micelle is oblate or prolate[1,2,9]. In the present discussion we will assume oblate ellipsoidal shapes. The value of f_{IS} would be equal to 1 for a sphere while it would be 1.8 for an ellipse with (a/b=4.5). The values of B_{MM} for non-ionic micelles increase with increase in f_{IS}.[1]

Interaction between macro-ions and small ions:

The charged surface of a micelle can be analyzed in a similar way as other macro-ions. In current literature, however, this consideration is generally neglected when analyzing micelles. The contribution of such electrostatic interactions, B_{MS} to B_2 (Equation (14) is given as:[1,7]

$$B_{MS} = \left(\frac{1000 v_1}{M_m^2}\right)\left[-\frac{\beta_{MS} \, m_s}{(4 + 2\beta_{SS} \, m_s)}\right] \quad (13)$$

where $\beta_{SS} = \left(\frac{\partial \ln \gamma_s}{\partial m_s}\right)$, where γ_s is the activity coefficient of the added electrolyte, β_{MS} arises from interactions at the micellar surface which is given as:[1]

$$\beta_{MS} = -\frac{(1.17(Z)^2)}{\sqrt{m_s}(1 + 0.328 \, R_m \sqrt{m_s})^2} \quad (14)$$

where R_m is the effective radius of the micelle.

This relation emphasizes that in any correct analysis, as regards ion-binding to micelles, the knowledge of R_m is essential.

RESULTS AND DISCUSSION

The relative magnitudes of the terms B_{Don}, B_{MM} and B_{MS} to the second virial coefficient have been studied in detail in previous reports[1,7,8,10,11]. In the case of proteins, Scatchard et.al.[10] presented a very detailed analysis. The interaction between added salt (m_s) and protein (bovine serum albumin) was exceptionally strong, such that the degree of chloride binding was estimated from the term B_{MS} (Equation (13)).

In another study,[7] it was shown that the magnitude of B_2 became generally zero for the isoelectric protein. Further, it was shown that when the valence (Z) of the protein was large, B_2 was large and positive, but B_2 decreases with increasing ionic strength (as expected from the B_{Don} term). It was obvious that a complete analysis of B_2 with the help of Equation (5) is not an easy matter, as pointed out by different investigators[1,7,8,10,11]. However, the dependence of each term on the right-hand side in Equation (4) on various parameters does indeed provide useful information.[1,7]

Before describing these results, it is worth mentioning that all the data on micellar N_w and N_n reported so far in the current literature indicate that $(N_w/N_n) \approx 1$.[1-5] This indicates that within the experimental accuracy, it is safe to accept that micelles (with $N_n=20$, for Na-deoxycholate[12] to 700, for SDS in high electrolyte solutions[4]) of a wide variety of systems show no significant poly-dispersity.

Non-ionic micelles:

In the case of non-ionic micelles, the terms B_{Don} and B_{MS} would be absent. In Figure 1 we present plots of B_2 versus temperature[13] for different systems. Further, it is seen that in the case of NPE_{10-18} series,[2] Table I, the value of B_2 increases by ~ 6 fold as the shape factor, f_{IS}, increases with an increase in number of ethylene oxide units. The data in Figure 1 show that B_2 decreases with temperature and becomes very small (>0) as micelle aggregation number increases.

Table I. The Magnitudes of Calculated (B_{MM}) and Measured B_2 of Non-ionic Micellar Solutions.[1]

Surfactant	R_m (Å)	M_m	f_{IS}	B_{MM} $(\frac{ml\ mol}{g^2})$	B_2 (measured) (ml mol/g^2)
NPE_{10}	30.5	100.000	1.6	$1.0\ 10^{-4}$	$0.36\ 10^{-4}$
NPE_{13}	36.0	60.000	1.8	$1.8\ 10^{-4}$	$1.2\ 10^{-4}$
NPE_{18}	38.2	55.000	3.0	$5.6\ 10^{-4}$	$1.9\ 10^{-4}$

(NPE_x=nonylphenol with x number of ethylene oxide units).

It is well known that non-ionic amphiphiles show inverse solubility in water (Cloud-point).[2,9,13] The micelles in a 'good-sol-
vent' should then be expected to exhibit larger B_2 values.[11] This
is also observed, Table I, since B_2 for $NPE_{18} > NPE_{13} > NPE_{10}$.

Further, as temperature increases the micelles are increasingly surrounded by a 'poor-solvent', since the system is getting closer to the 'cloud-point'. This leads one to expect that $B_2 \to 0$ as temperature increases ($\to$ cloud-point). This is also observed from the data
in Figure 1. It is also apparent that in order to provide thorough
analysis of any macromolecule solution, the knowledge of R_m is essential. In the current literature, the number of studies where both
N_n or N_w and R_m are given are very few.[2]

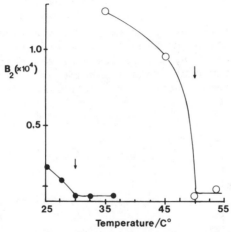

Figure 1. Plots of B_2 (ml mol g^{-2}) versus temperature (°C) for nonionic micelles in aqueous phase: $C_{12}H_{25}O\,(C_2H_4O)_5C_2H_4OH$, ($\bullet$);
$C_{16}H_{33}O(C_2H_4O)_{15}C_2H_4OH$; ($\circ$).   [Data from Reference (13)]

Ionic micelles:

In general the ionic micelles are more interesting than the
non-ionic micelles, since the effect of added electrolytes has much
more drastic effect on B_2 of the former than that of the latter.[12,14]
It is observed that in the relation as given in Equations (4) and
(5), if the addition of electrolyte, m_s, has none or negligible effect on β_{MM} or B_{MS}, and in case N_n (i.e., R_m) remains almost constant, then B_2 should vary linearly with $(1/m_s)$. This behavior has
indeed been reported in the case of various macroions.[7,11] However,

it is also obvious that if the addition of m_s gives rise to an increase in N_n (i.e., R_m) then B_2 will not vary linearly with $(1/m_s)$
as is also observed.[1,4]

   In Figure 2 and Table II are given B_2 data for various micellar systems in the regions m_s where the change in N_n is negligible.
These results show convincingly that even though the absolute values of B_{Don} are much larger in magnitude (both for polymers[7,10,11]
and micelles[1,4]) than B_2 measured, the plots in Figure 2 are linear
in all cases. It is of interest to mention that this linear dependence of B_2 on $(1/m_s)$ is valid for micelles with $N_n \approx 20$ (Na-deoxycholate) to ~ 200. The data in Figure 2 thus show that this dependence is valid for both anionic and cationic micelles, even though
the counter-ions are different in these two systems.

   At high m_s concentrations B_2 becomes zero or negative[1,4] in
most ionic micellar systems. We have ascribed this to a change in
micellar shape from spherical to ellipsoidal. The last argument has
been verified with the help of intrinsic viscosity $|\eta|$ data.[1]
However, light-scattering[14] and ultra-sonic measurements[15] also support these observations convincingly. In order to be able to provide
a more detailed analysis, the radius, R_m, and shape, f_{IS}, data are
needed. The latter measurements are being carried out in this laboratory and will be the subject of a future report.[9]

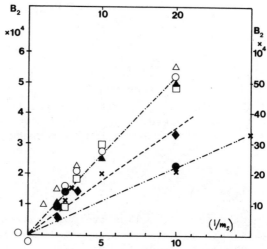

Figure 2. Variation of B_2 versus $(1/m_s)$ in the range of m_s (concentration of added electrolyte) where N_n (i.e., R_m) is found to show
a negligible change. See text for details. (SDS; (O) 30 °C, (□)
40 °C; SUS (●) 40 °C; DTAB; (▲) 30 °C, (△) 40 °C; TTAB (◆) 40 °C;
NaDC (X) 25 °C).

Table II. Dependence of B_2 on $(1/m_s)$ in the Region of m_s (Added Electrolyte) where N_n (and R_m) Show Negligible Change.[1,4,12]

Surfactant	temp.	$B_2{}^{*)} = \quad a_1 + b_1 \ (1/m_s)$	Range of m_s
DTAB	$30^{\,\circ}$	$= -5.5 \ 10^{-6} + \dfrac{4.9 \ 10^{-5}}{m_s}$	0.1-0.8
	$40^{\,\circ}$	$= 5.4 \ 10^{-5} + \dfrac{4.94 \ 10^{-5}}{m_s}$	"
TTAB	$30^{\,\circ}$	$= -2.6 \ 10^{-4} + \dfrac{4.9 \ 10^{-5}}{m_s}$	"
	$40^{\,\circ}$	$= -1.2 \ 10^{-4} + \dfrac{4.6 \ 10^{-5}}{m_s}$	"
SUS	$40^{\,\circ}$	$= 4.9 \ 10^{-5} + \dfrac{1.9 \ 10^{-5}}{m_s}$	0.1-0.7
SDS	$30^{\,\circ}$	$= 3.6 \ 10^{-5} + \dfrac{4.8 \ 10^{-5}}{m_s}$	0.1-0.4
	$40^{\,\circ}$	$= 3.9 \ 10^{-6} + \dfrac{4.94 \ 10^{-5}}{m_s}$	0.1-0.4
NaDC	$25^{\,\circ}$	$= 1.47 \ 10^{-3} + \dfrac{4.9 \ 10^{-5}}{m_s}$	0.03-0.1

DTAB = dodecyl trimethyl ammonium bromide;
TTAB = tetradecyl trimethyl ammonium bromide;
SUS = sodium undecyl sulfate;
SDS = sodium dodecyl sulfate;
NaDC = sodium deoxycholate.
*) Correlation coefficients in all systems were 0.95-0.999.

CONCLUSION

In the case of macroions it has been reported that B_2 varies linearly with $(1/m_s)$ as given in Equation (4). An analogous conclusion can be reached from the B_2 data analyzed here, Figure 2, for the case of different ionic micelles (and under varying experimental conditions). This observation leads us to conclude that micelles should be analyzed by a similar thermodynamic procedure as that used for the macromolecular solutions. The variation of B_2 in the case of non-ionic micelles allows us to conclude that the shape factor, f_{IS}, (Equation (10)) describes the data satisfactorily. The ion-binding term, B_{MS}, is not easily analyzed at this stage. However, as more data become available, the counter-ion binding, β_{MS}, can be estimated from B_2 data (unpublished data). The ratio $N_w/N_n \approx 1$ is found to be valid for a variety of micelles with $N_n=20$ to 700. The values of B_2 as found from light-scattering and osmometry are almost of the same magnitude.

REFERENCES

1. D. K. Chattoraj, K. S. Birdi and S. U. Dalsager, in "Solution Behaviour of Surfactants: Theoretical and Applied Aspects", K. L. Mittal and E. J. Fendler, Editors, Vol. 1 pp. 505-520, Plenum Press, N.Y., 1982.

2. K. S. Birdi, Koll. A. u. Z. Polym. 250, 731 (1972); K. S. Birdi Koll. Z. u. Z. Polym. 252, 551 (1974).

3. K. S. Birdi, S. Backlund, K. Sørensen, T. Krag and S. U. Dalsager, J. Colloid Interface Sci. 66, 118 (1978).

4   K. S. Birdi, S. U. Dalsager and S. Backlund, J. C. S. Faraday I 76, 2035 (1980).

5. S. Backlund, K. Rundt, K. S. Birdi and S. U. Dalsager, J. Colloid Interface Sci. 79, 578 (1981).

6. K. S. Birdi, M. Meyle and E. Stenby, These Proceedings.

7. J. T. Edsall, H. Edelhoch, R. Lontie and P. R. Morrison, J. Am. Chem. Soc. 72, 4641 (1950).

8. A. Isihara and T. Hayashida, J. Phys. Soc., Japan, 6, 46 (1951)

9. K. S. Birdi, to be published.

10. G. Scatchard, J. Am. Chem. Soc. 68, 2315 (1946); G. Scatchard, A. C. Batchelder, A. Brown and M. Zosa, J. Am. Chem. Soc. 68, 2610 (1946).

11a. C. Tanford, "Physical Chemistry of Macromolecules" Wiley, New York, 1961; b) H. Morawetz, "Macromolecules in Solution", Interscience, New York, 1965.

12. K. S. Birdi, Finnish Chem. Lett., 142 (1982).

13. D. Attwood, P. H. Elworthy and S. B. Kayne, J. Phys. Chem. 74, 3529 (1970).

14. P. J. Missel, N. A. Mazer, G. B. Benedek, C. Y. Young and M. C. Carey, J. Phys. Chem. 84, 1044 (1980).

15. H. Høiland, personal communication.

THERMODYNAMICS OF MICELLIZATION AND SOLUBILIZATION OF PENTANOL

IN THE SYSTEM WATER + SODIUM N-OCTANOATE + N-PENTANOL AT 25°C

Part 2. Solubilization of Pentanol

Jarl B. Rosenholm[1] and Per Stenius[2]

[1]Department of Physical Chemistry, Åbo Akademi
 Porthansgatan 3-5, 20500 Åbo, Finland

[2]Institute for Surface Chemistry
 Box 5607, 114 86 Stockholm, Sweden

Excess chemical potentials, enthalpies and entropies have been determined for each of the components in solutions of sodium octanoate and pentanol in water at 298K. The experimental methods include calorimetry, vapour pressure osmometry and head-space gas chromatography. The solubility of pentanol in octanoate solutions is much higher than expected from normal solubilizing capacities of micellar surfactants. The thermodynamic data indicate that this is due to interactions between octanoate and pentanol at concentrations considerably below the cmc. The interaction apparently does not lead to the formation of well-defined micelles. Thus, the system is intermediate between typical "hydrotropic" systems in which increased miscibility without micelle formation is achieved by the addition of an amphiphilic compound and typical micellar systems in which solubility is increased by solubilization.

INTRODUCTION

Sodium octanoate forms micelles in aqueous solutions, although the critical micelle concentration (cmc) is quite high (0.35 mol/dm^{-3} at 298K).[1] Pentanol can be solubilized in the octanoate micelles, but the solubility of pentanol is considerably higher than the solubility of alcohols with longer chains. This can be clearly seen by comparing the ternary phase diagrams for water, sodium octanoate and different n-alcohols.[2]

The pentanol/octanoate system thus is intermediate between systems in which the micellar effects completely predominate (as will be the case for longer-chain homologues) and systems in which the mutual solubility of water and a weakly polar compound is increased by addition of a third amphiphilic compound that does not form typical micellar aggregates, such as would be the case for shorter-chain homologues ("hydrotropy").[3,4] The former effects predominate at high surfactant concentrations while the latter effects are of importance in the concentration region around and just above the cmc.

In this paper we report on fairly detailed experimental investigations of the excess chemical potential, enthalpy and entropy of each of the components in aqueous solutions of the system water/ pentanol /sodium octanoate. We will show that the dualism in the solubilization effects is clearly observed in the partial molar thermodynamic properties of the components.

EXPERIMENTAL

Pentanol activities were determined by measuring the relative vapour pressure of pentanol in the solution with a gas-chromatographic system equipped for head-space analysis (Perkin-Elmer F-11, FID detector, Porapak Q column). Pure pentanol was used as reference. The reproducibility of peak areas was ±2%. The pentanol activity was calculated from

$$a(C_5OH) = \frac{A(\text{sample})}{A(\text{reference})} \tag{1}$$

where A(sample) is the mean value of at least five separate determinations of GC peak areas and A(reference) is the correspondingly calculated area for pure pentanol.

Water activities were measured in a Perkin-Elmer "Molecular Weight Apparatus 15" vapour pressure osmometer. This measurement is based on the heat evolved when water vapour condenses on a solution droplet. Sodium chloride solutions were used as reference. It was checked by separate measurements on NaCl solutions containing pentanol that the heat of condensation of pentanol did not cause systematic errors. Each activity is the mean value of at least five measurements with a standard deviation of about 2%.

Chemicals. Sodium octanoate was prepared by neutralization of octanoic acid (Fluka AG, puriss) with NaOH.[5] The purity was better than 99.5% (checked by titration). Other chemicals were of analytical grade. The water was doubly distilled and ion exchanged with a conductivity < 0.7 μS cm^{-1}.

CALCULATION OF THERMODYNAMIC QUANTITIES

The excess quantities were calculated with an ideal solution of water (component 1), fully dissociated sodium octanoate (component 2) and pentanol (component 3) as the reference solution. Thus, the mole fraction x_1 of component i is given by

$$x_i = \frac{n_i}{n_1 + 2n_2 + n_3} \tag{2}$$

where n_i is the amount of component i.

Excess chemical potentials. These were calculated from

$$G_i^E = RT \ln f_i \tag{3}$$

where $f_i = a_i/x_i$ and a_i is the activity of component i. The activity was directly measured for water and pentanol. The activity of sodium octanoate was calculated by integration of the Gibbs'-Duhem equation.

For solutions of octanoate without pentanol two forms of the integrated equation were used:

$$\ln a_2 = \ln a_2(s) - \int_{a_1'}^{a_1''} (x_2 f_1)^{-1} da_1 =$$

$$= \ln a_2(s) - 2 \int_{\sqrt{a_1'}}^{\sqrt{a_1''}} x_1 x_2^{-1} a_1^{-\frac{1}{2}} d\sqrt{a_1} \tag{4}$$

$\ln a_2(s)$, the lowest value of octanoate activity, was calculated by fitting data for low octanoate concentrations to the extended Debye-Hückel equation [6]

$$\log \gamma_+ = - \frac{A_o \sqrt{m_2}}{1 + B a_o \sqrt{m_2}} \tag{5}$$

using $A_o = 0.511$ $mol^{-\frac{1}{2}}$ $kg^{\frac{1}{2}}$, $B = 3.3$ $mol^{-\frac{1}{2}}$ $kg^{\frac{1}{2}}$, and a_o as an[6] adjustable parameter. The best least squares' fit was obtained for $a_o = 1.25$ nm giving $\ln a_2(s) = -13.9197$ at $x_2 = 0.00110$. Activ-

ities calculated from the two forms of Equation (4) deviated by
about 2.5%. Deviations from previously reported data[7],[8] are also
about 2,5%.

The octanoate activity in ternary solutions was calculated
from the integrated Gibbs'-Duhem equation written in the form

$$\ln a_2 = \ln a_2(s) - \int_{a_1'}^{a_1''} x_1 x_2^{-1} d(\ln a_1) - \int_{a_3'}^{a_3''} x_3 x_2^{-1} d \ln a_3 \qquad (6)$$

The starting value for a_2 ($a_2(s)$) was taken from binary solutions and the lower limit of a_3 (a_3') was determined by extrapolation of plots of a_3 versus x_3 to $x_3 = 0$ for each ratio x_1/x_3. It
is estimated that the uncertainty in $\ln a_2$ is a few percent.

Excess enthalpies and entropies. Partial molar enthalpies
were determined calorimetrically and are reported in ref.9. Excess
entropies were calculated from the Gibbs-Helmholtz equation. The
uncertainty in the entropies is about ± 5%.

A full account of the details of the calculations and the
experimental results will be given in a subsequent paper.

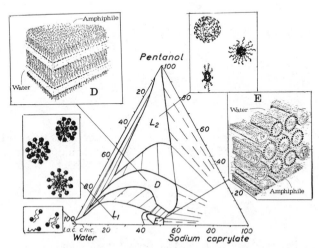

Figure 1.   Phase diagram at 293 K for the system water/sodium
octanoate (caprylate)/pentanol. Concentrations in weight-%.

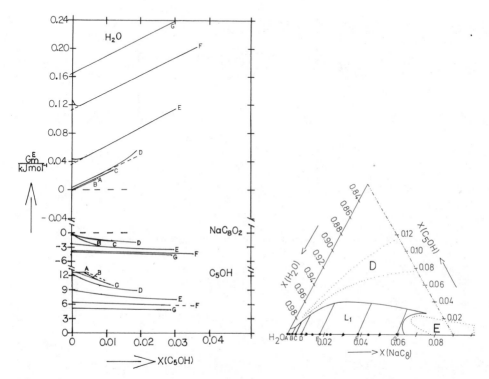

Figure 2.   Excess chemical potentials for water, sodium octanoate
and pentanol as a function of the mole fraction of added pentanol
at different molar ratios $n(H_2O)/n(NaC_8)$= A: pure water, B: 368;
C: 221; D: 134; E: 55.2; F: 24.1; G: 14.9.   The insert gives the
location of the lines.

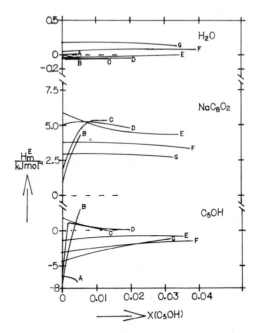

Figure 3.   Excess partial molar enthalpies for water, sodium octanoate and pentanol as a function of the mole fraction of added pentanol.   Molar ratios $n(H_2O)/n(NaC_8)$ as in Figure 2.

RESULTS

Figure 1 shows the complete phase diagram for the three components.[2]   The aqueous solution L_1 at very low surfactant concentrations (below the cmc of octanoate) is in equilibrium with pentanol solution; at higher concentrations lamellar phase is formed. Saturated octanoate solution is in equilibrium with hexagonal phase.   This phase is able to incorporate some pentanol.

Experimental determinations of water and pentanol activities were made in such a way that the thermodynamic quantities could easily be calculated for constant water/octanoate ratios.   These were chosen over the whole range of concentrations.   The results are given in Figures 2, 3, and 4.

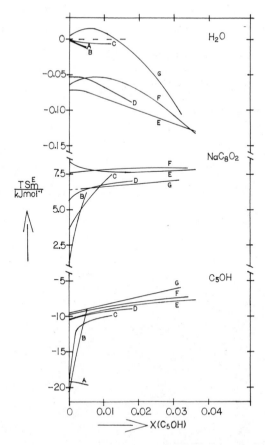

Figure 4.   Partial molar excess entropies (given as TS_m^E) for water, sodium octanoate and pentanol as a function of the mole fraction of added pentanol. Molar ratios $n(H_2O)/n(NaC_8)$ as in Figure 2.

DISCUSSION

Solubilization of Pentanol

Figure 5 compares the solubility of pentanol with the solubility of decanol in sodium octanoate solutions. The concentrations are given in weight-%. In order to compare solubilizing capacities in moles/surfactant the decanol concentrations should

therefore be multiplied by the molecular weight ratio M(pentanol)
/M(decanol) = 0.551; the result is shown in Figure 5.

Decanol is insoluble in pure water and very slightly soluble
in octanoate solutions below the cmc. The increase in solubility
above the cmc is entirely due to solubilization in the octanoate
micelles. Pentanol is much more soluble (about 2.5 weight-%) in
water but it is nevertheless obvious that the solubility above the

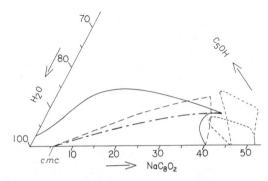

Figure 5. Solubilization (given as a partial three-component phase
diagram) of pentanol (————) and decanol (-------) in sodium
octanoate solutions. (-·-·-·-) solubilization of decanol multiplied
by molecular weight ratio M(pentanol)/M(decanol).

cmc is much larger than would be expected from the larger solubility
in water or the solubilizing capacity of the micelles for the
longer-chain homologue. It can be suggested that this is due to
either a strongly increased tendency to form micelles as pentanol
is added or to interactions between pentanol and octanoate without the formation of typical micelles. The continuous increase
in pentanol solubility as octanoate is added far below the cmc
shows that the latter effect must be important.

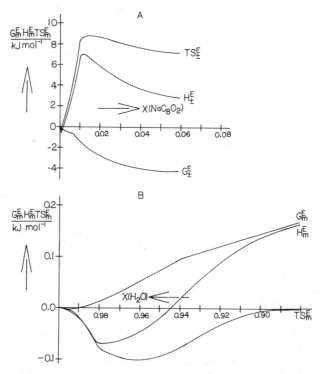

Figure 6. Excess chemical potentials, partial molar enthalpies and partial molar entropies in aqueous solutions of sodium octanoate A: NaC_8; B: water. [10]

Further evidence is given by the solubility in concentrated octanoate solutions. The solubility is gradually reduced and when the solution is saturated with octanoate the solubilizing capacities for pentanol and decanol are equal. At this concentration there will be very little free intermicellar solution into which pentanol can dissolve.[5,10-13]

The addition of an amphiphilic solubilizate to micelles is expected to lower the cmc[14], but it is nevertheless obvious that the very drastic increase in pentanol solubility at low octanoate concentrations can hardly be explained by this effect.

Aqueous Solution of Sodium Octanoate

It is of interest to compare the partial molar quantities illustrated in figures 2-4 with the corresponding quantities for the binary system sodium octanoate/water. To facilitate such comparision, figure 6 shows excess chemical potentials, entropies and enthalpies for this system[10]. A considerable number of investigations of various surfactants (e.g. ref. 30-34) show that this behaviour is typical of surfactants. The concentration dependence of the excess quantities has been interpreted in terms of hydrophobic interactions.

Excess Quantities at Low Concentrations

In Figures 2-4, curves B-C represent concentrations which are below the cmc of sodium octanoate. Curve A refers to solutions of pentanol in water. Such solutions (within the experimental accuracy of our measurement) show Henry's law behaviour; the enthalpy effects are very small (note that the standard state for pentanol is pure pentanol). As soon as octanoate is added to the solutions, however, the situation changes completely. Marked effects cannot be seen in the excess quantities of water. This is not surprising considering the high dilution of the solutions. The excess enthalpy and entropy of octanoate increases in a way very similar to the behaviour illustrated in Figure 5 (curves B and C, Figures 3 and 4). The excess entropy and enthalpy of pentanol changes in much the same way as that of octanoate. The result is a decrease in the chemical potential which is also very similar to that of octanoate in binary solutions (cf. curves B and C for octanoate in Figure 2 with the free energy curve in Figure 5). In particular, note that the excess enthalpy of pentanol which is added to octanoate concentrations below the cmc shows a maximum (curve C, Figure 3) around $x(C_5OH) \approx 0.002$.

The slope of the excess entropy curve (curve C, Figure 4) changes rapidly in the same concentration range. Thus, both quantities change in much the same way as the corresponding quantities for octanoate at somewhat higher concentrations. An obvious interpretation is that addition of pentanol promotes association in the octanoate solution and that the pentanol takes part in this association process.

Excess Quantities at Higher Concentrations

When pentanol is added to solutions above the cmc the excess quantities change much less than below the cmc. The enthalpies

of all three components as well as the entropies and free energies
of octanoate and pentanol are almost independent of the pentanol
concentration (with the possible exception of curve E for pentanol)
(figure 2). This must be due to an almost ideal mixing of pentanol
with the surfactant in the micelles. Pentanol is mainly incorporated between the octanoate chains. The insensitivity of the
partial molar quantities to the concentration of pentanol in the
aggregates implies that the thermodynamic properties, once large
aggregates have been formed, cannot be strongly dependent on local
interactions between the molecules or on intermicellar interactions.
In spectroscopic studies of counter ion binding[7],[21] the conclusion
has been drawn that binding is determined above all by the alcohol/
surfactant ratio in the micellar surface. It is somewhat surprising
that no effects of changes in counterion binding are seen in the
excess quantities of either pentanol or octanoate.

The excess entropy of water (curves E-G, Figure 4) goes through
a broad maximum. The interpretation is not straightforward. An
increased incorporation of octanoate and pentanol into aggregates
would lead to an increase in the entropy of water. Incorporation
of large amounts of pentanol into the octanoate micelles must lead
to a marked change in their surface charge density and, as a
result, changing intermicellar interactions (binding of counter
ions). At the highest concentrations of octanoate and pentanol
all of the water in the system is bound in hydration layers[5],[10],
[11-13] All these effects will influence the entropy of water.

Obviously, intermicellar effects are mainly reflected in the
thermodynamic properties of water. This is consistent with our
interpretation of the thermodynamic properties of octanoate and
pentanol and also with recently published theoretical calculations
of phase diagrams for ionic surfactant systems.[22],[23]

Thermodynamic Properties of Pentanol at Infinite Dilution

Extrapolation of the results in Figures 2-4 to infinite dilution will show the effect of dissolving pentanol in pure sodium
octanoate solutions. The result is shown in Figure 7.

These curves are very similar to the curves for sodium
octanoate (Figure 5). The enthalpy increases rapidly with the
concentration of octanoate and goes through a maximum at a concentration slightly lower than the corresponding maximum for pure
sodium octanoate.

The entropies can be compared in the same way. Pentanol
obviously interacts strongly with octanoate at concentrations con-

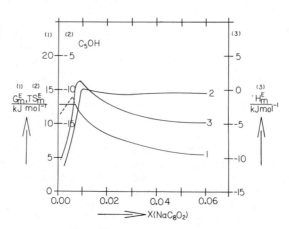

Figure 7.   Values of G_m^E, H_m^E and TS_m^E of pentanol at infinite dilution in sodium octanoate solutions.   Standard state: pure pentanol.

siderably below the cmc.   The thermodynamic properties of pentanol
in these solutions – in contrast to the behaviour in binary solutions of pentanol in water – are those generally associated with
"hydrophobic hydration".   However, micelle formation apparently
is not induced at concentrations much below those corresponding
to the enthalpy maximum in Figure 7.   This is clearly indicated
by the chemical potential of pentanol, shown in greater detail in
Figure 8.   This figure also gives an indication of the accuracy
of calculations based on activity measurements (which were made
for a larger number of octanoate concentrations than the determinations of enthalpies).

The free energy of pentanol below the cmc does not depend on
the octanoate concentration.   As expected, the free energy of
pentanol strongly decreases upon incorporation into the micelles;
the cmc of NaC_8 is ≈ 0.007.

Around $x(NaC_8)$ ≈ 0.03 - 0.04 the slope of the G_m^E curve changes.
A very weak minimum in the excess entropy can also be observed
(Figure 7, curve 2).   The change of the solution properties in this
concentration range have been interpreted as a change towards
elongated micelles. 24-26

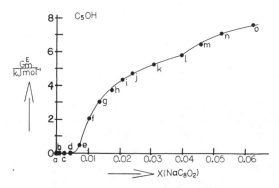

Figure 8. A comparison of free energies G_m^E of pentanol at infinite dilution in sodium octanoate solutions. Each point was determined by extrapolation of pentanol activities measured as a function of the concentration of pentanol at constant molar ratio water/octanoate. The free energy of pentanol in pure water is arbitrarily taken as zero.

CONCLUSIONS

The solubility of pentanol in sodium octanoate solutions around and just above the cmc is much higher than normal micellar solubilizing capacities or the solubility of pentanol in water. This increased solubility is accompanied by interactions between pentanol and octanoate that cannot be directly associated with the formation of well-defined micellar aggregates with a constant solubilizing capacity (such as is the case for, e.g., decanol).

Recent studies show that mixtures of highly mobile, short-chain ionic and nonionic surfactants may form highly flexible interfaces associated with the formation of aggregates much less well-defined than typical micelles [11,26]. Stilbs [27] compared the self-diffusion of long- and short-chain alcohols in sodium dodecyl sulphate solutions. The self-diffusion of long-chain alcohols was consistent with the incorporation of these alcohols into micelles. At low concentrations of butanol, data were also entirely consistent with a distinct distribution of butanol between water and micelles.

At approximately equal volumes of butanol and SDS, however, the micellar structure breaks down. On the basis of our thermodynamic data one may draw the conclusion that the octanoate/pentanol system behaves in a similar way. It may represent a system in which addition of a cosurfactant to a micellar solution creates the proper conditions for the formation of highly flexible interfaces that seemingly are a prerequisite for the formation of typical microemulsions with high concentrations of water and oil ("surfactant phase").[26,28,29]. A statistical thermodynamical analysis of the properties of these types of solutions thus probably cannot be done in terms of the commonly used models of micelle formation. A possible first approach would be the calculation of intermolecular affinities by application of the Kirkwood-Buff theory of solutions [34]. The use of this method for the study of hydrophobic interactions in aqueous solutions has been discussed by Ben-Naim[35,36]. The analysis, however, requires a knowledge of the heat capacities or compressibilities of the solutions which are not available for our system. Our results thus stress the need for more complete thermodynamic data in order to obtain a better understanding of solubilization phenomena.

ACKNOWLEDGEMENTS

This work was supported by grants from the National Science Research Council of the Finnish Academy, the Neste Foundation and the Foundation of Åbo Akademi.

REFERENCES

1. P. Ekwall and P. Holmberg, Acta Chem. Scand., 19, 455 (1965).
2. P. Ekwall, in "Liquid Crystals", G. Brown, Editor, Vol. 1, p. 1, Academic Press, New York, 1975.
3. C. Neuberg, Biochem. Z., 76, 107 (1916).
4. R. Durand, C. R. Acad. Sci., Paris, 223, 898 (1946).
5. J. B. Rosenholm and B. Lindman, J. Colloid Interface Sci., 57, 362 (1976).
6. R. A. Robinson and R. H. Stokes, "Electrolyte Solutions", 2nd revised ed., ch. 2, Butterworths, London, 1965.
7. E. R. B. Smith and R. A. Robinson, Trans. Faraday Soc., 38, 70 (1942).
8. P. Ekwall, H. Eikrem and P. Stenius, Acta Chem. Scand., 21, 1639 (1967).
9. M.-R. Hakala, J. B. Rosenholm and P. Stenius, J. Chem. Soc. Faraday Trans. I, 76, 473 (1980).
10. I. Danielsson, J. B. Rosenholm, P. Stenius and S. Backlund, Progr. Colloid Polymer Sci., 61, 1 (1976).
11. D. H. Smith and S. A. Templeton, J. Colloid Interface Sci., 68, 59 (1979).

12. K. Larsson, Chem. Phys. Lipids, 9, 181 (1972).
13. B. Lindman and B. Brun, J. Colloid Interface Sci., 42, 388 (1973).
14. K. Shinoda, J. Phys. Chem., 58, 1136 (1954).
15. C. Tanford, "The Hydrophobic Effect", 2nd ed., Wiley, New York, 1973.
16. P. Ekwall and O. Harva, Finska Kemistsamf. Medd., 52, 257 (1943).
17. P. Ekwall and P. Stenius, Acta Chem. Scand., 21, 1767 (1967).
18. S. G. Cutler, P. Meares and D. G. Hall, J. Chem. Soc. Faraday Trans. I, 74, 1758 (1978).
19. J. B. Rosenholm, Colloid Polymer Sci., 259, 1116 (1981).
20. P. Ekwall, I. Danielsson and P. Stenius, in "Surface Chemistry and Colloids", M. Kerker, Editor, MTP Int. Rev. Sci. Phys. Chem. Ser. 1, Vol. 7, p. 97, Butterworths, London, 1972.
21. J. B. Rosenholm, T. Drakenberg and B. Lindman, J. Colloid Interface Sci., 63, 538 (1978).
22. G. Gunnarson, B. Jönsson and H. Wennerström, J. Phys. Chem., 84, 3114 (1980).
23. B. Jönsson, H. Wennerström and B. Halle, J. Phys. Chem., 84, 2179 (1980).
24. I. Danielsson, Finska Kemistsamf. Medd., 72, 108 (1963).
25. P. Ekwall and P. Holmberg, Acta Chem. Scand., 19, 573 (1965).
26. B. Lindman, P. Stilbs and M. E. Moseley, J. Colloid Interface Sci., 83, 569 (1981).
27. P. Stilbs, J. Colloid Interface Sci., 87, 385 (1982).
28. S. Kislalioglu and S. Friberg, in "Theory and Practice of Emulsions", A. L. Smith, Editor, p. 257, Academic Press, New York, 1976.
29. E. Sjöblom and U. Henriksson, in these proceedings.
30. S. Lindenbaum, J. Chem. Thermodyn. 3, 625 (1971).
31. R. De Lisi, G. Perron and J.E. Desnoyers, Can. d. Chem. 58, 959 (1980).
32. J.E. Desnoyers, R. De Lisi and G. Perron, Pure Appl. Chem. 52, 433 (1980).
33. R. De Lisi, G. Perron, J. Paquette and J.E. Desnoyers, Can. J. Chem. 59, 1865 (1981).
34. J.G. Kirkwood and F.P. Buff, J. Chem. Phys. 19, 774 (1951).
35. A. Ben-Naim, "Hydrophobic Interactions", Plenum Press, New York, 1980, p. 78f.
36. A. Ben-Naim, J.Chem. Phys., 67, 4884 (1977).

DISSOLUTION AND MICELLIZATION OF LONG-CHAIN ALKYLSULFONIC ACIDS AND THEIR SODIUM SALTS IN WATER

M. Saito[*], Y. Moroi[**] and R. Matuura[**]

[*]
Department of Chemical Science, Faculty of Education
Yamaguchi University, Yoshida, Yamaguchi 753, Japan
[**]
Department of Chemistry, Faculty of Science, Kyushu
University, Hakozaki, Fukuoka 812, Japan

The properties of aqueous solution of alkylsulfonic acids and their sodium salts from 10 to 22 carbon atoms were studied. For both surfactants, Krafft points increase with increasing alkylchain, while the CMC values decrease with increasing alkylchain indicating a linear relation between logCMC and carbon number. Solubilities of sodium alkylsulfonates also decrease with increasing alkylchain with rate of the decrease diminishing with carbon numbers in the chain. The thermodynamic parameters for dissolution of sodium salts, $\Delta G°$, $\Delta H°$ and $\Delta S°$ were evaluated from the solubility change with temperature at 20°C. Since it was found that the $\Delta S°$ values decreased almost linearly with alkylchain length, water molecules around the alkylchain play a very important role in the dissolution behavior, and the dissolved alkylchain seems to remain stretched, not coiled. Comparison was made on micellar and dissolution behaviors between sodium alkylsulfonates and corresponding sulfonic acids. Differences of Krafft points between sulfonic acids and their sodium salts could be attributed mainly to the energy difference in their solid states. The slight difference in CMC value between them could be due to the difference in degree of the micellar dissociation between H^+ and Na^+ ions.

771

INTRODUCTION

The shape of dissolved amphiphilic molecules in dilute aqueous solution has been discussed not only from a colloid chemical point of view but also from a physiological interest. Mukerjee[1] analyzed the data obtained by Goodman[2] on the distribution of long-chain fatty acids between heptane and an aqueous phosphate buffer to evaluate the dimerization constants of ionized monomer species. He suggested the possibility of a sudden transition of the alkyl-chain from an extended structure to a coil between 16 and 18 carbons in the chain. On the other hand, from more careful study by Smith and Tanford[3] on the partitioning of the acids, it was shown that a linear relationship exists between the difference of the standard chemical potentials of the acids in water and in hydrocarbon for acids up to behenic acid (22 carbon atoms). The result by Smith and Tanford is inconsistent with the former suggestion by Mukerjee.

A few investigations concerning this problem have been made from the view-point of micelle formation of surfactants[4-6] However, usual ionic surfactants with a long alkyl-chain of more than 16 carbon atoms have a high Krafft point. This disadvantage not only makes it difficult to do experiment at an ordinary temperature but also leads to chemical change of the surfactants. Nonionic surfactants with a long alkyl-chain, on the other hand, must have a very long hydrophilic portion in the molecule in order to dissolve in water around room temperature. Hence, it becomes extremely difficult for such nonionic surfactants to be mono-disperse with respect to the hydrophilic portion. Thus, the use of nonionic surfactants will render the investigation of the above matter meaningless, too.

In the light of the above discussion, it is truly desirable to prepare ionic surfactants with more than 16 carbon atoms which have a Krafft point near room temperature. We could synthesize the ionic surfactants, alkylsulfonic acids, which have 10 (decyl sulfonic acid) to 22 (docosyl sulfonic acid) carbon atoms. Their CMC measurements were easily made on account of their relatively lower Krafft points. The most important point is that these surfactants are very stable chemically against both H^+ or OH^- ions and temperature. Alkylsulfonic acids were prepared by exchanging the counterion of sodium salts with hydrogen ion.

Few reports concerning the properties of long chain sodium alkylsulfonates have been published in spite of their importance in practical use as typical and stable anionic surfactants[7,8] Tartar et al. reported the solubilities and CMC values in aqueous solution of the sodium salts of 10 to 18 carbon atoms[9-10]

However, they did not measure the solubilities at temperatures far below the corresponding Krafft points and, in addition, did not discuss the results from thermodynamic point of view. Since the solubility and the Krafft point are specified with a solid - solution equilibrium, the solid structure must be taken into account for discussion of these properties of surfactants in aqueous solution. Therefore, we investigated properties of the solid state, in addition to those of aqueous solution, of sodium alkylsulfonates by the differential thermal analysis (DTA) and X-ray diffraction analysis.

Further, we measured the Krafft point, CMC, and the degree of dissociation of micelles of both surfactants and, comparing these results, discussed on possible shapes of dissolved amphiphilic molecules. We also measured the solubilities of the sodium salts in water by considering their solid structure and have compared these results with literature data. Also we have discussed dissolution behavior of amphiphilic molecules paying attention to the thermodynamic parameters for dissolution.

EXPERIMENTAL

Materials

Alkylsulfonic acids and their sodium salts were prepared by the method described before[12,13] and purified by ether extraction and by recrystallization from water.

Krafft Point

Conductance and temperature of the aqueous surfactant solution, containing the suspended solid, were determined with conductivity cell and thermocouple, respectively. The Krafft points were determined as the temperature at the intersection of the two straight lines obtained by plotting the conductance of the solution against reciprocal of the absolute temperature.

DTA

Thermal transitions of the solid sodium alkylsulfonates were analyzed with an Al_2O_3 reference and at a heating rate of 2.3°C/min using Shimadzu DTA apparatus. Sodium salts used for DTA analysis were kept in a desiccator containing pure water for two days and were equilibrated with water vapor.

X-ray Diffraction

X-ray diffraction analysis of the solid sodium alkylsulfonates were made by Rigaku X-ray Diffractometer of Model-DF,
with Ni-filtered Cu-α radiation and with a scan-speed of 2°/min.

Solubility

The 10ml injection cylinder containing the surfactant as a
suspension was dipped into a thermostat controlled within ± 0.01
°C, and the solution in the cylinder was stirred by means of a
disk rotor. Saturated solution was filtered by applying a
pressure on the injector and the filtrate was transferred into a
preweighed flask and then diluted to suitable concentration with
water for its conductance measurement. Mechanical agitation for
less than one hour was required to reach equilibration.
Conductance of the solution was measured and concentration was
determined using calibration curves. Membrane filters used were
of 0.22 μm pore size (Millipore. FGLP 01300).

CMC

The CMC's of sodium alkylsulfonates were determined by the
conductivity method as the concentration at the intersection of
the two lines obtained by plotting the specific conductance
against concentration. The CMC's of alkylsulfonic acids were
determined by the conductivity method as well as by the pH method.
In the pH method it was taken as the concentration at the intersection of the two lines obtained by plotting the pH against
logarithm of concentration.

RESULTS AND DISCUSSION

Krafft Point

The Krafft points of alkylsulfonic acids and their sodium
salts are plotted against number of carbons in the alkylchain in
Figure 1. The Krafft points of alkylsulfonic acids are lower
than those of the corresponding sodium salts, and the shorter
the alkylchain of surfactants, the difference of their Krafft
points becomes greater. This means that the hydrophilic portion
of a surfactant, a gegenion in particular, plays a more
important role than the hydrophobic portion for shorter chain
surfactants and vice versa for longer surfactants, because the
homologous hydrophobic and the same hydrophilic ionic portions
are involved.

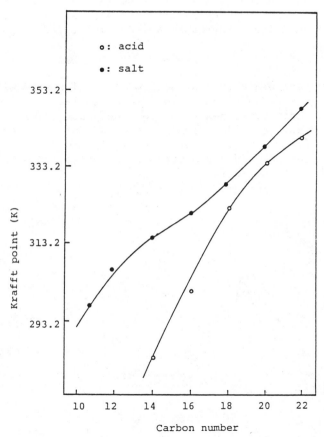

Figure 1.   Comparison of Krafft points of alkylsulfonic acids with those of sodium salts.

The Krafft point is defined as the temperature at which the solubility of ionic surfactants becomes high enough for singly dissolved molecules to aggregate to micelles[14] Evidently the solubility depends on free energy difference between a solid and a dissolved states of molecules. Difference in Krafft point between sulfonic acids and their sodium salts can be attributed mainly to the energy difference in their solid states, because their CMC difference is rather small. Thus, the energy state of sodium alkylsulfonates is lower in a solid states than that of corresponding acids, which leads to a smaller solubility of the former at a definite temperature and to higher Krafft point.

DTA

The typical DTA curves of sodium alkylsulfonates with C_{12} is shown in Figure 2. Every curve for C_{10} up to C_{22} is composed of three sharp endothermic peaks due to some phase transitions over a broad band due to evaporation of adsorbed water. Three transition temperatures were numbered as T_1, T_2 and T_3 from low to higher temperature, and are plotted in Figure 3 against the carbon number of the surfactants.

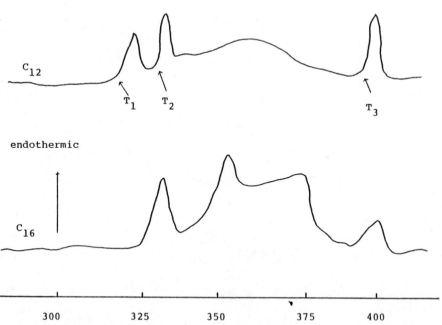

Figure 2. Phase transitions and corresponding endothermic peaks for sodium alkylsulfonate with C_{12} on DTA curves.

All these transition temperatures increase with increasing
carbon number of the alkylchain. What these transitions mean
crystallographically is not clear, but it is evident that the

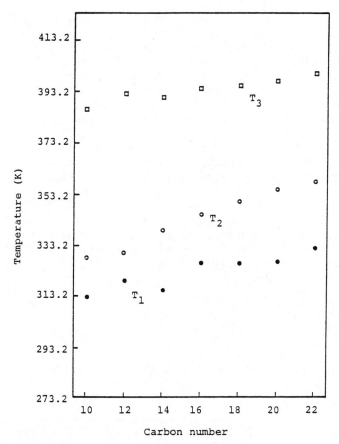

Figure 3.   Transition temperatures with carbon number of sodium
alkylsulfonates.

phase below the first transition temperature (T_1) can be assumed
to have the same type of lattice. A phase transition between
solid states, i.e., a change of crystal structure, has a strong

influence on solution behavior such as solubility; so it is
important to make solubility determination at temperatures below
T_1. The mole ratio of adsorbed water to surfactant at 296.2K
increases from 0.3 to 1.7 with decreasing alkychain from C_{22}
down to C_{10}.

X-ray Diffraction

    X-ray diffraction patterns were analyzed by the method of X-
ray crystallography to yield a long-spacing in a direction perpendicular to the plane of the crystal. The postulated model of
sodium alkylsulfonates in the solid state is shown in Figure 4.
The length (l) of the solid surfactants with carbon number (n_c)
is calculated from the following expression, $l = 4.58 + 1.265n_c$,
where 4.58 (Å) is obtained from an atom-to-atom distance of the
head group and van der Waals radius of the terminal methyl group,
and 1.265 (Å) is obtained from the distance of 2.53 Å between
alternate carbon atoms of a fully extended alkylchain. The angles
(θ) of alkylchain of surfactant molecule against the plane made
of head groups was calculated from the long-spacing. The long-
spacings ($L= 2l\sin\theta$) and the angles (θ) thus obtained are summarized in Table I. The long-spacing increases with increasing
carbon atoms up to C_{18} surfactant, but the angle does not show
any regularity. As a result, information which is useful enough
to discuss the influence of the solid structure on dissolution
behavior could not be obtained from X-ray diffraction data.

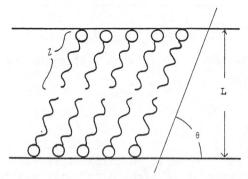

Figure 4. Model of sodium alkylsulfonate in the solid state.

Table I. Long-Spacing and Inclination of Alkylchain in Solid State.

Carbon number	Long-spacing	θ
C_{10}	29.24(Å)	58° 3'
C_{12}	31.84	53°44'
C_{14}	37.68	57°41'
C_{16}	38.38	50°38'
C_{18}	40.70	48°40'
C_{20}	32.19	32°35'
C_{22}	36.24	33°59'

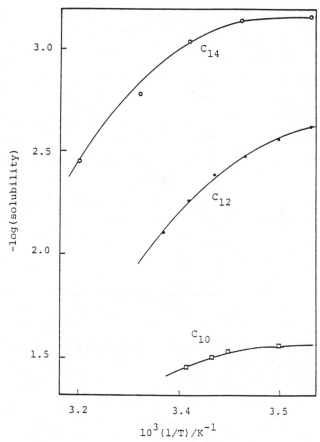

Figure 5.   Solubility change with temperature for C_{10}, C_{12}, and C_{14} salts.

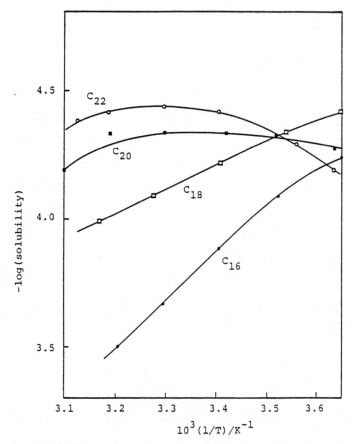

Figure 6. Solubility change with temperature for C_{16}, C_{18}, C_{20} and C_{22} salts.

Solubility

Solubilities at temperatures higher than the first phase transition (T_1) were not determined since a change of dissolution behavior would have occurred above T_1, and the suspended solution became so viscous that the filtration was difficult above T_1. Solubility data for sodium alkylsulfonates from 10 up to 18 carbon atoms obtained by Tartar and Wright[9] are in agreement with our data. In order to calculate the enthalpy change of the dissolution, values of log(solubility) are plotted against reciprocal of the absolute temperature in Figures 5 and 6.

Below the Krafft point, a singly-dispersed sodium alkyl-sulfonate can be assumed to be completely dissociated[15,16] So we can calculate the thermodynamic parameters for dissolution by using the following equations,

$$\Delta G° = -2RT\ln S \tag{1}$$

$$\Delta H° = -2RT[\partial\ln S/\partial(1/T)]_p \tag{2}$$

$$\Delta S° = (\Delta H° - \Delta G°)/T \tag{3}$$

where S refers to solubility. The values of $[\partial\ln S/\partial(1/T)]_p$ are obtained from the solpes of Figures 5 and 6 at 293.2 K. The parameters thus obtained are summarized in Table II, where values for C_{10} surfactant are omitted because of its relatively high concentration. The thermodynamic parameters for dissolution are plotted against carbon number of alkylchain and are shown in Figure 7.

The table and the figure show that the $\Delta G°$ values increase, whereas $\Delta H°$ and $\Delta S°$ decrease with increase in the carbon number of alkylchain. The increase of $\Delta G°$ with increasing chain length indicates that solubility becomes less with increasing alkyl-chain.

Only from the thermodynamic point of view, dissolution of surfactant molecule in the solid state must lead to an entropy increase. However, the results are contrary to expectation for C_{18}, C_{20} and C_{22} surfactants. Therefore, it can be considered that water molecules around a dissolved hydrocarbon chain become structured, i.e., entropy decreases on its dissolution[17] The longer the alkylchain, the greater the number of structured water molecules. This effect plays a more important role than positive entropy change due to its dissolution. At any rate, entropy decrease accompanied by a transfer of hydrophobic solute from hydrocarbon phase to an aqueous phase cannot be explained without considering the contribution of water molecules around hydrophobic solute molecules in aqueous phase to the total entropy.

Table II. Thermodynamic Parameters for Dissolution of Sodium
Alkylsulfonates in Water at 293.2K.

Carbon number	$\Delta G°$ (kJ mol^{-1})	$\Delta H°$ (kJ mol^{-1})	$\Delta S°$ (JK^{-1}mol^{-1})
C_{12}	25.2	102	261
C_{14}	33.7	70.9	127
C_{16}	43.6	67.0	80
C_{18}	47.6	32.6	−51
C_{20}	48.7	≈ 0	−166
C_{22}	49.8	−11.5	−209

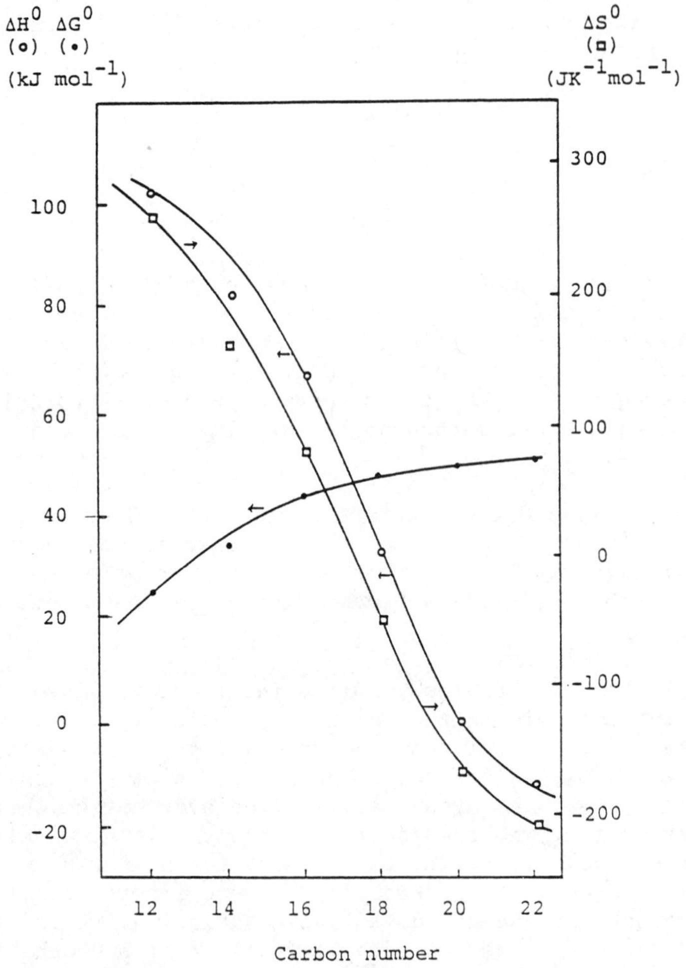

Figure 7. Thermodynamic parameters for dissolution with carbon
number.

CMC

The CMC's of the alkylsulfonic acids and their sodium salts are summarized in Tables III and IV. The CMC values of the alkyl-sulfonic acids are lower than those of the sodium salts. This is due to the lower degree of dissociation of H^+ ion than Na^+ ion at the micellar surface, which is also reflected in the lower Krafft temperatures of alkylsulfonic acids. A similar difference between H^+ and Na^+ is often seen in a difference in colligative properties of aqueous electrolyte solution between H^+ and Na^+ ions. At any rate, the difference in dissociation of the head group is due to a larger radius of hydrated sodium ion than that of the hydrated hydrogen ion.[18,19]

Variations of CMC values with carbon number of alkylchain are shown in Figure 8. A linear relation between log (CMC) and carbon number of alkylchain is shown for both surfactants. From this result, it cannot be said that any sudden change in the shapes of the dissolved amphiphilic molecules truly occurs in an aqueous solution by increasing the alkylchain length of these molecules.

Degree of Dissociation of Micelles

We have investigated properties such as Krafft point, CMC, and solubility of aqueous solution of alkylsulfonic acids and their sodium salts, and suggested that the difference in the properties of the two surfactants is due to the difference in the degree of dissociation of their micelles. Because of its importance, we have estimated α values for both surfactants using two methods and have discussed the results.

Table III. CMC Values of Alkylsulfonic Acids in mol dm^{-3}.

carbon number	303.2 K*	303.2 K	323.2 K	343.2 K
C_{10}	3.0×10^{-2}	—	—	—
C_{12}	7.3×10^{-3}	0.78×10^{-2}	1.02×10^{-2}	1.30×10^{-2}
C_{14}	2.4×10^{-3}	2.38×10^{-3}	2.80×10^{-3}	4.05×10^{-3}
C_{16}	5.8×10^{-4}	5.92×10^{-4}	8.05×10^{-4}	1.11×10^{-3}
C_{18}	—	—	2.26×10^{-4}	3.17×10^{-4}
C_{20}	—	—	—	1.02×10^{-4}
C_{22}	—	—	—	3.24×10^{-5}

* by pH measurement

Table IV. CMC Values of Sodium Alkylsulfonates in mol dm^{-3}.

carbon number	323.2 K	328.2 K	338.2 K	348.2 K
C_{10}	3.85 x10^{-2}	—	—	—
C_{12}	1.15 x10^{-2}	1.20 x10^{-2}	1.52 x10^{-2}	2.12 x10^{-2}
C_{14}	3.45 x10^{-3}	3.90 x10^{-3}	4.00 x10^{-3}	5.45 x10^{-3}
C_{16}	1.01 x10^{-3}	1.05 x10^{-3}	1.26 x10^{-3}	1.62 x10^{-3}
C_{18}	—	2.21 x10^{-4}	3.47 x10^{-4}	5.20 x10^{-4}
C_{20}	—	—	1.66 x10^{-4}	1.90 x10^{-4}
C_{22}	—	—	—	6.10 x10^{-5}

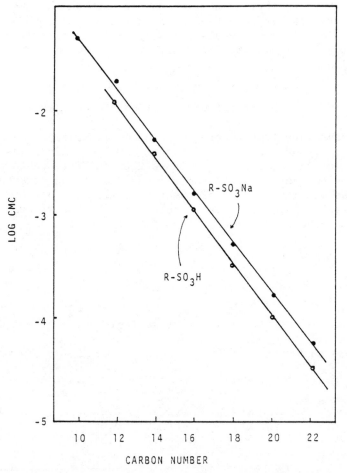

Figure 8.   Linear relationship between log CMC and carbon number.

i) <u>From conductivity measurement.</u>   Ionic equilibria in micellar solutions are discussed on the basis of micelle-monomer equilibrium, mass-conservation, and electroneutrality. One obtains Equation (4) from mass balance,

$$C = C_s + nC_m \qquad\qquad (4)$$

where C is the total (analytical) concentration of surfactant, C_s and C_m are the concentrations of monomer and micelle, respectively, and n is the aggregation number of a micelle. Electroneutrality also requires that

$$C_i = C_s + \alpha nC_m \qquad\qquad (5)$$

where C_i is the concentration of free counterions, and α is a degree of counterion dissociation from a micelle. Combining Equations (4) and (5), one obtains[20]

$$C_i = C_s + \alpha(C - C_s) \qquad\qquad (6)$$

Specific conductivity κ_{sp} is expressed as follows

$$\kappa_{sp} = C_i\lambda_i + C_D\lambda_D + C_m\lambda_m \qquad\qquad (7)$$

where C_D is the concentration of surfactant monomer anion, λ_i, λ_D and λ_m are the equivalent conductivities of counterion, monomer surfactant anion and micelle, respectively. Concentration of free counterion C_i is expressed as

$$C_i = \alpha C + (1-\alpha)C_0 \qquad\qquad (8)$$

where C_0 is CMC ; and above the CMC, C_D is assumed to be constant (= C_0). Combining Equations (8) and (7), one obtains

$$\kappa_{sp} = (\lambda_i + \lambda_D)C_0 + (\alpha\lambda_i + \lambda_m/n)(C - C_0)$$

$$= \kappa_{sp}^{cmc} + (\alpha\lambda_i + \lambda_m/n)(C - C_0) \qquad\qquad (9)$$

Equation (9) indicates a straight line between conductance and concentration above CMC as is shown in Figure 9, and the slope gives the value of $(\alpha\lambda_i + \lambda_m/n)$. Limiting conductivities of gegen ion[18] are used for values of λ_i assuming a complete dissociation of the surfactants. Thus, α is obtained from the slope by assuming $\alpha\lambda_i \gg \lambda_m/n$. The values of α for both surfactants thus obtained are presented in Table V.

ii) <u>From pH measurement.</u> As an approximation, Shirahama[21] expressed the pseudo-phase model for 1:1 type ionic surfactants forming a micellar solution in the absence of extraneous electrolyte as

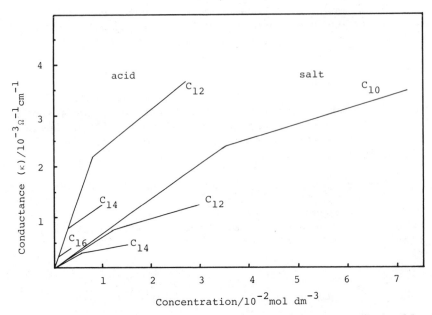

Figure 9. Conductance (κ) versus concentration of alkylsulfonic acids and corresponding sodium salts.

Table V. Degree of Dissociation of Micelles from Conductance vs. Concentration Plots at 323.2 K.

carbon	degree of dissociation (α)	
number	acid	salt
C_{10}	—	0.482
C_{12}	0.300	0.404
C_{14}	0.286	0.333
C_{16}	0.240	0.310

Table VI. Values of the Degree of Dissociation of Alkylsulfonic Acid Micelles Depending on Analytical Method.

carbon number	Botre's	Shirahama's
C_{10}	0.42	0.53
C_{12}	0.25	0.36
C_{14}	0.22	0.29

$$c_i - c_0^2/c_i = \alpha[c - (c_0^2/c_i)] \tag{10}$$

The plots of c_i against $(c - c_0)$ according to Equation (6) (Botre's method), and of $(c_i - c_0^2/c_i)$ against $(c - c_0^2/c_i)$ according to Equation (10) (Shirahama's method) give a linear relation with the slope α. As mentioned before, pH measurements of alkylsulfonic acids give c_i and c_0. The values of α for alkylsulfonic acids are given in Table VI.

From the result in Table V, one can see that the degree of dissociation of H^+ ions is lower than that of Na^+ ions at the micellar surface, and that the degree of dissociation of counterions from micelle decreases with increase in the carbon number of surfactants. Table VI shows the variation of α with the method of analysis and calculation for alkylsulfonic acids. It seems that the analysis using Equation (10) gives results which are in agreement with the results obtained by conductivity method. At any rate, the difference in the dissociation of the head groups is due to larger radius of hydrated sodium ion as compared with that of the hydrated hydrogen ion,[18,19] and this causes a difference in properties of aqueous solution, such as CMC and Krafft point, between alkylsulfonic acids and their sodium salts.

REFERENCES

1. P. Mukerjee, J. Phys. Chem., 69, 2821 (1965).
2. D. S. Goodman, J. Am. Chem. Soc., 80, 3887 (1958).
3. R. Smith and C. Tanford, Proc. Nat. Acad. Sci. USA 70, 289 (1973).
4. H. Kolbel, D. Klemann, and P. Kuzendorfer, in "Proceedings of the 3rd International Congress on Surface Activity, Cologne, 1960", Vol. 1, p.1. Verlag der Univeritatsdruckerei Mainz GmbH (1960).
5. J. E. Carless, R. A.Challie, and B. A. Mulley, J. Colloid Sci., 19, 201 (1964).
6. C. H. Evans, J. Chem. Soc., 579 (1965).
7. A. P. Brady and H. Huff, J. Colloid Sci., 3, 511 (1963).
8. J. K. Weil, F. D. Smith, A. J. Stirton, and R. G. Blistline, J. Am. Oil Chem. Soc., 40, 538 (1963).
9. H. V. Tartar and K. A. Wright, J. Am. Chem. Soc., 61, 539 (1939).
10. K. A. Wright and H. V. Tartar, ibid., 61, 544 (1939).
11. K. A. Wright, A. D. Abbott, and H. V. Tartar, ibid., 61, 549 (1939).
12. M. Saito, Y. Moroi, and R. Matuura, J. Colloid Interface Sci., 76, 256 (1980).
13. M. Saito, Y. Moroi, and R. Matuura, ibid., 88, 578 (1982).

14. K. Shinoda, Y. Nakagawa, B. Tamamushi, and T. Isemura, "Colloidal Surfactants", Chap. 1., Academic Press, N. Y., 1963.

15. I. Satake, T. Tahara, and R. Matuura, Bull. Chem. Soc. Jpn., 42, 319 (1969).

16. T. Sasaki, M. Hattori, and K. Nukina, ibid., 48, 1397 (1975).

17. C. Tanford, "The hydrophobic Effect: Formation of Micelles and Biological Membranes", John Willy & Sons, New York, 1973.

18. R. A. Robinson and R. H. Stokes, "Electrolyte Solutions", Butterworths, London, 1955.

19. J. O. M. Bockris and A. K. N. Reddy, "Modern Electrochemistry", Plenum/Rosetta Edition, New York, 1973.

20. C. Botre, V. L. Crescenzi, and A. Mele, J. Phys. Chem., 63, 650 (1959).

21. K. Shirahama, Bull. Chem. Soc. Jpn., 47, 3167 (1974).

THE EFFECT OF COSOLVENTS ON THE FORMATION OF MICELLES OF

CETYLTRIMETHYLAMMONIUM BROMIDE IN AQUEOUS SOLUTIONS

Lavinel G. Ionescu*, L. Stelian Romanesco*
and Faruk Nome
Laboratório de Química de Superfícies
Universidade Federal de Santa Catarina
Florianópolis, S.C., Brazil

and

Sarmisegetuza Research Group*
Las Cruces and Santa Fe, N. Mexico, U.S.A.

In terms of their effect on the micellization of cetyltrimethylammonium bromide (CTAB) in aqueous solutions the cosolvents studied can be divided into four categories: 1) Short-chain alcohols such as methanol, ethanol, n-propanol and isopropanol that enhance micelle formation at very low mole fraction and totally inhibit it at a slightly higher concentration ($X \approx 0.05$); 2) Compounds that form relatively strong hydrogen bonds with water such as acetone, acetonitrile, tetrahydrofuran and dioxane. They have a slight inhibitory effect on micelle formation at very low mole fraction and totally inhibit the process at an approximate mole fraction of 0.1; 3) Cosolvents that form well-defined stoichiometric hydrates with water. This group includes dimethyl sulfoxide; N,N-dimethylformamide; N,N-dimethylacetamide. They have a more pronounced effect on micelle formation and inhibit it totally at $X_{cosolvent} \approx 0.3$; and 4) Compounds similar to water such as glycerol, ethylene glycol and formamide that allow micelle formation over the entire range of aqueous solutions and also in the pure liquid. Study of the CTAB-H_2O- and CTAB-DMSO-H_2O systems by means of tensiometric, viscosity, NMR spin lattice relaxation and quasielastic light scattering techniques suggest a new picture for the CTAB micelle consisting of three well defined regions: a) a rigid center containing the terminal methyl groups b) a fluid region containing most of the methylene groups and c) a relatively rigid surface area containing essentially the N-methyl groups and the Br^- counterions; the bromide ions and the quaternary ammonium head groups being hydrated in aqueous solution. The addition of small amounts of dimethyl sulfoxide leads to a slight increase in the rigidity of the surface area and an increase in the size of the micelle. Recent surface tensiometric studies of the $CTAB-NaCl-H_2O$ system have shown that the apparent critical micellar concentration (CMC) of CTAB changes markedly as a function of added salt. Experimental parameters such as ΔH^o_{mic} indicate that the $CTAB-NaCl-H_2O$ system undergoes a phase transition at 25° C when the NaCl concentration is about 0.5 M . Viscosity and quasielastic light scattering studies of the same system also suggest the existence of structural changes as a function of NaCl concentration. Structural changes similar to lamellar and liquid crystalline forms have also been observed for solutions of CTAB in water containing NaOH, NaBr, NaOTs and other salts.

INTRODUCTION

As part of a systematic study of the process of micellization in mixed aqueous[1-7] and non-aqueous[8-10] solvent systems we have investigated the formation of micelles by means of spectroscopic, surface tensiometric, conductometric, NMR spin lattice relaxation, viscosity and quasielastic light scattering techniques.

Some of the surfactants studied were cetyltrimethylammonium bromide (CTAB), cetyltrimethylammonium chloride (CTACl), sodium lauryl sulfate (NaLS), cetylpyridinium chloride (CPCl) and various acylcarnitines. The solvent systems used included pure water, glycerol, ethylene glycol, formamide and aqueous solutions of dimetyl sulfoxide; N,N-dimethylformamide; N,N-dimethylacetamide; formamide, glycerol, ethylene glycol, acetone, acetonitrile, dioxane, tetrahydrofuran, methanol, ethanol, isopropanol, n-propanol, sodium hydroxide, various salts and others.

The present paper will review some of the experimental results that we have obtained for the surfactant cetyltrimethylammonium bromide (CTAB) in aqueous solutions containing different cosolvents and will deal principally with data obtained by means of surface tension, NMR spin lattice relaxation and quasielastic light scattering techniques.

   The   study of the effect of cosolvents on micelle formation
in aqueous solutions is relatively new and it will not be discussed here. It has originally been treated by Ray and Nemethy[11],[12]
and it has been reviewed recently in the literature[13]. The formation of normal micelles in non-aqueous polar solvents and in
aqueous solutions containing cosolvents is of interest because it
may result in a·better understanding and practical applications of
the process of micellization. In addition, some of the solvents
such as ethylene glycol and glycerol have been widely used in
protein conformation studies and as membrane simulators. These
dense and viscous liquids approximate portions of membranes in
terms of the anhydrous environment that they may provide[14],[15].

   For micelles in water, the process of micellization is generally explained in terms of hydrophobic interactions between water
and the surfactant. "Hydrophobic interactions" is in many ways a
convenient term that is used to describe an entire array of inter-
and intramolecular interactions involved in micellization and in
a certain way, the term disguises our ignorance about the actual
molecular dynamic processes that take place[9],[10]. The formation of
micelles in water is believed to take place by association of the
hydrophobic parts of the surfactants molecules and the repulsion
of the water molecules from their immediate environment. The overall process of micellization involves a decrease in the free energy of the system. For aqueous solutions it is generally regarded as
an entropy-directed process; the preponderant contribution of the
entropy term being explained by disordering of the water structure
and the breakup of the "Frank-Evans icebergs" by the surfactant
molecules.

   The kind of interactions occuring in the formation of micelles in polar solvents other than water are called solvophobic.
The understanding of solvophobic interactions and micellization
in non-aqueous and mixed aqueous media is considerably more
nebulous[9]. The driving force for micellization in such systems is
less than that for water and the increase in ΔG^{o}_{mic}. is believed to
be mainly the result of a decrease of the entropic contribution.
Few solvent systems are as highly ordered as water.

   The presence of the OH group as part of the structure of the
solvent is not an absolute prerequisite for the presence of solvophobic interactions. In fact, micelles for also in solvents like
formamide, $HCONH_2$, and ethylene diamine, $NH_2CH_2CH_2NH_2$. A comparison between ethylene diamine, 2-aminoethanol and ethylene glycol
shows that substitution of the OH groups by NH_2 groups gradually
decreases the solvophobic effect[11]. This effect has been quanti-
fied per unit of $-CH_2-$ and is approximately -2.89 kJ mol^{-1} for
water, -0.75 kJ mol^{-1} for glycerol and -0.71 kJ mol^{-1} for
ethylene glycol[8-11]. There seems to be no doubt that micellization is the result of a solvophobic effect and that this effect

is not due to water alone, but many other solvents. Its exact nature, however, could perhaps be understood by a detailed analysis and an experimental study of such factors as the number of OH and NH_2 groups, polarizability, dielectric constant, hydrogen bonding, surface tension and other physical chemical properties.

EXPERIMENTAL PROCEDURE

Materials

Cetyltrimethylammonium bromide, CTAB, $CH_3(CH_2)_{15}\overset{+}{N}(CH_3)_3\ Br^-$, was obtained from Aldrich Chemical Company, Milwaukee, Wisconsin, USA. It was recrystallized twice from ethanol and dried under vacuum. All of the cosolvents or salts used were of analytical reagent or spectro grade and were employed without any further purification. The water was deionized and distilled

Methods

The critical micellar concentration (CMC) was determined in most cases by means of plots of surface tension of solution versus the concentration or the logarithm of the concentration of CTAB. The inflection point in the graph was considered equivalent to the CMC. The experiments were usually performed at two or three different temperatures, i.e., 25°, 32° and 40° C. Periodically, other experimental methods such as conductance and spectroscopy were used to determine the CMC. The agreement between the various methods was usually within experimental error. The surface tension measurements were performed using a Du Nouy manual or Fisher Model 21 semi-automatic tensiometer.

The proton spin lattice relaxation study of the CTAB-H_2O and CTAB-H_2O-DMSO systems was performed employing D_2O from Bio-Rad Laboratories and $(^2H_6)$DMSO from Stohler Isotope Chemicals. Values of $1/T_1$ for three different kinds of protons were determined by means of inversion recovery method using a Jeolco Model JNM-PFT-100 Fourier transform NMR spectrometer with deuteron frequency-field lock system[3,5].

Viscosity and quasielastic light scattering measurements were performed using solutions of surfactant and cosolvent or salt prepared with much care to avoid contamination due to dust particles. The viscosity was measured at 25° C using Ostwald viscometers placed in a constant temperature bath. The quasielastic light scattering measurements were performed using an optical light source consisting of an argon laser operating at 4145 Å with an output power of about 100 mW. The intensity of the light scattered was measured at an angle of 90º at 15º, 25º or 45ºC.

The diffusion coefficient (D) and hydrodynamic radius (R_h) were calculated for different systems using a multi-channel correlator.

RESULTS AND DISCUSSION

Formation of Micelles of CTAB in Aqueous Solutions Containing Various Cosolvents

Most of the experimental values obtained for the critical micellar concentration (CMC) were determined from plots of the surface tension of the solutions versus the concentration or the logarithm of the concentration of CTAB. The abrupt change in the plots was taken as an indication of micelle formation and the inflection point as corresponding to the CMC. Linear behavior was considered as evidence for the absence of micelle formation. We have previously shown by means of NMR spin lattice relaxation that such is indeed the case for CTAB in aqueous solutions of DMSO[3,5]. Periodically the CMC was also determined by means of other experimental tecniques, particularly electrical conductance , spectroscopy and NMR and in almost all cases the agreement between the various methods was within experimental error. Careful observation of the surfactant solutions shows that in all cases where there is micelle formation, the systems tend to form stable foam and a stable film on the ring of the tensiometer.

The cosolvents so far studied include methanol, ethanol, n-propanol, isopropanol, acetone, acetonitrile, dioxane, tetra-hydrofuran, dimethyl sulfoxide (DMSO); N,N-dimethylformamide (DMF); N,N-dimethylacetamide (DMA); formamide, ethylene glycol, glycerol and others. Some typical results obtained for aqueous solutions of DMF at 25° C are given in Figure 1.

The variation of the critical micellar concentration of CTAB in water at 25° C as a function of the concentration of cosolvent for some representative cases is illustrated in Figure 2. As can be seen, for some cosolvents the process of micellization takes place over the entire concentration range, while for others it may be inhibited over short range. For the particular case of short-cahin alcohols, exemplified by ethanol, there is a slight enhancement of micellization and the CMC actually decreases at very low mole fraction of alcohol.

The thermodynamics of micellization in water has been dis-cussed extensively in the literature[16,17]. One treatment supposes the presence of two distinct phases, i.e., an aqueous and a mi-cellar phase at the CMC and a constant concentration of monomers in solution, once micelles are formed.

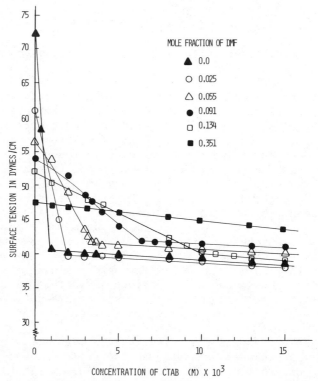

Figure 1. Dependence of surface tension on the concentration of cetyltrimethylammonium bromide (CTAB) in aqueous solutions of N,N-dimethylformamide (DMF) at 25° C.

The standard free energy of micellization, ΔG^o_{mic}, as a good approximation is given by Equation (1).

$$\Delta G^o_{mic} = RT \ln CMC \qquad\qquad (1)$$

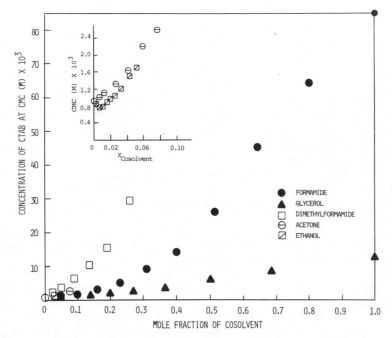

Figure 2. Dependence of the critical micellar concentration of
        cetyltrimethylammonium bromide (CTAB) in aqueous
        solutions at 25° C as a function of cosolvent
        concentration for some representative cases.

    Assuming that the aggregation number and the degree of ioni-
zation are temperature independent, the enthalpy, ΔH^{o}_{mic}, and
entropy of micellization, ΔS^{o}_{mic}, can be determined from the
temperature dependence of the CMC by means of Equation (2) and
Equation (3), respectively.

$$\Delta H^o_{mic} = - RT^2 \frac{d}{dT} \ln CMC \tag{2}$$

$$\Delta S^o_{mic} = (\Delta H^o_{mic} - \Delta G^o_{mic})/T \tag{3}$$

Using these equations, derived for solutions of surfactants in pure water, we have calculated thermodynamic parameters for micellization in ternary systems, CTAB-H_2O-Cosolvent, being well aware that in addition to water-surfactant interactions, they also include interactions of the type water-cosolvent and surfactant-cosolvent. The values thus calculated for ΔG^o_{mic} are very reliable, since they depend on very accurate determinations of CMC, often by more than one experimental technique. This is however not the case for ΔH^o_{mic}, that depends on the variation of CMC with temperature. The method of direct mesurement of thermodynamic properties of surfactants in aqueous solutions developed by Desnoyers and his associates[18-20] is more precise and more widely applicable, since it does not depend on micelle size, shape, and a well-defined CMC. However, it requires more sophisticated techniques and instrumentation not readily available.

Figure 3 illustrates some experimental results that we have obtained for ΔG^o_{mic} at 40° C for CTAB in aqueous solutions containing representative cosolvents. As can be noted, glycerol (G), ethylene glycol (EG) and formamide (F) allow micelle formation over the entire concentration range of aqueous solutions. In fact, CTAB also forms micelles in pure glycerol, ethylene glycol and formamide at 40° C, the process having the following order of spontaneity : H_2O > G > EG > F . Other cosolvents such as isopropanol and acetone inhibit micelle formation at a low concentration ($X_{cosolv} \approx 0.1$), while in the case of DMF, CTAB forms micelles in aqueous solutions up to $X_{DMF} \approx 0.3$.

A careful consideration of the experimental results obtained so far leads us to classify the cosolvents in terms of their effect on the micellization of CTAB in aqueous solutions into four categories or groups:

1) Short chain alcohols such as methanol, ethanol, n-propanol and isopropanol that enhance micelle formation at very low mole fraction and totally inhibit it at slightly higher concentration ($X_{alcohol} \approx 0.05$ or 10-15% by volume).

2) Compounds such as acetone, acetonitrile, tetrahydrofuran and dioxane that form relatively strong hydrogen bonds with water, have a slight inhibitory effect at very low concentration and totally inhibit the process at an approximate mole fraction of 0.1 or 15-20% by volume.

3) Cosolvents that form well-defined stoichiometric hydrates with water. This group includes dimethyl sulfoxide; N,N-dimethyl-formamide and N,N-dimethylacetamide. They have a more pronounced effect on micelle formation and totally inhibit it when X_{cosolv} approaches 0.3 or about 60-70% by volume.

4) Compounds similar to water such as glycerol, ethylene glycol and formamide that allow micelle formation over the entire range of aqueous solutions and also in the pure liquid.

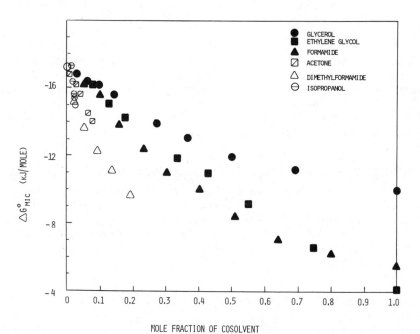

Figure 3. Dependence of the free energy of micellization of cetyl-trimethylammonium bromide (CTAB) in aqueous solutions containing various cosolvents at 25° C.

The effect of alcohols to enhance micelle formation at low concentration has been explained in terms of solubilization of the alcohol molecules inside the micelle[21]. The inhibitory effect at higher concentration follows the order $(CH_3)_2CHOH > CH_3CH_2CH_2OH > CH_3CH_2OH > CH_3OH$ and can be understood by considering the interactions between water and the alcohols that result in a decrease of the solvophobic effect. These interactions consist of the destruction of the original structure of water and the formation of new hydrogen bonds between water and the alcohols[22,23]. Preliminary quasielastic light scattering measurements for the $CTAB-H_2O-CH_3OH$ system indicate that the diffusion coefficient (D) of micellar aggregates is significantly higher in this system than in water and that the micelles are smaller in size[24]. It is important to note that quasielastic light scattering measurements in aqueous solutions containing cosolvents are considerably much more difficult to perform than the corresponding experiments in the presence of salts. This is normally because the solutions scatter either too much or too little light.

The inhibitory effect in the second group follows the order dioxane > acetone > tetrahydrofuran > acetonitrile and has been explained in terms of the formation of hydrogen bonds between the cosolvents and water. This conclusion is supported by experimental parameters measured such as ΔG°_{mic}, ΔH°_{mic} and ΔS°_{mic}[7,25].

DMSO, DMF and DMA form stoichiometric hydrates with water of the type Cosolvent·$2H_2O$ and the presence of hydrogen bonding has been confirmed by a variety of techniques[26-28]. Their dissolution in water is a highly exothermic process. The highly ordered array of the hydrates substantially restricts the motion of the surfactant molecules and essentially eliminates hydrophobic interactions[2-6]. The inhibitory effect follows the order DMA > DMF > DMSO.

A comparison of the spontaneity of micellization of CTAB in pure solvents shows that the process is thermodynamically more favored in water, followed by glycerol, ethylene glycol and formamide, respectively[29,30]. The diffusion coefficient (D) decreases and the size of CTAB micelles increases considerably as a function of ethylene glycol concentration[24].

Proton Spin Lattice Relaxation Study of $CTAB-H_2O$ and $CTAB-H_2O-$DMSO Systems

Proton spin lattice relaxation time (T_1) for terminal methyl, N-methyl and methylene groups of CTAB in water-dimethyl sulfoxide (DMSO) solutions was determined at concentrations above and below the CMC in an attempt to elucidate the nature of the intra- and intermolecular interactions and the role of solvent molecules in the formation of micelles. The mole fraction of DMSO used were 0.098 and 0.366 and the experiments were performed at 28.2^o and 40.0^o C. Average rotational correlation times $\tau^s(R)$ and effective activation energies of the various relaxation processes for the three groups were also determined. The mathematical analysis used in the treatment of the experimental data and the results obtained have been described extensively in the litera- ture[3,5]. Some representative average rotational correlation times, $\tau^s(R)$ in units of 10^{-12} s for CTAB in water below the CMC at 28.2^o C are: $\tau^s(CH_2)=0.31$, $\tau^s(CH_3)=4.17$ and $\tau^s(N-CH_3)=2.71_s$. For the system CTAB-DMSO-H$_2$O at $X_{DMSO}=0.098$, $\tau^s(CH_2)=0.69$, $\tau^s(CH_3)=9.86$ and $\tau^s(N-CH_3)=11.3$. At $X_{DMSO}=0.366$, $\tau^s(CH_2)=15.0$, $\tau^s(CH_3)=16.0$ and $\tau^s(N-CH_3)=15.2$. The spin lattice relaxation times are diffe- rent for solutions of CTAB below and above the CMC and also vary as a function of DMSO concentration.

A detailed analysis of the experimental results indicates that the increased structuring of the H$_2$O-DMSO liquid system, resulting in the formation of the stoichiometric hydrate DMSO.2H$_2$O overcomes the hydrophobic effect of the long alkyl chain of CTAB and inhibits totally micelle formation at $X_{DMSO} \approx 0.33$ in accord- ance with the results obtained by surface tensiometry. Under these conditions, plots of surface tension for CTAB-DMSO-H$_2$O versus the concentration of CTAB are linear[2].

The generalized picture that emerged for CTAB micelles is illustrated in Figure 4 and is somewhat different from models proposed later in the literature by Aniansson[31,32], Menger and associates[33,34] and Fromherz[35]. The CTAB micelle consists of three well delineated regions: a) a rigid center containing the terminal methyl groups b) a fluid region containing most of the methylene groups which are as mobile as those of monomers dispersed in water or hydrocarbons and c) a relatively rigid surface area containing essentially the N-methyl groups and Br$^-$ counterions; the quaternary ammonium head groups and the bromide ions being hydrated in aqueous solution[3,5,36,37].

The addition of DMSO leads to a slight increase in the rigidity of the surface area and an increase in the size of the micelle. Preliminary viscosity and quasielastic light scattering measurements suggest that DMF and DMA also decrease the diffusion coefficient(D) of micellar aggregates and that they also cause an increase in the size of the micelles. For example, D of micellar aggregates of 0.02 M CTAB at 25° C decreases from 5.5 x 10^{-7} cm^2 s^{-1} ($X_{DMA}=0.021$) to 4.0 x 10^{-7} cm^2 s^{-1} for ($X_{DMA}=0.162$)[24] compared to 11.3 x 10^{-7} cm^2 s^{-1} for the same concentration of CTAB and 0.02 M NaBr in water.

Study of the CTAB-NaCl-H$_2$O and Other Related Systems

During the past few years we have also analyzed the formation of micelles of CTAB in aqueous solutions containing salts such as NaCl, NaBr, NaOTs, NaI and NaOH by means of surface tension, viscosity and quasielastic light scattering techniques[38-43]. In all cases studied, the apparent critical micellar concentration, as measured by surface tension, decreases markedly as a function

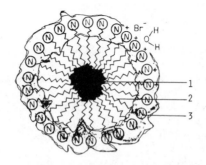

1 - RELATIVELY RIGID CENTER CONTAINING TERMINAL
    METHYL GROUPS
2 - FLUID REGION CONTAINING MOST OF THE METHYLENE
    GROUPS
3 - RELATIVELY RIGID SURFACE CONSISTING MAINLY OF
    THE N-METHYL GROUPS AND BROMIDE COUNTERIONS

Figure 4. Structure proposed for micelles of cetyltrimethyl-
          ammonium bromide (CTAB) on the basis of NMR spin lattice
          relaxation time measurements[3,5,36,37]

of added salt. Eventually, as the concentration of salt is in-creased all these systems form liquid crystalline mesophases at room temperature. Sodium p-toluene sulfonate (NaOTs) shows the most pronounced effect [39,40].

For CTAB-NaCl-H$_2$O at 25° C the apparent CMC varies from 9.2 x 10^{-4} M in pure water to 9.0 x 10^{-6} M for surfactant solutions containing 1.5 M NaCl[38,40]. Experimental values of ΔH^o_{mic} indicate that the system CTAB-NaCl-H_2O undergoes a phase transition at 25° C when the concentration of NaCl is about 0.5 M[38,40]. Viscosity and quasielastic light scattering studies of the same system suggest the existence of structural changes to lamellar and liquid crystalline forms as a function of NaCl. For fixed concentrations of NaCl, the diffusion ceofficient (D) for solutions of CTAB-NaCl-H_2O at 25° C has three types of behavior[24,43]. For low values of NaCl (0.01 - 0.10 M) D increases as a function of CTAB. For concentrations between 0.50 M and 1.0 M, D is practically constant and for solutions containing more than 1,5 M NaCl the diffusion coefficient decreases, indicating liquid crystalline growth[44]. Typical values obtained for D at 25° C vary between 1.5×10^{-7} cm^2 s^{-1} and 19.5×10^{-7} cm^2 s^{-1}.[38] Results obtained for the hydrodynamic radius (R_h) at limit surfactant concentration indicate that micelles of CTAB in the presence of NaCl have a size somewhat different from micelles of CTAB and CTACl in aqueous solutions containing salts with the same counterions[24,43,45].

Additional studies of the formation of micelles of CTAB[40] in the systems H_2O-NaCl-CH_3CH_2OH, H_2O-NaCl-DMSO, H_2O-NaOTs-CH_3CH_2OH and H_2O-NaOTs-DMSO showed that in all cases NaCl and NaOTs tend to counterbalance the inhibitory effect of ethanol and dimethyl sulfoxide and favor micelle formation.

ACKNOWLEDGEMENTS

Support received from CNPq-National Research Council of Brazil, FINEP and Sarmisegetuza Research Group, Las Cruces and Santa Fe, N. Mexico, USA is gratefully acknowledged.

REFERENCES

1. L.G. Ionescu and J. K. Tsang, Rev. Roum. Biochim.,15, 211 (1978).
2. L. G. Ionescu, T. Tokuhiro, B. J. Czerniawski and E. S. Smith, in "Solution Chemistry of Surfactants", K. L. Mittal. Editor, Vol. 1, p. 487, Plenum Press, New York, 1979.
3. T. Tokuhiro and L. G. Ionescu, in "Solution Chemistry of Surfactants", K. L. Mittal, Editor, Vol. 1, p. 507, Plenum Press, New York, 1979.
4. L. G. Ionescu, T. Tokuhiro and B. J. Czerniawski, Bull. Chem. Soc. Jpn., 52, 922 (1979).
5. T. Tokuhiro, D. S. Fung and L. G. Ionescu, J. Chem. Soc. Faraday Trans. II, 75, 975 (1979).
6. L. G. Ionescu and B. J. Czerniawski, Rev. Roum. Biochim., 18, 103 (1981).

7. L. G. Ionescu and V. T De Favere, in "Solution Behavior of Surfactants: Theoretical and Applied Aspects", K. L. Mittal and E. J. Fendler, Editors, Vol. 1, p. 407, Plenum Press, New York, 1982.

8. L. G. Ionescu, Areh. Biol. Med. Exp., 12(2), 272 (1979).

9. L. G. Ionescu and D. S. Fung, Bull. Chem. Soc. Jpn., 54, 2503 (1981).

10. L. G. Ionescu and D. S. Fung, J. Chem. Soc. Faraday Trans. I, 77, 2907 (1981).

11. A. Ray, Nature, 231, 313 (1971).

12. A. Ray and G. Nemethy, J. Phys. Chem., 75, 809 (1971).

13. L. Magid, in "Solution Chemistry of Surfactants", K. L. Mittal, Editor, Vol. 1, p. 427, Plenum Press, New York, 1979.

14. S. Damjanovich, B. Somogyi and J. Bot, Stud. Biophys., 59, 229 (1976).

15. T. Yubisui, M. Takeshita and Y. Yoneyama, Experientia, 32, 3989 (1976).

16. D. G. Hall, Trans Faraday Soc., 66, 1351 (1970).

17. N. Muller, in "Reaction Kinetics in Micelles", E. H. Cordes, Editor, p. 1, Plenum Press, New York, 1973.

18. J. E. Desnoyers, R. De Lisi, C. Ostiguy and G. Perron, in "Solution Chemistry of Surfactants", K. L. Mittal, Editor, Vol. 1, p. 221, Plenum Press, New York, 1979.

19. R. De Lisi, C. Ostiguy, G. Perron and J. E. Desnoyers, J. Colloid Interface Sci., 71, 147 (1978).

20. J. E. Desnoyers, R. De Lisi and G. Perron, Pure & Appl. Chem., 52, 443 (1980).

21. B. D. Flockhart, J. Colloid Interface Sci., 12, 557 (1957).

22. L. G. Ionescu and F. De Paula Soares Mol Filho, Supl. Cienc. Cult. 32(7), 438 (1980).

23. F. De Paula Soares Mol Filho, M. S. Thesis, Universidade Federal de Santa Catarina, Florianópolis, S.C., Brazil,1980.

24. L. G. Ionescu and L. S. Romanesco, unpublished results, 1982.

25. V. T. De Favere, M. S. Thesis, Universidade Federal de Santa Catarina, Florianópolis, S.C., Brazil, 1980.

26. W. Drinkard and D. Kivelson, J. Phys. Chem., 62, 1494 (1962).

27. J.M. G. Cowie and P. M. Toporowski, Canad. J. Chem., 39, 2240 (1961).

28. J. Bougard and R. Jadot, J. Chem. Thermodyn., 7, 1185 (1975).

29. L. G. Ionescu and S. M. H. Probst, Arq. Biol. Tecnol.,25, 106 (182).

30. S. M. Hickel Probst, M.S. Thesis, Universidade Federal de Santa Catarina, Florianopolis, S.C., Brazil, 1982.

31. G. E. A. Aniansson, J. Phys. Chem., 82, 2805 (1982).

32. G. E. A. Aniansson, Ber. Buns. Ges. Phys. Chem., 82, 988 (1978).

33. F. M. Menger, J. M. Jerkunica and J. M. Johnston, J. Am. Chem. Soc., 100, 4676 (1978).

34. F. M. Menger, Acc. Chem. Res., 12, 111 (1979).

35. P. Fromherz, Chem. Phys. Lett., 77, 460 (1981).

36. D. S. Fung, Ph. D. Dissertation, University of Detroit, Detroit, Michigan, 1978.
37. T. Tokuhiro, D. S. Fung and L. G. Ionescu, Paper presented at the 174th National Meeting of the American Chemical Society, Chicago, Illinois, August 28 - September 2, 1977, Abstr. Phys. 39.
38. L. G. Ionescu and T. H. M Do Aido, Arq. Biol. Tecnol.,24, 75 (1981).
39. L. G. Ionescu and T. H. M. Do Aido, Supl. Cienc. Cult. 33(7), 436 (1981).
40. T. H. M. Do Aido, M. S. Thesis, Universidade Federal de Santa Catarina, Florianópolis, S.C., Brazil, 1981.
41. L. G. Ionescu and T. H. M. Do Aido, Supl. Cienc. Cult.,34(7), 458 (1982).
42. L. G. Ionescu, C. A. Bunton, D. F. Nicoli and R. Dorshow, unpublished results, 1982.
43. L. G. Ionescu, E. Teixeira and L. S. Romanesco, unpublished results, 1982.
44. P. J. Missel, N. A. Mazer, G. B. Benedek, C. Y. Young and M. C. Carey, J. Phys. Chem., 84, 1044 (1980).
45. J. Briggs, R. B. Dorshow, C. A. Bunton and D. F. Nicoli, J. Chem. Phys., 76, 775 (1982).

THE SPHERE TO ROD TRANSITION OF CPX AND CTAX MICELLES IN HIGH IONIC STRENGTH AQUEOUS SOLUTIONS : THE SPECIFICITY OF COUNTERIONS

G. Porte[*] and J. Appell[**]

[*] Centre Dynamique Phases Condensées, LA 233 C.N.R.S.
[**] Lab. Spectr. Rayleigh-Brillouin, ERA 460 C.N.R.S.
Université des Sciences & Techniques du Languedoc
34060 Montpellier Cédex, France

The specificity of some counterions (X) to induce the elongation of cetylpyridinium (CPX) or cetyltrimethylammonium (CTAX) micelles in aqueous solutions has been investigated experimentally using in particular quasi-elastic light scattering spectroscopy to follow the size evolution of the micelles. F^- as well as $C\ell^-$ were found inefficient in promoting the elongation of CPX and CTAX micelles while Br^-, NO_3^-, $C\ell O_3^-$ do promote their elongation with increasing efficiency from Br^- to $C\ell O_3^-$. Addition of $C\ell^-$ (mixed salt solutions) results in no modification or even in some cases in a small decrease of the micellar size.

A crude model taking into account both chemical and electrostatic aspects of the ionic environment of the micelles allows for a semi-quantitative interpretation of the experimental facts with as few adjustable parameters as possible.

INTRODUCTION

The micellization of ionic surfactants in aqueous solutions is well known to result from a balance between the hydrophobic interaction between the alkyl chains of the surfactant and the electrostatic interaction between the ionic heads as well as with the counterions. The addition of simple electrolytes to the solution affects this balance and is thus expected to influence the critical micellar concentration (c.m.c.) and the evolution in size and shape of the micelles with temperature and concentrations as well as other properties of the solution. That this is indeed the case has been known for many years but many aspects of the "electrolyte effect", as it was named by Anacker[1] in his recent review, are still open to question.

In the present paper our interest lies in one particular aspect of this electrolyte effect namely the specificity of counterions to induce the elongation of micelles. That this specificity is related to the binding of the counterions on to the micellar surface is consistent with the accepted picture that elongation can occur if the counterions are sufficiently effective in lowering the electrostatic repulsion between the ionic surfactant heads so that a closer packing of these heads on the micellar surface becomes possible as is the case in a sphere – to – rod transition.[1,2]

We have studied previously the influence of added NaBr or NaCl on the micellization of cetylpyridinium bromide or chloride (CPX, $X = Br^-$ or Cl^-). We have shown[3] that CPBr micelles undergo a sphere-to-rod transition, while CPCl micelles remain spherical up to 1 M added NaCl. This is in excellent agreement with the earlier results of Anacker and Ghose.[4] A number of studies[5,6] have shown that a similar difference exists for the influence of Br^- and Cl^- on cetyltrimethylammonium (CTAX) micelles. Other anions display similar specificity as shown on CPX[4] or on CTAX.[7,8] This prompted us to examine the CPX and CTAX micelles with $X = F^-$, NO_3^-, ClO_3^- or $Br^- + Cl^-$, $NO_3^- + Cl^-$ and $ClO_3^- + Cl^-$ and compare their properties to those observed for CPBr micelles.[3,9,10,11]

As previously quasi-elastic light scattering spectroscopy (QELS) (QELS) was used here to follow the evolution in overall size of the micelles through their measured mean hydrodynamic radius $\overline{R_H}$.[3,11] F^- as well as Cl^- was found inefficient in promoting the elongation of CPX or CTAX micelles. The other anions Br^-, NO_3^- and ClO_3^- do promote the elongation of CPX and CTAX micelles with increasing efficiency from Br^- to ClO_3^-. The overall size of the micelles decreases in the same fashion with increasing temperature for both surfactants (which differ only by their ionic heads) with the three anions.

The extra addition of $C\ell^-$ to solutions initially containing Br^-, NO_3^- or $C\ell O_3^-$ (and thus containing large micelles) results in either no modification or in some cases in a small decrease in size of the micelles. This rather unexpected result rules out the hypothesis that the total ionic strength could be the determinant parameter for elongation.

In order to interpret these results a crude estimation of the difference in the free energy of monomers located either in spherical or rod-like micelles is made ; the basic assumption is that specific adsorption of the counterions takes place in the Stern layer at the interface between the micelle and the surrounding solution. This estimation allows for the observed facts. In particular, the paradoxical effect of $C\ell^-$ in mixed salts solutions is well accounted for if the binding constant of $C\ell^-$ is small as compared to that of the other anions. This conclusion is consistent with the results obtained on ionic exchange constants on CTAX by Gamboa *et al.*[8]

For the three anions, evidence has been obtained that the shape and structure of the CPX or CTAX micelles is similar to that of CPBr micelles which were shown previously[9,10] to be semi-flexible rods with a persistence length of 200 ± 50 Å .

EXPERIMENTS AND RESULTS

Reagents and Solutions

CPBr was synthesized as described earlier.[3] CTABr was the commercial product from Merck. The added NaBr, NaCℓ, NaF, NaNO$_3$ and NaCℓO$_3$ were reagent grade (Merck). The solutions were prepared by weight in tridistilled water and filtered through a 0.22 µm Millipore filter directly into the experimental cells. We neglect in what follows the contribution of the Br^- counterions brought by the surfactant ($\sim 6 \times 10^{-3}$M) as compared to the counterions brought by the added salt (> 0.2 M). This is reasonable in view of the identical preliminary results that we obtained for solutions prepared with CPBr or with CPCℓ.

Quasi Elastic Light Scattering Spectroscopy (QELS)

QELS is used here to measure the mean hydrodynamic radius $\overline{R_H}$ of the micelles of CPX or CTAX under different experimental conditions. The experimental set-up and data analysis have been described previously.[11]

Some of the results obtained for the overall size of micelles

of CPX and CTAX as reflected in their $\overline{R_H}$ are shown in Figures
1 to 3. Examination of the curves of Figures 1 and 2 leads to the
following remarks. 1) The evolution of $\overline{R_H}$ with temperature is
similar for CPX and CTAX micelles. 2) For both surfactants the
anions promote elongation with more and more efficiency in
the order Br^-, NO_3^- and ClO_3^- in agreement with the results of

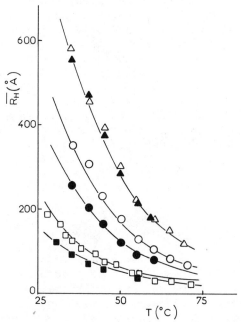

Figure 1. $\overline{R_H}$ versus temperature for CPX micelles. With CPBr
6×10^{-3} M and 0.2 M NaX : open signs;or 0.2 M NaX + 0.2 M NaCl :
full signs ; X = ClO_3^- : $\triangle$, $\blacktriangle$; NO_3^- : o,$\bullet$; and Br^- : $\square$,$\blacksquare$.
The solid lines are computed from the quantitative model as ex-
plained in the text.

Anacker and Ghose[4] for CPX. 3) Addition of Cl^- has but little influence on the micellar size : either not noticeable (CPClO$_3$/Cl ,
CTAClO$_3$/Cl and CTABr/Cl) or even slightly inhibiting the micellar elongation (CPNO$_3$/Cl , CPBr/Cl and CTANO$_3$/Cl). This is
rather unexpected and clearly indicates that the ionic strength
of the solution is not the determinant factor for elongation.

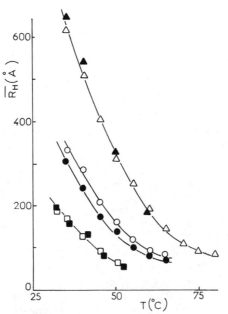

Figure 2. $\overline{R_H}$ versus temperature for CTAX micelles. With CTABr 6×10^{-3} M and 0.2 M NaX : open signs;or 0.2 M NaX + 0.2 M NaCℓ : full signs ; X = CℓO$_3^-$: △,▲; NO$_3^-$: ○,●; and Br$^-$: □,■. The solid lines are computed from the quantitative model as explained in the text.

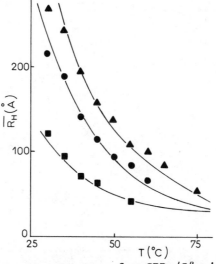

Figure 3. $\overline{R_H}$ versus temperature for CPBr/Cℓ micelles with CPBr 6×10^{-3} M, Br/Cℓ = 1/1 and total salt concentration 0.8 M : ▲ ; 0.6 M : ●; and 0.4 M : ■ . The solid lines are computed from the quantitative model as explained in the text.

In Figure 3 the size evolution of CPBr/Cℓ micelles is shown
as a function of temperature ; the ratio of Br$^-$ to Cℓ^- is kept
constant and equal to 1 while the total salt concentration varies
from 0.4 to 0.8 M . The variation of size with salt concentration and temperature is analoguous to that observed for pure NaBr
(cf Figure 7 of ref. 11). Similarly we found that the evolution in
size with the detergent concentration is similar to that observed
when pure NaBr is added.[11] From these results we thus infer that
the salt mixture promotes the micellar elongation in the same way
as a pure salt provided the ratio of salts is kept constant.

Angular Dissymmetry of Scattered Light

The ratio d (50°) of the intensities scattered by the micellar solutions at angles of 50° and 130° has been measured using
the technique described earlier.[9,10] We can then compare the measured evolution of d (50°) versus $\overline{R_H}$ to that observed earlier[10]
for CPBr micelles. This comparison is shown in Figure 4. Within
experimental errors the d (50°) versus $\overline{R_H}$ evolution is identical for CPBr , CPNO$_3$, CPCℓO$_3$ and CTABr micelles. This evidence
leads us to the conclusion that CPX and presumably CTAX micelles
have the same geometrical shape. From previous model calculations[10]
we have shown this geometrical shape to be semi-flexible rods with
a persistence length $\ell \simeq 200 \pm 50$ Å . The experimental points in
figure 4 can, indeed, be seen to closely follow the curve calculated previously[10] for polydisperse micelles of this shape.

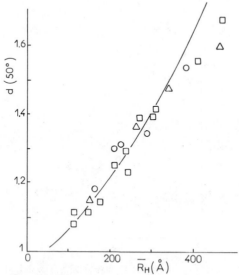

Figure 4. d (50°) versus $\overline{R_H}$. CPBr , □ from ref. 9; CPNO$_3$, O;
and CPCℓO$_3$, Δ . The curve is that calculated for polydisperse
semi-flexible rod-like micelles with ℓ = 200 Å from ref. 10.

INTERPRETATION OF THE VARIATIONS OF THE MICELLAR $\overline{R_H}$

We have previously analyzed the evolution of the $\overline{R_H}$'s of
CPBr micelles with a model based on a thermodynamic theory of multiple equilibrium (see e.g. ref. 13-16). Before analyzing the present results in the same manner we first recall the main hypothesis underlying this model and its prediction on the size evolution
of the micelles. This model was initiated by Mazer et al[15] and
Missel et al[16] who used it to analyze the size evolution of sodium
dodecyl sulfate (SDS) micelles.

The micelles are assumed to be spherocylinders with n_0 monomers in the two hemispherical ends (n_0 is a constant :
$n_0 \simeq 100$ for CPX micelles) and with n monomers in the cylindrical part (n can have any value between 0 and ∞). The micellar
solutions are assumed ideal (i.e. no intermicellar interactions)
and the ionic strength is high (i.e. the intramicellar interactions between the constitutive monomers are short-ranged). Under
these conditions the contribution to the free energy of the micelle from one of its monomer is entirely determined by the local
environment of the monomer : corresponding to its localization
either in the spherical ends or in the cylindrical part of the micelle. Under these assumptions the application of the multiple
equilibrium between the N-micelles ($N = n + n_0$) and the monomers in
solution leads to the following prediction : the size distribution
of the micelles is determined by the product KC where C is the
detergent concentration and K is a thermodynamic factor independent of C . K measures the tendency for n_0 monomers to aggregate into the cylindrical part of a micelle rather than to form
a minimum spherical micelle ; ℓnK can be written as :

$$\ell nK = (G^{0S} - G^{0C})/kT   , \tag{1}$$

where G^{0S} and G^{0C} are the standard free energies for n_0 monomers in the spherical or cylindrical part of a micelle respectively.

The variation of the $\overline{R_H}$'s of micelles as a function of KC
is shown in Figure 5. The details for the derivation of this plot
are given in ref. 11 where we had shown that the size evolution of the CPBr micelles with C was well described by this model. Furthermore from the evolution of $\overline{R_H}$'s with temperature
and salt concentration we found[11] that the corresponding K's
were given by the empirical expression:

$$\ell nK = a/T + b . \tag{2}$$

within experimental errors <u>a</u> was found independent of temperature and salt concentration while <u>b</u> only deepens on salt concentration. Of course both a and b are a priori expected to de-

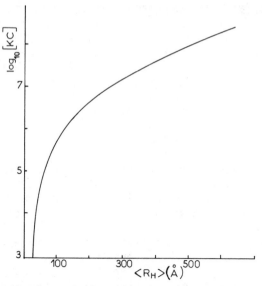

Figure 5. Prediction of the quantitative model log KC versus
$\langle R_H \rangle$ (from ref. 11).

pend on the nature of the detergent and of the counterion. In a
similar analysis of the present experimental results we hope to
find some systematic variations of $\underline{a}$ and $\underline{b}$'s with the nature
of the detergent and of the counterion; however one must re-
member that these parameters are derived from experimental results
interpreted in the light of an approximate quantitative model so
that the errors in $\underline{a}$ and $\underline{b}$ may well be the magnitude of the
of the effects which we are looking for.

    An illustration of the analysis of our experimental results
on CPX micelles is given in Figure 6. In fact the empirical expression
(2) is found to hold for all the present results within the limit
of validity of the quantitative model ($\overline{R_H}$ < 300 Å cf. ref. 11) so
that $\underline{a}$ and $\underline{b}$ are, as previously found, independent of temperature. Furthermore the points obtained for all CPX micelles are
found within experimental errors to line up on parallel lines with
a slope $\underline{a} \simeq 15000 \pm 2000$ °K, $\underline{a}$ is thus found independent of the
concentration and also of the nature of the counterion X^- . The
dependence of K on the nature and concentration of X^- is then
entirely reflected by the dependence of b on these parameters.
The $\underline{b}$'s obtained here are given in Tables I and II. Such a forced fit to,parallel lines is reasonable as shown in Figures 1 and
3 where the solid lines give $\overline{R_H}$ versus T derived from the
plot of Figure 5 with $\underline{a} = 15000$ °K and the appropriate $\underline{b}$'s
(from Tables I and II).

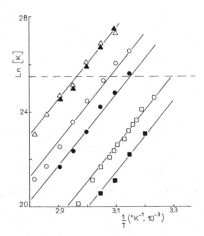

Figure 6. LnK as a function of 1/T for CPX micelles with
CPBr 1.1×10^{-4} mole fraction and 0.2 M NaX : open signs; or
0.2 M NaX + 0.2 M NaCℓ blackened signs ; with X = CℓO$_3^-$: △▲ ;
NO$_3^-$: ○● ; and Br$^-$: ◇♦. The solid lines have the mean slope
a ∿ 15000 °K .(The dotted horizontal line at LnK = 25.5 is the
limit of validity of the quantitative model for the present CPBr
concentration.)(Note that here K is expressed in (mole fraction)$^{-1}$
whereas it was expressed in M^{-1} in ref. 11).

Table I. The empirical parameter b and the computed values of
$((\Delta G_3 - \Delta G_4)/kT + b_0) = b'$ as a function of salt concentration for
CPBr and CPBr/Cℓ micelles (with CPBr 1.1×10^{-4} mole fraction).

Salt Concentration	CPBr		CPBr/Cℓ	
(M)	b[*]	b'[**]	b[***]	b'[**]
0.2	− 25.5	− 26		
0.4	− 22.5	− 22.5	− 26.5	− 26.3
0.6	− 21	− 20.5	− 24	− 24
0.8	− 19.5	− 19.3	− 23	− 22.5

[*] From the results given in reference 11 ; the b's have been
  reevaluated with the mean value of a=15000 °K .
[**] Computed (see text) with the K$_X$'s given in Table VI and
  $b_0 = -15$.
[***] Present results with a = 15000 °K .

    Similarly for CTAX micelles a mean a value is found :
a ≃ 17000 ± 2000 °K ; the corresponding b's are given in Table III.
Comparing the value of a found for CPX and CTAX micelles does
not allow for any decisive conclusion as to the influence of the
polar heads on this parameter. This is, however, not the purpose

of the present paper where a parallel study of CTAX micelles has
been conducted in order to test whether the peculiarities observed
in mixed salt solutions were not solely due to the particular chemical nature of the pyridinium ion.[17] The similarity observed here
for the behavior of CPX and CTAX micelles seems to rule out such
an hypothesis.

Table II. The empirical parameter b and the computed values of
$((\Delta G_3 - \Delta G_4)/kT + b_0) = b'$ for different CPX micelles.

Solutes X $\rightarrow$	Br^-		NO_3^-		$C\ell O_3^-$	
$\downarrow$						
CPBr +	b^*	b'^{**}	b^*	b'^{**}	b^*	b'^{**}
NaX 0.2 M	− 25.5	− 26	− 22.5	− 23.2	− 20.5	− 20.7
" + NaCℓ 0.2 M	− 26.5	− 26.3	− 23.5	− 23.7	− 20.5	− 20.4

* With a = 15000 °K .
** Computed (see text) with the K_X's given in Table VI.

Table III. The empirical parameter b for different CTAX micelles.

Solutes X $\rightarrow$	Br^-	NO_3^-	$C\ell O_3^-$
$\downarrow$			
CTABr +	b^*	b^*	b^*
NaX 0.2 M	− 31	− 28.5	− 26.5
NaX 0.2 M + NaCℓ 0.2 M	− 31	− 29	− 26.5

* With a = 17000 °K .

The counterion influence on micellar association has been introduced in the quantitative model either in terms of an electrostatic model : the Gouy-Chapman double layer approximation by
Mazer et al[15] and Missel et al[16-18] in their study on SDS or in
terms of the counterion binding approximation by us,[11] following
the suggestion of Mukerjee[13], in our study of CPBr. Surprisingly,
enough both descriptions, one purely electrostatic and the other
purely chemical, gave equally good interpretations of the evolution
of K or equivalently b with the counterion concentration. In the
first description the counterion specificity can hardly be introduced while the second lacks realism. Therefore we develop , in
what follows, a model taking into account both chemical and electrostatic aspects of the ionic environment of the micelles. This
crude model aims at a qualitative interpretation of the experimental facts with as few adjustable parameters as possible.

MICELLAR IONIC ENVIRONMENT : AN ELECTROCHEMICAL MODEL

The first basic assumption of the present model is that the micellar / solvent interface consists of a Stern layer containing the ionic surfactant heads and a certain amount of adsorbed counterions which lowers the apparent charge density of the micelle and of a diffuse layer where the distribution of ions is described classically by the Poisson-Boltzmann distribution. Such a description, classical for a polyion-solvent interface, has been used extensively by Stigter[12] for micelles. The second basic assumption is that the micelles retain constant internal structures (for both the spherical and cylindrical parts) which are independent of the nature and concentration of the counterion.

With these basic assumptions, we aim at evaluating the quantity which determines the variations of K with the salinity of the solution. We consider the hypothetical cycle of transformations shown in Figure 7. Starting with n_0 (monomers + counterions) arranged in a cylindrical portion of a micelle. ΔG_1 is the change in free energy during the rearrangement to a sphere at zero overall charge. If we assume that the adsorption energy of the counterions does not depend on the micellar shape (as is done below) and with the assumption of a constant internal structure stated above, ΔG_1 is independent of the nature and concentration of the counterions. ΔG_3 is the change in free energy when $n_0 \alpha_s^e$ counterions escape from the Stern layer into the solution leaving the spherical micelle with its equilibrium degree of ionization α_s^e; and ΔG_4 is the corresponding term for the cylindrical part of a micelle with its equilibrium degree of ionization, α_c^e. Then going from the equilibrium cylindrical to spherical packing of n_0 monomers with their appropriate number of counterions the change in free energy is ΔG_2 :

$$\Delta G_2 = \Delta G_1 + \Delta G_3 - \Delta G_4 \quad , \tag{3}$$

clearly $\Delta G_2 = G^{0S} - G^{0C}$ defined in equation (1) and can be split, in agreement with the experimental results (cf. Equation (2)) in a term ΔG_1 independent of the nature and concentration of the counterion and a term $\Delta G_3 - \Delta G_4$ which depends on it and that we now proceed to compute.

To derive ΔG_3 and ΔG_4, we consider the Stern layer and the rest of the solution (including the diffuse layer) as two distinct phases which remain separately in internal equilibrium. The variation in free energy of the interface upon desorption of one counterion, if $n = n_0 (1 - \alpha)$ counterions are adsorbed, can be written :

$$- \frac{\partial G}{\partial n} = \frac{1}{n_0} \frac{\partial G}{\partial \alpha} = \mu_X^0 - \mu_{BX}^0 + kT \ln N_X + e\phi_0 + KT \ln \frac{\alpha}{1 - \alpha} \quad , \tag{4}$$

where μ_X^0 and μ_{BX}^0 are the standard chemical potentials for X^-

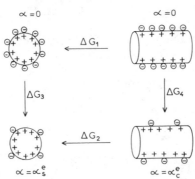

Figure 7.   Cycle of transformations used to compute the change in free energy for n_0 monomers in the spherical or cylindrical part of a micelle. α is the degree of ionization of the micelle with α_s^e and α_c^e the equilibrium values on the spherical and cylindrical parts respectively (see text).

in the solution or adsorbed in the Stern layer, where N_X is the mole fraction of salt in the bulk of the solution and where $e\phi_0$ is the electrical work to bring the counterion from the Stern layer where the potential is ϕ_0 to the bulk where, by convention the potential is zero. ϕ_0 , the potential on the Stern layer, is a function of the surface charge density and thus depends on α . In Equation (4), the term $kT\ln(\alpha/(1-\alpha))$ accounts for the contribution of the exchange entropy of $(1-\alpha)n_0$ bound ions on n_0 available adsorption sites. Doing this we thus assume that the density of available sites is identical to the density of polar heads on the micellar surface. The conditions for validity of this equality will be discussed further below.

At equilibrium, where $\partial G/\partial n = 0$, we write :

$$kT\ln K_X + kT\ln N_X + e\phi_0^e + kT\ln(\alpha^e/(1-\alpha^e)) = 0 \quad , \tag{5}$$

where ϕ_0^e and α^e are the equilibrium values of ϕ_0 and α and K_X is the equilibrium constant for the desorption-adsorption reaction of the counterion in the Stern layer defined by :

$$kT\ln K_X = \mu_X^0 - \mu_{BX}^0 \quad . \tag{6}$$

If two counterions X^- and Y^- are present with concentrations C_X and C_Y , the equilibrium will be the result of the competition between the two desorption-adsorption reactions. If K_X and K_Y are the corresponding equilibrium constants, the ratio of adsorbed X^- to adsorbed Y^- will be equal to $N_X K_X/N_Y K_Y$. If we define a mean constant $\overline{K}_T$ as

$$\overline{K}_T = (N_X K_X + N_Y K_Y)/N_T \quad , \tag{7}$$

with $N_T = N_X + N_Y$, it is readily shown that Equations (4) and (5) are valid provided K_X and N_X are replaced by $\overline{K}_T$ and N_T .

To compute ΔG_3 or ΔG_4 we simply integrate the differential Equation (4). For example :

$$\Delta G_3 = \int_0^{\alpha_s^e} \frac{\partial G_3}{\partial \alpha} \, d\alpha \quad , \tag{8}$$

which gives :

$$\Delta G_3 = n_0 \left[kT(\alpha_s^e \ln(\overline{K}_T N_T) + \alpha_s^e \ln \alpha_s^e + (1 - \alpha_s^e) \ln(1 - \alpha_s^e)) + \int_0^{\alpha_s^e} e\phi_0(\alpha) d\alpha \right] . \tag{9}$$

After having integrated $e\phi_0(\alpha) d\alpha$ by parts and remembering that when $\alpha = \alpha_s^e$ Equation (5) holds, we obtain the very simple expression for ΔG_3 :

$$\Delta G_3 = n_0 \left[kT\ln(1 - \alpha_s^e) - e \int_0^{\phi_0^e} \alpha d\phi_0 \right] \tag{10}$$

To proceed further we must now establish a relationship between α and ϕ_0 . This is derived from the Poisson-Boltzmann distribution for a uniformly charged plane (Gouy-Chapman model). This is justified under our experimental conditions ($C_T > \cdot 2M$) where the Debye screening length (< 7 Å) is small compared to the radius of curvature of the micellar surface (~ 25 Å) ; this approximation is certainly less severe than those listed above. We then write (see e.g. references [19-20]) :

$$e\phi_0 = 2 \, kT \, \sinh^{-1}(\alpha/a) \quad , \tag{11}$$

with $\quad a = \left[1.2 \, 10^{-3} C_T/\pi\ell_b \right]^{1/2} \left[e/\sigma_0 \right]$

where ℓ_b is the Bjerrum length ($\ell_b \sim 7$ Å in water at room temperature), C_T is the salt concentration in mole/ℓ, and σ_0/e is the ionic head density : $\sigma_0/e = 1/S_{TP}$ with S_{TP} the surface per polar head (Å^2).

Combining Equations (5) and (10) we can compute $\alpha_{s(c)}^e$ for the sphere (or the cylinder) as a function of $\overline{K}_T$ and N_T from :

$$(1 - \alpha_s^e)/\alpha_s^e = \overline{K}_T N_T \exp(2 \, \sinh^{-1}(\alpha_s^e/a_s)) \quad . \tag{12}$$

And using Equation (11), ΔG_3 is obtained from Equation (10) :

$$\Delta G_3 = n_0 kT \left[\ell n(1 - \alpha_s^e) - 2(\alpha_s^{e2} + a_s^2)^{1/2} + 2a_s \right] \quad , \tag{13}$$

and a similar expression holds for ΔG_4 .

The surface area per polar head on the Stern layer are computed

assuming the micellar geometry described in ref. 11 and the
Stern layer to be a sphere or cylinder with a radius $\sim$ 25 Å; [12,21,22]
these surfaces are assumed independent of the nature of the counterion consistent   with the initial assumption of a constant internal
structure of the micelles. We verified in the course of the calculations that with different values (within ± 20 %) for the Stern
layer radius, qualitatively similar results were obtained.

Table IV. Variations of the apparent charges and difference in
electrochemical energy for one monomer in the spherical or cylindrical part of a micelle with K_X and C_X .

K_X (mole fraction)$^{-1}$	C_X (M)	α_s^e	α_c^e	$\left((\Delta G_3 - \Delta G_4)/kTn_0\right)$
	0.2	0.856	0.752	− 0.60
1	0.4	0.852	0.749	− 0.54
	0.8	0.844	0.744	− 0.46
	0.2	0.564	0.450	− 0.32
10	0.4	0.552	0.442	− 0.27
	0.8	0.530	0.430	− 0.21
	0.2	0.287	0.222	− 0.12
100	0.4	0.264	0.208	− 0.08
	0.8	0.226	0.186	− 0.05

With Equations (12) and (13) and their analogues for the cylindrical part, we can now compute the electrochemical energy difference : $\Delta G_3 - \Delta G_4$ as a function of the salt concentration with K_X
(or $\overline{K}_T$) as an adjustable parameter. The results obtained for typical values of C_X and K_X are shown in Table IV ; these illustrate
the fact that α_s^e and α_c^e depend more noticeably on K_X than on C_X.
Thus upon addition of a second ion Y^- with $K_Y \ll K_X$ the effective $\overline{K}_T$ is lowered and the increase in ionic strength may be insufficient to compensate for this lowering; this is illustrated in
Table V. Addition of large amounts of a second ion results in a
decrease of $\Delta G_3 - \Delta G_4$, and the inbition of the micellar elongation
observed experimentally is actually predicted by the model.

Table V. Variations of the electrochemical energy difference with
ionic strength upon addition of a second salt (K_X = 100 , K_Y = 5).

C_X	C_Y	$\overline{K}_T$	$(\Delta G_3 - \Delta G_4)/kTn_0$
0.2	0	100	− 0.116
0.2	0.2	52.5	− 0.121
0.2	0.6	28.75	− 0.118

COMPARISON WITH EXPERIMENTS

We now turn back to the experimental results and seek for a set of K_X's for the ions Cl^-, Br^-, NO_3^-, ClO_3^- which yield a set of $(\Delta G_3 - \Delta G_4)/kT$ related to the b's (the empirical parameter of Equation (2)) by :

$$b = (\Delta G_3 - \Delta G_4)/kT + b_0 \quad , \tag{14}$$

where b_0 is a constant which depends on the internal structure of the micelle and is thus a part of ΔG_1 (see above Equation (3)). As can be seen in Table I the dependence of b on the salt concentration (in this case NaBr) is correctly reflected by the computed values of $(\Delta G_3 - \Delta G_4)/kT$ assuming $b_0 = -15$ and $K_{Br} = 110$. Assuming this value for b_0 a set of K_X's can be calculated, as shown in Table VI, with which a fair agreement is obtained between the experimental and computed b's as illustrated in the Tables I and II. The inhibiting influence of Cl^- on the elongation of micelle is indeed found for $K_{Cl} < K_{Br}/10$. From the K_X's the non-electrical part of the standard free energy of transfer of the anions from the CPX micelles to the water can be evaluated using Equation (6); these are shown in Table VI to be positive indicating that the anions tend to bind to the micelles. They are of the same order of magnitude as the total free energy of transfer found by Gamboa et al^8 for CTAX micelles.

Table VI. The set of adsorption-desorption equilibrium constants K_X's and the standard free energy of transfer of the anions from the micelles to the water.

X^-	K_X	$(\mu_X - \mu_{BX})$ (kcal/mole)
ClO_3^-	300	3.4
NO_3^-	180	3.1
Br^-	110	2.8
Cl^-	5	0.96

DISCUSSION

The quasi-quantitative agreement between the experimental and calculated b values is striking if one recalls the crudeness of our model. It is thus worthwhile reexamining the main hypothesis of the model in order to discuss whether the found agreement is sound or just fortuitous.

The first assumption is that the internal structure of the spherical and cylindrical parts of the micelle do not depend, at least in a first approximation, on the ionic environment of the micelle.

This allows for the separation of the free energy difference in two
parts as is done in Equation (3), one part depending only on the
internal structure and thus on the nature of the detergent ion and
the other depending also on the nature and concentration of the
counterion. This assumption is sustained by some experimental evi-
dences namely the constancy of the persistence length with respect
to the nature and concentration of the counterions (cf. Figure 4
and reference 10) and also the invariance of the empirical parameter a in Equation (2) with respect to the same factors which has
been observed for CPX and CTAX (see above). Moreover the same invariance was also observed in the case of SDS by Missel *et al.*[16, 18]

The second assumption concerns the spatial distribution of the
counterions in the neighbourhood of the charged micelle. The Poisson-Boltzmann approach of the diffuse layer, although widely used,
has been often questioned (see e.g. reference 23 and reference
therein). In particular in a recent paper, Fixman[24] discusses the validity of this description for polyions - small ions solutions and
he arrives at the conclusion that this approach is especially valuable when the charge density of the polyion and the ionic strength
of the solution are large; and these two conditions are met in the
present case.

A few remarks are now necessary concerning the specific adsorption of ions and the adjustable parameter of the model : K_X . The
adsorption of ions with increasing efficiency in the order :
$Cl^- < Br^- < NO_3^-$ has received experimental support in the case of
CTAX micelles[7,8] and evidence for a competitive adsorption of Cl^-
and Br^- on CTA^+ or CP^+ has been found by Mirallas[25] from
NMR measurements. These experiments, however, yield no information
on the nature of the interactions responsible for this specific adsorption. In the model described above, an assumption is made about
this interaction : it is of short enough range to leave the counterion distribution in the diffuse layer unaltered ; but no assumption is made about its nature. It probably consists of several contributions either localized in the vicinity of the polar heads
(chemical interaction) and/or unlocalized such as that due to the
polarization of the counterion in the very strong gradient of dielectric constant at the micellar interface.

In this frame the identification of the density of available
adsorption sites to the density of polar heads on the micellar surface is valid only if the unlocalized attraction is not too large :
the density of adsorbed counterions remains smaller than the density of polar heads and each adsorbed ion stands close to a polar
head in order to cancel its charge.

On the contrary, if the unlocalized adsorption energy is extremely high an inversion of the charge of the micelles could be obser-

ved as is observed for colloids.[26] The density of adsorbed ions is
then greater than the density of polar heads so that the density of
adsorption sites can no longer be identified with the latter. Apparently, such an inversion of the apparent micellar charge does not
occur in the present study ; it could however arise with other ions
so that we reconsider the above calculation with a different definition of the adsorption site in the Appendix.

There is no a priori reason for assuming that K_X is identical
on the spherical and cylindrical part of the micelle. But although
we could easily have introduced two distinct values, we prefer making this assumption in order to minimize the number of adjustable
parameters. This choice seems a posteriori justified by the good
agreement found with one parameter only.

Clearly a number of points have been neglected in the model
such as the finite size of the ions, their hydration, the variation
of the dielectric constant etc... We feel that the good agreement
which is observed between the model calculation and experiment,
despite these approximations, is probably due to the validity of the
two main assumptions and to the fact that the adjustment of K_X com-
pensates efficiently for the minor approximations. Thus the $\bar{K}_X$'s ,
as determined herein, clearly cannot be considered as   true equilibrium constants for the adsorption-desorption equilibrium of the
X^-'s .

APPENDIX

The density of adsorption sites is now assumed to be equal to
$1/S_X$ where S_X is the surface occupied by one adsorbed ion and we
define f_X by :

$$f_X = S_{TP}/S_X - 1 \quad , \qquad\qquad\qquad\qquad (A1)$$

(f_X/S_{TP}) is the excess density of adsorption sites compared to the
density of polar heads. Thus f_X will have different values f_X^S
and f_X^C on the spherical and cylindrical parts of the micelle.

In order to compute $\Delta G_2 = G_s^0 - G_c^0$ we consider the same cycle
of transformation as above (cf. Figure 7). The two uncharged states
are obtained by assuming that n_0 counterions are adsorbed so that
a different number of adsorption sites will remain free on the cylindrical and spherical micelles. Then ΔG_1 will be the sum of $\Delta G_i^\bullet$
dependent solely on the internal structure and of an entropic term :

$$- TS_{cs}^{\alpha=0} = kTn_0 \int_0^1 \left(\mathrm{Ln} \, \frac{1-\alpha}{\alpha+f_s} - \mathrm{Ln} \, \frac{1-\alpha}{\alpha+f_c} \right) d\alpha \quad . \qquad (A2)$$

Equation (3) is then written as :

$$\Delta G_2 = \Delta G_1^0 - TS_{cs}^{\alpha=0} + \Delta G_3 - \Delta G_4 \quad , \qquad (A3)$$

which can again be split into two terms one independent of the counterions ΔG_1^0 and the other dependent on the counterions namely
$\left[\Delta G_3 - \Delta G_4 - TS_{cs}^{\alpha=0}\right]$ or $\left(\Delta G_3' - \Delta G_4'\right)$ with

$$\Delta G_3' = \Delta G_3 + kTn_0 \int_0^1 Ln \frac{1-\alpha}{\alpha+f_s} \, d\alpha \quad ,$$

$$\Delta G_4' = \Delta G_4 + kTn_0 \int_0^1 Ln \frac{1-\alpha}{\alpha+f_x} \, d\alpha \quad , \qquad (A4)$$

ΔG_3 or ΔG_4 are computed as previously by integration of the free
energy change upon desorption of one ion from a micelle with
$n = n_0(1-\alpha)$ adsorbed counterions which now is written as :

$$-\frac{\partial G}{\partial n} = \frac{1}{n_0}\frac{\partial G}{\partial \alpha} = kTLnK_X N_X + kTLn \frac{\alpha+f_X}{1-\alpha} + e\phi_0 \quad . \qquad (A5)$$

As before K_X is given by Equation (6) and ϕ_0 as a function of
α by Equation (11). From the equilibrium condition $\partial G/\partial n = 0$ at
$\alpha = \alpha^e$ the relationship giving α^e for the sphere (or equivalently for the cylinder) is given by :

$$\frac{1-\alpha_s^e}{\alpha_s^e + f_X^s} = K_X N_X \exp\left[2 \sinh^{-1}(\alpha_s^e/a_s)\right] \qquad (A6)$$

And we finally obtain :

$$\Delta G_3' = kTn_0 \left[Ln(1-\alpha_s^e) + f_X^s Ln(f_X^s + \alpha) - (1+f_X^s)Ln(1+f_X^s)\right.$$
$$\left. - 2(\alpha_s^{e2} + a_s^2)^{1/2} + 2a_s\right] \quad , \qquad (A7)$$

and an equivalent relationship for $\Delta G_4'$. It is easy to verify that
if f_X^s is set equal to zero, Equation (A7) transforms to Equation
(3) as it should.

In the case of mixed salts solutions we assume two different
specific surfaces S_X and S_Y for the two counterions; correspondingly f_X's and f_Y's are defined by Equation (A1). At equilibrium the ratio (R) of adsorbed X and Y counterions is readily
shown to be :

$$R = \frac{\left(1 + f_X\right)n_X K_X}{\left(1 + f_Y\right)n_Y K_Y} = \frac{S_Y n_X K_X}{S_X n_Y K_Y} \qquad (A8)$$

from Equation (A8) R is seen to be independent of the micellar
structure considered so that the cycle of transformations from Figure 7 can still be considered provided R is held constant and
equal to its equilibrium value given by Equation (A8). It is then
easy to show that the relationships written above remain valid

provided K_X, N_X and f_X are replaced by $\overline{K}_T$, N_T and f_T defined by :

$$N_T = N_X + N_Y \ , \ \overline{K}_T = \frac{N_X K_X + N_Y K_Y}{N_T} \ , \ f_T = \frac{f_X N_X K_X + f_Y N_Y K_Y}{N_T \overline{K}_T} \quad (A9)$$

REFERENCES

1. E.W. Anacker in "Solution Chemistry of Surfactants" K.L. Mittal Editor, p. 247, Plenum Press, New York, 1979.
2. H. Wennerström and B. Lindman, Phys. Reports, 52, 1 (1979).
3. G. Porte, J. Appell and Y. Poggi, J. Phys. Chem., 84, 3105 (1980).
4. E.W. Anacker and H.M. Ghose, J. Amer. Chem. Soc., 90, 3161 (1968).
5. U. Herniksson, L. Odberg, J.C. Eriksson and L. Westman, J. Phys. Chem., 81, 76 (1977).
6. J. Ulmius, B. Lindman, G. Lindblom and T. Drakenberg, J. Colloid Interface Sci., 65, 88 (1978).
7. D. Bartet, C. Gamboa and L. Sepulveda, J. Phys. Chem., 84, 272 (1980).
8. C. Gamboa, L. Sepulveda and R. Soto, J. Phys. Chem., 85, 1429 (1981).
9. J. Appell and G. Porte, J. Colloid Interface Sci., 81, 85 (1981).
10. J. Appell and G. Porte, J. Colloid Interface Sci., 87, 492(1982).
11. G. Porte and J. Appell, J. Phys. Chem., 85, 2511 (1981).
12. D. Stigter, J. Phys. Chem., 68, 3603 (1964) ; idem, 79, 1008 (1975) ; idem, 79, 1015 (1975).
13. P. Mukerjee, J. Phys. Chem., 76, 765 (1972).
14. J.N. Israelachvili, D.J. Mitchell and B.W. Ninham, J. Chem. Soc. Faraday Trans 2, 72, 1525 (1976).
15. N.A. Mazer, M.C. Carey, G.B. Benedek, J. Phys. Chem., 80, 1075 (1976) ; in "Micellization Solubilization and Microemulsions", K.L. Mittal,Editor, Vol.1, p. 359, Plenum Press, New York, 1977.
16. P.J. Missel, N.A. Mazer, G.B. Benedek, C.Y. Young and M.C. Carey, J. Phys. Chem., 84, 1044 (1980).
17. P. Murkerjee and A. Ray, J. Phys. Chem., 70, 2149 (1966).
18. P.J. Missel, N.A. Mazer, M.C. Carey and G.B. Benedek in "Solution Behavior of Surfactants - Theoretical and Applied aspects", K.L. Mittal and E.J. Fendler,Editors, Vol.1, pp. 373-388, Plenum Press, New York, (1982).
19. J.O'M. Bockris, A.K.N. Reddy "Modern Electrochemistry", Plenum Press, New York, 1977.
20. G. Weisbuch, M. Guéron, J. Phys. Chem., 85, 517 (1981).
21. C. Tanford, J. Phys. Chem., 78, 2469 (1974).
22. R. Zana, J. Colloid Interface Sci., 78, 330 (1980).
23. G. Gunnarsson, B. Jönsson and H. Wennerström, J. Phys. Chem., 84, 3114 (1980).
24. M. Fixman, J. Chem. Phys., 70, 4995 (1979).
25. Ph. Mirallas, Thèse de Spécialité, USTL, Montpellier, France,1980.
26. J.T. Davies and E.K. Rideal, "Interfacial Phenomena", Academic Press, New York, 1963.

SALT-INDUCED SPHERE-ROD TRANSITION OF IONIC MICELLES

Shoichi Ikeda

Department of Chemistry, Faculty of Science
Nagoya University
Chikusa, Nagoya 464, Japan

Micelle size and shape of surfactants can be most directly determined by conventional light scattering measurements. This method has been applied for aqueous salt solutions of several ionic surfactants having a dodecyl group, and the conditions for stability of spherical and rodlike micelles have been examined by changing concentration and ionic species of added salt as well as by modifying polar head group of surfactants. At low salt concentrations the micelle molecular weight is low and the micelle shape should be globular or spherical. When the salt concentration exceeds a certain threshold value characteristic of surfactant species and counterion species, the spherical micelles formed at the critical micelle concentration associate together into rodlike micelles with increasing micelle concentration, and the micelle molecular weight becomes very large in concentrated salt solutions. The double logarithmic relation between micelle molecular weight and ionic strength is linear for both spherical and rodlike micelles. The threshold ionic strength for the sphere-rod transition is higher, as the ionic head group is bulkier and the counterion binding is less tight. Thus the main factors determining micelle shape are the geometrical adaptability of surfactant molecules or ions in a micelle and the intra- and inter-micellar interactions of their polar head groups. Co-ion species of added salt do not alter the threshold ionic strength but influence the micelle molecular weight.

INTRODUCTION

Two kinds of micelles have been proposed for ionic surfactants in dilute aqueous solutions. One is spherical or globular, and the other is rodlike or cylindrical. The former is due to Hartley,[1] and the latter was proposed by Debye and Anacker.[2] However, there have been suggested no clear way to ascertain whether a given surfactant forms spherical micelles or rodlike micelles under given conditions. Most ionic surfactants appear to form spherical micelles in water, and there has been serious objection to the idea of rodlike micelles.[3,4]

During the last several years we have been investigating the micelle size and shape of ionic surfactants in aqueous salt solutions, in order to elucidate the stability of spherical and rodlike micelles under different conditions. In this article we present an overview of the results of our investigation, mainly based on the light scattering measurements. We have confined ourselves to aqueous solutions of ionic surfactant having a dodecyl group, which contain simple salt having common counterion species.

LIGHT SCATTERING RESULTS

Light scattering from micellar solutions can be represented by the Debye equation

$$\frac{K(c - c_o)}{R_\theta - R_\theta{}^o} = \frac{1}{M\, P(\theta)} + 2\,B\,(c - c_o) \tag{1}$$

where R_θ is the reduced intensity of light scattered in the direction of scattering angle, θ, by a surfactant solution of concentration, c (g cm^{-3}), and K is the optical constant

$$K = \frac{2\,\pi^2 \tilde{n}_o{}^2 (\partial \tilde{n}/\partial c)_{C_s}{}^2}{N_A\, \lambda^4} \tag{2}$$

where N_A: the Avogadro number, λ: the wavelength of light, $\tilde{n}_o$: the refractive index of the solution at the critical micelle concentration (cmc), and $(\partial \tilde{n}/\partial c)_{C_s}$: the specific refractive index increment of solutions at constant salt concentration, C_s (M). The subscript, o, refers to the cmc. Then M is the (apparent) micelle molecular weight, and B is the (apparent) second virial coefficient. If the micelle has a radius of gyration, R_G, larger than $\lambda/20$, light scattered by a micelle undergoes interference, and the particle scattering factor, $P(\theta)$, differs from unity and depends on the angle in the following way:

$$\frac{1}{P(\theta)} = 1 + \frac{16 \ \pi^2 n_o^2}{3 \ \lambda^2} \ R_G^2 \ \sin^2 \frac{\theta}{2} + \cdots \qquad (3)$$

If the micelle is a rigid rod, its length is given by $L = 3.464 \ R_G$.

The (apparent) micelle aggregation number, m', is derived simply by dividing M by the molecular weight of surfactant molecule. The 'true' values of molecular weight and aggregation number of micelles can be obtained from M and B, if the micellar solution behaves ideally so that the second virial coefficient arises from the direct effect of effective micellar charge.[5,6] It proves that these values differ from M and m', respectively, by less than 10 % in the presence of more than 0.01 M salt. However, this theory does not take into account the long-range electrostatic interaction as well as the excluded volume effect.

In order to present typical light scattering behavior, we illustrate here the results on aqueous NaCl solutions of dodecyldimethylammonium chloride (DDAC).[7-9]

Figure 1 shows the Debye plots for micellar solutions of DDAC in water and in the presence of NaCl less than 1 M. No dissymmetry is observed for these solutions, so that $P(\theta) = 1$. In this range of NaCl concentrations the micelles strongly repel one another, but the repulsion decreases with increasing NaCl concentration, as expressed by the decrease in B. Values of M and m' obtained from the reciprocal of the intercept and of B from the slope are given in Table I.

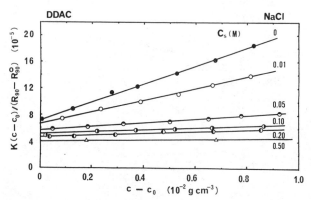

Figure 1. The Debye plot for aqueous NaCl solutions of DDAC at 25 °C. Low NaCl concentrations. (reprinted with permission from ref. 7).

Table I.   Size and Repulsion of Spherical Micelles of DDAC.

C_s (M)	Cmc (10^{-3} M)	M	m'	B (10^{-3} cm^3 g^{-1})
water	14.9	13,900	55.6	6.81
0.01	11.6	15,200	60.8	4.47
0.05	6.8	17,600	70.4	1.43
0.10	5.0	19,400	77.6	0.75
0.20	3.5	21,500	86.0	0.70
0.50	2.5	25,200	101	0.25

The observed aggregation number suggests that the micelles are spherical. However, the micelle shape cannot be exactly a sphere but will be an oblate spheroid such as proposed by Tartar[10] and others.[11,12] The aggregation number changes by a factor two with NaCl concentration.

When the NaCl concentration increases further, scattering of light becomes tremendously stronger with increasing micelle concentration, and its dissymmetry, $z_{45} = P(45)/P(135)$, also exceeds 1.05, i. e., $P(\theta) < 1$. The results are shown in Figure 2.

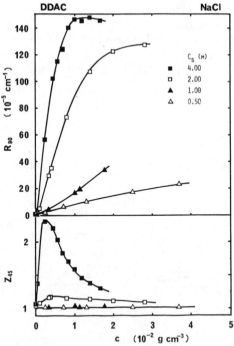

Figure 2. Reduced intensity (upper) and dissymmetry (lower) for aqueous NaCl solutions of DDAC at 25 °C. High NaCl concentrations. (reprinted with permission from ref. 7).

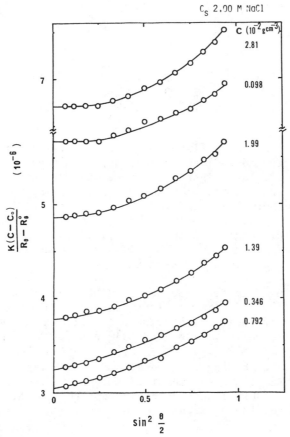

Figure 3. Angular dependence of light scattering for micellar solutions of DDAC in 2.00 M NaCl at 25 °C.

The angular dependence of light scattering from micellar solutions in 2.00 M NaCl is shown in Figure 3. With increasing micelle concentration, the intercept decreases rapidly and reaches a minimum. If the contribution from the second virial coefficient is negligible, this means that the small micelles formed at the cmc associate into larger micelles. The small micelles will be spherical, but the large micelles are rodlike, as shown below. At the finite micelle concentration where the Debye plot has a minimum value, the reciprocal intercept and the initial slope of angular dependence essentially correspond to the molecular weight and radius of gyration of the large micelles, respectively. At lower micelle concentrations, the reciprocal intercept gives the weight-average molecular weight of two kinds of micelles, spherical and rodlike.

Table II. Molecular Weight and Length of Rodlike Micelles of DDAC.

C_s (M)	Cmc (10^{-3} M)	M (10^3)	m'	R_G (A)	L (A)
1.00	0.92		224		
2.00	0.64	333	1,330	174	600
3.00	0.32	1,200	4,800	515	1,780
4.00	0.14	2,940	11,800	884	3,060

We have obtained similar plots for micellar solutions of DDAC
in more concentrated NaCl solutions, and from these plots we have
derived the molecular weight and radius of gyration of the large
micelles. Table II summarizes these results. The molecular weight
increases from 50,000 to 3,000,000 on changing NaCl concentration
from 1 M to 4 M. As the radius of gyration is roughly proportional
to the molecular weight, so the large micelles must be rodlike,
having a uniform thickness. Table II also includes values of the
length of rodlike micelles, assuming rigid rods.

We have investigated aqueous salt solutions of other ionic surfactants and observed similar light scattering behavior for aqueous
NaBr solutions of dodecyldimethylammonium bromide (DDAB)[13] and dodecyltrimethylammonium bromide (DTAB)[14] and for aqueous Na halide
solutions of sodium dodecyl sulfate (SDS).[15,16] Kushner et al.[17]
measured light scattering from aqueous NaCl solutions of dodecylammonium chloride (DAC) and reported values of micelle molecular
weight. Their results can also be interpreted in terms of the
sphere-rod conversion of micelle shape, as shown below.

SPHERE-ROD CONVERSION OF MICELLES

We may assume that the micellization of ionic surfactants
proceeds as follows: below the threshold salt concentration spherical micelles, D_m, are formed from monomers, D, in such a way as

$$m\ D \ \rightleftarrows\ D_m \tag{4}$$

and above the threshold salt concentration the spherical micelles
further associate together to form rodlike micelles, D_{mn}, as given
by

$$n\ D_m \rightleftarrows D_{mn} \tag{5}$$

The sphere-rod conversion of micelle shape occurs as a shift of a
concentration-dependent equilibrium at a fixed salt concentration.
At the threshold salt concentration, the intermicellar repusion is
compensated by the intermicellar attraction, so the micellar solution behaves ideally, i. e., B = 0.

On the other hand, the sphere-rod conversion of micelles occurs as a salt-induced (or counterion-induced) transition, if the micelle concentration is kept finite. Since spherical micelles are at first formed above the critical micelle concentration, the rod-like micelles begin to form only after the spherical micelles are present. Then the threshold salt concentration can be observed at finite micelle concentrations.

A theory of micellar growth beyond spherical micelles was given by Mukerjee,[18] and it has been elaborated by Tausk and Over-beek[19] and others.[20],[21] This theory postulates a series of association equilibria of micelles with monomer, which are isodesmic for all steps (other than the formation of the smallest or spherical micelle). It leads to a shift of the series of stepwise association equilibria towards higher aggregation with increasing micelle concentration but predicts a broad distribution of micelle size. There is, however, some evidence suggesting that rodlike micelles are not very polydisperse in size.

The salt-induced transition of ionic micelles is most clearly revealed when the micelle molecular weight is plotted against the ionic strength, where the ionic strength is given by the monomer and the added salt, i. e., $C_O + C_s$. (The monomer concentration is equal to the cmc, C_O (M)). Figure 4 shows the double logarithmic plots of M against $C_O + C_s$ for DDAC and its head group homologs, DAC and DTAC,[17],[22] all in aqueous NaCl solutions, where DTAC is dodecyltrimethylammonium chloride.

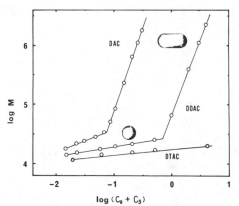

Figure 4. The double logarithmic plots of micelle molecular weight and ionic strength for DAC, DDAC and DTAC.

   Two linear portions having gradual and steep slopes are present for DDAC and DAC micelles. The portion with gradual slope
corresponds to the spherical micelles, and the portion of steep
slope represents the rodlike micelles. DTAC micelles, however,
follow a single straight line, indicating formation of only spherical micelles even in the saturated NaCl solutions. Generally, the
spherical micelles have size slightly dependent on the salt concentration, whereas the rodlike micelles alter their length much more
sensitively with the change in ionic strength, keeping their thickness constant.

   Figure 5 shows that similar linear double logarithmic plots
are obtained for DDAB and DTAB micelles in aqueous NaBr solutions.
13,14

   From these results we may conclude that electrostatic repulsion, which is stronger at low ionic strengths, stabilizes spherical micelles of ionic surfactants; and, if it is compensated by the
shielding effect of the added salt, the spherical micelles associate
together to form rodlike micelles.

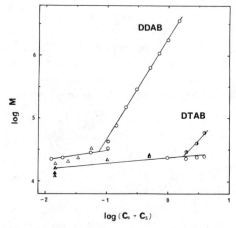

Figure 5. The double logarithmic plots of micelle molecular weight
and ionic strength for DDAB and DTAB at 25 °C.

FACTORS DETERMINING THE SPHERE-ROD TRANSITION

The threshold salt concentration for the sphere-rod transition of ionic surfactants can be readily idintified from the intersection of the two linear portions on the double logarithmic plot of M vs. $C_0 + C_s$. The threshold NaCl concentration is 0.07 M for DAC, 0.80 M for DDAC and is higher than 5.5 M for DTAC. Similarly, the threshold NaBr concentration is 0.07 M for DDAB and 1.8 M for DTAB.

From these results one can see that an ionic surfactant undergoes the sphere-rod transition at a lower concentration of added salt, as the ionic head group is smaller. Since the surfactant ions having a bulkier head group are accommodated more comfortably in a spherical micelle than in a rodlike micelle, we can assign the geometrical adaptability of the shape of surfactant molecules or ions to a micelle as one of the important factor for the sphere-rod transition.

Comparing the threshold salt concentration for different counterion species, Cl^- and Br^-, one can see the effect of specific binding of counterions on the sphere-rod transition. In spite of the same surfactant ion species, the threshold salt concentration for the Cl^- micelles in aqueous NaCl solutions is much higher than that for the Br^- micelles in aqueous NaBr solutions. These results clearly indicate that the specific binding of counterions to the micelles plays a major role in reducing intramicellar electrostatic repulsion and destabilizing the spherical micelles: Br^- binds more tightly to the micelles than Cl^-.

We further find that linear double logarithmic relations between M and $C_0 + C_s$ also hold for SDS micelles in aqueous NaX solutions, where X^- is F^-, Cl^-, Br^- or I^-.[15,16] They are shown in Figure 6. In these plots two linear portions intersect at 0.45 M, irrespective of co-ion species, X^-. Thus the co-ion species does not alter the threshold salt concentration for the sphere-rod transition. However, it influences the micelle molecular weight, as was observed for DTAB micelles in aqueous NaBr[14] and KBr[23] solutions and for dodecylpyridinium bromide micelles in aqueous LiBr, KBr and RbBr solutions.[24] In most cases, except for spherical micelles of SDS, the micelle aggregation number is larger as the size of the co-ion of added salt becomes larger.

From these results we can attribute the effect of added salt on the sphere-rod transition to that of counterion species, i. e., to the specific binding of counterions to the micelles.

The effect of the size of the ionic head group on the micelle aggregation number was examined for decylammonium bromide and its N-methyl-substituted homologs in 0.500 m NaBr solutions.[25] These results show that the micelle size of decylammonium bromide and its

methyl derivative is more than ten times larger than that of its
dimethyl and trimethyl derivatives; the latter two have size as-
signable to spherical micelles. The large aggregation numbers for
the former two can be adequately explained if it is postulated that
they have the threshold NaBr concentrations lower than 0.500 m, so
that they can form rodlike micelles in 0.500 m NaBr solutions.

The effect of counterion species on the micelle size was investigated for various cationics, especially on dodecylpyridinium
halide micelles in aqueous K halide solutions,[24] and DTAB[26] and
cetylpyridinium bromide[27] micelles, both in various Na salt solutions. The large differences observed in the aggregation number
dependent on counterion species can be interpreted in terms of the
differences in micelle shape, i. e., spherical vs. rodlike; DTAB
micelles are spherical in most of Na salt solutions, whereas they
have aggregation number as large as 1100 and should be rodlike in
0.500 m NaSCN solutions.[27]

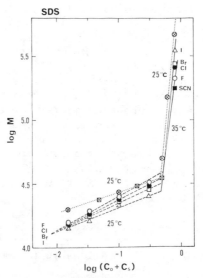

Figure 6. The double logarithmic plots of micelle molecular weight
and ionic strength for SDS at 25 and 35 °C. (including data of
other investigators)   (reprinted with permission from ref. 16).

COUNTERION BINDING AND MICELLE STRUCTURE

Counterion species of added salt influence not only the sphere-rod transition in micelles but also affect the micelle aggregation number, as they bind to micelles to different degrees.

Table III lists the aggregation number of spherical micelles of DDAC and DDAB as a function of concentration of added salt having common counterion species. In spite of the identical surfactant ion species constituting charged micelles, the aggregation number differs considerably at a given salt concentration. A similar difference in aggregation number is observed between spherical micelles of DTAC and DTAB. Generally, the Br^- micelles have a size about 20 surfactant ions larger than the Cl^- micelles. Thus the Br^- micelle must deviate from a sphere to a larger extent than the Cl^- micelle, if the micelle is closely packed by the surfactant ions. The deviation would be towards the distorted oblate spheroids.[10-12]

Table III. Aggregation Number, m', of Spherical Micelles.

C_s (M)	0	0.01	0.05	0.10	(Na salt)
DDAC	55.6	60.8	70.4	77.6	(NaCl)
DDAB	77.1	79.9	95.8	107	(NaBr)

Table IV gives the molecular pitch of rodlike micelles of DDAC nad DDAB at different aggregation numbers. The molecular pitch of both micelles decreases with increasing aggregation number, suggesting flexibility of the micelles. Viscosity measurements have shown that the rodlike micelles of DDAC behave like wormlike chains in concentrated NaCl solutions.[9]

Table IV. Molecular Pitch, L/m', of Rodlike Micelles (in A).

m'	985	1330	2185	3700	4800	5980	11800	12100
DDAC		0.45			0.37		0.26	
DDAB	0.39		0.38	0.35		0.28		0.23

Table IV also indicates large difference in molecular pitch between DDAC and DDAB micelles. DDAC has a longer pitch than DDAB at a given aggregation number, in spite of the identical surfactant ion species. If the surfactant ions are packed in the micelles with an equal density in the direction of contour length, the difference in molecular pitch gives different numbers of surfactant ions in a layer of cross-section of the rodlike micelle. Let us assign values of 0.40 and 0.35 A as the molecular pitch of DDAC and DDAB micelles, respectively. If the extended surfactant ion

is 6.0 A thick, then 15 and 17 surfactant ions must be incorporated
in the cross-section of DDAC and DDAB micelles, respectively. This
difference means that the cross-section deviates from a circle to
different degrees: the section of the Br$^-$ micelles should be more
distorted from a circle than that of the Cl$^-$ micelles, both being
elliptical.

From these results we can now suggest models for surfactant
micelles such as illustrated in Figure 7. Spherical micelles would
be distorted oblate ellipsoids, and rodlike micelles are wormlike
and have elliptical cross-sections.

TWO-STEP MICELLIZATION

We have found that micellization is followed by the sphere-rod
equilibrium, when the concentration of added salt exceeds a certain
threshold value. This is expressed by Equations (4) and (5). We
can expect that under certain conditions these two processes occur
in two steps on increasing the surfactant concentration. Around
the threshold salt concentrations, DTAB and SDS exhibit, besides
the first cmc, the second cmc above which rodlike micelles are
formed.[14,15] Figure 8 shows the Debye plots of DTAB micelles in
concentrated NaBr solutions. Slightly above the threshold NaBr
concentration, the second cmc is more than ten times higher than the
first cmc. With increasing NaBr concentration further, the second

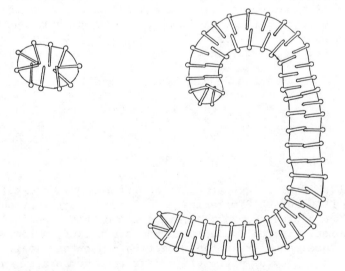

Spherical micelle   Rodlike micelle

Figure 7. Shapes and structures of two kinds of micelles.

cmc lowers much more rapidly than the first cmc, and eventually the second cmc becomes indistinguishable from the first cmc.

Both DTAB and SDS have bulky ionic head groups so that spherical micelle are much more stable because of geometrical reason. Formation of rodlike micelles is hampered by the relatively high stability of these spherical micelles, until certain extent of stabilization is attained by their mixing with solvent.

It should be noted, however, that the sphere-rod conversion discussed here occurs only when the added salt is present beyond a certain threshold concentration. Consequently, the sphere-rod equilibrium found by us must be distinguished from that reported by Reiss-Husson and Luzzati even without added salt.[28] The latter was usually observed for more concentrated surfactant solutions by small angle x-ray scattering.

ANGULAR DEPENDENCE OF LIGHT SCATTERING

It is relevant to discuss characteristic features of angular dependence of light scattering from micellar solutions containing spherical and rodlike micelles together. Light scattered by such a solution behaves differently at low and high scattering angles. This feature can be seen in aqueous NaCl solutions of DDAC shown in Figure 3, and it is more enhanced in aqueous NaI solutions of SDS, as shown in Figure 9.

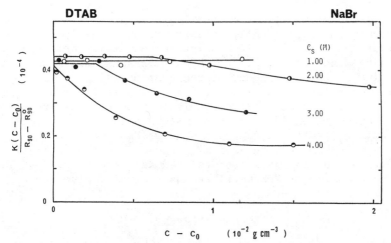

Figure 8. The Debye plots for aqueous NaBr solutions of DTAB at 25 °C. High NaBr concentrations. (reprinted with permission from ref. 14).

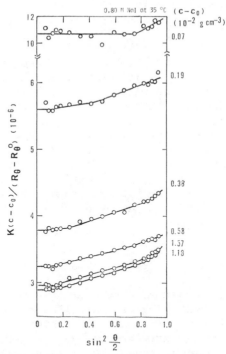

Figure 9. Angular dependence of light scattering for micellar solutions of SDS in 0.80 M NaI at 35 °C. (reprinted with permission from ref. 16).

It is known that light scattering from a solution of macromolecules consisting of optically anisotropic units exhibits anomalous angular dependence.[29-34] In a solution containing both spherical and rodlike micelles, light scattered by a spherical micelle has a uniform intensity independent of the scattering angle, but light scattered by a larger rodlike micelle undergoes internal interference. The interference effect consists of a usual isotropic effect as given by Equation (3) and an anomalous anisotropic effect. This latter effect leads to a negative angular dependence at low angles and to a positive dependence at high angles, so that a minimum may appear around 90° in the reciprocal reduced intensity.[35,36] This would be stronger for SDS micelles than for DDAC micelles.

ACKNOWLEDGEMENT

I am greatly indebted to my collaborators, especially to Dr. S. Ozeki, Mr. S. Hayashi and Dr. T. Imae, for their careful experimental work and stimulating discussion.

REFERENCES

1. G. S. Hartley, "Aqueous Solutions of Paraffin-Chain Salts," Hermann, Paris, 1936.
2. P. Debye and E. W. Anacker, J. Phys. Colloid. Chem., 55, 644 (1951).
3. L. M. Kushner and W. D. Hubbard, J. Colloid Sci., 10, 428 (1955).
4. C. Tanford, J. Phys. Chem., 78, 2469 (1974).
5. W. Prins and J. J. Hermans, Proc. K. Ned. Akad. Wet., B 59, 162 (1956).
6. L. H. Princen and K. J. Mysels, J. Colloid Sci., 12, 594 (1957).
7. S. Ikeda, S. Ozeki and M. Tsunoda, J. Colloid Interface Sci., 73, 27 (1980).
8. S. Ikeda, S. Ozeki and S. Hayashi, Biophys. Chem., 11, 417 (1980).
9. S. Ozeki and S. Ikeda, J. Colloid Interface Sci., 77, 219 (1981).
10. H. V. Tartar, J. Phys. Chem., 59, 1195 (1955).
11. C. Tanford, J. Phys. Chem., 76, 3020 (1972).
12. J. N. Israelachvili, D. J. Mitchell and B. W. Ninham, J. Chem. Soc., Faraday Trans. II, 72, 1525 (1976).
13. S. Ozeki and S. Ikeda, to be published.
14. S. Ozeki and S. Ikeda, J. Colloid Interface Sci., 87, 424 (1982).
15. S. Hayashi and S. Ikeda, J. Phys. Chem., 84, 744 (1980).
16. S. Ikeda, S. Hayashi and T. Imae, J. Phys. Chem., 85, 106 (1981).
17. L. M. Kushner, W. D. Hubbard and R. A. Parker, J. Res. Nat. Bur. Stand., 59, 113 (1957).
18. P. Mukerjee, J. Phys. Chem., 76, 565 (1972).
19. R. J. M. Tausk and J. Th. G. Overbeek, Biophys. Chem., 2, 175 (1974).
20. P. J. Missel, N. A. Mazer, G. B. Benedek, C. Y. Young and M. C. Carey, J. Phys. Chem., 84, 1044 (1980).
21. G. Porte and J. Appell, J. Phys. Chem., 85, 2511 (1981).
22. S. Ozeki and S. Ikeda, Bull. Chem. Soc. Jpn., 54, 552 (1981).
23. H. J. L. Trap and J. J. Hermans, Proc. K. Ned. Akad. Wet., B 58, 97 (1955).
24. W. P. J. Ford, R. H. Ottewill and H. C. Parreira, J. Colloid Interface Sci., 21, 522 (1966).
25. R. D. Geer, E. H. Eyler and E. W. Anacker, J. Phys. Chem., 74, 369 (1971).
26. E. W. Anacker and H. M. Ghose, J. Phys. Chem., 67, 1713 (1963).
27. E. W. Anacker and H. M. Ghose, J. Am. Chem. Soc., 90, 3161 (1968).
28. F. Reiss-Husson and V. Luzzati, J. Phys. Chem., 68, 3504 (1964).
29. P. Horn, H. Benoit and G. Oster, J. chim. phys., 48, 530 (1951).
30. P. Horn, Ann. phys., 10, 386 (1955).
31. M. Nakagaki, Bull. Chem. Soc. Jpn., 34, 834 (1961).
32. H. Utiyama and M. Kurata, Bull. Inst. Chem. Res., Kyoto Univ., 42, 128 (1964).
33. K. Nagai, Polymer J., 3, 67 (1972).

34. G. C. Berry, J. Polymer Sci., C 65, 143 (1978).
35. T. Imae, K. Okahashi and S. Ikeda, Biopolymers, 20, 2553 (1981).
36. T. Imae and S. Ikeda, to be published.

QUASIELASTIC LIGHT SCATTERING STUDIES OF THE MICELLE TO VESICLE

TRANSITION IN AQUEOUS SOLUTIONS OF BILE SALT AND LECITHIN

P. Schurtenberger [+], N.A. Mazer [*], W. Känzig [+]
and R. Preisig [†]

Laboratorium für Festkörperphysik, ETH Zürich [+];
Department of Physics, MIT Cambridge and Gastro-
enterology Division, Brigham and Women's Hospital,
Boston [*]; Department of Clinical Pharmacology,
University of Berne [†]

In recent years detailed light scattering studies
of bile salts (BS) and lecithin (L) in aqueous solutions[*]
have shown that thermodynamically stable aggregates
(mixed micelles) are formed. Their size distribution depends strongly upon the total lipid concentration C, the
molar ratio (L/BS) and the temperature T. If C decreases
at constant (L/BS) and T, the micellar size increases
and appears to diverge at a critical value C_c. At concentrations below C_c the disc-shaped mixed micelles
undergo a transition to monodisperse versicles with a
mean hydrodynamic radius $\bar{R}_h$ in the range of 120-600 Å,
depending on (L/BS), C and T. Similar phenomena can be
observed by varying (L/BS) and T at constant C.

By serially diluting aqueous solutions of the trihydroxy bile salt glycocholate and egg yolk lecithin in
0.15 M NaCl/Tris buffer (pH 8) with different (L/BS)-
ratios, we have delineated in a phase diagram the two
regions where mixed micelles and where vesicles are present. The size of the vesicles decreases monotonically
with increasing dilution and approaches a minimum radius
of about 120 Å independent of (L/BS). A simple theoretical analysis of this behavior suggests that the vesicle

---

*) N.A. Mazer, G.B. Benedek and M.C. Carey,
   Biochemistry 19, 601 (1980)

size is determined by the partitioning of bile salt
molecules between the aqueous and the bilayer phase.

In the micellar region of the phase diagram the
aggregates are in thermodynamic equilibrium, and changes in the size distribution induced by varying C, (L/
BS) or T are reversible. However, when the micellar
phase boundary is crossed and vesicles are formed,
their size not only depends on the values of C, (L/BS)
and T, but can also depend on how these parameters are
varied to reach the final state of the solution. This
behavior indicates that solutions containing vesicles
can be brought into metastable states.

INTRODUCTION

Bile salts, lecithin and cholesterol, the major lipid components of bile, show a variety of aggregation phenomena in aqueous
solutions. While the biologically important long-chain lecithins
form liquid crystalline bilayer structures in water they can be
solubilized in the presence of bile salt molecules by forming thermodynamically stable mixed micelles [1,2,3]. Recent studies [4] using
quasielastic light scattering have led to a model of discshaped
mixed micelles consisting in essence of a lecithin bilayer. The
bile salts form a belt around the perimeter and thus prevent the
contact between water and the lecithin hydrocarbon chains. In
addition they are incorporated within the disc in a fixed stoichiometry depending upon the bile salt species and the temperature only
(see Figure 1).

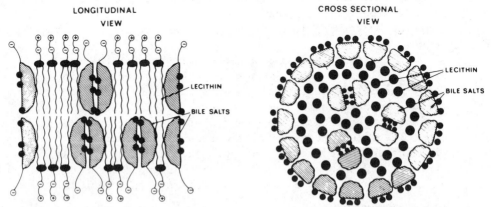

Figure 1.   Schematic model for the structure of the bile salt-
lecithin mixed micelles according to   ref. 4.

Systematic investigations of the aggregative properties of aqueous micellar solutions of bile salt and lecithin [4] have shown the existence of three distinct regimes in a phase diagram where one plots the bile salt concentration versus the lecithin concentration (see Figure 2). At low molar ratios of lecithin to bile salt (L/BS) simple micelles appear to be in coexistence with mixed micelles. At higher L/BS ratios only mixed micelles are present in solution. Their size strongly depends upon the total lipid concentration and the L/BS ratio. By diluting a mixed micellar solution one can observe a marked increase of the mean size and the polydispersity with decreasing concentration. At a critical dilution the size of the micelles seems to diverge. One has reached the mixed micellar phase limit. Upon further dilution a transition from disc-shaped mixed micelles to nearly monodisperse vesicles occurs [5].

The concentration dependence of the micellar size and the existence of a phase limit can be explained with the mixed disc model by taking into accout the equilibrium between mixed micelles and bile salt monomers in the intermicellar solution [4]. The system tends to conserve the intermicellar concentration (IMC) of bile salt monomers, and thus the relative amount of bile salts available for solubilizing the lecithin decreases upon dilution. At concen-

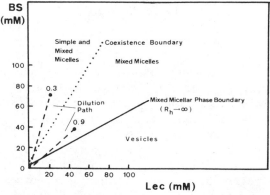

Figure 2. Phase diagram for the taurocholate-lecithin system in 0.15 M NaCl at 20°C (from ref.4). Also shown are dilution paths at two different (L/BS) molar ratios (0.3 and 0.9). The initial total lipid concentration for each path was 50 mg/ml. See text for explanation.

trations below the phase limit, the remaining bile salt molecules
are too few to solubilize the lecithin in disc-shaped mixed micelles
with a finite size and a transition to closed bilayer structures
(vesicles) occurs. An analogous transition from mixed micelles to
vesicles occurs if the detergent molecules are removed by dialysis
or by gel filtration [6,7].

Although vesicles have become of considerable importance as
model membranes and drug delivery systems [8] there is still a lack
of basic quantitative knowledge about the formation of these bilayer structures. Even the question as to whether vesicles are
thermodynamically stable is not yet answered. For these reasons we
have employed static and dynamic light scattering methods to study
the micelle to vesicle transition in bile salt-lecithin solutions.
We have determined the mean size and polydispersity of vesicles
produced by diluting mixed micellar solutions as functions of the
(L/BS) ratio, the total concentration and the temperature. To study
the question of whether such vesicles are thermodynamically stable
we have designed several experiments to see if the vesicle size
depends only upon the values of these physical and chemical parameters or whether it shows a dependence upon the path in the field
of these parameters leading to the final state of the solution
(i.e. on the particular sequence of events leading to the final
temperature and composition of the system). Such a path dependence
would indicate that solutions containing vesicles can be in a metastable state, in contrast to micelles which are in thermodynamic
equilibrium.

EXPERIMENTAL

Materials and Solutions

Egg yolk lecithin was obtained from Lipid Products (South
Nutfield,Surrey, U.K. (Grade I)) and the sodium salt of glycocholic
acid was from Calbiochem. The glycocholate was dissolved in ethanol,
filtered and recrystallized according to Cortese [9]. The NaCl was
obtained from Merck (analysis grade) and the Tris(hydroxymethyl)
aminomethane was obtained from Fluka (puriss p.a.).

Mixed micellar bile salt-lecithin solutions with different
(L/BS) ratios were prepared by the method of coprecipitation [10].
After dissolving an appropriate amount of each lipid in ethanol,
the mixture was dried in vacuo (Büchi rotavapor) until the dry

weight was constant. An appropriate amount of a 0.15 M NaCl/Tris
buffer (pH 8) was then added in order to obtain stock solutions
with a total lipid concentration of 50 mg/ml. The final concentrations were then prepared from the stock solutions by a number of
dilution steps. Each sample was flushed with purified N_2, sealed
and then incubated at the desired temperature until the light
scattering properties were time independent. In contrast to micellar systems which reached equilibrium within minutes the incubation
time needed for solutions containing vesicles was usually at least
48 hours due to the very slow kinetics of the micelle to vesicle
transition.

Apparatus and Methods

Our light scattering apparatus for both static and dynamic
measurements consists of an argon ion laser (Spectra Physics, model
171, λ = 5145 Å), a temperature controlled scattering cell holder,
a digital Malvern autocorrelator (K 7023, 96 channels) and an "on-
line" data analysis performed by a Nova 3 computer. Details are
given elsewhere [11].

Aqueous bile salt-lecithin solutions (0.5 ml) were pipetted
into acid washed cylindrical quartz scattering cells (inner diameter 8 mm). These were sealed and centrifuged at 15000 g for 10
minutes to sediment dust. Measurements were usually performed at
90° scattering angle at temperatures of 20°C and 40°C.

The autocorrelation function of the scattered light intensity $C(\tau)$ was measured at 48 channels equally spaced in time. The
remaining 48 channels were delayed to provide a direct measurement
of the baseline (B). The B values agreed within experimental error
with the base line computed from the monitor channels, indicating
the absence of dust and other sources of long decay times. A cumu-
lant analysis [12] was used to deduce the mean diffusion coefficient
$\bar{D}$ and the polydispersity V (as defined in Ref. 13). From $\bar{D}$ an
apparent mean hydrodynamic radius $\bar{R}_H$ as defined by Equation (1)
was calculated in analogy with the Stokes-Einstein relation [13,14]

$$\bar{R}_H = \frac{kT}{6\pi\eta\,\bar{D}} \tag{1}$$

where kT is the thermal energy and η the viscosity of the solvent.

Results and Discussion

a) Equilibrium properties of solutions containing bile salt-lecithin vesicles

        In the first series of experiments we have measured the mean
hydrodynamic radius as a function of dilution for bile salt-lecithin solutions having different L/BS ratios at 20°C. In each dilution series a characteristic concentration dependence of the size
and polydispersity of the aggregates was observed. In Figure 3 the
mean hydrodynamic radius $\bar{R}_H$ and the polydispersity index V are
plotted versus the dilution factor for a L/BS ratio of 0.9. Initially $\bar{R}_H$ and V correspond to the micellar stock solution. With increasing dilution the observed micellar size and polydispersity

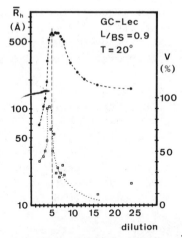

Figure 3. Mean apparent hydrodynamic radius $\bar{R}_H$ (— ● —) and polydispersity index V (·· □ ··) versus the dilution factor for a dilution series of a glycocholate-lecithin stock solution with L/BS=
0.9 and a total concentration of 50 mg/ml. A dilution factor of 1
corresponds to the stock solution concentration of 50 mg/ml, and
a dilution factor of n to 50 mg/ml divided by n.

first increases as predicted by the mixed disc model and then passes through a maximum. At this critical dilution we have reached the mixed micellar phase limit. At higher dilutions we can observe nearly monodisperse vesicles whose size decreases with increasing dilution from approximately 600 Å to 150 Å at high dilutions. A similar aggregation behaviour of biliary lipids has been recently observed in native dog bile collected at different excretion rates [15].

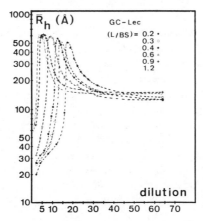

Figure 4. Mean apparent hydrodynamic radius $\bar{R}_H$ versus the dilution factor for six dilution series differing in their L/BS ratio. The total lipid concentration of the stock solutions was always 50mg/ml.

In Figure 4 we have plotted the hydrodynamic radius of the aggregates versus the dilution factor for six dilution series differing in their L/BS ratio. At higher L/BS ratios the critical dilution is shifted towards lower values. This is in agreement with the predictions of the mixed disc model (4) and is understandable from the phase diagram in Figure 2. The critical dilution is reached at the point where the dilution path intersects with the

micellar phase limit. One can see that for the higher of two different L/BS ratios (0.3 and 0.9) this intersection is located at a
higher total lipid concentration and thus requires a smaller dilution factor. For all L/BS ratios investigated the hydrodynamic
radius decreases with increasing dilution and approaches values between 120-150 Å. This is close to the minimum size of detergent free
unilamellar lecithin vesicles prepared with sonication [8]. For a
qualitative understanding of the characteristic concentration dependence of the vesicle size one has to take into account the influence of dilution on the equilibrium between the bile salt molecules in the aqueous solution and within the aggregates. The micellar system tends to conserve the IMC. Therefore dilution decreases the amount of bile salt molecules in the mixed micelles and
necessitates micellar growth according to the mixed disc model [4].
At the critical dilution there is no bile salt left to solubilize
the lecithin by the formation of discs of finite size and the transition to vesicles occurs. Because of the partition equilibrium for
the bile salt between the vesicles and the monomers in solution the
amount of bile salt molecules in the vesicles is expected to decrease with further increasing dilution. The decreasing vesicle
size with increasing dilution led us therefore to the hypothesis
that the vesicle size is mainly determined by the remaining amount
of bile salt molecules incorporated in the bilayer. A knowledge of
this relation between the vesicle size and the bilayer composition
is essential for a quantitative understanding of the aggregative
properties of the system. Therefore we have tried to analyse our
data using a simple model based on the following three assumptions:

1.) The radius of the vesicles R at a fixed temperature is a mono-
    tonic function of the bile salt to lecithin ratio in the bi-
    layer $(\frac{BS}{L})_b$

$$R = f\ [(\frac{BS}{L})_b]\qquad\qquad(2)$$

2.) The bile salt molecules "dissolved" in the lecithin bilayer are
    in equilibrium with those dissolved as monomers $((BS)_{aq})$. This
    equilibrium at a fixed temperature involves an equilibrium con-
    stant K which is assumed to be independent of the bilayer cur-
    vature.

$$(\frac{BS}{L})_b = K \cdot (BS)_{aq}\qquad\qquad(3)$$

3.) Because of the extremely small solubility of lecithin in water all lecithin molecules are incorporated in the vesicles.

Taking into account the conservation of bile salts,

$$(BS)_{tot} = (BS)_{aq} + (\frac{BS}{L})_b \cdot (L)_{tot} \qquad (4)$$

where $(BS)_{tot}$ and $(L)_{tot}$ are the total molar concentrations of bile salt and lecithin, respectively, one finds

$$(L)_{tot} = (\frac{BS}{L})_b^{-1} \cdot (BS)_{tot} - K^{-1} \qquad (5)$$

Therefore the model yields a simple relation between the solution composition, the postulated equilibrium constant K and the vesicle size (as a consequence of assumption 1). This relation contains several predictions that can easily be tested with our experimental data. In a plot of $(BS)_{tot}$ versus $(L)_{tot}$ vesicles of equal size and therefore equal $(BS/L)_b$ (Equation 2) should be on straight lines. The slope is given by the ratio $(\frac{BS}{L})_b$ and the intercept with the abscissa by the value $-(1/K)$. An analysis of our dilution series is presented in Figure 5. The linear relationship predicted by the model is indeed observed over a wide range of solution composition and vesicle radii. Moreover all lines intercept the "lecithin-axis" at nearly the same point, leading to a value for K^{-1} of 16 mM.

According to Equations (2) and (5) $\bar{R}_H$ should depend solely on the quantity $\frac{(L)_{tot} + K^{-1}}{(BS)_{tot}}$ which corresponds to the ratio of lecithin to bile salt in the bilayer of the vesicles. In Figure 6 we have replotted $\bar{R}_H$ of the vesicles measured in the different dilution experiments versus $\frac{(L)_{tot} + K^{-1}}{(BS)_{tot}}$ using the value of 16 mM for K^{-1} as determined from Figure 5. The finding that all vesicle sizes in Figure 4 fall on a single functional dependence offers strong support in favour of a general relationship between the vesicle size and the bilayer composition. It is noteworthy to point out the strong composition dependence of the vesicle size for values of $\frac{(L)_{tot} + K^{-1}}{(BS)_{tot}}$ between 3.5 and 7 while at higher values $\bar{R}_H$ is nearly independent.

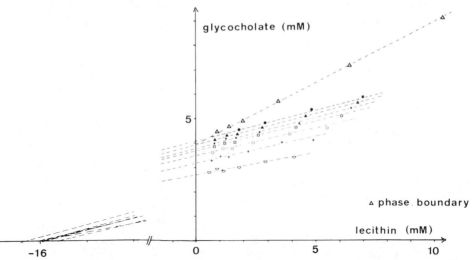

Figure 5.   Phase diagram for glycocholate-lecithin systems at 20°C
in 0.15 M NaCl/Tris buffer. The phase boundary separates mixed
micellar systems which lie above from vesicle systems which lie
below. Compositions that produced vesicles of the following sizes
are indicated by the corresponding symbols: 500 Å (●), 450 Å (▲),
400 Å (★), 350 Å (□), 300 Å (○), 250 Å (+) and 200 Å (▽). The
broken lines are derived from a least squares analysis of the data
for each vesicle size and are seen to intercept the "lecithin-axis"
at approximately the same value in agreement with Equation (5).

b) Dependence of the properties of the solution upon the path of
preparation

     The equilibrium model together with the results discussed
above permits a prediction of the vesicle size as a function of
the final total concentration and the L/BS ratio using Figure 5.
This makes it possible to investigate the dependence of the system
upon variations in the preparation method.

     According to the phase diagram in Figure 5 addition of bile
salts to a solution of small vesicles should cause a marked growth
of the vesicles and could even result in the formation of mixed
micelles if the final composition crosses above the phase boundary.
We have investigated the readjustment of the size distribution in
a solution of small glycocholate-lecithin vesicles (mean hydro-

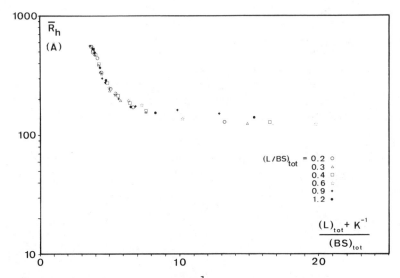

Figure 6. $\bar{R}_H$ versus $\dfrac{(L)_{tot} + K^{-1}}{(BS)_{tot}}$ for the glycocholate-lecithin

dilution series presented in Figure 4. A value of 16 mM was used for K^{-1} according to the value determined from Figure 5.

dynamic radius 175 Å, C_{tot} = 4.17 mg/ml, L/BS = 1.2) after adding aqueous solutions of glycocholate with different concentrations. The composition of the final solutions, the predicted values for $\bar{R}_H$ and the experimental results are presented in Table 1. Although the preparation method is quite different from the simple dilution method, the final sizes of the vesicles and micelles are in good agreement with the predictions based on the model, indicating that the bile salt molecules can influence the size of preformed leci- thin vesicles and even solubilize them into mixed micelles. This ability of bile salts and other detergents [16] to influence vesicle properties can have important consequences in the use of liposomes as drug carriers [18]

Changes in temperature have a marked influence on the micel- lar phase limit [5,17] and cause a shift of the characteristic curve $\bar{R}_H$ versus dilution (see Figure 7). Temperature changes are therefore

Table I. Mean apparent hydrodynamic radius $\bar{R}_H$ of a vesicle solution of glycocholate and lecithin after the addition of aqueous
solutions of glycocholate at different concentrations. The total
concentration of the stock solution was 4.17 mg/ml, the L/BS ratio
equal to 1.2 and the hydrodynamic radius $\bar{R}_H$ = 175 Å. The final L/BS
ratios and the total concentrations are presented in the first and
second columns, respectively. The expected aggregate size determined from the dilution series (Figure 5) are presented in column 3.

L/BS	C_{tot} [mg/ml]	$\bar{R}_H$ (predicted) [Å]	$\bar{R}_H$ (measured) [Å]
(0.2	4.67)*	40	40
(0.37	4.51)*	90	80
0.9	3.68	195	170
0.6	3.47	250	250
0.6	3.68	290	280
0.4	3.29	395	360
0.65	4.35	400	420
0.3	3.47	500	515

*) These compositions correspond to mixed micellar
   solutions (see Figure 5).

a convenient method for the observation of the dynamics of the rearrangement of the vesicle size distribution and of the existence
of possible path dependent metastable states. We have equilibrated
two dilution series of the same micellar stock solution (L/BS=0.3,
C_{tot} = 50 mg/ml) at 20°C and at 40°C. As presented in Figure 7 the
two dilution curves show the same characteristic concentration dependence of the aggregate size. However, the critical dilution for
40°C is shifted towards a lower value and the maximum value of $\bar{R}_H$
is smaller. Upon lowering the temperature from 40°C to 20°C in the
series equilibrated at the higher temperature we would expect that
in the absence of path dependent behaviour the size distribution
of both micelles and vesicles would readjust until the equilibrium
values for 20°C (as measured in the other dilution series) were
reached. Figure 8 A shows the agreement between expectation and

experiment in both the micellar and the vesicle regions. A completely different behaviour is observed however after increasing the
temperature from 20° to 40°C in the other dilution series.
In the micellar region the rearrangement does take place, consistent with the thermo-reversibility of micellar system [4,13]. However,
in the samples for which the micellar phase limit has already been
crossed at 20°C the equilibrated vesicles show no indication of

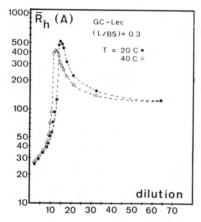

Figure 7. $\bar{R}_H$ versus the dilution factor for two dilution series
of the same glycocholate-lecithin stock solution (L/BS = 0.3, total
concentration 50 mg/ml) equilibrated at 20°C and 40°C respectively.

shrinking and their mean size remains at the values observed for
an equilibration temperature of 20°C (Figure 8 B).

    This clear indication of path dependence is confirmed by
other experiments. For example when previously equilibrated solutions containing large vesicles (i.e. $\bar{R}_H$ 200-500 Å) were diluted
to concentrations at which the "equilibrium" vesicle size should
be ~120 Å (according to Figure 5) they failed to show any decrease
in size during observations extending over three months. While our

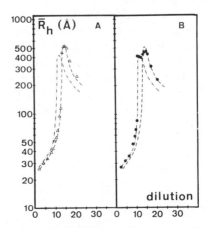

Figure 8. Mean hydrodynamic radius $\bar{R}_H$ versus the dilution factor for the dilution series presented in Figure 7 after changing the temperature from (A) 40°C to 20°C and from (B) 20°C to 40°C. The curves correspond to the equilibrium values at 20°C (---) and 40°C (·—·—·) determined from Figure 7.

previous data in table 1 and Figure 8 A demonstrate that vesicles are capable of growing we have as yet never seen the shrinking of vesicles equilibrated beforehand. Thus it would appear that vesicle solutions can become trapped in a metastable state with a larger mean size than would be predicted from the phase diagram given in Figure 5.

CONCLUSIONS

In summary the results presented in this paper show that the size of vesicles formed by diluting mixed micellar solutions of lecithin and bile salts can vary from ~120 to ~500 Å depending upon the total lipid concentration, the lecithin-to-bile salt molar ratio and the temperature. On the basis of a simple partition equilibrium model we can interpret these observations as due to the influence of the bilayer composition on the vesicle size. The model has enabled us to deduce from the experimental data a general relation between the size of the vesicles and the composition of the solution. Although this "equilibrium" relation could be verified in numerous epxeriments, we also found cer`ain paths of sample preparation where the expected vesicle sizes were not obtained. Therefore we conclude, that solutions containing vesicles can be brought into metastable states, in contrast to micellar solutions which are in thermodynamic equilibrium.

ACKNOWLEDGEMENT

This work was supported by The Swiss National Science
Foundation (Grant No. 3.614.80). The authors thank Dr. Ph. Huguenin
for purifying the bile salt material used in this study and
M. Hürlimann for the important contributions made during his
diploma work.

REFERENCES

1. D. M. Small, J. Lipid Res., $\underline{8}$, 551 (1967).
2. J. Ulmius, G. Lindblom, H. Wennerström, L.B.-Å. Johannson,
    K. Fontell, O. Söderman and G. Arvidson, Biochem., $\underline{21}$, 1553
    (1982).
3. M. C. Carey and D. M. Small, Am. J. Med., $\underline{49}$, 590 (1970).
4. N. A. Mazer, G. B. Benedek and M. C. Carey, Biochem., $\underline{19}$, 601
    (1980).
5. N. A. Mazer, Ph.D. Thesis, Massachusetts Institute of Technology
    (1978).
6. J. Brunner, P. Skrabal and H. Hauser, Biochim. Biophys. Acta,
    $\underline{455}$, 322 (1976).
7. M. Milsmann, R. Schwendener and H. G. Weder, Biochim. Biophys.
    Acta, $\underline{512}$, 147 (1978).
8. F. Szoka and D. Papahadjopoulos, Ann. Rev. Biophys. Bioeng.,
    $\underline{9}$, 467 (1980).
9. F. Cortese, J. Am. Chem. Soc. $\underline{59}$, 2532 (1937).
10. D. M. Small, S. A. Penkett and D. Chapman, Biochim. Biophys.
    Acta, $\underline{176}$, 178 (1969).
11. H. R. Haller, Dissertation (6604), ETH Zürich (1980).
12. D. E. Koppel, J. Chem. Phys., $\underline{57}$, 4814 (1972).
13. N. A. Mazer, G. B. Benedek and M. C. Carey, J. Phys. Chem. $\underline{80}$,
    1075 (1976).
14. N. A. Mazer, M. C. Carey and G. B. Benedek, in "Micellization,
    Solubilization and Microemulsions", K. L. Mittal, Editor, Vol.1,
    p. 359, Plenum Press, New York 1977.
15. N. A. Mazer, P. Schurtenberger, W. Känzig, M. C. Carey and
    R. Preisig, paper presented at the 4th International Conference
    on Surface and Colloid Science, Jerusalem, Israel, July 1981.
16. D. Lichtenberg, C. F. Schmidt, M. Jackson and B. J. Litman,.
    paper presented at the 4th International Conference on Surface
    and Colloid Science, Jerusalem, Israel, July 1981.
17. P. Schurtenberger, N. A. Mazer and W. Känzig, manuscript in
    preparation.
18. R. N. Rowland and J. F. Woodley, Biochim. Biophys. Acta, $\underline{620}$,
    400 (1980).

THE APPLICABILITY OF MICELLAR MODELS TO THE ACTIVITY COEFFICIENTS

OF SODIUM CARBOXYLATES IN AQUEOUS SOLUTION

G. Douhéret and A. Viallard
Laboratoire de Thermodynamique et Cinétique Chimique
E.R.A. C.N.R.S. 1005. Université de Clermont-Ferrand 2
F - 63170, Aubière, France

The work reported here was carried out in the context of research on enhanced recovery of mineral oil. Its aim was to seek a predictive theoretical model for the description of the variation with concentration of the mean activity coefficients $\gamma_{\pm s}$ of ionic amphiphiles in aqueous solution, using in particular the pseudophase model (p.p.model). We have established that with this model, assuming no interactions between phases and uncharged micelles, close-to-theoretical behaviour is only observed with long-chain amphiphiles. Increasingly marked departure from theoretical behaviour occurs as the chain shortens. This departure may be explained qualitatively by assuming that the micelles are charged, and become increasingly so as the chain length decreases. The assumption that the micellar charge does not vary with the overall concentration of the amphiphile proves to be incompatible with the model. However, assuming an invariant degree of counter-ion binding, a theoretical law may be developed which closely fits experimental data.

The mass action law model (m.a.model) was also examined. We had already shown that this model could satisfactorily describe the experimental $\gamma_{\pm s}(m_s)$ relationship for short-chain carboxylates. In order to remain consistent with the assupmtions made using the p.p.model, it was assumed that interactions between aggregates and the surrounding medium are negligible. A mixed model then leads to a theoretical law which fits experimental data.

In the case of charged species, the assumption of the invariance of the micellar charge with changing overall concentration of the amphiphile is incompatible with the model. If the ratio of the anionic and cationic stoichiometric coefficients is assumed to be constant, the theoretical law found to be rigorous with the p.p.model becomes only an approximate one.

This approach remains open to improvement, but does nevertheless describe quite well a certain amount of experimental data. The approximations introduced and limits of applicability are discussed, and the directions in which further work in this area might usefully go are indicated.

INTRODUCTION

In the course of previous work, it was found that the variation of apparent molal volume[1] and apparent molal heat capacity[2]vs. concentration of homologous sodium carboxylates in aqueous solution could be accurately described using the p.p.model in conjunction with an original method of data treatment. The present work set out to see whether and to what extent this model could describe the variation of the activity coefficients of these species with concentration. This was felt to be of particular importance since these values determine the thermodynamic properties of the solution. Robinson and Stokes[3], Davies[4], Hall[5] and Koshinuma[6] among others have called attention to the difficulties encountered in expressing the activity coefficients of electrically charged dissociated species. This is the case when a micellar charge is introduced with the p.p.model, or when charged aggregates are introduced with the m.a.model. In addition, it must be stressed that very few approaches have yet been attempted to describe the variation of the activity coefficients of surfactants over a very wide concentration range including monomer and micelle ranges and beyond (see Pytkowicz[7]). Nevertheless, the interesting tentative work of Leyendekkers and Hunter[8], the comprehensive paper of Vikingstad[9] and particularly the promising studies of Stenius et al[10] warrant mention here. Recently, we also developed a method for the calculation of the mean activity coefficient $\gamma_{\pm s}$ of sodium carboxylates in aqueous solution using the m.a.model[11]. This model is being used by Desnoyers et al[12] to analyse micellar aggregation processes in the course of a systematic investigation of thermodynamic parameters (ϕ, V, H, C_p, α_p and K_T). The m.a.model[13], like the p.p.model[14], is widely used for the description of amphiphile aggregation and micelle formation, though in both cases, a number of crude assumptions are necessary. The present work aimed to investigate the applicability of these two models to the description of activity coefficients of sodium carboxylates in aqueous solution, by thorough examination of the assumptions made explicitly and implicitly. Electrically neutral micelles (p.p.model) and aggregates (m.a.model) were first examined, followed by charged micelles and aggregates.

EXPERIMENTAL DATA

A number of measurements pertaining to the activity coefficients of sodium carboxylates in aqueous solution are documented in the literature. Some, such as sodium octanoate, have been studied in great detail[15,16]. Systematic studies have also been carried out on short and long-chain carboxylates[17,18,19,20,21,22]. However, irrespective of their quality, the available data are seldom comparable. Firstly, different ranges of concentration have been studied using a variety of different experimental techniques. Secondly, the data fall into two distinct main categories, the first corresponding to studies at variable ionic strength of the solute, while the second concerns measurements at constant ionic strength, maintained at a generally high level by a background electrolyte according to the constant ionic medium principle[23]. This implies complete lack of comparability. In this study, we chose to consider only the former class.

NON-CHARGED SPECIES

Pseudophase Model

The p.p.model is often regarded as a rather rough, though easily applied approximation[24], particularly when used to model short-chain amphiphiles of low cooperativity. Nevertheless, some of our studies led us to the conclusion that the p.p.model can be useful, and when applied concurrently with the m.a.model, which is regarded as having poor applicability to high-cooperativity surfactants and at high concentrations, that it can cover a wide range of amphiphiles and surfactants[1].

In its basic form, the p.p.model assumes that a solute s dissolved in a solvent 0 is present in two separate phases, namely the solvent containing phase α and a phase composed solely of micellar species, β. Equilibrium of s between α and β implies that the chemical potentials of the solute in each phase are the same, i.e.,

G. DOUHÉRET AND A. VIALLARD

$$\mu_s^{\alpha} = \mu_s^{\beta} \tag{1}$$

In addition, the Gibbs-Duhem rule applied to phase β gives μ_s^{β} to be constant. It therefore follows from Equation (1) that the mean activity of s in α, $a_{\pm s}^{\alpha}$ is constant whence, since $a_{\pm s}^{\alpha} = m_s^{\alpha}.\gamma_s^{\alpha}$, the corresponding mean molal activity coefficient $\gamma_{\pm s}^{\alpha}$ may be written as a linear function of the reciprocal molality of s in α, $(m_s^{\alpha})^{-1}$:

$$\gamma_{\pm s}^{\alpha} = C/m_s^{\alpha} = C.\xi_s^{\alpha} \tag{2}$$

where C is a constant.

Assuming no interactions between α and β, the free enthalpies G_{α} and G_{β} are additive. Hence, the experimental chemical potentials of solute s, μ_s and solvent 0, μ_0 may be equated to their values in α. Consequently, for the same reference state of s in α and in the whole solution,

$$a_{\pm s} = a_{\pm s}^{\alpha} = \text{constant} \tag{3}$$

where $a_{\pm s}$ is the experimental activity of s in the solvent. Thus alongside Equation (2), we may write :

$$\gamma_{\pm s} = C'/m_s = C' . \xi_s \tag{4}$$

since $m_{\pm s}$, the mean molality of s, is equal to the stoichiometric molality, m_s, whose reciprocal is ξ_s. If the p.p.model is valid, a plot of $\gamma_{\pm s}$ against ξ_s should yield a straight line with a zero ordinate intecept. As shown in Figure 1, plots of experimental values of $\gamma_{\pm s}$ against ξ_s are close to linear for all the carboxylates studied over concentration ranges which are increasingly wide as n increases. In addition, the ordinate intercepts A(n) tend to zero as n increases. This same pattern has also been observed with other types of surfactants, e.g. n-alkyl sulfates and n-alkyltrimethyl-ammonium bromides[25]. Thus a law of the type :

$$\gamma_{\pm s} = A(n) + B(n).\xi_s \tag{5}$$

accurately accounts for the experimental data. The coefficients A(n) and B(n) specific to various carboxylates are given in Table I.

The departures from the linear law (Equation (4)) may be ascribed either to inappropriateness of the model, or to unwarranted assumptions made, such as the absence of interactions between phases. At high concentrations, different structures may be present also[1]. For the time being, we decided to further examine the implications of the model itself. From Equation (3), we can obtain an expression for the activity coefficients $\gamma_{\pm s}$ and $\gamma_{\pm s}^{\alpha}$:

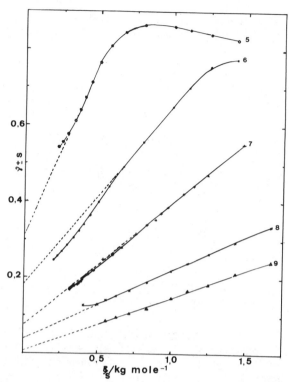

Figure 1 - Mean molal activity coefficient for carboxylates Na$^+$ H(CH$_2$)$_n$COO$^-$ in aqueous solution at 298,15 K (from Smith and Robinson[19] (n = 5, 6, 8, 9) and Ekwall et al[15] (n = 7) as a function of the reciprocal molality (various curves refer to different values of n).

$$\gamma_{\pm S} = \gamma_{\pm S}^{\alpha} (1 - N \frac{m_{NM}}{m_S})$$

where m_{NM} represents the molality of the micelles, assumed to be uncharged. Equation (3) also implies that $\gamma_{\pm S}^{\alpha} (m_S - Nm_{NM})$ is constant. Still assuming no interactions between phases, $\gamma_{\pm S}^{\alpha}$ is solely a function of $m_{\pm S}^{\alpha}$, whence $m_{\pm S}^{\alpha}$ is also constant. Consequently, for neutral micelles,

$$m_+^{\alpha} = m_-^{\alpha} = constant = c.m.c.$$

where m_+^{α} and m_-^{α} are respectively the cationic and anionic molalities in phase α. It may therefore be concluded that the c.m.c. has a real meaning only when all the assumptions introduced above are warranted, i.e. uncharged micelles, no solvent in phase β and no interactions between phases.

Table 1 - Coefficients of Equation (5) applicable to the mean molal activity coefficients of sodium carboxylates $Na^+H(CH_2)_nCOO^-$, in aqueous solution, at 298.15 K.

n	Reference	concentration range (mole kg^{-1})	Number of experimental points	A(n)	B(n)
5	19	2.0 - 3.5	6	+ 0.3202	+ 0.8834
6	19	0.8 - 1.5	5	+ 0.1904	+ 0.4552
7	15	0.7 - 1.3	9	+ 0.0639	+ 0.3311
8	19	0.5 - 2.5	12	+ 0.0475	+ 0.1719
9	19	0.5 - 1.8	10	+ 0.0091	+ 0.1397

The possibility of a change in the aggregation number N as a function of m_S may be envisaged. Relatively low values of N might be expected to lead to a variation of the chemical potential μ_S^β of the solute in β with N. It is here that the pseudophase model differs from the phase separation model (invariance of μ_S^β with droplet size in an emulsion). If, therefore, variations of N cause variations of μ_S^β, the phase rule, which imposes invariance at constant T and P for a binary two-phase system, is not obeyed. Hence, the p.p.model and its assumptions imply invariance of N. Clearly, then, the assumptions introduced into this model restrict its scope. In a previous semi-empirical approach, the m.a.model was found to provide a close fit for the activity coefficients of the same compounds (n ⩽ 9). We therefore endeavoured to compare the two models, making the same assumptions.

Mass Action Law Model

This model assumes that a single aggregate with aggregation number N is in equilibrium with the parent monomer in a homogeneous medium :

$$N M^+ + N A^- \rightleftarrows (M_N A_N)^0 \qquad (6)$$

where M^+ and A^- are the cation and anion resulting from dissociation of the solutes, and $(M_N A_N)^0$ is the electrically neutral N-mer.

The equilibrium constant $K_N^{\ominus}$ (T) is given by

$$K_N^{\ominus}(T) = \frac{a_{NM}}{(a_+)^N \cdot (a_-)^N} \qquad (7)$$

where a_i is the activity of species i. Applying Euler's theorem, at equilibrium (null affinity) we obtain:

$$\mu_S = \mu_+ + \mu_-$$

and :

$$a_{\pm S} = (a_+ \cdot a_-)_{NA}^{1/2}$$

whence :

$$\gamma_{\pm S} = \frac{(a_+ \cdot a_-)_{NA}^{1/2}}{m_S} \qquad (8)$$

After rearranging :

$$\gamma_{\pm S} = \left| \frac{\gamma_{NA}}{K_N^{\ominus}(T)} \right|^{1/2N} \cdot \frac{(m_{NA})^{1/2N}}{m_S} \qquad (9)$$

where γ_{NA} is the activity coefficient of the neutral aggregate and m_{NA} is its molality. Clearly, knowledge of $\gamma_{\pm S}(m_S)$ alone is not sufficient in theory to determine the parameters $K_N^{\ominus}(T)$ and N. An additional relationship is needed. However, a semi-empirical method for evaluating $K_N^{\ominus}(T)$ and N has been developed by us[11]. This method gives a good fit of experimental data and also enables the limits of the premicellar and micellar ranges of aqueous solutions of sodium carboxylates to be determined. Stenius[26], from highly detailed and accurate emf measurements, suggested that, for certain types of surfactants, very small premicellar aggregates form within a narrow concentration range where the micelle concentration starts to rise. The lower experimental accuracy involved in the determination of activity coefficients does not evidence such aggregates which the model may therefore neglect. Stenius[26] considers that this type of model gives a good description of the variation of activity coefficients with concentration.

If m_{NA} is replaced by $(m_S - m_1)/N$ in Equation (8), where m_1 is the molality of the monomer, the expression for $\gamma_{\pm S}$ becomes:

$$\gamma_{\pm S} = \left| \frac{\gamma_{NA}}{NK_N^{\ominus}(T)} \right|^{1/2N} \cdot \xi_S \cdot (\frac{1}{\xi_S} - m_1)^{1/2N} \qquad (10)$$

Unlike the p.p. model, this model does not imply that $\gamma_{\pm S}$ is a linear function of ξ_S, since γ_{NA} and m_1 depend on T, P and m_S. However, some linearity may be observed in certain ranges of concentration if appropriate assumptions are introduced. First, as was done with the p.p.model, let us assume that the interactions between aggregates are sufficiently weak for γ_{NA} not to vary, i.e. $\gamma_{NA}(T, P, N)$ has a constant value. Next, since $m_1(m_s)$ is unknown, let us assume that m_1 is almost constant and close to the c.m.c. This corresponds to a mixed "m.a.-p.p." model. Equation (10) then becomes :

$$\gamma_{\pm s} \sim A(N,T) \cdot (\frac{1}{\xi_s} - CMC)^{1/2N} \cdot \xi_s \qquad (11)$$

where $A(N,T)$ represents the quantity $\left| \dfrac{\gamma_{NA}}{N.K_N^\Theta(T)} \right|^{1/2N}$, and γ_{NA}

is considered as constant.

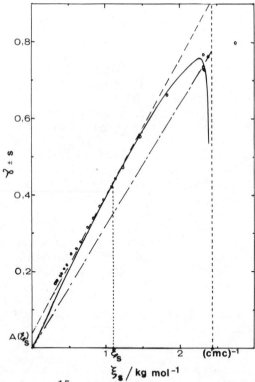

Figure 2 - Experimental[15] (o) and calculated (drawn curve : mixed model ; dotted curve : p.p.model) mean molal activity coefficient for sodium octanoate in aqueous solution at 298.15 K, as a function of ξ_s.

Figure 2 compares results using the two models for solutions of sodium octanoate for which detailed, though not very accurate, data are available[15]. The values of $K_N^0(T)$ and N (respectively 7×10^6 and 11) are reported[11]. The curve obtained with the mixed model using a value of 0.40 mole kg^{-1} for the CMC fits the experimental data fairly well except in the premicellar range where the approximation is obviously inappropriate. At the highest concentrations, changes in the structure of the micellar phase occur[1]. The ordinate intercept $A(\xi_s)$ of the tangent at point $(\xi_s, \gamma_{\pm s})$, deduced from Equation (10) is given by

$$A(\xi_s) = f(N,T) \cdot (\frac{1}{\xi_s} - CMC)^{(\frac{1}{2N} - 1)} \qquad (12)$$

Increase in the length of the hydrocarbon chain means an increase in N. It can be shown that $A(\xi_s)$ decreases as N increases. Thus, experimentally observed behaviour is accounted for and, as remarked by Lindman and Wennerström, "the larger the value of n, the more cooperative is the association and the more one approaches phase separation behaviour"[27].

At this stage, then, the choice of model depends as much on the nature of the amphiphile as on the assumptions involved. Neither the m.a.model nor the p.p.model is entirely satisfactory, but both have something to offer. A better agreement might be expected through a more realistic description of the systems. Thus, the assumption of uncharged micelles, though warranted as a rough approximation, may physically be an oversimplification. Accordingly, the assumption of a micellar charge was introduced into each model.

However, before going further, it might be worthwhile to add some comments about the so-called "mixed" models and their use. Common practice is to assume a_{NA} approximately constant[28] in Equation (7) giving :

$$a_+ \cdot a_- \backsim constant = a_{\pm} = a_{\pm s}$$

This approximation apparently makes this into a p.p.model, but it in fact results from confusion in the definitions of a_{NA} in the two models. In the p.p.model, a_{NA} (in fact a_{NM}) represents the activity of the N-mer in the micellar phase which is indeed constant under well-defined conditions. In the m.a.model, a_{NA} is the activity of the N-mer in the homogeneous ternary medium (water/monomer/N-mer). It thus varies with m_{NA}, the concentration of the N-mer in the solution. At most, γ_{NA} might be assumed constant if the interactions with the micelle are assumed negligible, which seems reasonable in the case of neutral micelles. This justifies à posteriori the distinction made between NA (m.a.model) and NM (p.p.model).

CHARGED MICELLES

Pseudophase Model

When charged micelles are considered, a difficulty arises in the definition of the solute in phases α and β. Thus, phase α contains in addition to the solvent, positive and negative ions at different concentrations. Hence, the composition of the mixture depends on m_s according to :

$$m_+^\alpha = m_s - q m_{CM} \tag{13}$$

$$m_-^\alpha = m_s - p m_{CM} \tag{14}$$

where m_i^α is the molality of species i, p and q are the numbers of negative and positive ions in the micelle of molality m_{CM} in the mixture. The system must be considered as a ternary diphasic system (two ions and a solvent). The equilibrium of the ions between phases gives :

$$\mu_+^\alpha = \mu_+^\beta \tag{15}$$

$$\mu_-^\alpha = \mu_-^\beta \tag{16}$$

where μ_i^ϕ is the chemical potential of i in phase ϕ. The Gibbs-Duhem rule is applied in phase β giving :

$$q(d \ln a_+^\alpha)_{T,P} = - p(d \ln a_-^\alpha)_{T,P} \tag{17}$$

Assuming no interations between phases (admittedly a bold assumption since charged micelles are being considered), from Equations (15) and (16), using Euler's theorem, we obtain the following expression for G :

$$G = n_s (\mu_+^\alpha + \mu_-^\alpha) + n_o^\alpha \, \mu_o^\alpha$$

and in general for the system (solute s + solvent 0)

$$G = n_s \mu_s + n_o \mu_o$$

for any values of n_s and n_o. It then follows that :

$$\mu_s = \mu_+^\alpha + \mu_-^\alpha \; ; \qquad \mu_o = \mu_o^\alpha$$

and $\qquad \mu_+ = \mu_+^\alpha \; ; \qquad \mu_- = \mu_-^\alpha$

With the same reference states in phase α and in the whole solution :

$$a_+ = a_+^\alpha \quad ; \quad a_- = a_-^\alpha$$

Thus,

$$a_{\pm s} = a_{\pm s}^\alpha \tag{18}$$

whence, with Equations (13) and (14), we deduce :

$$\gamma_{\pm s} = \gamma_{\pm s}^\alpha \cdot (1 - q \cdot \frac{m_{CM}}{m_s})^{1/2} \cdot (1 - p \frac{m_{CM}}{m_s})^{1/2} \tag{19}$$

Unfortunately, the parameters p and q cannot be extracted from this relation without further equations. In particular, the expression of $\gamma_{\pm s}^\alpha$ (m_+^α and m_-^α) requires modeling the influence of the ionic concentrations on $\gamma_{\pm s}^\alpha$ in phase α, in which electrical neutrality is not obtained.

At the present stage, the simplest treatment of Equation (19) would involve assuming p and q to be constant, i.e. that the charged micelles are monodisperse and that their shape and size do not vary. Hence, addition of solute causes only an increase in the number of micelles. This assumption implies that γ_+^β and γ_-^β are invariant, provided there are no interactions between phases and between micelles and that there is no solvent in phase β. Phase β then being a binary mixture, one of its constituents can be considered as the "solvent", giving :

$$a_+^\beta = \frac{q}{p} \cdot \frac{10^3}{M_-} \cdot \gamma_+^\beta \; (p, q)$$

$$a_-^\beta = \frac{10^3}{M_-} \cdot \gamma_-^\beta \; (p, q)$$

Consequently, a_+^β = constant and a_-^β = constant and, from Equations (15) and (16), it results in a_+^α = constant and a_-^α = constant, so that, for neutral micelles, we have :

$$\gamma_{\pm s} = \text{constant}/m_s = C' \cdot \xi_s$$

Equation (4) is thus valid for both charged and uncharged micelles, and so, the comments already made regarding its fit with the experimental data still apply.

However, since this system is a ternary diphasic one, the phase rule would require that, at constant T and P, it be invariant, i.e., when m_s varies, the chemical potential of any of the constituents in either of the phases at equilibrium may be arbitrarily chosen. Now, this is incompatible with the invariance of the activities a_i^β established above assuming p and q constant. Accordingly, this model, with the crude treatment applied and with the assumptions made above, is incompatible with invariant p and q, i.e. with a constant micellar charge.

If now, we assume p and q to be variable but such that p/q = r = constant(T,P), Equation (17) becomes :

$$(a_+^\alpha)^r \cdot (a_-^\alpha) = \text{constant}$$

and as above, assuming no interactions between phases :

$$a_+^r \cdot a_- = \text{constant (T, P)} \tag{20}$$

Experimentally, if this law holds, a plot of $\ln a_-$ against $\ln a_+$ should give a straight line with slope = - r which would give directly r, the degree of counterion binding. Clearly, $a_+(m_s)$ would have to be known. In Figure 3, for aqueous solutions of sodium octanoate, we can see two regions of linearity, one of which has a slope giving r = 0.59 which agrees closely with the value deduced from self-diffusion measurements reported by Lindman and Brun[30]. The other region of linearity with slope of opposite sign corresponds to an entirely different phenomenon which will be discussed below. The Equation (19) does not fit for short-chain carboxylates (n = 3, 4, 5) for which values of single ion activities are available[31]. This finding is consistent with spectroscopic data for these species reported by Umemura et al[32], suggesting that extensive pre-

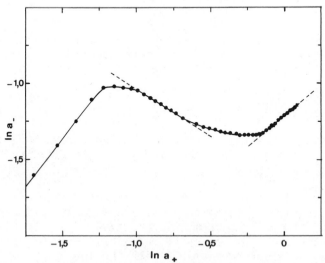

Figure 3 - Octanoate *vs* sodium activity, in aqueous solution, at 298.15 K (from Ekwall and Stenius[29]).

micellar aggregation occurs rather than true micellization. Lack of
directly available data for sodium decanoate (n = 9) prevented us
from seeing what sort of fit could be had with Equation (19), but
experimental data for other types of surfactants[33,34] fitted well.
Evidently, then, the assumption of micellar charge with constant r
is not only "thermodynamically" possible, but agrees well with available experimental data. It may be added that a constant r implies
variation of the aggregation number (here p), i.e. of the size of
the micelles and therefore of their charge $\eta = p(r - 1)$.

It is found experimentally that, for sodium carboxylates[35], r
tends to 1 as n increases, which reduces the micellar charge.
Unfortunately, it has been found[1,11] that the values of N vary
considerably according to the technique used to measure them and
are not very reproducible. In addition, at least for carboxylates,
their determination has always assumed the micelles to be neutral.
Error calculations show that it is not yet possible to know, using
the above expression for η, how the micellar charge varies with n.
It might be expected, for any particular carboxylate, that p would
increase with m_s, in which case η would increase. Assuming that the
set of micelles or the micelle itself are only stable within a
certain range of micellar charge, then beyond a critical value of
m_s, some new structure would be expected to appear. The existence
of successive micellar phases of different nature has indeed been
clearly observed for sodium octanoate[1] and sodium dodecyl sulfate[36]
among others.

Of the various assumptions made, that of no interactions between
phases may be examined, though no model for such interactions is yet
completely developed. To a first approximation, a_+^α and a_-^α might be
assumed to differ only slightly from a_+ and a_- and r = q/p to remain
roughly constant, so that Equation (20) would be approximately valid.
If $a_{\pm s}^\alpha \sim a_{\pm s}$ and a_1^ϕ vary only slightly with m_s, then :

$$a_{\pm s} \sim B(n) \cdot m_s + A(n)$$

i.e. $a_{\pm s}$ is no longer constant, but interactions only induce a first-
order perturbation in m_s. Hence :

$$\gamma_{\pm s} \sim B(n) + A(n) \cdot \xi_s \qquad (21)$$

Experimentally, as shown in Figure 1, such a law fits for small n,
with B(n) tending to zero as n increases. In other words, the per-
turbation term of $a_{\pm s}$ tends to zero and $\gamma_{\pm s}$ tends to $A(n) \cdot \xi_s$.
Now, $\gamma_{\pm s} = A(n) \cdot \xi_s$ is the expression obtained with the p.p.model
for neutral micelles assuming no interaction between phases. This
observation therefore supports the idea of a lowering of micellar
charge and, with it, a diminution of interactions between phases as
the chain length of carboxylates increases.

Mass Action Law Model

The equilibrium between the charged species is written as :

$$q \, M^+ + p \, A^- \rightleftarrows (M_q \, A_p)^{q-p}$$

The equilibrium condition (reaction affinity null) gives

$$\mu_{CA} = q\mu_+ + p\mu_-$$

i.e.

$$K^\ominus_{pq}(T) = \frac{a_{CA}}{a_+^q \cdot a_-^p} \tag{22}$$

where a_i is the activity of species i in the solution. As previously established, the relationship between experimental and theoretical activities, applying the equilibrium condition and Euler's theorem, will be :

$$a_{\pm S} = (a_+ \cdot a_-)^{1/2} \tag{23}$$

Equations (22) and (23) lead to :

$$\gamma_{\pm S} = \left| \frac{a_{CA}}{K^\ominus_{p,q}(T)} \right|^{\frac{1}{2p}} \cdot a_+^{\frac{1}{2}(1 - \frac{q}{p})} \cdot \xi_S \tag{24}$$

Equation (9) is obtained (neutral micelles) for $p = q = N$.
The equivalent to Equation (20) in logarithmic form is obtained :

$$\ln a_- = \frac{1}{p} \ln \left| \frac{a_{CA}}{K^\ominus_{p,q}(T)} \right| - \frac{q}{p} \ln a_+ \tag{25}$$

As with the p.p.model, several options can be envisaged.

If p and q are constant (which would seem a reasonable assumption a priori) and if a_{CA} is approximately constant, then :

$$\ln a_- \sim \text{constant} - r \ln a_+$$

Equation (20) is thus obtained, which was found to fit for octanoate (Figure 3) and other amphiphiles[34,35]. Unfortunately, the activity of the micelle cannot remain constant if m_S varies, since at least its concentration, m_{CA}, varies with m_S. In this case, the m.a.model with constant micellar charge cannot be checked against experimental data.

If $r = q/p =$ <u>constant</u>, the relationship above can only be checked against experimental data if

$$\frac{1}{p} \ln \left| \frac{a_{CA}}{K^{\ominus}_{p,q}(T)} \right|$$

varies only slightly with m_s. This cannot be ascertained as yet. Finally, Equation (20), which is experimentally supported, though rigorous for $r = $ constant with the p.p.model assuming no interaction between phases and no solvent in the micellar phase, is with the m.a.model only a possible relationship.

CONCLUDING REMARKS

Evidently, the pseudophase and mass action models used in the habitual way can both help interpret the behaviour of the mean activity coefficients of sodium carboxylates in aqueous solution. These models may be regarded as complementary and, at present, neither can justifiably be discarded in favour of the other, as too many unverifiable assumptions are introduced in both. Complementary measurements are required, in particular of single ion activities and micellar charge. Application of the models to charged micelles,as described here, is clearly only a first approximation but does nevertheless fit experimental data. Further refinement is clearly required, involving in particular the introduction of electrochemical potentials. This implies knowledge of interface potentials which is unfortunately lacking as yet. The introduction of interactions between phases[10,37,38,39] requires knowledge of parameters the values of which are not readily accessible.

Finally, both models,as used here, remain useful for certain parameters, but the accuracy of the available experimental data is still insufficient to allow more sophisticated models to be tested. The relative success of the approaches used here may well be due to the low micellar charges present in the systems studied.

ACKNOWLEDGMENTS

Pertinent remarks of Professor P. Stenius, Institute for Surface Chemistry, Stockholm, Sweden, relative to the m.a.model, are gratefully acknowledged. Thanks are also due to R. Ryan, University of Clermont-Ferrand 1, for his assistance in correcting the English text.

REFERENCES

1. G. Douhéret and A. Viallard, J. Chim. phys., 78, 85 (1981)
2. G. Douhéret, G. Roux-Desgranges and A. Viallard, in preparation
3. R.A. Robinson and R.H. Stokes, "Electrolyte Solutions" 2nd ed, pp. 37-38, Butterworths, London, 1968.
4. C.W. Davies, "Ion association", Butterworths, London, 1962.
5. D.G. Hall, J. Chem. Soc., Faraday I, 77, 1121 (1981).
6. M. Koshinuma, Bull. Chem. Soc. Japan, 54, 3128 (1981).
7. R.M. Pytkowicz, "Activity Coefficients in Electrolyte Solutions", Vols. I and II, C.R.C. Press, Boca Raton, 1979.
8. J.V. Leyendekkers and R.J. Hunter, J. Electroanal. Chem., 81, 123 (1977).
9. E. Vikingstad, J. Colloid and Interface Sci., 72, 68 (1979).
10. F. Eriksson, J.C. Eriksson and P. Stenius, in "Solution Chemistry of Surfactants", K.L. Mittal, Editor, Vol. 2, pp. 297-310, Plenum Press, New York, 1979.
11. G. Douhéret and A. Viallard, Fluid Phase Equil., 8, 233 (1982).
12. J.E. Desnoyers, G. Caron, R. de Lisi, D. Roberts, A.H. Roux and G. Perron, J. Phys. Chem., in press.
13. E.R. Jones et C.R. Bury, Phil. Mag., 4, 841 (1927).
14. K. Shinoda and E. Hutchinson, J. Phys. Chem., 66, 577 (1962).
15. P. Ekwall, H. Eikrem and P. Stenius, Acta Chem. Scand., 21, 1639 (1967).
16. Z. Bedö and E. Berecz, Acta Chim. Acad. Sci. Hung., 103, 199 (1980).
17. M. Randall, J.W. McBain and A. McLaren White, J. Amer. Chem. Soc., 48, 2517 (1926).
18. J.W. McBain and M. Barker, Trans. Faraday Soc. 31, 149 (1935).
19. E.R.B. Smith and R.A. Robinson, Trans. Faraday Soc., 38, 70 (1942).
20. P. Stenius, Acta Chem. Scand., 27, 3435 (1973).
21. S. Backlund, F. Eriksson, R. Friman, K. Rundt and J. Sjöblom, Acta Chem. Scand., A 34, 381 (1980).
22. S. Backlund, F. Eriksson, R. Friman, J. Sjöblom and B. Thyhn, Acta Chem. Scand., A 35, 521 (1981).
23. G. Biedermann and L.G. Sillén, Arkiv Kemi, 5, 425 (1953).
24. P. Mukerjee, Adv. Colloid Interface Sci., 1, 241 (1967).
25. G. Douhéret and A. Viallard, unpublished data (1982).
26. P. Stenius, personal communication (1982).
27. B. Lindman and H. Wennerström, Topics Current Chem., 87, 1 (1980).
28. T. Sasaki, M. Hattori, J. Sasaki and K. Nukina, Bull. Chem. Soc. Japan, 48, 1397 (1975).
29. P. Ekwall and P. Stenius, Acta Chem. Scand. 21, 1767 (1967)
30. B. Lindman and B. Brun, J. Colloid Interface Sci., 42, 388 (1973).
31. S. Backlund, F. Eriksson and R. Friman, Finnish Chem. Letters, 277 (1974).

32. J. Umemura, H.H. Mantsch and D.G. Cameron, J. Colloid Interface Sci., 83, 558 (1981).
33. S.G. Cutler, P. Meares and D.G. Hall, J. Chem. Soc. Faraday I, 74, 1758 (1978).
34. K.M. Kale, E.L. Cussler and D.F. Evans, J. Phys. Chem., 84, 593 (1980).
35. E. Vikingstad, A. Skauge and H. Høiland, J. Colloid Interface Sci., 66, 240 (1978).
36. A.H. Roux, unpublished data (1980).
37. E. Ruckenstein and R. Nagarajan, J. Phys. Chem., 79, 2622 (1975).
38. C. Tanford, "The Hydrophobic Effect : Formation of Micelles and Biological Membranes", 2nd ed., Wiley, New York, 1980.
39. G. Gunnarsson, B. Jönsson and H. Wennerström, J. Phys. Chem., 84, 3114 (1980).

KINETIC APPLICATIONS OF BILE SALT AMPHIPHILES

Charmian J. O'Connor, Robert G. Wallace and Beng Tatt Ch'ng

Chemistry Department
University of Auckland
Private Bag, Auckland, New Zealand

The nomenclature and mechanism of formation of bile salt amphiphiles have been reviewed. The effects of pH and of anaerobic bacteria in the small bowel and the intestine have been highlighted to show the importance of internal environment to bile salt composition during the enterohepatic circulation.

The aggregation properties of bile salts do not always conform to the simple phase separation model of micellization. Sub-micellar aggregates of well defined size have been identified and interactions with these smaller oligomers have contributed to the complexities observed in physical measurements.

Kinetic studies reported for reactivity of bile salt amphiphiles are surveyed and recent results on the activity of p-nitrophenylacetate, N-ethyl-4-nitrotrifluoroacetanilide and methyl orthobenzoate in conjugated and unconjugated bile salt solutions at pH <7 and T = 310 K are reported. These results suggest that the association process of bile salts is one of multiple association of monomers or small oligomers rather than a monomer $\rightleftharpoons$ n-mer equilibrium.

INTRODUCTION

In the past decade there have been increasing interest and awareness of the importance of bile salts in both health and disease.   Bile salts are important surfactants in the biological system, for they solubilise lecithin and cholesterol to form mixed micelles in bile,[1,2] regulate a number of cholesterol metabolizing enzymes in the liver and intestine[3] and facilitate absorption of dietary fats by emulsifying large droplets, making them more accessible to pancreatic lipase.[4]   Further, bile salts remove the products of pancreatic hydrolysis, *e.g.* fatty acids, through micelle formation.   In the light of their importance, it is therefore rather surprising that the number of physical organic kinetic studies of their reactivity is limited, but the explanation for this lack may lie in the relative insolubility of the bile acids and their apparent inability to catalyse the rates of hydrolysis of biological substrates in the absence of enzymes. Nevertheless, the micellar properties of bile salts and their chemical structure lead to interesting chemistry, for they offer an attractive opportunity for correlating chemical composition with micellar properties in a series of molecules differing only slightly in their chemical structure.

Bile Salt Formation

The common naturally occurring bile salts are steroids derived from tetracyclic triterpenes.[5]   They are usually composed of a saturated cyclopentanophenanthrene nucleus with alpha oriented polar hydroxyl groups attached at specific positions within the nucleus.   Bile salts, particularly in the entero-hepatic circulation, are derivatives of the primary bile acids, cholic (CA) and chenodeoxycholic (CDCA) acids ($3\alpha,7\alpha,12\alpha$-trihydroxy- and $3\alpha,7\alpha$-dihydroxy-cholanic acids respectively).   CA is synthesised at about twice the rate of CDCA.   These acids, while still in the hepatocyte, are conjugated with either glycine or taurine.   Scheme (1) shows events in the liver cell in the formation of primary bile acids.

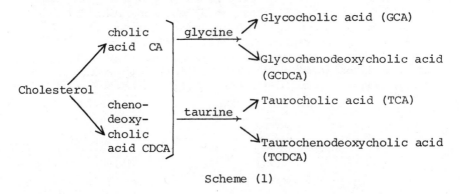

Scheme (1)

Anaerobic bacteria in the small intestine produce a 7α-dihydroxylase, which converts the primary bile acids to secondary bile acids, deoxycholic (DCA) (3α,12α) and lithocholic (LCA) (3α) acids.   These events are summarized in Scheme (2).

GCA }   ⎡ glycodeoxycholic acid (GDCA)
     }   7α-dehydroxylase   ⎢
TCA }   ────────────────────→   ⎢
                                                    ⎣ taurodeoxycholic acid (TDCA)

GCDCA }   7α-dehydroxylase   ⎡ glycolithocholic acid (GLCA)
      }   ────────────────────→   ⎢
TCDCA }   ⎣ taurolithocholic acid (TLCA)

Scheme (2)

Figure 1 summarises the structures and nomenclature of these bile acids.   The pK$_a$'s of the bile acids are profoundly affected by the presence of the conjugate base and are slightly dependent

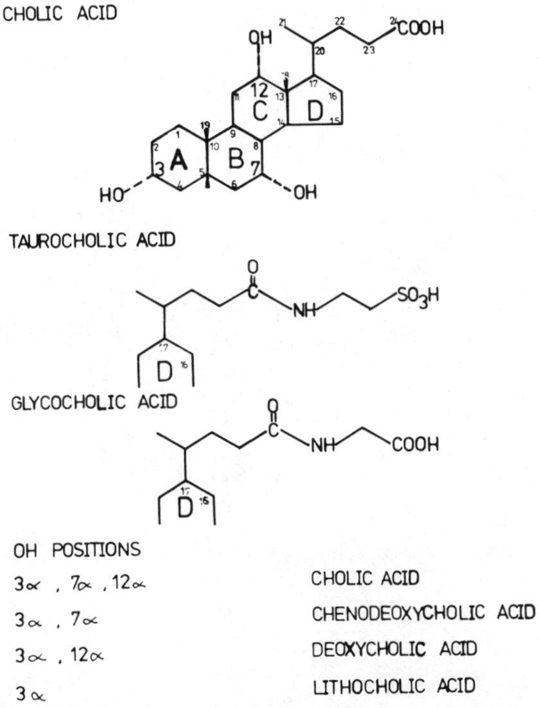

Figure 1.   The structure and nomenclature of the bile acids.

on the concentration of the acid itself.   Values for CA, DCA and
CDCA range[6] from 5.0-5.5, 5.3-6.3 and 5.9-6.5 respectively, those
for the glycine conjugates GCA, GDCA and GCDCA from 3.8-4.1, 4.7-
4.8 and 4.2-4.3 respectively and those for the taurine conjugates
have pK_a values less than 2.   The presence of the unprotonated
hydrophilic side chain enhances the solubility of the conjugated
bile salts at intestinal pH 6-7.LCA and its salts are relatively
insoluble and 80% of them are excreted from the body.[7]   That
portion absorbed is conjugated but few chemical studies have been
made on lithocholates.

The other bile salts are much too valuable to be lost from
the body, so after completing their major intestinal digestive and
absorptive functions they are conserved by absorption in the small
bowel and returned *via* the portal drainage system to the liver for
recycling - the enterohepatic circulation (Figure 2).   Bile salts
are produced in the foetus by the age of 12 weeks but the only one
of these passing easily over to the placenta is DCA.Up to 26 weeks
of life the taurine conjugates are the most abundant, whereas in
an adult the glycine conjugates are most abundant.   By the age of
2-3 months after birth an adult composition pattern has started to
emerge.   The total bile salt content of the body is c. 3 to 5g

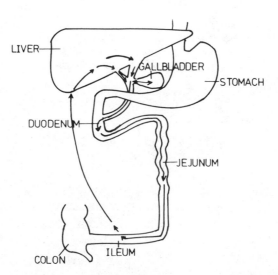

Figure 2.   Enterohepatic circulation of bile salts.

and this pool is recycled *via* the bowel, in the enterohepatic circulation, 6-12 times per day.   More than 95% of the bile salts reaching the ileum are reabsorbed and the remaining 2 to 5% (about 500 mg daily) are excreted.

Bile salts are continually secreted by day and night but during fasting most of the bile salt pool is concentrated in the gallbladder.[7,8]   Attempts have been made to study the formation of gallstones under fasting conditions.[9]

When food passes from the stomach to the duodenum the bile salts in the gallbladder and the bile duct system are released. The pH of the duodenojejunal contents, after eating a fatty meal, has been variously measured as 6.0 ± 0.5[8] and between 6 to 7.7,[10] and the bile salt concentration as 9 ± 2 m mol[8] and 8.6 ± 1.4 m mol[11] for adults and 5.3 ± 0.6 m mol[12] for full term infants.

Figure 3 shows the relative distribution of the bile salts upon entering the intestine.   The precise percentages will be affected by diet, hormonal levels and numerous other factors[5] and have been averaged from data in several sources.[5,7,10,13]

Most of the conjugated primary bile salts are reabsorbed unchanged, but up to 25% are deconjugated by enzymes derived from bacteria normally resident in the lower small bowel.   Only about 30% of the secondary bile acid DCA is reabsorbed, but after re-absorption the unconjugated recycled bile acids are reconjugated in the liver.   Figure 4 shows the bile salt composition just before reabsorption.   In calculating these percentages no allowance was made for absorption[10] of the protonated bile acids before they reach the ileum.   A small fraction of the unionized glycocholic acids is absorbed by diffusion in the duodenum and jejunum but the rest of the conjugated bile acids remain until they reach the ileum.   The enterohepatic circulation cycles the bile salts at least twice during the digestion of a meal.[7,10,13]   Should this circulation be interrupted severe malabsorption of fat and fat soluble vitamins can result.[13]   The bile salt micelles (*vide infra*) transport the insoluble products of fat digestion across the unstirred aqueous layer of the intestinal mucosal surface and present these products at the epithelial cell for final absorption.[7,8,13]

Bile Salt Aggregation

The junction of the A and B rings in the steroidal nucleus (Figure 1) is kinked so that the rings lie in a *cis* configuration. This skeletal modification, caused by the liver, of the bile salt precursor, cholesterol, brings the hydroxyl groups at the 3 and 7 positions closer together.   Thus the bile acid molecule exhibits

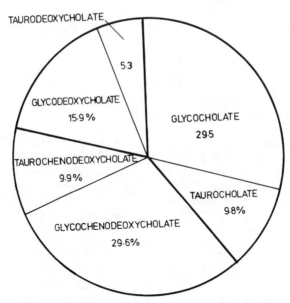

Figure 3. Bile salt composition upon entering the intestine
           (lithocholate excluded)

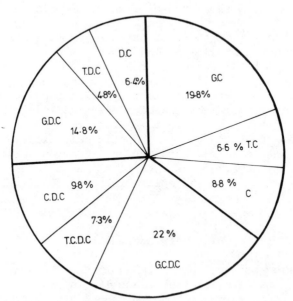

Figure 4. Bile salt composition just before reabsorption
           (lithocholate excluded)

planar polarity, with the hydrophilic groups being situated below
the equator of the molecule and most of the steroid skeleton with
its protruding methyl groups lying above it. The mobility of the
side chain allows its polar groups to lie in the same plane as the
hydroxyl groups, thus providing a major contribution to the
polarity of the under surface of the molecule.[14]

The molecular structure does not therefore allow bile salts to
exhibit the distinct end-to-end polarity which characterises many
aqueous surfactants. Nevertheless, the presence of both hydrophobic and hydrophilic portions and the planar polarity allow them
to aggregate at moderate concentrations, (10^{-4}-10^{-2} mol dm^{-3}), to
form micelles which dissolve sparingly water soluble compounds.[15]

Many of the early works and formulations of the basic
theories of micellisation were carried out by McBain[16,17] and Hartley.
[18,19] Bury et al.[20-22] showed that a fairly abrupt transition
was to be expected from application of the mass action law to
equilibrium between simple ions (or molecules) and the "micelles"
composed of a fairly large number of them. The more or less well
defined concentration at which the transition occurred (normally
over the range 2-3x10^{-3} mol dm^{-3}) they called "the critical
concentration for micelles" (CMC). Later IUPAC[23] defined CMC thus:
"There is a relatively small range of concentrations separating the
limit below which virtually no micelles are detected and the limit
above which virtually all the additional surfactant forms micelles.
Many properties of surfactant solution, if plotted against the
concentration, appear to change at a different rate above and below
this range.".

Any increase in detergent concentration above CMC serves only
to increase the number, but not the nature, of micelles. Moreover,
the solubility of hydrophobic compounds should increase sharply at
the CMC, owing to the new, hydrophobic environment being formed in
the interior of the micelles, and then increase steadily with
increasing numbers of micelles present.

But it appears that bile salts do not always conform to this
simple model. Thirty years ago Ekwall et al.[24] found that the
lipophilic solubility in solutions of sodium cholate (NaC),
glycocholate (NaGC), deoxycholate (NaDC) and taurocholate (NaTC)
increased, with increasing concentration, in definite steps,
indicating that micelle formation passed through several successive
distinct stages, each one being characterized by constant
properties of the solute.

Later investigations have confirmed abnormal behaviour.
Norman[25] considered that the changes he observed in the conductance
of solutions of NaC and NaDC and their taurine and glycine
conjugates were in agreement with a stepwise association process.

[1]H n.m.r. and fluorescent probe studies[26] gave evidence of a
continuous increase in the size of the aggregates, and also of
altered molecular packing, with increasing NaDC concentration;
Mayer *et al.*[26] also reported increased micelle size with increasing
sodium taurodeoxycholate (NaTDC) concentration.   Ultrasonic and
density studies caused Djavanbakht *et al.*[27] to state, "These
results lead to the conclusion that the association takes place in
the whole range of concentration and not at some critical micelle
concentration, and that the distribution curve of aggregation
sizes must be wide and shifting upward as the concentration is
increased.".

     Thus, it seems that, instead of a simple transition from
monomers to a complete and fully developed micelle over a
relatively small concentration range, there is a wide concentration
range in which the aggregation appears to start usually first with
dimers[6] and then, with increasing size of the oligomer, there is
increasing development of the internal micellar environment.   When
the dye Rhodamine 6G is dissolved in solutions of NaTC and NaTDC,
λ_{max} shifts[28] as the concentration of bile salt increases from 1 to
8×10^{-3} mol dm^{-3};   this shift is dependent both on temperature and
concentration of added sodium chloride but is most influenced by
the larger oligomers.

     Thus the criterion of change in property of a solute as a
determinant of CMC is invalidated if this property depends not only
on the number but also on the size of the micelles.   Indeed, it
has been stated,[29] "While CMC values reported in the literature are
dependent on pH and ionic strength of the medium, the differences
in these values do not lend themselves to easy rationalisation.".
In this context, the comments of Mukerjee and Cardinal[30] are
important.   They observed great differences in behaviour when they
compared the solubility of naphthalene in NaC with that of a dye in
a typical anionic surfactant.   A plot of Δ(solubility) against
Δ(detergent concentration) showed a sharp vertical increase for the
latter case, whereas for the bile salt showed a smooth curve from
which it was possible to choose many values of CMC and for each of
these it was possible to extrapolate linearly the data points
above and below the chosen value.   They concluded that the
significance of a CMC value for NaC was clearly limited and that
"a complex pattern of association, the formation of dimers and one
or more higher oligomers is indicated".   The composition of the
aggregates, even at NaC concentrations 5 to 10 times greater than
that of apparent CMC, was still changing markedly.   Similar
smooth transitions have been found[31] for the solubilization of
20-methylcholanthrene in solutions of NaC, sodium chenodeoxycholate
(NaCDC) and NaDC and the glycine and taurine conjugates of these
salts.

     Serious attempts have been made to distinguish between the

phase separation model, which treats micellization as the formation of a separate hydrocarbon phase in the bulk aqueous phase and the mass-action model of micellization, which predicts a smooth change in physical properties as the concentration of the bile salt passes through the CMC.   Studies on the thickness of the surface layer of NaDC solutions led Thomas and Christian[32] to favour the mass-action model.   Moreover, several sub-CMC interactions have been identified.

Although the measured CMC for NaDC is equal to $2.6^{\pm}0.3$ m mol dm^{-3}, binding[33] between virus membranes and C-14 labelled NaDC began at <0.1 m mol dm^{-3} and the amount of bound virus increased with increasing NaDC concentration until lysis of the viral membrane occurred at [NaDC] = $0.9^{\pm}0.1$ m mol dm^{-3}.   (The use of different bile salts for solubilising components of cell membranes has been extensively investigated.[34])

Similar sub-CMC interactions have been found between solutions of NaDC and a cellulose acetate millipore filter.[35]   A surfactant layer-liquid membrane formed at the interface, with the membrane becoming progressively more covered as the bile salt concentration increased.   At [NaDC] = 0.5 CMC the area was half covered and coverage was complete at [NaDC] = CMC.

The apparent CMC of a mixture of two different bile salt species[36] lies closer to the value of the less soluble species, probably because that species lies predominantly in the micellar phase while the more soluble species lies preferentially in the solution phase.

Several workers have investigated the ability of bile salts to solubilize their corresponding bile acids and the related property of the pH at which precipitation occurs.[28,37,38]   The trihydroxy bile salts are less efficient than the dihydroxy ones, for they require a higher ratio of deprotonated:protonated species and the glycine conjugates tend to be more efficient than their unconjugated equivalents.   The pH at which precipitation occurs generally decreases with decreasing pK_a of the bile acid, but Igimi and Carey[37] observed that the efficiency of NaCDC and sodium glycochenodeoxycholate (NaGCDC) to solubilise their corresponding acids increased with increasing concentration of the bile salt and levelled off at high concentration.   This trend, which reflects an increase in the micelle size, runs parallel to that of the measured values[28,38] of pK_a of the bile acids, which showed that the activity of the acid molecules was lowered when they were solubilised in a micellar environment, thus leading to an increase in pK_a.

Bile salt solubility increases with temperature[28] and NaC aggregates dissociate almost completely to monomers[39] at 343 K.

Increasing temperature increases the mobility of the anionic side chain and this increases the mutual repulsion between neighbouring members of the aggregates.

By using partial molar volumes, Murata *et al.*[40] were able to show that the structure of NaC micelles and the mechanism of their formation were quite different at pH 7.17 from those at the higher pH values of 8.20 and 9.20. Hydrogen bonding by the protonated carbonyl groups apparently contributes to micelle formation and micellization occurs at concentrations between 10^{-4} and 10^{-5} mol dm^{-3}, well below those required for helical formation, 4×10^{-3} mol dm^{-3}.

The effect of protonation on bile salt properties has important biological implications as the pH of the duodenum is often in the acid region and the presence of protonated bile acids will affect the structure of the mixed micelles present and hence their ability to solubilise substrates and assist in chemical decomposition.

Another possible influence on bile salt structure and reactivity is the buffer used in solution. Paul and Balaram[41] found that, in Tris HCl, the fluorescent probe dansyl cadaverine bound strongly to DCA and phospholipid mixed micelles but in sodium phosphate little binding could be detected. Separation of NaTDC on Sephadex G-100 columns was considerably slowed[42] in the presence of phosphate buffer compared with that in the presence of NaCl.

Although there is now considerable evidence that bile salts do not always conform exactly to the requirements of the pseudo phase model of micellization, many physical measurements have been made[14,28,32] to search for a distinct value of the CMC and interpretation of many of the kinetic results has depended on the ability of the investigators to identify such a value. An account of the kinetic applications of bile salt amphiphiles follows.

Kinetic Applications of Bile Salt Amphiphiles

Although catalysis by amphiphiles was first reported[43] in 1906 and the field of micellar catalysis has experienced such an enormous growth that today there are literally thousands of publications concerning micellar catalysis in the literature,[44-46] we can identify only five such publications for bile salts in aqueous solution in the absence of enzymes,[47-51] one in dimethyl-sulfoxide as solvent[52] and four for reactivity of bile salts in the presence of bile salt stimulated lipase.[53-56] These last four papers do not fall within the scope of this survey - suffice to say that in the presence of bile activated lipase there is both esterase and lipase activity which apparently occur, at least partially, through formation of a ternary complex involving enzyme, bile salt and substrate.

Borgstrom and Erlanson[57] have made a comprehensive study on the interactions and effects of bile salts on the activity of rat and porcine pancreatic lipase and co-lipase. Co-lipase overcomes the inhibition of lipase caused by bile salts alone and recent studies by Lombardo et al.[58,59] have confirmed that bile salts play an important part in the activity of human carboxyl ester hydrolase. There are two sites of bile salt recognition. One site, specific to the $3\alpha,7\alpha$ hydroxyl groups induces activation of the enzyme; the other site, unspecific towards bile salt hydroxylation is located at the active centre and is implicated in substrate recognition. The general inhibitory effects of bile salts on enzyme activity has been confirmed in a number of studies.[60-65]

One further kinetic study[66] should be noted. The kinetics of pyrene fluorescent decay in the presence of nonionic and ionic quenchers have yielded information about the permeability of NaTC micelles. Positively charged quenchers are strongly absorbed at the surface of the micelles and the charge leads to destruction of the micelle. Hydrated electrons have a low reactivity towards monomer NaTC and the reactivity decreases further upon micellization.

Hydrolysis of phenyl esters has been the subject of most of the kinetic papers. Fendler et al.[52] investigated the displacement of p-nitrophenoxide ion from bis-p-nitrophenyl phenylphosponate as a function of [NaC] and [H$_2$O] in dipolar aprotic dimethyl sulfoxide at 298 K. The observed rate constant, k_ψ, as a function of [NaC], exhibited sigmoidal dependence followed by a plateau and at constant [NaC] k_ψ decreased curvilinearly with increasing [H$_2$O]. Similar profiles were obtained for reactivity of N-tert-butyl-2,4,6-trinitrobenzamide in the presence of NaC and NaDC in this medium. The amide underwent facile nucleophilic aromatic substitution at C-1 forming a highly stable σ, or Meisenheimer complex.

These reactions in DMSO are quite rapid, $k_\psi \sim 10^{-3} - 2.0$ s^{-1}, and the bile salt acts as an efficient catalyst. By contrast, in water $k_\psi \sim 10^{-3} - 10^{-5}$ s^{-1} for hydrolysis of PNPA and the bile salt is frequently an inhibitor.

In the presence of conjugated or free bile salt, Menger and McCreery[49] found that the p-nitrophenyl esters of acetic, hexanoic, octanoic and dodecandioic acids invariably displayed diminished reactivity towards hydroxide ion in the pH range 10.5 - 12.0 at 298 K. On the other hand, p-nitrophenyldodecanoate, PNPD, hydrolysis was accelerated by NaC and NaCDC. Both the inhibitions and accelerations became appreciable only above the CMC of the bile salts with the breaks in the curves corresponding

to known CMC values.   Without exception, the dihydroxy bile salts
showed a greater inhibitory effect (or smaller catalysis in the
case of PNPD) than their trihydroxy analogues.   Conjugation of
the bile salts lessened the inhibitory effect but did not change
the order NaDC > NaCDC > NaC.   These results showed that both the
number and position of the hydroxy group substituents on the
steroid nucleus are important and they are likely to be a
reflection of the closeness of the packing in an aggregate held
together by both hydrophobic and hydrophilic bonding.   Lengthening
the p-nitrophenyl carboxylate chain from 8 to 18 carbon atoms
caused no change in the value of k_ψ in 0.1 mol dm^{-3} NaC showing
that the environment of the ester functionality within the micelles
is not greatly dependent upon chain length.   The chains probably
exist in a folded conformation inside the micelles.   The micelle
interior is less conducive to reaction than the surface of a
steroid monomer or dimer.   The large rate constant calculated for
adsorbed PNPA in NaDC solution suggests that this smaller ester is
"probably adsorbed near the micelle surface".   The longer the
ester chain, the larger is the binding constant.

    Cairns-Smith and Rasool[50] confirmed that, contrary to the usual
inhibition of alkaline hydrolysis of p-nitrophenyl esters by
anionic micelles, the hydrolysis of PNPD was catalysed by NaC and
this was seen as reflecting a suppression of the self aggregation
of the long chain ester by its incorporation within the micelle.
These authors also made a molecule consisting of two cholic acid
residues joined by a flexible amido oxyethylene oligomer.   This
molecule can halve the CMC of NaC when present in the proportion
of one part in 50.   It was postulated that the double molecule
spent much of its time in a conformation resembling the two-
molecule nucleus from which cholate micelles must presumably grow
and that these double molecules form nuclei to which the cholate
ions can add at a concentration about half that required for net
formation of two-molecule nuclei by NaC molecules themselves.
Such a 'tied-micelle' is likely to mimic the solid-like structure
of globular protein interiors more closely than normal surfactant
micelles and the modest rate enhancements found, for the
hydrolysis of PNPD, confirmed that reducing the degrees of
freedom was the right step towards modelling membrane reactivity.

    An earlier study[48] revealed that the spontaneous hydrolysis of
both cationic and anionic esters in the presence of CA at pH8 and
at 303 K was decreased upon binding of the substrates to the bile
salt.   The variation of the apparent binding constant, 7 mol^{-1}
dm^3 for cationic m-acetoxy-N-trimethylanilinium iodide to 3650
mol^{-1} dm^3 for anionic p-nonanoyloxybenzoic acid was considered to
reflect the magnitude of hydrophobic forces between CA and
substrates.   Hydrolysis of these same substrates was catalysed by
a histamine amide of cholic acid and the rate data obeyed the
Michaelis-Menten equation for enzyme catalysis.   However, the

nature of the binding site was not altered by the histamine moiety and it seemed that the steroid ring provided a non-specific hydrophobic binding site for the long methylene chain in the substrate molecules used.

An attempt to synthesise[49] an enzyme 'mimic' by attaching a good nucleophile, an oxime group, to the 7 position of deoxycholate produced a catalyst which increased the rate of hydrolysis of PNPD nearly 100-fold above background hydrolysis, at pH 12.0, at concentrations of steroid oxime below CMC.   Beyond CMC the rate of reaction increased only slightly.   Under compaable conditions there was only a 7-fold linear increase in rate by acetone oxime.   This contrast in behaviour between steroidal and free oxime was not observed in the hydrolysis of shorter chain length esters and the difference was attributed to the nature of the substrate.   When PNPD complexes with a steroid monomer or dimer (below CMC) or with a bile salt (above CMC), the long chain ester unravels and exposes its carbonyl group to attack but it reacts relatively slowly with acetone oxime owing to its coiling and/or self aggregation.

The effects of bile salts upon the hydrolysis of benzyl-penicillin[47] differ from those with phenyl esters.   Below CMC, NaDC decreases benzylpenicillin resistance to the hydrolytic activity of penicillinase while NaC has no effect.   Above CMC, both NaDC and, to a lesser extent, NaC helical complexes increase the resistance to hydrolysis.   As with phenyl esters, the interaction between antibiotic and bile salt was not merely electrostatic and seemed to be stereospecific in nature.   No interaction and therefore no decrease in hydrolytic activity of penicillinase was found with the α-aminophenylacetyl (ampicillin) and dimethoxyphenyl derivatives, in which the structure is modified by the presence of a polar group or total absence of a side chain, respectively.

Sugimoto *et al.*[67] confirmed that host-guest fitting plays an important role in asymmetric reactions by studying the reduction of ketones within the chiral environment of NaC micelles.   When alkyl phenyl ketones of acetophenone, isobutyrophenone and tert-butyl phenyl ketones were reduced in the presence of NaC, no chirality was introduced into the alcohols.   However, in the reduction of naphthyl ketones optically active alcohols were obtained.   The binding site of the aromatic ketones in the micelle was firmly determined to be the chiral hydrophobic interior and the greater ^{1}H n.m.r. chemical shift of the C-19 methyl protons compared with those at C-18 suggested the preferential binding site to be near the A/B ring junctures of the steroid.   The steroselectivity of asymmetric induction is strongly dependent upon the geometrical fitting between the micellar chiral interior and an aryl group in aromatic ketones.

The differing behaviour[68] of NaDC and NaC micelles towards benzene and naphthalene substrates was also observed in the deuterium spin-lattice relaxation time (T_1) measurements of perdeuterobenzene and perdeuteronaphthalene. Values for perdeuterobenzene were larger than expected and the authors concluded that the benzene molecules had a limited degree of vibrational motion in the molecular plane and a C_6 oscillation. In contrast, the T_1 values for perdeutero naphthalene were decreased in the presence of micelles indicating that, while there remained a vibrational motion within the micelle, the molecules were not able to undergo rotation in their molecular plane.

C-13 n.m.r. measurements[69] showed that, within NaDC micelles, p-xylene molecules underwent little restriction of their molecular motion while 2-methyl naphthalene molecules were subject to considerable restriction.

In the light of this evidence on the differing effects of micellar environment on benzene and naphthalene derived substrates, it is not surprising that chiral effects were observed during the hydrolysis of p-nitrophenyl N-lauroylalaninate.[51] The L-ester was hydrolysed more rapidly than the D-isomer. However, the hydrolysis of p-nitrophenyl N-acetyl-phenylalaninate was inhibited by the micelles and the stereo-specific effect was not observed. The presence of the extra phenyl group in the latter ester (and the long alkyl chain in the former) is likely to affect the host-guest cooperativity and hence the reactivity.

These literature studies on reactivity have been concentrated, for the most part, on bile salt concentrations greater than CMC and in solutions of alkaline pH where all the bile acids are converted to the anionic form. But these conditions do not necessarily hold *in vivo* and our studies have been concentrated on solutions of pH ⩽7 where a considerable fraction of the steroid molecule (with the exception of the taurocholates) is present as the bile acid. Inevitably, this has placed a relatively low limit on the maximum concentration that can be solubilised and as a result many of our data refer to concentrations well below CMC. The very small interactions found between PNPA and bile acid/bile salt mixtures make this ester an excellent probe to investigate the effect of varying bile salts on bile salt stimulated esterase activity. These enzyme modified reactions will be reported later.[70]

RESULTS AND DISCUSSION

We have confirmed that bile salts are generally inhibitors of the hydrolysis of esters and amides. Figure 5 shows rate profiles for hydrolysis of PNPA in the presence of deoxycholic,

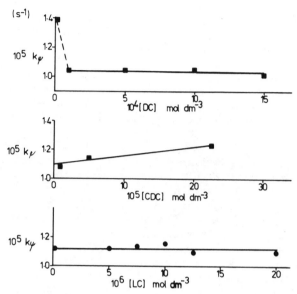

Figure 5.   Plots showing the effect of unconjugated bile salts on
the hydrolysis of PNPA in 0.01 mol dm^{-3} phosphate
buffer, pH 7, T = 310 K.

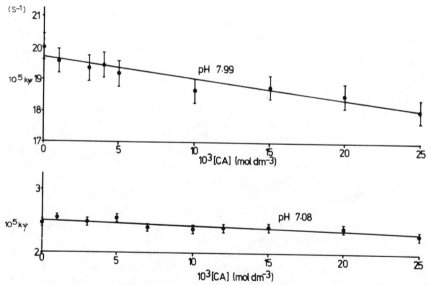

Figure 6.   Plots showing the effect of cholic acid on the
hydrolysis of N-ethyl-4-nitrotrifluoroacetanilide in 0.05 mol dm^{-3}
trizma base at pH 7.99 and pH 7.08, T = 310 K.

chenodeoxycholic and lithocholic acids in 0.01 mol dm^{-3} phosphate buffer at pH 7 and Figure 6 those for hydrolysis of N-ethyl-4-nitrotrifluoro acetanilide in the presence of cholic acid in 0.05 mol dm^{-3} trizma base at pH 7.99 and pH 7.08. We tested the reactivity, at pH 7, of all the bile acids and their glycine and taurine conjugates with PNPA; with the exception of taurodeoxycholate, the taurine and glycine conjugates were virtually ineffective, as was lithocholate; deoxycholate was slightly inhibitory and chenodeoxycholate and taurodeoxycholate showed 10 and 20% rate enhancement at 0.2 and 5 m mol dm^{-3} respectively. Cholate inhibited the rate up to 1.5 m mol dm^{-3} and then the rate gradually increased curvilinearly up to a concentration of 40 m mol dm^{-3}. This last result contrasts with that obtained[49] at pH 12.04 which showed almost linear inhibition up to [NaC] = 100 m mol dm^{-3} but such a contrast illustrates the importance on reactivity of the presence of undissociated bile acid. At pH 12 all the cholic acid is deprotonated whereas at pH 7 c. 10% is present in the protonated form. 25 m mol dm^{-3} Cholic acid inhibits the hydrolysis of N-ethyl-4-nitrotrifluoro acetanilide c. 10% at both pH 7 and pH 8.

Catalysis in the presence of anionic aqueous micelles is rare,[71,72] so these results were not surprising. However anionic micelles are known to catalyse the hydrolysis of methyl ortho-benzoate[73] and other ortho esters[74] and we hoped for similar results with the bile salt aggregates. In 0.03 mol dm^{-3} acetate buffer at pH 4, a quite complicated picture emerges for the hydrolysis of methyl orthobenzoate. The rate profiles in the presence of taurocholates are characterised by quite sharp peaks and troughs and in each case these coincide with discontinuities or short plateaux in the plots of surface tension $vs.$ log[bile salt concentration] measured under the same conditions of pH and temperature. Figure 7 shows typical comparative plots for NaTC.

At a pH corresponding to 44% protonation of the glycocholates, the rate profiles show sharp discontinuities and again these correspond to changes in slope in the surface tension plots. Figure 8 shows comparative plots for GDCA. As the pH increases, so that the percentage of the undissociated acid form present decreases, the rate profiles in the presence of NaGC/GCA change distinctly, Figure 9, but in each case the trends in the rate profiles can be mirrored as discontinuities in the surface tension plots. It appears that, at the relatively low pH values we have used, the association process in bile salts is one of multiple association of monomers or small oligomers rather than a monomer ⇌ n-mer equilibrium and that these small aggregates have their own characteristic substrate binding constants. The size of the aggregates and the concentration of bile salt at which they are formed also depends on the proportion of bile salt anion present. The "bumpy" profiles we observe are similar to those recorded for

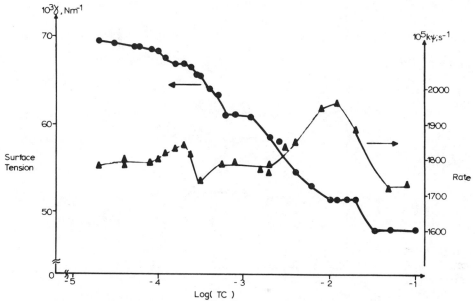

Figure 7. Observed rate constants, k_ψ, for hydrolysis of methyl-orthobenzoate, ▲, and observed surface tension, γ, ●, as a function of sodium taurocholate, TC, concentration in 0.03 mol dm^{-3} acetate buffer, pH 4, T = 310 K.

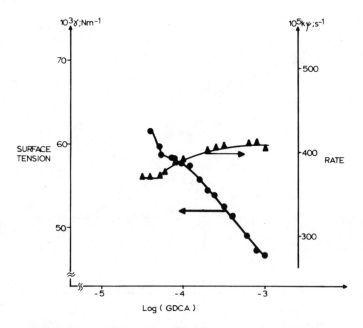

Figure 8. Observed rate constants, k_ψ, for hydrolysis of methyl
orthobenzoate, ▲, and observed surface tension, Υ, ●, as a
function of glycodeoxycholic acid, GDCA, concentration in 0.03 mol
dm^{-3} acetate buffer, pH 4, T = 310 K.

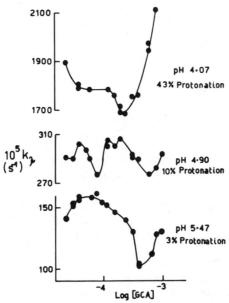

Figure 9. Observed rate constants, k_ψ, for hydrolysis of methyl orthobenzoate as a function of glycocholic acid, GCA, concentration at various pH values.

the reactivity[75] of tris(oxalato)chromate(III) anion in dodecyl-ammonium propionate and ocytlammonium tetradecanoate in benzene. At that time, no satisfactory explanation was forthcoming to explain the complex rate profiles. However, we now believe that the answer lies in changing aggregation patterns as the concentration of the surfactant increases, even at concentrations well below 'CMC', and that each aggregate has its own kinetic identity as well as an unique ability to modify the surface tension properties of the solvent. The complexity of these present profiles is dependent on the buffer medium used, the pH, the nature of the substrate and of the bile salt amphiphile.

EXPERIMENTAL

Bile salts and acids were purchased from Sigma and were used without purification except for cholic acid which was recrystallised to remove the 1% chenodeoxycholic acid impurity. Water used as solvent was triply distilled.

All kinetic data were monitored spectrophotometrically on a Cary 219 Spectrophotometer using wavelengths of 275 nm for PNPA, 408 nm for N-methyl-4-nitrotrifluoroacetanilide, and 228 nm for methyl orthobenzoate.

Surface tensions were recorded on a White Tensiometer.   All
solutions were maintained at 310 K using Grant Thermocirculators.

ACKNOWLEDGEMENTS

Financial assistance from the Research Committees of the New
Zealand Universities' Grants Committee and the University of
Auckland Research Committee (to C.J.O'C) is gratefully
acknowledged.

REFERENCES

1.   D.M. Small, M. Bourges and D.G. Derivichian, Nature, $\underline{211}$, 816
     (1966).
2.   D.M. Small, Advan.Int.Med., $\underline{16}$, 243 (1970).
3.   H. Danielsson and J. Sjovall, Ann.Rev.Biochem., $\underline{44}$, 233 (1975).
4.   D.M. Small, Gastroenterology, $\underline{52}$, 607 (1967).
5.   W.G.M. Hardison, in "Progress in Liver Diseases", H. Popper
     and F. Schaffner, Editors, Vol.VI, Chapter 6, Grune and
     Stratton, New York, 1979.
6.   D.M. Small, in "The Bile Acids", P.P. Nair and D. Kritchevsky,
     Editors, Vol.I, pp.249-254, Plenum Press, New York, 1971.
7.   C. Tasman-Jones, Patient Management, 138 (1980).
8.   D.R. Saunders and G.B. McDonald, in "Recent Advances in
     Gastroenterology", I.A.D. Bouchier, Editor, No.3, Chapter 2,
     Churchill Livingstone, Edinburgh, 1976.
9.   R.N. Redinger and D.M. Small, Arch.Intern.Med., $\underline{130}$, 618
     (1972).
10. H.W. Davenport, "The Physiology of the Digestive Tract", 3rd
     Edition, Year Book, Medical Publishers Inc., 1971.
11. H.P. Porter and D.R. Saunders, Gastroenterology, $\underline{60}$, 997
     (1971).
12. J.B. Watkins, P. Szezepanik, J.B. Gould, P. Klein and
     R. Lester, Gastroenterology, $\underline{69}$, 706 (1975).
13. W. Admirand and L.W. Way, in "Gastrointestinal Disease",
     M.H. Sleisenger and J.S. Forattran, Editors, Chapter 26,
     W.B. Saunders Co., Philadelphia, 1973.
14. M.C. Carey and D.M. Small, Arch.Intern.Med., 130, 506 (1972).
15. U. Wosiewitz and S. Schroebler, Experientia, $\underline{35}$, 717 (1979).
16. J.W. McBain, M.E. Laing and A.F. Tetley, J.Chem.Soc., $\underline{115}$,
     1279 (1919).
17. J.W. McBain, in "Third Report on Colloid Chemistry", British
     Association for the Advancement of Science, London, 1920.
18. G.S. Hartley, "Aqueous Solutions of Paraffin-Chain Salts",
     Hermann et Cie, Editors, Paris, 1936.
19. G.S. Hartley, in "Progress in the Chemistry of Fats and
     Other Lipids", Pergammon Press, London, 1955.

20.   J. Grindley and C.R. Bury, J.Chem.Soc., 679 (1929).
21.   D.G. Davies and C.R. Bury, J.Chem.Soc., 2263 (1930).
22.   E.R. Jones and C.R. Bury, Phil.Mag., 4, 841 (1927).
23.   Manual of Symbols and Terminology, Appendix II, Part I,
      International Union of Pure and Applied Chemistry, Pure Appl.
      Chem., 31, 612 (1972).
24.   P. Ekwall, E.V. Lindström and K. Setala, Acta Chem.Scand., 5,
      990 (1951).
25.   A. Norman, Acta Chem.Scand., 14, 1300 (1960).
26.   N.A. Mayer, R.F. Kwasnick, M.C. Carey and G.B. Benedek, in
      "Micellization, Solubilization and Microemulsions" K.L. Mittal,
      Editor, Vol 1, pp.383-402, Plenum Press, New York, 1977.
27.   A. Djavanbakht, K.M. Kale and R. Zana, J.Colloid Interface
      Sci., 59, 139 (1977).
28.   M.C. Carey and D.M. Small, J.Colloid Interface Sci., 31, 382
      (1969).
29.   R. Paul, M.K. Mathew, R. Narayanan and P. Balaram, Chem. and
      Phys.of Lipids, 25, 345 (1979).
30.   P. Mukerjee and J.R. Cardinal, J.Pharm.Sci., 64, 882 (1975).
31   A. Norman, Acta Chem.Scand., 14, 1295 (1960).
32.   D.C. Thomas and S.D. Christian, J.Colloid Interface Sci., 82,
      430 (1981).
33.   A. Helenius, E. Fries, H. Garoff and K. Simons, Biochim.
      Biophys.Acta, 436, 319 (1976).
34.   A. Helenius, D.R. McCaslin, E. Fries and C. Tanford,
      Methods in Enzymology, LVI, 734 (1979).
35.   R.C. Szivastava and S. Yadav, J.NonEquilib.Thermodyn., 4, 219
      (1979).
36.   R.H. Dowling and D.M. Small, Gastroenterology, 54, 1291 (1968).
37.   H. Igimi and M.C. Carey, J.Lipid Research, 21, 72 (1980).
38.   P. Ewall, T. Rosendahl and N. Löfman, Acta Chem.Scand., 11,
      590 (1957).
39.   D.M. Small, S.A. Penkett and D. Chapman, Biochim.Biophys.
      Acta, 176, 178 (1969).
40.   Y. Murata, H. Akisada, M. Yoshida, G. Sugihara and M. Tanaka,
      Fukuoka University Science Reports, 9, 41 (1979).
41.   R. Paul and P. Balaram, Biochem.Biophys.Research Commun.,
      81, 850 (1978).
42.   B. Borgström, Biochim.Biophys.Acta, 106, 171 (1965).
43.   E. Twitchell, J.Am.Chem.Soc., 28, 196 (1906).
44.   F.M. Menger, Accts.Chem.Res., 12, 111 (1979).
45.   J.H. Fendler and E.J. Fendler, "Catalysis in Micellar and
      Macromolecular Systems", Academic Press, New York, 1975.
46.   J.H. Fendler, "Membrane Mimetic Chemistry", Wiley-Interscience,
      1982.
47.   F. Alhaique, C. Botré, G. Lionetti, M. Marchetti and
      F.M. Riccieri, J.Pharm.Sci., 56, 1555 (1967).
48.   S. Shinkai and T. Kunitake, Bull.Chem.Soc.Japan, 44, 3086
      (1971).

49.   F.M. Menger and M.J. McCreery, J.Am.Chem.Soc., <u>96</u>, 121
      (1974).
50.   A.G. Cairns-Smith and S. Rasool, J.Chem.Soc.Perkin Trans.II,
      1007 (1978).
51.   S. Miyagishi and M. Nishida, Yukagaku, <u>28</u>, 923 (1979).
52.   E.J. Fendler, D.J. Koranek and S.N. Rosenthal, in "Solution
      Chemistry of Surfactants", K.L. Mittal, Editor, Vol.2, p.575,
      Plenum Press, New York, 1979.
53.   O. Hernell and T. Olivecrona, Biochim.Biophys.Acta, <u>369</u>, 234
      (1974).
54.   O. Hernell, Europ.J.Clin.Invest., <u>5</u>, 267 (1975).
55.   B. Fredrikzon, O. Hernell, L. Blackberg and T. Olivercrona,
      Pediat.Res., <u>12</u>, 1048 (1978).
56.   P.-S. Wang, J.Biol.Chem., <u>256</u>, 10198 (1981).
57.   B. Borgström and C. Erlanson, Eur.J.Biochem., <u>37</u>, 60 (1973).
58.   D. Lombardo, J. Fauvell and O. Guy, Biochim.Biophys.Acta,
      <u>611</u>, 136 (1980).
59.   D. Lombardo and O. Guy, Biochim.Biophys.Acta, <u>611</u>, 147 (1980).
60.   B. Borgström, J. Donner and C. Erlanson, in "Adv. Bile Acid Res.
      3rd Bile Acid Meeting", S. Matern, J. Hackenschmidt, P. Back ,
      Editors, pp. 213-217, Schattauer, Stuttgart, Germany, 1975.
61.   F. Takao, Y. Sugimoto and H. Fuwa, Dempun Kagahu, <u>25</u>, 12
      (1978).
62.   D. Lairon, G. Nalbone, H. Lafont, J.Leonardi, N. Domingo,
      J.-C. Hauton and R. Verger, Biochemistry, <u>17</u>, 5263 (1978).
63.   D. Lairon, G. Nalbone, H. Lafont, J. Leonardi, N. Domingo,
      J.-C. Hauton and R. Verger, Adv.Exp.Med.Biol., <u>101</u>, 95 (1978).
64.   A. Poulos and K. Beckman, Clin.Chim.Acta, <u>101</u>, 277 (1980).
65.   D. Lairon, M. Charbonnier-Augeire, G. Nalbone, J. Leonardi,
      J.-C. Hauton, G. Pieroni, F. Ferrato and R. Verger, Biochim.
      Biophys.Acta, <u>618</u>, 106 (1980).
66.   M. Chen, M. Grätzel and J.K. Thomas, J.Am.Chem.Soc., <u>97</u>,
      2052 (1975).
67.   T. Sugimoto, Y. Matsumura, T. Imanishi, S. Tanimoto and
      M. Okano, Tetrahedron Lett., <u>37</u>, 3431 (1978).
68.   B.M. Fung and L. Thomas, Chem.Phys. of Lipids, <u>25</u>, 141 (1979).
69.   D. Liebritz and J.D. Roberts, J.Am.Chem.Soc., <u>95</u>, 4996 (1973).
70.   C.J. O'Connor and R.G. Wallace (1982), unpublished data.
71.   E.J. Fendler and J.H. Fendler, Adv.Phys.Org.Chem., <u>8</u>, 271
      (1970).
72.   C.J. O'Connor, E.J. Fendler and J.H. Fendler, J.Chem.Soc.
      Perkin Trans.II, 1901 (1973).
73.   R.B. Dunlap and E.H. Cordes, J.Phys.Chem., <u>73</u>, 361 (1969).
74.   E.H. Cordes and C. Gitler, Progr.Bioorganic Chem., <u>2</u>, 1
      (1973).
75.   C.J. O'Connor and R.E. Ramage, Aust.J.Chem., <u>33</u>, 695 (1980).

ELECTROSTATICS OF MICELLAR SYSTEMS

Jens Frahm[1] and Stephan Diekmann[2]

[1]Max-Planck-Institut für biophysikalische Chemie
D 3400 Gottingen, West Germany

[2]Biochemical Laboratories, Harvard University
Cambridge, Mass. 02138 USA

In recent years we have developed an electrostatic description of micellar systems in aqueous solutions based on a modified Poisson-Boltzmann equation. The influence of these modifications introduced were analysed using a model system. The theory has been applied to various experimental conditions and has been shown to provide a general explanation for the experimental findings.

Two metal complex formation reactions have been studied: one taking place entirely at the micellar surface, the other in the bulk phase between the micelles. The apparent rate changes of both reactions due to the presence of micelles can be quantitatively described by our electrostatic model. By combining the theory with surface potential measurements the electrostatic properties of micellar sodium-alkylsulfates have been determined. Recently, the rate constants of electron transfer reactions taking place at micellar surfaces were found to be changed not only due to an electrostatic shift of the local ion concentrations but in addition to that, by an "environmental effect".

INTRODUCTION

Micelles built up by sodium alkylsulfates are highly negatively charged systems. In aqueous solution the existence of hydrophobic pockets (the micellar interior) and charged surfaces gives rise
to catalytic effects which are of great chemical interest. In addition, since hydrophobic and electrostatic effects are the basis of
biochemical reactions and enzymatic activity, the understanding of
"micellar catalysis"[1] is of biochemical importance.

In recent years, we have developed an electrostatic description of micellar systems in aqueous solutions based on a modified
Poisson–Boltzmann equation.[2] This theoretical model has been
applied to various experimental conditions and has been shown to
provide a general explanation for the experimental findings. In
particular, by combining the theory with surface potential measurements the electrostatic properties of micellar sodium alkylsulfates
could be determined.[3] Recently, the rate constants of electron
transfer reactions taking place at micellar surfaces were found to
be influenced not only by the electrostatic shift of the local ion
concentrations but in addition to that by an "environmental effect"
[4].

THEORETICAL MODEL

Micelles are not rigid but highly mobile in spite of their
aggregated structure[5]. For concentrations close to the critical
micelle concentration (cmc) they form spheres with up to 100 monomers which are in dynamic equilibrium with the free monomers in
solution. Kinetic studies indicate that the lifetime of a monomer
in the micelle is about 10^{-5} sec.[6] The intramolecular motions of
monomers within the micelle can be assumed to be rather unrestric-
ted with an enhanced mobility of the terminal CH_3-group predominantly oriented within the micellar core and of the ionic headgroups situated at the surface of the micelle.[5] Correlation times
of these motions are considerably shorter than residence times of
individual monomers in the micelle.

Thus for any process or measurement with a characteristic
time range slower than that of the ionic headgroup mobility, the
micelle appears to be a sphere of uniform smeared-out surface
charge. As a result, a constant surface potential can be assigned
to the micellar surface. This, of course, applies only to counterions which do not chemically react with the headgroup of a micellar monomer.

In our model the ion distribution in the micellar solution is
described by the Poisson–Boltzmann equation (PBE)

$$\text{div} \, (\varepsilon \text{grad} \, \phi) = - \rho/\varepsilon_o \qquad (1)$$

(ε the dielectric constant, ε_0 the dielectric constant of free
space). Since we are dealing only with spherical micelles, the lo-
cal potential ϕ is simply a function of the distance r from the
micellar surface.
For a constant surface potential ϕ_0, the boundary condition
$\phi_{r = r_m} = \phi_0$ (r_m the micellar radius) has to be used.
Sodium alkylsulfate micelles are expected to have high surface potentials. Thus, the PBE is not allowed to be linearized which ex-
cludes an analytical mathematical treatment.

Due to the high surface charges, two physical effects are
taken into account which appear as modifications of the PBE:

1. the dielectric saturation, and 2. the volume saturation at the
micellar surface.

1. Within the surface region of the micelle the dielectric constant is changing from the high value in water ($\varepsilon \cong 80$) to a
   low value in the hydrophobic micellar core ($\varepsilon \cong 5$ to 10). Outside the micellar surface, for $r > r_m$, this variation can be
   described by the dependence of the dielectric constant ε on the
   local electrical field strength E with $E = -$ grad (ϕ (r)). With
   ε (E) given by Booth[7], the dielectric saturation is corrected
   for without introducing an adjustable parameter in the PBE.

2. The attraction of counterions to micellar surfaces with high
   potentials, in some cases, causes extremely high "surface concentrations" of counterions. This surface concentration is limited by the total volume of a surface shell around the micelle
   and the volume of the hydrated counterions.
   Again, as for the dielectric saturation, the PBE is corrected
   for the ion volume saturation[8] without introducing an adjustable parameter in the PBE.
   This correction should be important especially when counterions
   with different charges are competing in the diffuse double
   layer.

The boundary conditions for the solution of the PBE are chosen
with respect to the cell model (see e.g. ref. 2 and further references therein) of micellar solutions: the total aqueous solution is
divided into as many identical cells as micelles are present in
solution assuming one micelle in each cell. Due to the attraction
to the charged surface, the concentration of counterions in the
bulk solution between the micelles might decrease markedly below
the appropriate mean concentration. Thus, the bulk concentration
of counterions should not be identified with the mean ion concentration in solution as is usually done in the Debye-Huckel theory
for simple electrolytes. In the cell model the mass balance is
calculated for each ion present yielding the correct bulk ion concentration (at the cell boundary).

The properties of the nonlinearized PBE with these two modifications are analyzed using a model system of charged spheres with
a given surface potential.[2] The ionic environment is taken to be
a mixture of mono- and divalent ions. Different theoretical treatments give rise to measurable differences especially when compared
at the charged surface; there the differences are largest. Therefore, the number of 2+ ions in the vicinity of the surface of the
charged spheres is calculated: the radius-dependent ("local") 2+ ion
density is integrated within a shell-like volume around the sphere
between two radii. The inner radius of this shell is the radius of
the sphere (i. e. the radius where the surface potential is defined). The outer radius is, in general, dependent on the experimental conditions. In this model system the outer radius is chosen
to be 8 Å larger than the inner radius.

In order to obtain a quantitative measure of the influence of
the two modifications upon the solution of the nonlinearized PBE,
calculations have been performed with and without them (constant
dielectric constant and the assumption of point charges). The calculations show good agreement for the case of small surface potentials ($\phi_0 \lesssim 100$ mV) and small 2-2 electrolyte concentrations
($c_{2-2} \lesssim 10^{-3}$ M). However, if one of these values is increased, the
deviation between the theoretical models becomes considerable. The
unmodified nonlinearized PBE always provides higher value for the
number of attracted ions. This obviously arises from neglecting
the ion sizes.

The surface potentials of sodium alkylsulfate micelles are in
the range of -100 mV to -200 mV. Thus, following Figure 1, any
electrostatic model should at least take into account the volume
saturation at the micellar surface. As will be shown in the next
section, both modifications are sufficient to explain the experimental results.

METAL COMPLEX FORMATION IN SODIUM ALKYLSULFATE SYSTEMS

The electrostatic model presented above has been applied to
metal complex formation reactions between Ni^{2+} and two organic
compounds. These compounds, PADA[9] (pyridine-2-azo-p-dimethyl-ani-
line) and murexide[10] change colour when bound. Thus, the reaction
can easily be followed by absorption spectroscopy. The apparent
rate changes of these reactions due to the presence of micelles
can be quantitatively explained by our electrostatic model.

The complex formation between PADA and Ni^{2+} has been previously studied in aqueous[11] and in various surfactant solutions
12,13,14. The reaction profile, the apparent reaction rate plotted

versus the surfactant concentration, increases sharply at the cmc
of the system by two to three orders of magnitude (see Figure 2).
Below the cmc only a slight increase in the apparent rate was observed which could be explained by the "catalysis" due to premi-

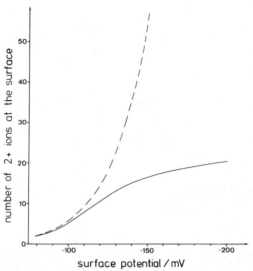

Figure 1.   Calculated number of 2+ counterions in a shell of 8 Å
thickness around one model sphere of 15 Å radius, plotted versus
the surface potential of the sphere. The 2-2 electrolyte ·concen-
tration is 10^{-3} M, the concentration of the spheres is 10^{10} spheres
/cm^3.
——— modified non-linearized PBE,
– – – – unmodified non-linearized PBE.
Reprinted with permission from ref. 2.

cellar aggregates.[9] Above the cmc a maximum is observed followed
by a steady decrease due to the dilution of metal ions (constant
overall concentrations) at the surface of micelles which grow in
number.

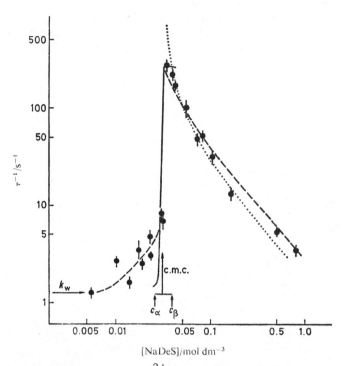

Figure 2.   Reaction rate of Ni^{2+}-PADA complexation with NaDeS present in solution. The cmc is indicated together with the appearence range of micelles (c_α to c_β);k_w is the reaction rate in water;
● experimental points with error bars;
--- above the cmc: theoretical model using the modified non-linearized PBE (ion diameter 5 Å, micelle radius 15 Å, aggregation number 50, surface potential -102 mV),
--- below the cmc: rate increase due to modelled premicellar aggregates[9],
··· simplified model (James and Robinson[12]),
── calculated rate increase making concrete assumptions[9] on the number increase of micelles at the cmc and the number of PADA molecules in a micelle (see text).
Reprinted with permission from ref. 9.

    The increase in apparent rate at the cmc is determined by the formation of micelles and the binding constant of the hydrophobic ligand (PADA) to the micellar core. Since the binding constant of the hydrophobic ligand in general can only roughly be estimated from the reaction profile, we have chosen a ligand strongly hydrophobic in character (PADA). This allows us to assume that each micelle can take up one or two PADA molecules instead of introducing a binding constant. The concentration of PADA was chosen to be small (2×10^{-5} M), so that slightly above the cmc all PADA molecules were already bound. The difference in the reaction profile assuming one or two PADA molecules per micelle was found to be negligible.[9]

It had already been concluded from James and Robinson[12] analysis of the system Ni^{2+}/PADA/NaDS (sodium dodecylsulfate) that the rate limiting step of the complex formation is not changed when the reaction takes place in the surface region of the micelle rather than in the bulk. This conclusion is reasonable since the rate limiting step is the rate of exchange of a water molecule from the inner coordination sphere of the Ni^{2+} which should not be influenced by an external potential change:

$$Ni(H_2O)_6^{2+} + \underset{N}{\overset{N}{N}}) \underset{fast}{\overset{K}{\rightleftarrows}}$$

$$((H_2O)_5NiOH_2 \underset{N}{\overset{N}{N}}))^{2+} \underset{k_b}{\overset{k_{ex}}{\rightleftarrows}} ((H_2O)_4Ni \underset{N}{\overset{N}{N}}))^{2+} + 2H_2O$$

outer-sphere complex   inner-sphere complex

For a very similar micellar system Ni^{2+}/PADA/NaDeS (sodium decylsulfate) we measured the surface potential and the reaction profile, i. e. the dependence of the apparent reaction rate on the surfactant concentration.

To compare the theoretical model with the measured rates of the latter system the thickness of the surface shell has to be determined over which the Ni^{2+} counterion density is integrated to give the "surface concentration". The thickness of this shell is solely determined by the reaction mechanism and the partners involved, it is a measure of the distance between the hydrated metal ion and the ligand before the outer-sphere-complex is formed, and should be independent of both, the nature of the surfactant and the surfactant concentration. The inner radius of the shell is determined by the surface at which the surface potential is measured. This radius is identified with the micellar radius, i. e. the mean position of the charged headgroups. The outer radius should be larger than this inner radius by 5 Å to 10 Å.

Using the measured surface potential, a thickness of 6.5 Å for the surface shell was found to yield good agreement between the theoretical model and the measured rates of the complex formation. Good agreement is found over the whole surfactant concentration range studied (see Figure 2) confirming that the rate limiting step k_{ex} is not changed when the reaction takes place at the micellar surface.

In addition to our calculations using the modified non-linearized PBE computations have been performed with the same numerical procedure neglecting effects due to finite ion sizes as well as due to a variable dielectric constant in order to estimate the influence of these modifications.[9] For the NaDeS system the comparison of the electrostatic parameters needed for a quantitative fit of the rate data using these two approaches gives deviations of ca. 10 %. For surfactants with longer chainlength this difference is expected to be larger.

Using the set of parameters thus determined, the temperature dependence of the complexation rate was predicted taking into account the temperature dependence of the reaction in water. Again, good agreement with the experiments was found. The apparent rate shifts measured in a variety of other surfactant systems (NaDS, NaTS (sodium-tetra-), and NaOS (-octylsulfate)) could also be described by our theory (see figure 3 for NaDS).

As pointed out in the theoretical section our model not only provides the concentration of ions at the surface, but also the ion concentration in the bulk of the solution (at the cell boundary). This value was verified experimentally by application to a complexation reaction between Ni^{2+} and the hydrophilic ligand murexide.[10] This reaction has been analysed by Fischer et al..[16] The murexide ion is hydrophilic and negatively charged so that it is located solely in the bulk aqueous phase. Above the cmc of the system the apparent equilibrium constant of the reaction drops dramatically (see Figure 4), since now the micelles attract Ni^{2+} ions to their surface and thus reduce the Ni^{2+} concentration in the bulk. The micellar parameters for this system were taken from the literature (the surface potential was interpolated between the measured values). Thus, the general shape of the curve in Figure 4 could be determined without any adjustable parameter. Since no cmc value was reported by Fischer et al.[16], we fitted the calculated decrease of the relative apparent equilibrium constant to the experimental data. Changing the cmc value, however, moves the computed curve to higher or lower NaDS concentrations without changing the slope of the curve. Acceptable agreement is obtained.

The applications presented and the good agreement between theory and experiments allows us to conclude that the observed rate enhancement (Ni^{2+}/PADA) or decrease in apparent equilibrium constant (Ni^{2+}/murexide) is solely due to the local change in the concentration of the metal ion Ni^{2+}. By chosing ligands with dif-

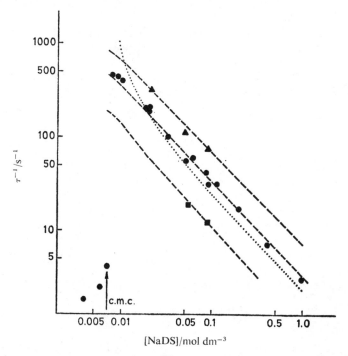

Figure 3. Reaction rate of Ni^{2+}-PADA complexation with NaDS present in solution[12],[13]. The cmc is indicated; experimental points for three temperatures are presented: ▲ 308° K, ● 298° K, and ■ 283° K;

- - - theoretical model (ion diameter 5 Å, micelle radius 18 Å, surface potential -106 mV, aggregation number 71) for the three different temperatures (the rate change in water has been taken into account)

..... simplified model (James and Robinson[12]), 298° K.
Reprinted with permission from ref. 9.

ferent properties (pos. - neg. charge, hydrophobic - hydrophilic) and thus different locations in the micellar solution, the Ni^{2+} ion concentrations in these regions of the solution are measured. The rate limiting step of the complex formation (the water exchange from the inner coordination sphere of the Ni^{2+} ion) is not changed when micelles are added to the solution. We suggest that this finding is not only true for the reactions analysed, but probably holds for all complex formations which follow the Eigen mechanism.[15] The catalytic effects are due to a local enrichment of the reaction partners with the intrinsic rate constants being unchanged. Thus, no "true catalysis" is observed.

ELECTROSTATIC SURFACE PROPERTIES OF SODIUM ALKYLSULFATES

    The knowledge of the surface potential (or equivalently: the
surface charge) is essential for the calculation of ion concentrations in polyelectrolyte systems. Thus, this value should not be
determined theoretically but should be measured. The experimentally
determined surface potential values should be taken as the basis of

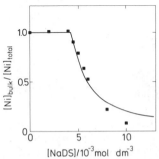

Figure 4. Ni_{bulk}/Ni_{total} as a function of NaDS concentration.
Ni_{total} = 2 · 10^{-4} mol/dm^3. Experimental points have been taken
from Fischer et al.[16]. Ni_{bulk}/Ni_{total} is identical to the apparent
equilibrium constant relative to the constant in water.
—— theoretical model (ion diameter 5 Å, micellar radius 18 Å,
aggregation number 71, surface potential –125 mV, cmc = 4.3 · 10^{-3}
mol/dm^3).
Reprinted with permission from ref. 10.

the calculations (as was mostly done in our calculations). Therefore, to provide a complete set of surface potentials for different
sodium alkylsulfates under various conditions (see below), extensive measurements were carried out by Haase[3] by means of a method
described by Fernandez and Fromherz.[17] Two fluorescent pH-indica-
tors (7-hydroxocoumarine (I), 7-aminocoumarine (II)) which are sub-

stituted with a long alkyl chain and therefore insoluble in water, are incorporated into micelles. NMR measurements show that for the pH–indicators used the center of the chromophore is located near the second CH_2-group of the surfactant, while the titrated amino- and hydroxygroups are in the plane of the micellar headgroups.[18]

Careful surface potential measurements show a strong increase of the potential with the monomer chain length (the micellar radius) and – within experimental uncertainties – independence on temperature and surfactant concentration (up to ten times the cmc). The logarithm of the potential was found to decrease linearly with the ionic strength.[3]

The measured micellar surface potentials are used to calculate the appropriate surface charges.[3] With the independent knowledge of the mean aggregation numbers from light scattering measurements the micellar dissociation constants are easily computed. The average dissociation constant for micelles formed by the sodium salts of different surfactant molecules is about 0.4, and is almost independent of micellar radius, ionic strength and temperature in the range analysed (see Table 1). The calculated surface charge of the homogeneously charged model sphere is attributed to the number of dissociated monomer headgroups of the real micelle. Since the assumed model of the micelle cannot account for different surface sites or states, the associated counterions can either be interpreted as inner-sphere complexes of the Na^+ ions with the sulfate groups or loosely bound outer-sphere complexes. Our theoretical model distinguishes only between bound states within the surface and free states in the diffuse double layer outside the mean micellar radius. Thus, no Stern layer [19] is considered. For micelles of lithium-, sodium-, and caesium dodecylsulfate a strong increase in the mean aggregation number was found.[20] This increase was thought to be due to the different radii of the counterions. Therefore, their effect or the micellar surface properties was investigated for these micellar systems.[3] However, only small differences were found for the cmc and the surface potentials. Therefore the free enthalpy of micellization as well as its electrostatic contribution should depend only slightly on the nature of the surfactant counterions. A small decrease of the surface charge observed in going from LiDS to CsDS corresponds to the radii changes of the hydrated cations (r_{Li} = 3.4 Å, r_{Na} = 2.5 Å, r_{Cs} = 2.28 Å). This effect, however, cannot account for the dramatic increase in the mean aggregation number.

Similar effects due to different monovalent counterions are not only observed in micellar systems but also in solutions containing DNA[21], a highly negatively charged polyelectrolyte of cylindrical geometry.

Table I. Micellar Surface Dissociation Constants α_{mic} (Surface
Charge/Aggregation Number m) for Different Surfactants and NaClO$_4$
Concentrations.
*) (NaCl) = 0.1M; NaDeS sodiumdecylsulfate; NaDS sodiumdodecylsul-
fate; NaTS sodiumtetradecylsulfate; NaHS sodiumhexadecylsulfate;
r_{mic} = micellar radius

Surfactant conc.	Temp. °C	NaClO$_4$ M	m	r_{mic} Å	α_{mic}
NaDeS	25	0	40	15.0	0.45
0.06 M		0.010	41	15.1	0.44
		0.033	43	15.4	0.48
		0.100	50	16.2	0.47
NaDS	25	0	55	18.0	0.49
2xcmc		0.010	58	18.3	0.48
		0.033	72	19.7	0.43
		0.100	111	23.7	0.35
		0.100*)	99*)	21.9	0.46
NaTS	35	0	73	20.0	0.38
0.003 M		0.010	84	21.0	0.43
		0.033	93	21.7	0.45
		0.100	120	23.6	0.41
NaHS 0.00083 M	35	0	105	22.0	0.35

ELECTRON TRANSFER REACTIONS IN SODIUM DODECYLSULFATE

After having determined the electrostatic properties of sodium alkylsulfates these data were used to analyse rate constants of
electron transfer reactions in micellar solutions. Metal complex
formation was ideal to demonstrate the capacity of the theoretical
model, because no change of the rate limiting step was observed
and thus, the shift of the measured rate would only be due to local concentration changes of the metal ion. In the case of outer-
sphere electron transfer reactions, however, changes in the rate
limiting step are expected at the surface due to the influence of
the high local potential on the electronic system of the organic
ligands involved. For a bulk reaction the rate constant should not
be altered by the presence of micelles (as for metal complex formations). Thus, the decrease or enhancement of electron transfer reaction rates in the bulk (above the cmc) should be predicted by
our theoretical model as well.

A suitable system to test the predictions was found by Pramauro and Pelizzetti in the reaction $Fe^{2+}/IrCl_6^{2-}/NaDS$[4]. $IrCl_6^{2-}$ is hydrophilic and negatively charged, and thus it should entirely be located in the bulk. The cmc and the surface potential as well as other micellar parameters could be interpolated from the measured values. Thus, the bulk concentration of Fe^{2+} and $IrCl_6^{2-}$ were calculated without any adjustable parameter. Above the cmc the bulk concentration of Fe^{2+} is decreased remarkedly due to the attraction of the Fe^{2+} ions to the micellar surface, while that of $IrCl_6^{2-}$ is increased only slightly as compared to the corresponding values in the absence of NaDS.

If the measured reaction rates are corrected for the calculated bulk concentration, their values are approximately independent of NaDS concentration and are also equal to the rates determined in the absence of surfactant[4]. This confirms that the reaction proceeds in the bulk of the micellar solution. In this case the measured rates can quantitatively be described by pure electrostatics.

The situation totally changes for electron transfer reactions taking place at the surface of micelles. Two systems were studied[4] in cooperation with Pramauro and Pelizzetti:

$$Fe^{2+}/Os(bpy)_3^{3+}/NaDS \quad \text{(bpy: 2,2 –bipyridine)} \quad \text{and}$$

$$Fe^{3+/2+}/Os(dmbpy)_3^{2+/3+}/NaDS \quad \text{(dmbpy: 4,4'-dimethylbipyridine).}$$

The presence of micelles causes an increase in the measured reaction rate of three orders of magnitude which reaches a maximum and then decreases again. This behaviour is expected[22] when both reactants are attracted to micelles.

Due to the high charge of the complexes and the hydrophobic character of the ligands, these can be assumed to be located entirely at the micellar surface. For the other reactant, Fe^{2+}, the concentration at the surface is calculated taking into account the presence of the charged Os-complexes. Since for this type of surface reaction, differences in the reaction rates between electrostatically calculated and experimentally determined values are expected, the thickness of the shell defining the surface reaction environment should not be determined by fitting the theoretical model to the experimental results. Two values for the shell thickness, 5 Å and 8 Å, are discussed which are supposed to be extrema with respect to the sizes of the ions involved in these reactions. Using these values a remaining non–electrostatic enhancement factor is found for the $Fe^{2+}/Os(bpy)_3^+/NaDS$-system which is independent of ionic strength and surfactant concentration[4] (see Figure 5). This qualitative behaviour cannot be altered by adjusting the micellar parameters (cmc, surface potential, shell thickness). In fact, by varying the shell thickness from 5 to 8 Å, the remaining non-elec-

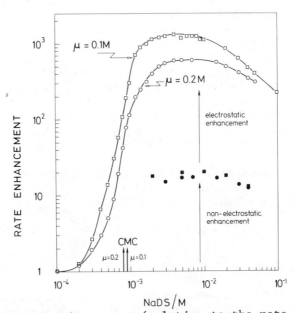

Figure 5. The rate enhancement (relative to the rate in water) of
the reaction $Fe^{2+} + Os(bpy)_3^{3+} \rightarrow Fe^{3+} + Os(bpy)_3^{2+}$ versus the concentration of NaDS for two ionic strengths: □ 0.1 M and o 0.2 M.
The cmc's of the two ionic strengths are indicated.
The experimental results are corrected for the electrostatic contribution to the rate enhancement using our theoretical model
(ionic strength: ■ 0.1 M (● 0.2 M), aggregation number 100 (140),
micellar radius 22 (24), surface potential –75 mV (–70 mV),
ion radius 5 Å (5 Å), shell thickness 5 Å (5 Å).
The electrostatic and the non-electrostatic enhancement are indicated by arrows.
Reprinted with permission from ref. 4;Copyright 1982 American Chem.
Soc.

trostatic factor grows from 17 ± 5 to 30 ± 5. The main part of the
non-electrostatic contribution may be ascribed to the ligand-mi-
celle interaction, which indeed should not be affected very much
by the ionic strength and the surfactant concentration.

The $Fe^{3+/2+}$/Os$(dmbpy)_3^{2+/3+}$/NaDS-system offered the possibility
to investigate the influence of the micelles upon the equilibrium
constant of a redox reaction together with the effect on the forward and backwards rates. The measured reaction rate is found to
be larger for the backward than for the forward reaction[4]. This
finding is opposite to what would be expected on the basis of pure
electrostatics, because the forward rate is increased due to the
attraction of the higher charged Fe^{3+} ion compared to Fe^{2+} for the
backreaction. For the backreaction, a non-electrostatic enhancement
factor is found which is comparable to that obtained for the previous system discussed above. For the forward reaction, however, a
non-electrostatic inhibition factor is experienced.

As a consequence of the different characteristics of the forward and back reaction rates, the observed equilibrium constant
varies with the surfactant concentration (see Figure 6) in a way
such that the apparent equilibrium constant drops by two orders of
magnitude[4]. For a discussion of the influence in the presence of
NaDS the apparent equilibrium constant can be written as

$$K_{app} = K_{water} \, F_1 \, F_2$$

(F_1: electrostatic factor relative to the water value,
 F_2: non-electrostatic factor relative to the water value).
Since K_{water} is known and F_1 can be calculated, F_2 can be obtained.
By this procedure the electrostatic influence can be separated from
the non-electrostatic. At the cmc the non-electrostatic factor decreases by two orders of magnitude and then stays constant independent of the surfactant concentration. The strong decrease at the
cmc is counterbalanced partly by an increase of the electrostatic
factor which then slowly decreases with increasing NaDS concentration.

The large differences between the experimental rates and the
theoretical electrostatic contribution are interpreted due to the
environmental influence. To our knowledge the presented method provides for the first time a reliable separation of electrostatic
and environmental effects on micellar catalysis.

CAN THE PRESENT APPROACH BE GENERALIZED?

The electrostatic model of micellar systems outlined above
is shown to be consistent and sufficient to explain various experimental findings. This is mainly due to the properties of micellar aggregation: the monomers have no fixed positions in the micelle

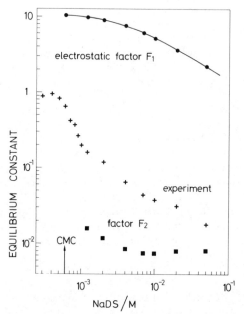

Figure 6. The apparent equilibrium constant (relative to the constant in water) for the system

$$Fe^{3+} + Os(dmbpy)_3^{2+} \overset{K}{\rightleftharpoons} Fe^{2+} + Os(dmbpy)_3^{3+}$$

is plotted (+) versus the concentration of NaDS. The cmc is indicated. (The ionic strength is 0.1 M).

The theoretical shift of the equilibrium constant due to the electrostatic contribution F_1 (●) is included. From experiment (+) and F_1 (●) the non-electrostatic shift F_2 (■) is obtained.

The parameters used in the calculations are:
aggregation number 100, ionic strength 0.1 M,
micellar radius 22, surface potential -75 mV,
ion radius 5 Å, and shell thickness 5 Å.

Adapted with permission from ref. 4; Copyright 1982 American Chemical Society.

and their charged headgroups, in general, can be assumed to reorient almost as freely at the micellar surface as in the monomeric
state[5]. Thus, for calculating the concentration of the reacting metal ions, the polyelectrolyte surface can be assumed to have a
smeared-out surface charge or, equivalently, a constant surface potential. Our theoretical approach should be used only for systems
where this general requirement is fulfilled.

Most biological polyelectrolyte systems, however, have a
crystal-like structure in solution: these macromolecules are built-
up by small molecular units in a well defined way. In general, the
small molecular units do not reorient freely within the macromolecules or the system (membranes are an exception) nor is it allowed
to take the mean value over a sequence of molecular units (for example nucleotides in DNA) when surface properties are discussed.

As an example, double helical DNA might be taken. The double
helix is built-up by two chains of nucleotides, the proper sequence
of which is essential for its biological function. The internal
motions of DNA - although very fast (in the nanosecond time range)
- are too small in amplitude[23] to allow a general smearing-out of
the regularly positioned charges in the chains. By smearing-out the
phosphate charges, parts of the biological function might be lost.

Within the diffuse double layer, some distance away from the
surface, the charges of the crystal-like macromolecule are partially screened by counterions. The discrete structure of the surface
charges might appear to be smeared-out at this imaginative surface
surrounding the macromolecule. At this "outer surface" the local
potential can be renormalized to a smaller value (being constant
within this surface) which now may enter a Poisson-Boltzmann equation describing the outer region of the diffuse double layer. For
DNA this was done with great success by Manning[24]. The bulk properties of the solutions containing these linear polyelectrolytes are
well described. However, the detailed surface properties which are
expected to have influence on the biological function, are averaged
by this procedure. The DNA double helix enclosed by the renormalization surface is defined by a line charge density and a number of
condensed counterions. In general, a more detailed picture of the
polymer is required for processes where the potential of the polyion enters directly (especially in physical studies like NMR,
spectroscopy, diffusion, conductivity, etc.)[25].

We suggest that the electrostatic surface properties of those
polyelectrolytes which exhibit a defined but irregular microscopic
structure may be described by calculating the potential of the discrete surface charges[25,26]. To determine the positions of counterions, pair correlation functions of the ions[27] and probably the
dipole character of the water molecules should be taken into account[28] at least within the renormalization surface.

REFERENCES

1. J. H. Fendler and E. J. Fendler, "Catalysis in Micellar and Macromolecular Systems", Academic Press, New York,]975.
2. J. Frahm and S. Diekmann, J. Colloid Interface Sci. 70, 440 (1979).
3. J. Frahm, S. Diekmann, and A. Haase, Ber. Bunsenges. Phys. Chem. 84, 566 (1980).
4. E. Pramauro, E. Pelizzetti, S. Diekmann, and J. Frahm, Inorg. Chem. 21, 2432 (1982).
5. E. Lessner and J. Frahm, J. Phys. Chem. 86, 3032 (1982).
6. S. Diekmann, Ber. Bunsenges. Phys. Chem. 83, 528 (1979).
7. F. Booth, J. Chem. Phys. 19, 391 (1951); 23, 453 (1953).
8. E. Wicke and M. Eigen, Z. Elektrochem. 56, 551 (1952).
9. S. Diekmann and J. Frahm, JCS Faraday I 75, 2199 (1979).
10. S. Diekmann and J. Frahm, JCS Faraday I 76, 446 (1980).
11. M. A. Cobb and D. N. Hague, JCS Faraday I 68, 932 (1972).
12. A. D. James and B. H. Robinson, JCS Faraday I 74, 10 (1978).
13. B. H. Robinson and N. C. White, JCS Faraday I 74, 2625 (1978).
14. N. C. White, personal communication.
15. M. Eigen and K. Tamm, Z. Elektrochem. 66, 93 (1962); Ber. Bunsenges. Phys. Chem. 66, 107 (1962).
16. M. Fischer, W. Knoche, B. H. Robinson, and J. Maclagan Wedderburn, JCS Faraday I 75, 119 (1979).
17. M. S. Fernandez and P. Fromherz, J. Phys. Chem. 81, 1755 (1977).
18. A. Haase, Ph. D. thesis, Giessen, 1980; A. Haase and P. Fromherz, in preparation.
19. D. Stigter, J. Phys. Chem. 79, 1008 (1975); 79, 1015 (1975).
20. G. Klar, Ph. D. thesis, Gottingen, 1978.
21. P. D. Ross and R. L. Scruggs, Biopolymers 2, 231 (1964); D. W. Grunwedel, C. H. Hsu, and D. S. Lu, Biopolymers 10, 47 (1971).
22. I. V. Berezin, K. Martinek, and A. K. Yatsimirskii, Russ. Chem. Rev. (Engl. transl.) 42, 787 (1973).
23. G. Lipari and A. Szabo, Biochem. 20, 6250 (1981); B. H. Robinson, L. S. Lerman, A. H. Beth, H. L. Frisch, L. R. Dalton, and C. Auer, J. Mol. Biol. 139, 19 (1980).
24. G. S. Manning, Quart. Rev. of Biophys. 2, 179 (1978); Acct. Chem. Res. 12, 443 (1979); Biophys. Chem. 9, 65 (1978).
25. D. M. Soumpasis, J. Chem. Phys. 69, 3190 (1978).
26. D. Perahia, A. Pullman, and B. Pullman, Int. J. Quant. Chem., Biol. Symp. 6, 353 (1979); A. Pullman and B. Pullman in "Molecular Electrostatic Potentials in Chemistry and Biochemistry", P. Politzer and E. G. Truklar, Editors, pp. 381, New York 1980.
27. D. M. Soumpasis, paper presented at the Symposium on "Polyelectrolytes and Charged Interfaces" in Strasbourg, June 10-11, 1981.
28. R. Lavery and B. Pullman, Nucleic Acid Res. 9, 3765 (1981).

Part IV
Solubilization

SOLUBILIZATION EQUILIBRIA STUDIED BY THE FT-PGSE-NMR MULTICOMPONENT

SELF-DIFFUSION TECHNIQUE

Peter Stilbs

Institute of Physical Chemistry
Uppsala University
Box 532, S-751 21 Uppsala, Sweden

The binding of small molecules (s) (or ions) to macromolecules in solution is directly reflected in the time-averaged self-diffusion coefficient of the small molecule (ion), according to the relation

$$D^S_{obs} = pD^S_{bound} + (1-p)D^S_{free}$$

which can be rewritten into the form

$$p = \frac{c^S_{bound}}{c^S_{total}} = \frac{D^S_{free} - D^S_{obs}}{D^S_{free} - D^S_{bound}} = \frac{D^S_{free} - D^S_{obs}}{D^S_{free} - D_{macromolecule}}$$

With the new possibility of rapid, convenient and accurate multicomponent self-diffusion measurements in complex mixtures by NMR, an application to micellar solubilization becomes straightforward. Typical micellar self-diffusion coefficients are of the order of $10^{-11} m^2 s^{-1}$, while the free diffusion of small molecules is characterized by values around $10^{-9} m^2 s^{-1}$. Solubilization (to a degree p) is therefore strongly and clearly reflected in D^S_{obs}. Recent studies with these new techniques are reviewed.

The partitioning of a third component (denoted s) between micelles and water in surfactant/water systems is a fundamental albeit difficult phenomenon to quantify experimentally or even to define. Within the phase separation model, one introduces a simple distribution coefficient between assumingly structureless micellar and aqueous phases. Hence

$$K_x = \frac{x^s_{micellar\ phase}}{x^s_{aqueous\ phase}} \tag{1}$$

or, alternatively

$$K_c = \frac{c^s_{micellar\ phase}}{c^s_{aqueous\ phase}} = \frac{p}{1-p} \cdot \frac{V_{aqueous\ phase}}{V_{micellar\ phase}} \tag{2}$$

where p represents the fraction of s-molecules in the micellar phase. This macroscopic description of the solubilization phenomenon inherently neglects that real micelles have a distinct degree of molecular order, that this order will lead to different regions or sites of solubilization and that the micellar structure and solubilization site necessarily must change in proportion to the amount of solubilized compounds. The arbitrary and imaginary nature of partition coefficients (K_c or K_x) becomes clearly apparent on considering the solubilization of molecules with different solubilization sites (*e.g.* one solubilized in the micellar surface region and another equally distributed over the whole micellar volume).

Several proven experimental techniques for the quantification of micelle-water partition equilibria already exist (see *e.g.* references 1-2). However, these are far from ideal. *Membrane osmometry* and *selective vapour pressure* data (from glc) can give information about the solubilization of low-molecular weight compounds. Apart from being demanding and difficult experimentally it should be noted that the quantity measured is related to solubilizate activity, not to concentration. *Fluorescence quenching* in the aqueous phase has a limited applicability for natural reasons. The increased *total solubility* of partly water-soluble compounds in the presence of surfactants is the most evident practical result of the solubilization phenomenon. Micelle-water partition coefficients can, in principle, be deduced from such data under certain conditions. An approach of this kind involves an assumption that the experimentally determined turbidity onset at some level of addition of solubilizates to a surfactant solution reflects the micellar saturation limit. Assuming that the aqueous solubility is known one can thus determine the fraction of

solubilizate in the micelles. At best this approach will give
information about highly perturbed micelles. For solubilizates
which form mesophases in these surfactant/water/solubilizate
systems (*e.g.* alcohols) the turbidity onset is evidently due to
a mesophase transition and not to "micellar saturation". This
complex situation has been pointed out by Ekwall *et al.*[3] and also
by Wennerström and Lindman in their recent review article[1] (pp 60-
61) but still seems to have been overlooked in many recent studies.

Certainly there is a need for an alternative and more
convenient and general method. One new approach, which was
suggested in a recent communication[4] is based on the very different
self-diffusion properties of micelles and nonaggregated constituents
in micellar solution. Typical micellar self-diffusion coefficients
are of the order of 10^{-11}-$10^{-10} m^2 s^{-1}$, while the free diffusion of
small to medium-sized molecules in aqueous solution is characterized
by values between 10^{-10} to $10^{-9} m^2 s^{-1}$. Assuming a two-site model
for the time-averaged self-diffusion behaviour one can deduce for
simple Brownian diffusion ($D=<x^2>/2\Delta t$) that

$$D^s_{obs} = pD^s_{micelle} + (1-p)D^s_{free} \tag{3}$$

where p is the fraction of s-molecules in the micellar phase.
Provided that an observation process monitors the diffusion over
a period longer than a millisecond or so, the micelles diffuse
many times their own diameter and solubilizate diffusion within the
micelles becomes unimportant and $D^s_{micelle}$ is then equal to $D_{micelle}$.
Hence equation (3) can be rewritten into the form

$$p = \frac{D^s_{free} - D^s_{obs}}{D^s_{free} - D_{micelle}} \tag{4}$$

and the degree of solubilization is obtained from a comparison
of experimentally determined self-diffusion coefficients. Evident
as it may seem,this has been an impractical and unrealistic
measurement approach until very recently, due to lack of a
suitable experimental technique. The situation has completely
changed with the recent practical realization of the Fourier
transform modification[5,6] of the proven pulsed-gradient[7] spin-
echo NMR experiment[8] at a new level of measurement convenience,
frequency resolution and sensitivity[9-11]. A typical experiment
(as illustrated in Figure 1) can usually be performed in about
10 minutes on a conventional pulsed FT-NMR spectrometer.

An identical measurement approach is applicable and has
been used for the study of micellar aggregation[12] and counterion

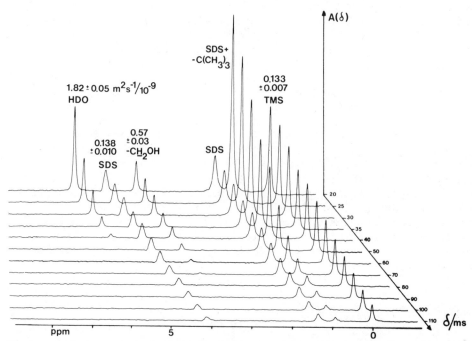

Figure 1. A sequence of 99.6 MHz pulsed-gradient spin-echo proton
NMR spectra on a dilute SDS/D$_2$O/neopentanol/TMS sample as a function
of the magnetic field gradient duration, δ. Peak heights (A$_i$)
directly reflect self-diffusion coefficients according to the
relation $A_i \alpha \exp(-\gamma^2 G^2 \delta^2 D_i (\Delta - \frac{1}{3}\delta))$ where γ, G and Δ are constants
of the experiment. See references 9-11 for further details on the
experimental procedures.

binding in polymeric polyelectrolyte-water systems[13]. In principle,
it can be used in any other situation where a small molecule (or
ion) binds to macromolecules or supramolecular aggregates of some
kind in solution.

Noteworthy results from the SDS solubilization study presented
in Reference 11 are

a) the degrees of solubilization (p) of short-chain alcohols are
0.32, 0.44, 0.77, 0.92, 0.96 and 0.985 from n-propanol to
n-octanol at 25°C, for a 0.25 M SDS concentration and a low
solubilizate/SDS ratio; these values correspond to a ΔG^o
increment of $\approx$2.6 kJ mole^{-1} for transfer of a $-CH_2-$group
from the aqueous to the micellar phase. A similar value
was determined for methyl-n-alkyl ketones.

b) branched alcohols are less solubilized than their linear counterparts.

c) crown ether SDS "solubilization" increases strongly with ring size in the series 1,4-dioxane to crown 18-6. The effect disappears in comparative studies with dodecyltrimethyl-ammonium bromide micelles which leads to the conclusion that the effect is due to crown ether Na^+ ion binding and subsequent electrostatic attraction to the SDS micellar surface.

In a continuation of this work the solubilization behaviour in all combinations of 6 different surfactants and 12 different solubilizates was studied[14]. The results show that, with the exception of the nonionic surfactant $C_{12}E_6$ (hexaethylene glycol dodecyl monoether), p-values do not change much with the type of surfactant. The nonionic surfactant solubilizes n-alkyl compounds only weakly, aromatics as strongly as the other surfactants and surprisingly, crown ethers and dioxane very poorly, despite their structural similarity to $C_{12}E_6$.

In summing up the particular advantages of the FT-PGSE multi-component self-diffusion approach to micellar partition equilibria one finds:

a) the method is very direct; only the micellar and aqueous phases contribute to the observable, D^S_{obs}.

b) studies on selected systems with one solubilizate molecule or less per micelle are possible.

c) arbitrarily concentrated isotropic systems can be investigated.

d) the measurements are rapid and easy to perform, once the spectrometer system has been set up properly.

e) the method monitors concentrations directly and is unaffected by non-ideal solution behaviour.

f) the quantity p is well-defined within the two-site diffusion model.

Structural aspects on surfactant/water systems in the presence of additives (alcohols, hydrocarbons etc.) can conveniently be monitored by self-diffusion measurements, as has been demonstrated in a series of papers by Lindman and coworkers.[15-19] The FT-PGSE technique, of course, is ideal in this multicomponent situation. Of immediate interest was to investigate possible structural changes in micelle/water-systems when alcohols are added. A study of this kind has recently been made.[20] Based on observations of very much enhanced self-diffusion rates for SDS, alcohols, as well as for trace amounts of hydrocarbon (TMS,

tetramethylsilane) above a ~ 1:1 total alcohol-surfactant volume
ratio it was concluded that structural changes in solution have
occured and that the distinct micelle/water structure has been
significantly altered into a less ordered state. The effects are
enhanced with increasing SDS concentration and alcohol chain length
from C_1 to C_5. Under such conditions it is evident that the concept
of micelle-water partition equilibria at high total solubilizate/
surfactant ratios is difficult to rationalize, a further argument
against the "total solubility approach" to micellar partition
equilibria.

ACKNOWLEDGEMENTS

This work has been supported by the Swedish Natural Sciences
Research Council. Thanks are due to Professor Björn Lindman for
many stimulating discussions.

REFERENCES

1. H. Wennerström and B. Lindman, Phys. Rep., 52, 1 (1979).
2. B. Lindman and H. Wennerström, Top. Curr. Chem., 87, 3 (1980).
3. P. Ekwall, L. Mandell and K. Fontell, Mol. Cryst. Liq. Cryst.
    8, 157 (1969).
4. P. Stilbs, J. Colloid Interface Sci., 80, 608 (1981).
5. R. L. Vold, J. S. Waugh, M. P. Klein and D. E. Phelps, J. Chem.
    Phys., 48, 3831 (1968).
6. T. L. James and G. G. McDonald, J. Magn. Reson., 11, 58 (1973).
7. E. O. Stejskal and J. E. Tanner, J. Chem. Phys., 42, 288 (1965).
8. E. L. Hahn, Phys. Rev., 80, 580 (1950).
9. P. Stilbs and M. E. Moseley, Chem. Scr., 15, 176 (1980).
10. P. Stilbs and M. E. Moseley, Chem. Scr., 15, 215 (1980).
11. P. Stilbs, J. Colloid Interface Sci., 87, 385 (1982).
12. P. Stilbs and B. Lindman, J. Phys. Chem., 85, 2587 (1981).
13. P. Stilbs and B. Lindman, J. Magn. Reson., 48, 132 (1982).
14. P. Stilbs, submitted
15. B. Lindman and B. Brun, J. Colloid Interface Sci., 42, 388 (1973).
16. H. Fabre, N. Kamenka and B. Lindman, J. Phys. Chem., 85, 3493
    (1981).
17. P.-G. Nilsson and B. Lindman, J. Phys. Chem., 86, 271 (1982).
18. B. Lindman, N. Kamenka, T.-M. Kathoupoulis, B. Brun and P.-G.
    Nilsson, J. Phys. Chem., 84 2485 (1980).
19. B. Lindman, P. Stilbs and M. E. Moseley, J. Colloid Interface
    Sci., 83, 569 (1981).
20. P. Stilbs, J. Colloid Interface Sci., in press.

SELECTIVE SOLUBILIZATION IN AQUEOUS SURFACTANT SOLUTIONS

R. Nagarajan and E. Ruckenstein[*]

The Pennsylvania State University
University Park, PA 16802
[*]State University of New York at Buffalo
Amherst, NY 14260

Experimental results as well as a thermodynamic
treatment of solubilization of mixtures of hydrocarbons
in aqueous surfactant solutions are presented. Binary
mixtures of benzene-hexane, benzene-cyclohexane and
hexane-cyclohexane have been solubilized in aqueous
solutions of some common anionic and cationic surfactants and of sodium deoxycholate. It is shown that
molecules can be selectively solubilized from binary
hydrocarbon mixtures. Large selectivity ratios of up
to about seven have been obtained for benzene compared
to hexane especially when the binary solubilizate phase
is hexane-rich. In general, the selectivity ratio of
one component over another depends on the difference in
the molar solubilization ratios of the single components.

A thermodynamic treatment is developed to examine
the origin of the selective solubilization and to predict the molar solubilization ratios of various compounds both singly and in mixtures. The treatment
takes into account (i) the change in the surface area
per amphiphile of the micelle due to the incorporation
of the solubilizates (this causes a change in the
steric and electrostatic interactions between the head
groups and in the micellar core-water interactions),
(ii) the hydrophobic effect associated with the transfer of the solubilizate from water to the micelle, and
(iii) the entropy of mixing of the surfactant and solubilizates in the micelle. It is found that the relative balance between the changes in the head group

interactions on the one hand and that in the micellar core-water interactions on the other   principally determine the nature and magnitude of solubilization. In general, the more polar solubilizates, those with smaller molecular volumes and those possessing low stereochemical flexibility, are preferentially solubilized.

The observed absence of any selectivity for benzene over hexane or cyclohexane in solutions of sodium deoxycholate is shown to be consistent with the comparable molar solubilization ratios of the two pure hydrocarbons. This markedly different solubilization behavior is interpreted in terms of the peculiar molecular structure of sodium deoxycholate and its ability to generate only very small primary aggregates.

Experimental solubilization data involving benzene + hexane mixtures are observed to display a synergistic effect. In the framework of the thermo-dynamic treatment, this synergism is shown to imply that at least some of the benzene molecules are located close to the micelle-water interface.

INTRODUCTION

Surface active molecules self-assemble as micelles or vesicles in dilute aqueous solutions so as to minimize the contact between their hydrophobic tails and water. As a result, the interior of micelles and the spherical shells of vesicles are highly non-polar, capable of accommodating other non-polar molecules. Consequently, hydrocarbon molecules which are only sparingly soluble in water can be solubilized in significant amounts in aqueous surfactant solutions[1-3]. Here, by solubiliza-tion we understand both (i) the molecular dispersion of the solubilizates between the hydrocarbon tails of the surfactant molecules forming the micelles or vesicles (Type I) as well as (ii) the formation of a core of solubilizates surrounded by a layer of surfactant molecules (Type II) (Figure 1). The extent of solubilization depends on the chemical characteristics of sur-factant and solubilizate as well as on the concentration of the solution and temperature. Experimental data regarding single component solubilizates show that the solubilization capacity of surfactants differs significantly for different solubilizates. Figure 2 shows the molar solubilization ratios (i.e., moles of solubilizate per mole of surfactant in micelles) for various solubilizates in 0.1M cetyl pyridinium chloride solution as a function of the molecular volume of the solubilizate. The same data are also represented against a combined function of both

molecular volume and polarity of the solubilizate, using the interfacial tension of the solubilizate against water σ_s as a measure of
the polarity. In general, the molar solubilization ratio is found
to decrease with increasing molecular volume of the solubilizate.
Also, the more polar molecules are solubilized in relatively larger

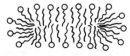

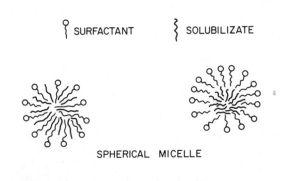

SPHERICAL MICELLE

CYLINDRICAL MICELLE

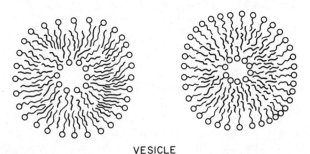

VESICLE

| TYPE I | TYPE II |
| SOLUBILIZATION | SOLUBILIZATION |

Figure 1.   Schematic representation of micelles and vesicles containing solubilizates. For Type I solubilization, the solubilizate
molecules are located between the surfactant tails in the aggregates. For Type II solubilization, the solubilizate molecules do
not mix with the surfactant tails in the aggregates and form a core
of their own in micelles or a shell in vesicles.

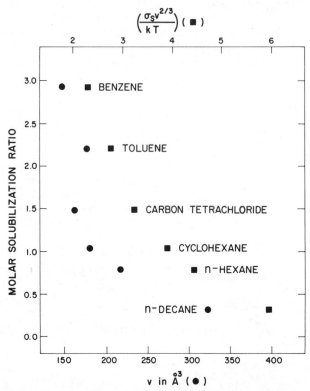

Figure 2. Molar solubilization ratio of hydrophobic solubilizates in 0.1M Cetyl Pyridinium Chloride solution against the molecular volume of the solubilizate. The same data are also plotted against the product of σ_s/kT (representing the polarity of the solubilizate) and $v^{2/3}$ (representing the molecular size of the solubilizate).

amounts compared to the less polar ones. Similar results have been reported in the literature for other surfactants[4-7]. A natural question is whether this significant variation in the molar solubilization ratios can lead to highly selective solubilization of one of the components when the surfactant solution is contacted with a binary or a multicomponent mixture of solubilizates. The only experimental study that has been reported in literature[8] involves the binary mixture of benzene + cyclohexane solubilized by sodium deoxycholate micelles. In this case, the composition of benzene + cyclohexane solubilized inside the micelles is found to be identical to that of the bulk organic phase in contact with the aqueous surfactant solution. This implies that there is no preference for one or the other of the compounds. It should be noted, however, that in this case, even for single

components, the molar solubilization ratios do not differ
appreciably. Consequently, there is no a priori reason to antici-
pate any selectivity when binary mixtures of benzene + cyclohexane
are solubilized in sodium deoxycholate solutions.

One goal of this paper is to investigate and explain why the
surfactant solutions exhibit high solubilization capacity for some
molecules and low capacity for others. In particular, one would
like to identify the role of the molecular volume of solubilizate,
its polarity, the nature of surfactant tail and of the polar head
group, in determining the molar solubilization ratios. For this
purpose, a thermodynamic treatment of solubilization is developed
linking measurable solubilization characteristics to the molecular
characteristics of the surfactant and solubilizate. Further, one
is especially interested in knowing whether the molecules with high
molar solubilization ratios as single components would also be
selectively solubilized when present in binary or multicomponent
mixtures. Consequently, the second goal of this paper is to determine, both experimentally and on the basis of a thermodynamic treatment, the molar solubilization ratios and the selectivity when
binary hydrocarbon mixtures are solubilized.

In the next section, the experimental results obtained in this
laboratory are presented. In Section III, a thermodynamic treatment for the solubilization of single components and binary mixtures
is developed. In Section IV, the experimental results are interpreted on the basis of the thermodynamic model and the origin of
high or low solubilization capacities for various solubilizates is
explained.

II. EXPERIMENTAL STUDY

The solubilization capacities for benzene, hexane and cyclohexane as single components and as binary mixtures in a number of
surfactant solutions have been measured. The surfactants used
include cetyl pyridinium chloride (CPC), sodium dodecyl sulfate
(SDS), sodium deoxycholate (SDC), dodecyl trimethylammonium
chloride (DTAC), Aerosol AY, and some of their binary mixtures.
The effect of added electrolytes has also been examined. Only
representative results are presented in this paper; more complete
experimental data will be published elsewhere[9].

The solubilization measurements have been carried out as
follows: equal volumes of aqueous surfactant solution and solubilizate phase have been contacted over a period of 24 hours to
insure equilibration. The resulting system was centrifuged to
break any emulsion formed and to separate the aqueous and organic
phases. The composition of the solubilizates in both phases have

been determined by gas chromatography using a flame ionization
detector.   In a few cases, a third emulsion phase has been observed.
The solubilizate composition in this emulsion phase was always
found to be identical to that of the organic phase.   From the gas
chromatographic analysis the total amount of solubilizate in the
aqueous phase was determined.   Assuming that the entire surfactant
is retained in the aqueous phase and accounting for the critical
micelle concentration (CMC) of the surfactant and for the solubility of the hydrocarbon in water, the molar solubilization ratio
(i.e., the moles of solubilizate per mole of surfactant in the
micelle) has been calculated.

Figure 2 presents the molar solubilization ratios in 0.1M
cetyl pyridinium chloride solutions for a number of organic compounds.   As mentioned before, the molar solubilization ratio is
found to bear a relation to the molecular volume v of the solubilizate.   This correlation improves further if in addition to the
molecular volume, one takes into account the polarity of the solubilizate as well.   It should be noted that while cyclohexane and
toluene have comparable molecular volumes, they give rise to
significantly different molar solubilization ratios.   However, if
the polarity of toluene via σ_s is taken into account, then the
figure shows that the correlation is improved.

Figure 3 contains results for the solubilization of benzene +
cyclohexane mixtures in sodium deoxycholate solutions.   The only
other experimental data available in literature[8] for this binary
system are also included for comparison.   The present results,
obtained using a direct gas chromatographic analysis method, are
in reasonable agreement with those from literature; the somewhat
lower values reported in literature[8] are probably due to the limitations of the method of extraction into carbon tetrachloride of
the solubilized micelles, prior to analysis.   Both sets of experimental data show that, in sodium deoxycholate solutions, the molar
solubilization ratios for single component benzene and cyclohexane
phases do not differ appreciably.   These results can be contrasted
to those shown in Figure 2 for cetyl pyridinium chloride solutions.

Figure 4 shows the molar solubilization ratios of benzene and
hexane in cetyl pyridinium chloride (CPC) and sodium dodecyl
sulfate (SDS) solutions as functions of the bulk phase composition.
The selectivity ratio for benzene over hexane in CPC, SDS and DTAC
are shown in Figure 5.   The selectivity ratio for component A in a
binary mixture of A and B is defined as:

$$\text{Selectivity Ratio for A} = \frac{(X_A/X_B) \text{ in micelles}}{(X_A/X_B) \text{ in bulk solubili-}} , \qquad (1)$$
$$\text{zate phase}$$

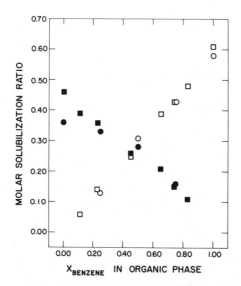

Figure 3.   Molar solubilization ratio of benzene and cyclohexane
in 0.1M sodium deoxycholate solutions as a function of the composition of the bulk solubilizate phase of benzene + cyclohexane.   Open
symbols are benzene and filled symbols are cyclohexane.   Squares
represent present experimental data and circles are from literature[8]
both at 0.01M NaOH and 0.15M NaCl.

where X is the mole fraction.   The experimental data show that
there is no selectivity for benzene or cyclohexane when sodium
deoxycholate is used as surfactant.   In marked contrast, DTAC, CPC
and SDS solutions show significant selectivity for benzene when
the organic phase is hexane-rich.   The selectivity ratio is as
large as seven when solutions of DTAC are contacted with hexane-
rich phases.   The selectivity for benzene is exhibited over the
entire range of composition, the selectivity ratio decreasing as
the mole fraction of benzene increases in the solubilizate phase.
These results are interpreted later on the basis of a thermodynamic
treatment of solubilization which is developed in the paper.

III.   THERMODYNAMICS OF SOLUBILIZATION

    At equilibrium, the aqueous solution consists of singly dis-
persed surfactant and solubilizate molecules, as well as surfactant
aggregates containing solubilizates.   The size distribution of the

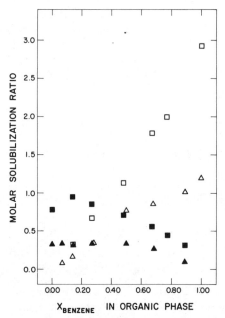

Figure 4. Molar solubilization ratio of benzene and hexane in 0.1M cetyl pyridinium chloride (squares) and in 0.1M sodium dodecyl sulfate (triangles) solutions as a function of the composition of the bulk solubilizate phase of benzene + hexane. Open symbols represent benzene and filled symbols represent hexane.

aggregates containing g surfactant molecules and j' and j" solubilizate molecules of kind s' and s" respectively, is given[10] by

$$X_{gj'j"} = X_1^g X_{1s'}^{j'} X_{1s"}^{j"} \exp\left(-\left(\frac{\mu^o_{gj'j"} - g\mu^o_1 - j'\mu^o_{1s'} - j"\mu^o_{1s"}}{kT}\right)\right).$$ (2)

In the above equation, X is the mole fraction, μ^o is the standard chemical potential in dilute solutions, k is the Boltzmann constant and T is the temperature. The subscripts 1, 1s' and 1s" refer to the singly dispersed surfactant and solubilizates s' and s", respectively. When there is only one solubilizate j"=0 and for surfactant solutions containing no solubilizate j'=j"=0. The molar solubilization ratio of solubilizate j'can be calculated as the ratio

$$\left(\sum_{g>1} \sum_{j'} \sum_{j"} j' X_{gj'j"} / \sum_{g>1} \sum_{j'} \sum_{j"} g X_{gj'j"}\right)$$

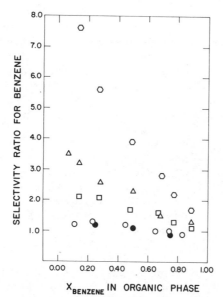

Figure 5.   Selectivity ratio for benzene against the mole fraction
of benzene in the bulk phase.   Open circles: benzene + cyclohexane
solubilized in 0.1M sodium deoxycholate; Filled circles: benzene +
cyclohexane solubilized in 0.1M sodium deoxycholate from literature[8],
Squares: benzene + hexane solubilized in 0.1M sodium dodecyl sulfate; Triangles: benzene + hexane solubilized in 0.1M cetyl
pyridinium chloride; Hexagons: benzene + hexane solubilized in 0.1M
dodecyl trimethyl ammonium chloride.

Since the surfactants used here form only micelles and not vesicles,
the thermodynamic treatment is developed only for micelles.

     The calculations are carried out, in case of Type I solubili-
zation, assuming the micelles to be spherical at small aggregation
numbers and cylindrical with hemispherical ends at large aggregation
numbers.   The transition from spheres to cylinders is considered to
occur when the radius of the micelle becomes equal to the extended
length of the surfactant molecule.   For micelles of intermediate
sizes in the transition region between spheres and cylinders, various
shapes such as ellipsoids, discs, etc., have been suggested[11].   Here,
the intermediate sized micelles are also treated as cylinders with
hemispherical ends.   For Type II solubilization, the micelles are
assumed to be spherical and to contain a core of solubilizate
molecules.

The difference in the standard chemical potential, $\Delta\mu^o =$ $(\mu^o_{gj',j''} - g\mu^o_1 - j'\mu^o_{1s'} - j''\mu^o_{1s''})$, contains the following contributions: (i) the free energy change due to the transfer of the surfactant tail and solubilizates from water to the micelle, $\Delta\mu^o_{Tr}$, (ii) the interfacial free energy at the micellar core-water interface, $\Delta\mu^o_{interface}$, (iii) the free energy due to steric interactions between the surfactant head groups at the micellar surface, $\Delta\mu^o_{steric}$, (iv) electrostatic interactions between the head groups (for ionic surfactants) at the micellar surface, $\Delta\mu^o_{electrostatic}$, and, (v) the free energy change due to the mixing of surfactant and solubilizate molecules inside the micelles, $\Delta\mu^o_{mixing}$ (in the case of Type I solubilization). Expressions for these contributions have been obtained in an earlier paper[10] by using a statistical thermodynamic approach, but are presented here in a simplified form so as to facilitate their use in calculations. A thermodynamic treatment of single component solubilization has been independently developed by Dadyburjor[12] along somewhat similar lines.

Transfer free energy. When singly dispersed surfactant and solubilizate molecules are incorporated into micelles, the hydrocarbon tail of the surfactant and the hydrocarbon solubilizate molecule are transferred from water to an almost liquid hydrocarbon-like environment. This change in environment produces a negative free energy contribution to $\Delta\mu^o$.

$$\Delta\mu^o_{Tr} = g\Delta\mu^o_{Tr,1} + j'\Delta\mu^o_{Tr,1s'} + j''\Delta\mu^o_{Tr,1s''} . \qquad (3)$$

For the transfer of a hydrocarbon molecule from water to a bulk hydrocarbon liquid, the transfer free energy $\Delta\mu^o_{Tr}$ has been experimentally estimated[3,13] to be about $-1.47kT$ per CH_2 group and $-3.25kT$ per CH_3 group for aliphatic hydrocarbons, $-1.83kT$ per CH_2 group for cyclohexane, and $-1.28kT$ per CH group for benzene. However, compared to a hydrocarbon molecule in a bulk hydrocarbon liquid, the hydrocarbon tails of the surfactant constituting the micelle are somewhat constrained, since their polar heads are anchored at the micellar surface[3,14-19]. To estimate the correction to the free energy change $\Delta\mu^o_{Tr,1}$ associated with these conformational constraints, a model was used in which the surfactant molecules in the micelle were free to rotate about their long axis while rotation about their two short axes were forbidden[14,15]. On the basis of such a physical model, the correction to $\Delta\mu^o_{Tr,1}$ due to the conformational constraints has been estimated to be $0.25kT/CH_2$ group and $0.42kT/CH_3$ group of an aliphatic hydrocarbon tail. Taking into account these corrections, $\Delta\mu^o_{Tr,1}$ for the surfactant tail is estimated for aliphatic hydrocarbons by using the following modified group contributions: $-1.22kT/CH_2$ group and $-2.83kT/CH_3$ group. Values of $\Delta\mu^o_{Tr,1}$ can also be estimated using the experimental values of CMC as a function of surfactant tail length for a homologous series of surfactants. On this basis, Tanford[3] has obtained $\Delta\mu^o_{Tr,1}$ of about -800 cal/mole (-1.32kT/CH$_2$ group) for methylene groups.

The estimation of the transfer free energy of the solubilizates depends on the type of solubilization. For Type II solubilization, the solubilizates are in an environment which is identical to their bulk liquid phase. Therefore, the value of $\Delta\mu^o_{Tr,1s}$ is taken equal to that for the transfer of a hydrocarbon molecule from water to liquid hydrocarbon. In contrast, for Type I solubilization, the solubilizates are in an environment somewhat different from that of a liquid hydrocarbon because they are located between the surfactant tails which possess a constrained conformation. This may impose some configurational restrictions on the aliphatic solubilizate molecules inside the micelles compared to their liquid hydrocarbon state. On the other hand, for the aromatic solubilizates which have a more rigid structure, the environment inside the micelles is probably not much different from that in their bulk liquid phase. Therefore, for Type I solubilization of aromatic molecules, $\Delta\mu^o_{Tr,1s}$ is taken equal to that for the transfer of a hydrocarbon molecule from water to liquid hydrocarbon. For Type I solubilization of aliphatic molecules, two extreme cases can be considered. In one, the solubilizate molecules experience inside the micelle an environment identical to that of a hydrocarbon liquid. In the other, the solubilizate molecules experience conformational constraints as severe as those experienced by the surfactant tails. For both extreme cases, the group free energy contributions needed to estimate $\Delta\mu^o_{Tr,1s}$ have been given above. However, since the solubilizates are not as constrained as the surfactant molecules whose polar heads are anchored at the interface, it is likely that their environment is closer to that described by the first extreme case.

Another important consideration for Type I solubilization concerns the location of the solubilizate[20-26]. Again, two limiting cases can be considered. In one, the solubilizate is assumed to be located near the interfacial region of the micelle such that a part remains in contact with a more polar environment. In this case, the magnitude of $\Delta\mu^o_{Tr,1s}$ has to be decreased to account for the incomplete transfer of the solubilizate into the hydrophobic region of the micelles. In the other limiting case, the solubilizate is assumed to be located farther away from the interfacial region of the micelle. In this case, the complete magnitude of $\Delta\mu^o_{Tr,1s}$ can be retained.

Interfacial Free Energy. This positive free energy contribution to $\Delta\mu^o$ arises because of the contact between a part of the hydrophobic core of the micelle and water. This interfacial free energy contribution can be written as the product of the interfacial tension and the area of the hydrocarbon core of the micelle exposed to water.

$$\Delta\mu^o_{interface} = \sigma g(a-a_o) \qquad (4)$$

In the above equation, a is the surface area of the hydrophobic core of the micelle per surfactant molecule and a_0 is the area per surfactant molecule shielded by the polar head from contact with water. σ is taken equal to the interfacial tension between the surfactant tail and water for Type II solubilization and also for those cases of Type I solubilization in which the solubilizate is assumed to be located farther away from the interface. However, when in Type I solubilization the solubilizate is assumed to be located near the interface, σ has to account for the presence of the solubilizate.

Steric Interactions. The micellar surface is crowded with polar head groups of the surfactants. Therefore, the steric repulsions between the head groups at the micellar surface has a positive free energy contribution to $\Delta\mu^o$. An expression for the free energy due to steric repulsions can be written by analogy with those used for excluded volume interactions. One of the simplest forms appears in the van der Waals equation where the volume occupied by the molecules themselves is unavailable for their translation. This excluded volume thus constitutes a measure of steric repulsion between the molecules. Using a two dimensional analog of the above physical picture, one can write the following expression for the free energy due to steric repulsions:

$$\Delta\mu^o_{steric} = -g \, kT \, \ln(1-a_p/a) \ . \tag{5}$$

Here, a_p is the effective cross-sectional area of the polar head group.

Electrostatic Interactions. In ionic surfactant solutions one has to consider also the electrostatic interactions between the head groups at the micellar surface. In the framework of the Debye-Hückel approximation, this repulsive contribution to $\Delta\mu^o$ can be written[3,27] as

$$\Delta\mu^o_{Electrostatic} = \left(\frac{e^2\beta^2g^2}{2\varepsilon r}\right) \left(\frac{1+\kappa a_i}{1+\kappa a_i+\kappa r}\right) \tag{6}$$

for spherical micelles and

$$\Delta\mu^o_{Electrostatic} = \left(\frac{e^2\beta^2g^2}{\varepsilon L}\right) \left(\frac{K_0(\kappa r)}{\kappa r K_1(\kappa r)} + \ln\left(\frac{r+a_i}{r}\right)\right) \tag{7}$$

for cylindrical micelles. In the above expressions, e is the electronic charge, β is the degree of dissociation of the ionic head groups at the micellar surface, ε is the dielectric constant of water, κ is the reciprocal Debye length, a_i is the radius of the counterion, r is the radius of the spherical (or cylindrical) micelle measured at the location of the charged head group, L is the length of the cylindrical micelle and K_0 and K_1 are the modified Bessel functions of order 0 and 1, respectively.

Internal Mixing in Micelles. For Type I solubilization, the surfactant tails mix with the solubilizate molecules, whereas for Type II solubilization, the solubilizates form a core and do not mix with the surfactant tail. Therefore, the internal mixing of surfactant tails and solubilizates provides, for Type I solubilization, another negative free energy contribution to $\Delta\mu^o$. This entropic contribution can be estimated from

$$\Delta\mu^o_{mixing} = kT(g\ \ln\frac{gv_1}{gv_1+j'v_{1s'}+j''v_{1s''}} + j'\ln\frac{j'v_{1s'}}{gv_1+j'v_{1s'}+j''v_{1s''}}$$

$$+ j''\ \ln\frac{j''v_{1s''}}{gv_1+j'v_{1s'}+j''v_{1s''}}) \qquad (8)$$

In the above expression, the entropy of mixing inside the micelles is calculated by taking into account the differences in the molecular volumes of the surfactant tail (v_1) and of solubilizates ($v_{1s'}$ and $v_{1s''}$).

Introducing the various contributions to $\Delta\mu^o$ in Equation (2), one can calculate the micellar size distribution and, consequently, the various solution properties of the system.

IV. RESULTS AND DISCUSSION

Some illustrative calculated results on solubilization in anionic sodium dodecyl sulfate (SDS) and cationic cetyl pyridinium chloride (CPC) solutions are now presented. In these calculations, it is assumed that the degree of dissociation of the ionic head groups at the micellar surface is unaffected by the extent of solubilization.

The concentration of the singly dispersed solubilizate molecules has an upper bound given by its solubility limit. It has been suggested that this solubility limit may be modified somewhat by the presence of the singly dispersed surfactant and its premicellar aggregates[28]. However, in the absence of quantitative solubility data, the solubility limit in the presence of the surfactant is assumed to be the same as that in water. The limiting solubilization capacity in the surfactant solution is calculated for this upper bound of the concentration of singly dispersed solubilizate molecule. For binary systems, the limiting concentrations in water of the singly dispersed solubilizate molecules are taken to be proportional to their mole fractions in the bulk solubilizate phase.

This simplification is satisfactory because of the ideal behavior of the mixtures considered here.

The values of the molecular parameters appearing in the expressions for the various contributions to $\Delta\mu^o$ are listed in Table I and in the Appendix.

For the surfactants and solubilizates used in the present calculations, only Type I solubilization, i.e., solubilization between the surfactant tails, appears to occur.

Table I. Properties of Surfactant and Solubilizate Molecules

	CPC	SDS	Hexane	Cyclo-hexane	Benzene
a_p $(\overset{o}{A}{}^2)$	30	17			
a_i $(\overset{o}{A})$	1.9	1.0			
K (litre/mole)	0.5	1.0			
β^*	0.8	0.7			
v $(\overset{o}{A}{}^3)$	460	352	216	180	162
$\Delta\mu_{Tr}^o$ (in units of kT)	-21.13	-16.26	-12.17	-10.6	-7.68
Solubility in water (mole fraction)	--	--	4.24×10^{-6}	1.7×10^{-5}	4.6×10^{-4}

Single Component Saturated Hydrocarbon Solubilizates. The solubilization capacity of saturated hydrocarbons hexane and cyclohexane in solutions of SDS and CPC have been calculated. Since the solubilizates are highly non-polar, they are assumed to be located farther away from the micellar interface. As mentioned earlier, $\Delta\mu_{Tr,1s}^o$ for these solubilizates is calculated for two extreme cases. If the solubilizates are considered to have in the micelles an environment similar to that of a bulk hydrocarbon liquid, then $\Delta\mu_{Tr,1s}^o$ is -12.38kT per molecule of hexane and -10.98kT per molecule of cyclohexane. On the other hand, if the solubilizates are assumed to be as constrained as the surfactant tails are, then $\Delta\mu_{Tr,1s}^o$ is -10.54kT for hexane and -9.48kT for cyclohexane (assuming a correction due to conformational constraints of 0.25kT/CH$_2$ group which has been estimated for aliphatic hydrocarbon molecule). As anticipated before on intuitive grounds, the calculations based on the latter assumption predict very low solubilization capacities

compared to the available experimental data. The former assumption,
namely, that the solubilizates are in a hydrocarbon liquid like
environment, predicts solubilization capacities which are only 10 to
20 percent larger than those experimentally observed. The modification of the transfer free energy to hydrocarbon liquid by 0.2kT and
0.4kT for hexane and cyclohexane, respectively (the resulting values
are listed in Table I), leads to excellent agreement between the
experimentally observed (Fig. 2) and calculated molar solubilization
ratios. These corrections are a measure of the slight conformational
constraints experienced by the solubilizates in the micelles. The
calculations thus suggest a predominantly hydrocarbon liquid like
environment for these solubilizates. Calculations of molecular conformations such as those of Dill[29] and Gruen[30], but incorporating
solubilizates inside micelles, may be used to examine theoretically
the above observations.

The calculations further show that the average number of surfactant molecules per micelle slightly increases in the presence of
solubilizates. The molar solubilization ratio increases from a low
value near the CMC, to reach its maximum asymptotic value at
sufficiently high concentration of surfactant. This ratio becomes
practically constant when the surfactant concentration is appreciably
higher than the CMC. The calculated values of CMC and the molar solubilization ratio are presented in Table II for both SDS and CPC solutions. The calculated molar solubilization ratios in CPC compare
well with the experimental data shown in Figure 2. The calculated
decrease in the CMC due to the presence of the solubilizates is about
30 to 40 percent, comparable to available experimental data in
literature[31].

Table II.   Solubilization Characteristics of CPC and SDS at 25°C.

(The molar solubilization ratio and the CMC are calculated for a
  surfactant solution saturated with the solubilizates)

	CPC		SDS	
Solubilizate	Molar Solubilization Ratio	CMC (mole fraction)	Molar Solubilization Ratio	CMC (mole fraction)
None	--	2.6×10^{-5}	--	1.55×10^{-4}
Hexane	0.79	1.8×10^{-5}	0.36	1.10×10^{-4}
Cyclohexane	1.03	1.47×10^{-5}	0.50	0.96×10^{-4}
Benzene	2.93	0.71×10^{-5}	1.10	0.68×10^{-4}

Single Component Aromatic Solubilizates. Calculations of the molar solubilization ratio of benzene in solutions of SDS and CPC have been carried out. Because they have some polarity (due to the aromatic ring) in addition to their hydrophobicity, the aromatic solubilizate may be located farther away from the polar interface of the micelle or close to this interface. When they are assumed located away from the interface, the calculations have been carried out using values for $\Delta\mu_{Tr,1s}^{o}$ given in Table I (which correspond to $\Delta\mu_{Tr}^{o}$ for the transfer to liquid hydrocarbon) and a value of 50 dyne/cm for the interfacial tension σ at the micellar core-water interface. The calculated molar solubilization ratios listed in Table II compare well with our experimental data plotted in Figure 2. The calculated CMC, for CPC and SDS with and without the solubilizate are also listed in Table II. Again, the depression in CMC in the presence of solubilizates is about 50 percent, comparable to typical experimental values available in literature.

The possibility that the solubilized benzene molecules are located near the micellar interface was also considered. In this case a part of the solubilizate molecules remains exposed to water instead of being in a highly hydrophobic environment. Consequently, the magnitude of $\Delta\mu_{Tr,1s}^{o}$ has to be decreased with respect to that indicated in Table I. Further, because of the location near the interface of the aromatic solubilizates, the value of the interfacial tension σ has also to be reduced somewhat from the value of 50 dyne/cm. The larger the reduction in the magnitude of $\Delta\mu_{Tr,1s}^{o}$, the lower should be the corresponding value of σ. Of course, the smallest value which can be taken for σ is 34 dyne/cm, i.e., the benzene-water interfacial tension. Since $\Delta\mu_{Tr,1s}^{o}$ becomes less negative and σ less positive, the two effects at least in part compensate one another.

Model calculations have been carried out here to examine the solubilization of benzene in solutions of CPC for arbitrary values of $\Delta\mu_{Tr,1s}^{o}$ and σ. The calculated molar solubilization ratio and the CMC of CPC in the presence of solubilized benzene are presented in Table III. It can be seen that for interfacial tensions less than 50 dyne/cm and values of $\Delta\mu_{Tr,1s}^{o}$ smaller in magnitude than those listed in Table I, one can obtain reasonable results for the molar solubilization ratio and for the depression in CMC caused by the solubilizates. These calculations show that the experimentally measured molar solubilization ratio can be satisfactorily explained by either of the assumptions regarding the location of benzene inside the micelles. Some calculations discussed later in the paper indicate that it is likely that benzene does occupy both locations. Such a model involving a part of the solubilized benzene near the interface and the remaining located away from the interface can predict satisfactorily both the observed molar solubilization ratio and the depression in the CMC. It is worthwhile to note that many experiments[20-24] indicate that aromatic molecules are solubilized

near the interface while some observations[26] suggest that aromatic
molecules are solubilized in the micellar interior.

Table III. Solubilization of Benzene in CPC for Various Assumed
            Values of $\Delta\mu^{o}_{Tr,1s}$ and σ at 25°C

σ (dyne/cm)	$\Delta\mu^{o}_{Tr,1s}$ (in units of kT)	Molar Solubilization Ratio	CMC (in mole fraction)
Benzene is assumed located farther away from the interface			
50	-7.68	2.93	0.71×10^{-5}
Benzene is assumed located near the interface			
45	-7.53	2.92	0.40×10^{-5}
45	-7.48	2.57	0.45×10^{-5}
40	-7.48	3.5	0.16×10^{-5}
40	-7.28	2.1	0.26×10^{-5}
35	-7.48	5.1	0.50×10^{-6}
35	-7.28	2.9	0.90×10^{-6}

    Binary Solubilizate Mixtures. The experimental results as well
as the calculated ones for single solubilizates show that the molar
solubilization ratio decreases in the order benzene > cyclohexane >
hexane. In order to examine whether these differences in solubilization capacities are retained in binary systems, calculations have
been carried out for benzene + hexane, benzene + cyclohexane, and
hexane + cyclohexane mixtures in CPC and SDS solutions. While
hexane and cyclohexane are assumed to be located in a highly non-
polar environment, for benzene, two extreme locations inside the
micelles as described earlier are considered. When benzene is
assumed located farther away from the interface, the calculations
have been carried out using $\Delta\mu^{o}_{Tr,1s}$ = -7.68 kT and σ = 50 dyne/cm.

    When benzene is assumed located near the interface, the calcu-
lations have been performed with $\Delta\mu^{o}_{Tr,1s}$ = -7.53 kT and σ = 45 dyne/
cm. One may note again that both of the above assumptions predict
satisfactory values of the molar solubilization ratio for pure
benzene (see Table III).

    The calculated molar solubilization ratios in CPC micelles are
plotted against composition, for the benzene + hexane mixture, in

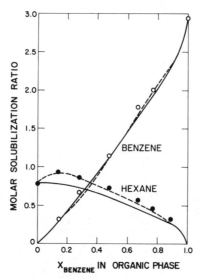

Figure 6.   Calculated molar solubilization ratio of benzene and
hexane in 0.1M cetyl pyridinium chloride.   The continuous line is
based on the assumption that benzene is located farther away from
the interface and the dotted line is based on the assumption that
some benzene molecules are located near the micellar interface.
Open circles are experimental molar solubilization ratio of benzene
and the filled circles are the experimental data for hexane.

Figure 6.   For comparison, the experimental values presented in
Figure 4 are also replotted in Figure 6.   One may observe that the
experimental data show a synergistic effect for the solubilization
of hexane in the presence of benzene, namely, the molar solubilization ratio of hexane does not decrease with decreasing mole fraction
of hexane in the bulk solubilizate phase.   While the molar solubilization ratio of hexane in CPC micelles is 0.79 when the organic phase
is 100% hexane, the ratio increases to about 0.89 when the organic
phase is 90% hexane and 10% benzene.   If one assumes that benzene is
farther away from the interface, it is difficult to explain this
synergistic effect.   For, in this case, there are deviations between
the calculated results and the experimental data.   However, if one
assumes that the solubilized benzene is located near the interface,
the calculations predict well the observed synergistic effect.
This supports the assumption that benzene is at least partly located
near the micellar interface.

A significant prediction of the present calculations is the selective solubilization of benzene compared to hexane or cyclohexane and of cyclohexane over hexane. The calculated results are plotted in Figure 7 for the three binary mixtures solubilized in 0.1M CPC and 0.1M SDS solutions. The data are represented as selectivity ratio for one component over another (defined by Equation (1)) against the composition of the organic phase. The selectivity ratio for benzene over hexane is comparable to the experimentally measured values presented in Figure 5. The selectivity ratio for benzene is greater than unity over the entire range of composition and particularly for hexane-rich conditions. If the three binary mixtures are compared, then the maximum selectivity ratio is shown by benzene + hexane and the smallest selectivity is shown by cyclohexane + hexane. Thus, the selectivity ratio parallels the molar solubilization ratios of single component solubilizates, the selectivity ratio being the largest when the two components have the widest difference in their molar solubilization ratios.

Origin of Selectivity in Solubilization. The experimental results on the solubilization of single component solubilizates shown in Figure 2 suggest that the ratio of solubilizate to surfactant in a micelle increases with decreasing molecular volume and increasing polarity of the solubilizate. In the framework of the thermodynamic treatment, the incorporation of a solubilizate in the micelle alters the area per surfactant molecule of the micelle giving rise to two opposing effects: the positive interfacial free energy contribution to $\Delta\mu^o$ increases (when σ does not change) whereas the steric and the electrostatic repulsions between the head groups at the micellar surface decrease. The same increase in the area per surfactant molecule of micelle can be caused either by a larger number of solubilizates with a low molecular volume or by solubilizing fewer solubilizates with larger molecular volumes. The former possibility also gives rise to a larger negative contribution to $\Delta\mu^o$ arising from the entropy of internal mixing in micelles. These explain why the smaller solubilizate molecules are preferentially solubilized. This would also explain the experimental observation[4] that the molar solubilization ratios of various isomers of a hydrocarbon molecule are comparable. When the solubilizations of two aliphatic straight chain hydrocarbons are compared, then the longer chain experiences a greater restriction on its conformational flexibility inside the micelles than the shorter chain does. This effect (in addition to the molecular volume effect mentioned above) modifies the magnitude of the transfer free energy $\Delta\mu^o_{Tr,1s}$ and significantly reduces the solubilization ratio for longer chain hydrocarbons. The experimental molar solubilization ratios for n-alkanes reflect the above trend.

Aromatic compounds are solubilized in larger amounts than the corresponding saturated hydrocarbon. The larger solubilization

capacity for benzene as compared to hexane is partly due to the
smaller molecular volume of the former. The molecular restrictions
inside the micelle compared to liquid hydrocarbon state reduces·
further the capacity of hexane to be solubilized. An additional contributing factor is the polarity of benzene which reduces somewhat

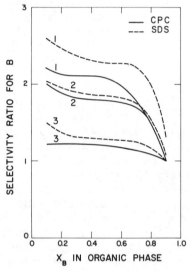

Figure 7. Calculated selectivity ratio of component B as a function
of the mole fraction of B in the bulk phase for cetyl pyridinium
chloride (———) and sodium dodecyl sulfate (----) solutions.
(1) Component A is hexane, B is benzene; (2) Component A is cyclohexane and B is benzene; (3) Component A is hexane and B is cyclohexane.

the micellar core-water interaction energy thus permitting larger
solubilization of benzene. The latter two factors can be invoked
to explain the higher molar solubilization ratio for toluene compared to cyclohexane which has about the same molecular volume.

Naturally, the factors affecting the molar solubilization ratios of single components are operative when binary mixtures are considered. The experimental results shown in Figure 4 and Figure 5 for the solubilization of benzene + hexane mixtures in CPC and SDS solutions clearly show the selective solubilization of benzene, especially for organic phase compositions that are hexane-rich. The calculated selectivity ratios plotted in Figure 7 indicate that benzene is preferred to hexane. In general, the selectivity for a component in binary systems closely parallels the trend observed in molar solubilization ratios of single components.

In solutions of sodium deoxycholate the molar solubilization ratios for cyclohexane and benzene are not too different from one another. Consequently, no selectivity is observed when their binary mixtures are solubilized. The solubilization behavior can be explained in this case in terms of the markedly different structure and micellization characteristics of sodium deoxycholate, compared to those of SDS and CPC. Sodium deoxycholate forms very small primary aggregates[32-35] with an aggregation number in the range of 4 to 10. These primary aggregates can further associate to form secondary aggregates of large aggregation numbers. Solubilization occurs inside the primary aggregates and not in the more polar region between the primary aggregates. Therefore, the factors governing the formation of primary aggregates rather than that of the secondary aggregates are relevant here. The primary aggregates are generated when the hydrophobic steroid surfaces of sodium deoxycholate are packed close to each other such that the polyhydroxylated surfaces are in contact with water. The forces responsible for such an association are relatively weak compared to those involved in the micellization of typical surfactants such as SDS and CPC. This is reflected in the very small aggregation numbers of the primary micelles of sodium deoxycholate. The incorporation of the solubilizates inside the primary aggregates hinders the close packing of the deoxycholate molecules thus exposing very large hydrocarbonaceous areas to water. As a result, the total number of molecules solubilized per micelle is rather small, irrespective of the nature of the solubilizate. Indeed, the experimentally observed molar solubilization ratio is 0.44 for cyclohexane and 0.56 for benzene which implies that each primary aggregate can incorporate only 2 to 4 molecules of benzene or cyclohexane.

V.   CONCLUSIONS

Using a direct gas chromatographic analysis method, the molar solubilization ratios for a number of solubilizates as single components and in binary mixtures have been determined experimentally. It is shown for the first time, that molecules can be selectively solubilized from binary hydrocarbon mixtures. Large selectivity ratios of up to about seven have been obtained for benzene compared to hexane especially when the binary mixture is hexane-rich. In general, the selectivity ratio of a binary mixture depends on the difference in the molar solubilization ratios of the single components, the selectivity is the largest when the two solubilizates have the widest difference in their molar solubilization ratios.

A thermodynamic treatment is developed and used to predict the molar solubilization ratio of various compounds singly and in mixtures. The treatment reveals the importance of the molecular volume and polarity of the solubilizate as well as the role of molecular constraints experienced by the molecules inside the micelles in determining the observed molar solubilization ratios.

Further, the absence of any selectivity for benzene or cyclohexane in solutions of sodium deoxycholate is shown to be consistent with the comparable molar solubilization ratios of the two pure components. This markedly different solubilization behavior is interpreted in terms of the peculiar molecular structure of sodium deoxycholate and its ability to generate only very small primary aggregates.

Finally the solubilization data involving benzene + hexane mixtures are used to explore the location of benzene inside the micelles. The synergistic effect observed experimentally is shown to imply, in the framework of the thermodynamic treatment, that at least some of the benzene molecules are located close to the interface.

ACKNOWLEDGEMENTS

We are indebted to Mr. Mark A. Chaiko for performing the experiments. This work was supported by the National Science Foundation.

REFERENCES

1. M.E.L. McBain and E. Hutchinson, "Solubilization and Related Phenomena," Academic Press, New York, 1955.
2. P. H. Elworthy, A. T. Florence and C. B. McFarlane, "Solubilization by Surface Active Agents," Chapman and Hall, London, 1968.
3. C. Tanford, "The Hydrophobic Effect," Wiley, New York, 1973.
4. J. W. McBain and P. H. Richards, Ind. Eng. Chem., 38, 642 (1946).
5. P. H. Richards and J. W. McBain, J. Am. Chem. Soc., 70, 1338 (1948).
6. H. B. Klevans, Chem. Rev., 47, 1 (1950).
7. R. C. Stearns, H. Oppenheimer, E. Simon and W. D. Harkins, J. Chem. Phys., 15, 496 (1947).

8. D. C. Thomas and S. D. Christian, J. Colloid Interface Sci.,
    82, 439 (1981).
9. M. A. Chaiko, R. Nagarajan and E. Ruckenstein, J. Colloid
    Interface Sci. (to be submitted).
10. R. Nagarajan and E. Ruckenstein, Sep. Sci. Technol., 16, 1429
    (1981).
11. J. N. Israelachvili, D. J. Mitchell and B. W. Ninham, J. Chem.
    Soc., Faraday Trans. II, 72, 1525 (1976).
12. R. Mallikarjun and D. B. Dadyburjor, J. Colloid Interface Sci.,
    84, 73 (1981).
13. G. Nemethy and H. A. Scheraga, J. Chem. Phys., 36, 3401 (1962).
14. R. Nagarajan and E. Ruckenstein, J. Colloid Interface Sci., 71,
    580 (1979).
15. R. Nagarajan and E. Ruckenstein, J. Colloid Interface Sci., 60,
    221 (1977).
16. K. A. Dill and P. J. Flory, Proc. Natl. Acad. Sci., 78, 676
    (1981).
17. K. A. Dill, J. Phys. Chem., 86, 1498 (1982).
18. D. W. R. Gruen, J. Colloid Interface Sci., 84, 281 (1981).
19. D. W. R. Gruen, Biochim. Biophys. Acta, 595, 161 (1980).
20. J. R. Cardinal and P. Mukerjee, J. Phys. Chem., 82, 1614 (1978).
21. P. Mukerjee, in "Solution Chemistry of Surfactants,"K.L. Mittal
    Editor, Vol. 1, pp. 153-173, Plenum Press, New York (1979).
22. P. Mukerjee, J. R. Cardinal and N. Desai, in "Micellization,
    Solubilization and Microemulsions", K. L. Mittal, Editor,
    Vol. 1, pp. 241-261, Plenum Press, New York (1977).
23. C. Hirose and L. Sepulveda, J. Phys. Chem., 85, 3689 (1981).
24. M. Almgren, F. Grieser and J. K. Thomas, J. Am. Chem. Soc.,
    101, 279 (1979).
25. D. W. R. Gruen and D. A. Haydon, Pure Appl. Chem., 52, 1229
    (1980).
26. S. A. Simon, R. V. McDaniel and T. J. McIntosh, J. Phys. Chem.,
    86, 1449 (1982).
27. C. Tanford, J. Phys. Chem., 78, 2469 (1974).
28. P. Somasundaran and B. M. Moudgil, J. Colloid Interface Sci.,
    47, 290 (1974).
29. K. A. Dill. These proceedings.
30. D. W. R. Gruen and Emma H. B. DeLacey. These proceedings.
31. K. Shinoda, in "Colloidal Surfactants," K. Shinoda, T. Nakagawa,
    B. Tamamushi and T. Isemura, Editors, Academic Press, New York,
    1963.
32. N. A. Mazer, M. C. Carey, R. F. Kwasnick and G. B. Benedek,
    Biochemistry, 18, 3064 (1979).
33. N. A. Mazer, G. B. Benedek and M. C. Carey, Biochemistry, 19,
    601 (1980).
34. M. C. Carey, J. C. Montet, M. C. Phillips, M. J. Armstrong and
    N. A. Mazer, Biochemistry, 20, 3637 (1981).
35. D. G. Oakenfull and L. R. Fisher, J. Phys. Chem., 81, 1838
    (1977).

APPENDIX

Notations and Definitions

a_p = cross-sectional area of the polar head group of the surfactant.

a_o = area per surfactant molecule shielded from contact with water = 21 $\mathring{A}^2$.

C_1 = molar concentration of singly dispersed surfactant molecules, moles/litre.

C_{add} = molar concentration of added salt, moles/litre.

e = electronic charge = 4.8×10^{-10} esu.

g = number of surfactant molecules in a micelle.

k = Boltzmann constant = 1.38×10^{-16} erg/$^\circ$K.

K = equilibrium constant for counterion binding.

l_o = length of the hydrocarbon tail of surfactant = $(1.50 + 1.269 \ N_c)$ in $\mathring{A}$.

N_c = number of CH_2 and CH_3 groups in the hydrocarbon chain.

T = absolute temperature, $=298^\circ$K in the present calculations.

β = degree of dissociation of the ionic head groups at the micellar surface, = $\beta^*/\{1 + K(C_1 + C_{add})\}$.

β^* = the value of β at zero ionic strength.

δ = distance of separation between the hydrophobic surface of the micelle and the location of charge on the polar head group of the surfactant.

ε = dielectric constant of water, = 80.

κ = reciprocal Debye length, = $(C_1 + C_{add})^{1/2}/(3.08 \times 10^{-8})$ cm^{-1} at 25°C.

Geometrical Properties of Micelles

For spherical micelles (Type I solubilization):

$$r = r_o + \delta = [3 \ (gv_1 + j'v_{1s'} + j''v_{1s''})/4\pi]^{1/3} + \delta$$

$$r_o \leq l_o$$

$$a = 4\pi r_o^2/g$$

For spherical micelles (Type II solubilization):

$$r = r_o + \delta = [3(gv_1 + j'v_{1s'} + j''v_{1s''})/4\pi]^{1/3} + \delta$$

$$a = 4\pi r_o^2/g$$

For cylindrical micelles (Type I solubilization):

$$r = 1_o + \delta$$

$$L = [(gv_1 + j'v_{1s'} + j''v_{1s''}) - \frac{4}{3} \pi 1_o^3]/\pi 1_o^2$$

$$a = (4\pi 1_o^2 + 2\pi L 1_o)/g$$

THERMODYNAMICS AND MECHANISMS OF SOLUBILIZATION OF ALCOHOLS IN AQUEOUS IONIC SURFACTANT SYSTEMS

H. Høiland[1], O. Kvammen[1], S. Backlund[2], and K. Rundt[2]

1. Department of Chemistry, University of Bergen, N-5014 Bergen-U., Norway. 2. Department of Physical Chemistry, Åbo Akademi, SF-20500 Åbo 50, Finland

Solubility, density, viscosity, ultrasound, electromotive force, and conductivity measurements at various temperatures have been used to study the properties of solutions containing ionic surfactant-alcohol micelles. The surfactants studied were sodium-dodecylsulfate (NaDDS), hexadecyltrimethylammonium bromide (HTAB), sodium decanoate (NaC_{10}), and di-sodium 2-carboxytetradecanoate (Na_2C_{15}). The alcohols were 1-pentanol to 1-decanol. The alcohols are distributed between the micellar and aqueous phases and the distribution coefficients have been determined. The distribution coefficients as well as the partial molar volumes show that the larger the alcohol the larger the proportion of alcohol in the micellar phase. The mean molar standard Gibbs' energy for the transfer of alcohol from the aqueous to the micellar phase was determined to be -2.78 kJ mol^{-1} per methylene group at 298.2 K. When pentanol or hexanol were added to NaDDS solutions most of the associated counter ions were released from the micelles. Both pentanol and hexanol promote structural changes in HTAB micelles as does hexanol in concentrated solutions of NaDDS micelles. Addition of alcohols lowered the critical micelle concentration of the surfactant.

INTRODUCTION

One of the most important properties of micellar solutions is their ability to solubilize substances that are otherwise insoluble or only slightly soluble in water. The primary question in this field concerns the amount of substance that can be solubilized by a micellar solution of a specific surfactant at a specified concentration. The first systematic studies of this kind were carried out in the 1940's by McBain and Richards[1] and Harkins et al.[2,3]

From a physico-chemical point of view one of the most interesting questions concerns the mechanisms of solubilization. In order to elucidate this question it appears necessary to look at the partitioning of the solubilizate between the micelles and their aqueous surroundings as well as the location of the solubilizate in the micelle. It is also important to investigate the effects of the solubilizate on the micelles, i. e. to what extent does the solubilizate cause structural changes of the micelles or influence the micellar aggregation number. Changes in the critical micelle concentration, c.m.c., and the counter-ion association are also of considerable importance.

In this paper we shall deal with the solubilization of n-alcohols in aqueous solutions containing ionic micelles, The attention is focused on the issues mentioned above.

MATERIAL USED

Pentanol, hexanol, heptanol, octanol, nonanol, and decanol were all supplied by Fluka (their highest purity). They were used without further purification. In the following they will be termed C_5OH for pentanol to $C_{10}OH$ for decanol (all normal alcohols).

Table I. Solubility of Hexanol in HTAB Solutions at the c.m.c. compared to the Solubility of Hexanol in Water.

T / K	m_{C_6OH} / mol kg^{-1} (water)		
	intercept	measured	literature
288.2		0.0626	0.0656
293.2	0.061		0.0611
298.2	0.058	0.0585	0.0581
308.2	0.054	0.0548	0.0541

Sodium dodecysulfate, termed NaDDS, was supplied by BDH-especially pure. Sodium decanoate, NaC_{10} was prepared as previously described.[4] Hexadecyltrimethylammonium bromide, HTAB, was supplied by Sigma. Disodium 2-carboxytetradecanoate, Na_2C_{15}, was prepared as previously described.[5]

ENERGETICS OF MICELLAR SOLUBILIZATION

The distribution coefficient of a solubilizate between the micellar and aqueous phases can be defined as:

$$K = \frac{x^{mic}_{solub}}{x^{aq}_{solub}} = \frac{\dfrac{n^{mic}_{solub}}{n^{mic}_{solub} + n^{mic}_{surf}}}{\dfrac{n^{aq}_{solub}}{n^{aq}_{solub} + n^{aq}_{surf} + n_{H_2O}}} \qquad (1)$$

Here x^{mic}_{solub} and x^{aq}_{solub} are the mole fractions of the solubilizate in the micellar and aqueous phases, respectively. The subscript surf refers to surfactant. By rearranging Equation (1) the following expression can be obtained:

$$m^{tot}_{solub} = \frac{K M_{H_2O} m^{aq}_{solub}}{1 - K M_{H_2O} m^{aq}_{solub}} (m^{tot}_{surf} - m^{cmc}_{surf}) + m^{aq}_{solub} \qquad (2)$$

m_i is the content of component i expressed as mol per kg water. The superscript tot and cmc refers to the total content and the content at the c.m.c., respectively.

In a previous work[6] Equation (2) was used to calculate the distribution coefficient of hexanol in NaDDS and HTAB solutions saturated with respect to hexanol. It was shown that the total content of hexanol was indeed a linear function of the total micellar surfactant content, $m^{tot}_{surf} - m^{cmc}_{surf}$. According to Equation (2) the intercept gives the solubility of the solubilizate in the aqueous phase. This provides a method to check the validity of the model. In Table I the intercepts are presented and compared with measured solubilities in water and with literature data. The agreement between the solubility of hexanol in HTAB solutions at the c.m.c. and the solubility in water is excellent as it should be.[7]

Assuming ideal behaviour of the solubilizate in the micelles and in the aqueous phase, the molar standard Gibbs' energy can be calculated for the process:

$$\text{solubilizate(aq)} = \text{solubilizate(mic)} \qquad (3)$$

For Process (3):

$$\Delta G_m^0 = - RT \ln K = - RT \ln \frac{x_{solub}^{mic}}{x_{solub}^{aq}} \qquad (4)$$

The results for n-alcohols in various surfactant solutions at 298.2 K are summarized in Figure 1. The linearity is remarkably good, and ΔG_m^0 appears to be rather independent of the surfactant. The slope of the straight line in Figure (1) is -2.78 kJ mol^{-1}. This is the standard Gibbs' energy per methylene group. This value can be compared to Stilb's value of -2.6 kJ mol^{-1} per methylene group for the transfer of n-alcohol molecules from the D_2O phase to the NaDDS micellar phase.[9] Wishnia[10] has given a value of -3.23 kJ mol^{-1} for gaseous alkanes. The results are seen to agree reasonably well. It is well established that the transfer of an amphiphile from the aqueous medium to the core of the micelle is, in principle, analogous to the transfer of an alkyl chain from an aqueous medium to a nonpolar solvent,[11] thus reflecting the liquid-like nature of the micellar interior.

Molyneux and Rhodes[12] prefer to look at the process:

Solubilizate(pure state) ⇌ Solubilizate(mic)   (5)

From the data of Bell[13] it is possible to calculate the molar standard Gibbs' energy per methylene group for the transfer of n-alcohol from water to the pure state. The result is -3.31 kJ mol^{-1}. For the corresponding process Tanford[11] gives -3.70 kJ mol^{-1} for n-alkanes.

For some of the systems studied ΔG_m^0 have been determined at several temperatures. The molar enthalpy and the molar entropy can thus be estimated:

$$\Delta H_m^0 = \partial(\frac{\Delta G_m^0}{T}) \Big/ \partial(\frac{1}{T}) \qquad (6)$$

and

$$\Delta S_m^0 = \frac{\Delta H_m^0 - \Delta G_m^0}{T} \qquad (7)$$

The thermodynamic parameters that control the solubilization equilibria are given in Table II.

The solubilities of alcohols in aqueous micellar solutions can be obtained from density measurements as described earlier.[6] In Figure 2 the total content of various alcohols in NaDDS

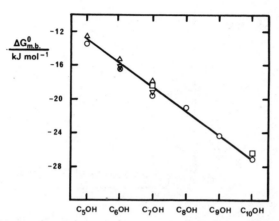

Figure 1. Molar standard Gibbs'energy of the Process(3) for n-alco--
hols in different surfactant solutions as a function of chain
length. NaDDS (O); NaC (Δ); HTAB (x); Na_2C_{15} (∇); and KC_{14} (□).
the latter from Reference 14. The temperature is 298.2 K.

Table II. Energetics of Solubilization of Pentanol, Hexanol, and
Heptanol in Different Surfactant Systems at 298.2 K.

Surfactant	Alcohol	ΔG_m^o kJ mol^{-1}	ΔH_m^o kJ mol^{-1}	ΔS_m^o kJ mol^{-1} K^{-1}
NaC_{10}	C_5OH	-12.5	3.8	55
	C_6OH	-15.2	9.6	83
	C_7OH	-17.9	4.4	75
NaDDS	C_6OH	-16.5	4.1	69
HTAB	C_6OH	-16.5	5.6	74
Na_2C_{15}	C_6OH	-16.3	---	--
	C_7OH	-19.2	---	--

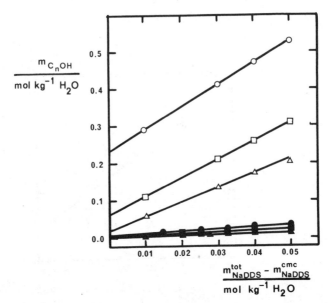

Figure 2. The solubilities of different alcohols as a function of the micellar content of NaDDS at 303.2 K. C_5OH (O); C_6OH (□); C_7OH (Δ); C_8OH (●): C_9OH (■); $C_{10}OH$ (▲)

solutions have been plotted versus m_{NaDDS}^{tot} according to Equation (2). It can be seen from Figure 2 that the solubilities of the alcohols C_8OH to $C_{10}OH$ in NaDDS solutions are small and that a marked increase appears going from octanol to heptanol. The same tendency can be observed from the data of Harkins and Oppenheimer,[3] for alcohols in 0.3m NaDDS. In our studies of the influence of alcohols on the aggregation number of the micelles,[14] it was found that water soluble alcohols (methanol to butanol) decrease the aggregation number. Moderately soluble alcohols (pentanol to heptanol) had little effect on the aggregation number, and sparingly soluble alcohols (octanol to decanol) increase the aggregation number. Thus it appears that the effects of alcohols on the aggregation number of micelles also depends on the solubility of the alcohol in water.

If we inspect Figure 2 with respect to the ratio of alcohol to surfactant, it turns out to be larger than one for pentanol, hexanol, and heptanol. For the systems NaDDS-C_6OH it appears that approximately 5 moles of hexanol can be solubilized per mole NaDDS. Such a high ratio has also been reported by other workers for aqueous NaDDS or HTAB systems.[15-19] The pertinent question is: Where is this large amount of alcohol situated? It appears as a possibility that a large portion of the alcohol is situated in the core of the micelles. This idea, which was loosely founded on the basis of the total solubility of these alcohols in alkanes, has recently found support in the light scattering work of Candau and Zana.[20] Thus, in our opinion the solubilization sites are not fixed. There may be an equilibrium between various sites in the micellar core and at the micellar surface. There is surely also a rapid equilibrium between alcohols in the micelle and in the surrounding aqueous medium, and hence the importance of the alcohol solubility in water.

For sodium decanoate and disodium 2-carboxytetradecanoate the ratio of alcohol to surfactant is only about one; as illustrated for the NaC$_{10}$ system in Figure 3. It thus appears that the polar group of the surfactant is important regarding the amount of alcohol that can be solubilized per mole of surfactant. Possibly the ability of the caboxylic group to form hydrogen bonds with the hydroxyl group of the alcohols is significant in this respect.

MOLAR VOLUMES OF ALCOHOLS

In order to obtain more information about the localization of moderately soluble alcohols in different media, the molar volumes of the alcohols have been evaluated. The apparent molar volumes can be calculated from the equation:

$$V_\phi = \frac{\rho^* - \rho}{m_{C_nOH}\rho\rho^*} + \frac{M_{C_nOH}}{\rho} \tag{8}$$

In Equation (8) ρ^* and ρ are the densities of the solvent and the solution, respectively.

For each alcohol V_ϕ was plotted as a function of m_{C_nOH}. V_ϕ was found to be nearly independent of m_{C_nOH}, and the value at infinite dilution V_2^∞, was determined for each alcohol in different micellar solutions. The partial molar volume of the alcohol infinitely diluted in a surfactant solution at the c.m.c., $V_{2(cmc)}^\infty$, could be evaluated by another plot of V_2^∞ against the surfactant molality extrapolated to the c.m.c.. This is illustrated in

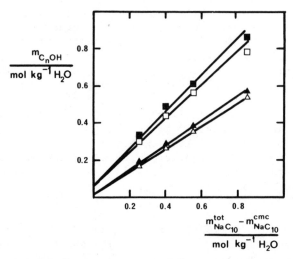

Figure 3. The solubility of hexanol (□) and heptanol (Δ) in micellar sodium decanoate at 298.2 K (open symbols) and at 308.2 K (filled symbols).

Figure 4. The partial molar volumes of the alcohols in water[21], $V^{\infty}_{2(aq)}$, and the molar volumes of the alcohols in the pure state, V^*_2 have been plotted in the same figure. From Figure 4 it turns out that $V^{\infty}_{2(aq)} < V^{\infty}_{2(cmc)} < V^*_2$. The difference between $V^{\infty}_{2(aq)}$ and $V^{\infty}_{2(cmc)}$ is smallest for pentanol and increases with chain length of the alcohol. It shows that the higher the alcohol, the larger the proportion of alcohol in the micellar state. This, of course, is in accordance with increasing distribution coefficients or decreasing molar standard Gibbs' energy of the Process (3).

CHANGES IN MICELLAR SHAPE FROM VISCOSITY AND ULTRASONIC STUDIES

         In a typical case the micelles are nearly spherical in shape over a rather wide concentration range above the c.m.c.. However, it is well known that addition of a third component, electrolyte or nonelectrolyte, can cause structural changes like the formation of rod-like micelles. Such changes lead to dramatic increase in the viscosity of the solutions.[18,22]

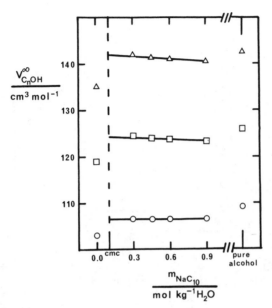

Figure 4. V_2^∞ plotted against the molality of sodium decanoate at 303.2 K. Pentanol (0), hexanol (□), heptanol (Δ). V_2^* and $V_{2(aq)}^\infty$ have also been plotted as the points to the far right and far left respectively.

    In order to elucidate the influence of hexanol on the shape of NaDDS micelles the viscosities of various NaDDS solutions, with and without hexanol, were measured by Ubbelohde viscometers. The results are presented in Figure 5 where the relative viscosity has been plotted versus the volume fraction of NaDDS for the binary system NaDDS in water and versus the volume fraction of NaDDS plus hexanol for the ternary system water-NaDDS-hexanol. The volume fractions have been calculated according to Clarke and Hall.[23]

    In pure NaDDS the micelles are spherical,[24] and the relative viscosity is a linear function of the volume fraction. The intercept is close to unity. The data for 0.05 and 0.1m NaDDS with hexanol added fall on the same straight line indicating no change

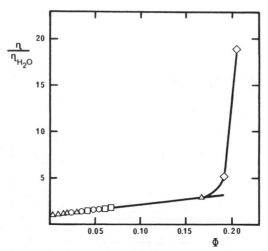

Figure 5.   Relative viscosity as a function of the volume fraction
of pure NaDDS solutions (△), 0.05m NaDDS + C_6OH (O), 0.1m NaDDS +
C_6OH (□), and 0.6m NaDDS + C_6OH (◇).   The temperature is 298.2 K.

in the micellar shape.   However, by increasing the NaDDS molality
to 0.6m a rapid increase in the relative viscosity is observed as
hexanol is added.   It appears that hexanol can cause structural
changes in NaDDS micelles, but only at relatively large molalities.

It has previously been shown[25] that ultrasound measurements can be used to monitor sphere to rod transitions in HTAB
solutions.   For HTAB in water a transition point around 0.3m was
observed.   In Figure 6 the relative speed of sound (relative to the
speed of sound of the solvent) in HTAB solutions with added hexanol
has been plotted versus the HTAB molality at 303.2 K.   The transition point is clearly shifted towards lower HTAB molalities as the
hexanol content increases.[26] This is in good agreement with the
results of Lindblom et al.[26]   For comparison, the data for HTAB in
0.01m pentanol have also been plotted in Figure 6.   It can be seen
that hexanol is clearly the more efficient additive in shifting
the sphere to rod transition point toward lower HTAB molalities.

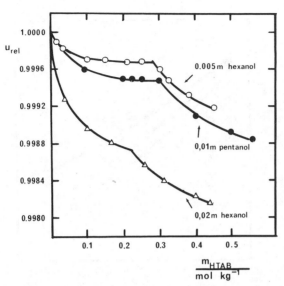

Figure 6. The relative speed of sound in HTAB + alcohol solutions
as a function of HTAB molality at 303.2 K.

EFFECTS OF ALCOHOLS ON THE COUNTER-ION ASSOCIATION AND THE C.M.C.

        The degree of counter-ion association in aqueous NaDDS can
be defined as:

$$\beta = \frac{m_{NaDDS}^{tot} - m_{Na^+}^{tot}}{m_{NaDDS}^{tot} - m_{DDS^-}^{aq}} \qquad (9)$$

The molality of free sodium ions in water, $m_{Na^+}^{aq}$, has been determined
from electromotive force measurements using a sodium responsive
glass electrode as previously described.[19] The molality of monomeric dodecylsulfate has been taken as equal to the molality at the
c.m.c. in a NaDDS solution containing a specified amount of added
alcohol. The results for NaDDS with added pentanol and hexanol at
298.2 K are shown in Figure 7. The observed decrease in β with
alcohol addition means that sodium ions are released from the
micellar surface. The higher the alcohol content, the more sodium
ions are released. The results also show that the higher alcohol
is the most efficient when it comes to counter-ion dissociation.

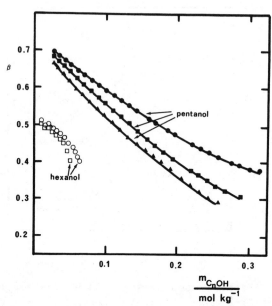

Figure 7. The degree of counter-ion association as a function of
alcohol molality in solutions containing NaDDS; 0.0133m (●); 0.0223m
(■); 0.0414m(▲); 0.064m (O); and 0.081m (□). The temperature was
298.2 K.

        The last point to be considered is the effects of alcohols
on the c.m.c. of the surfactants. Numerous investigations have already shown that there is an almost linear decrease in the c.m.c.[27]
with increasing alcohol content, see for instance Shinoda. It
therefore seems sufficient to state that we find the same. For
instance, the c.m.c. of NaC_{10} has been found to decrease linearly
from 0.1 mol kg^{-1} in water to 0.051 mol kg^{-1} for a solution saturated with respect to hexanol.[28] The c.m.c. was measured both by the
conductivity method and by electromotive force.[19]

CONCLUDING REMARKS

        Results from several experimental techniques have been
combined in order to elucidate the properties of aqueous micellar
solutions containing alcohol. The following factors have emerged
as being of importance:

1) The amount of surfactant and alcohol
2) The length of the hydrocarbon chain both of the surfactant and the alcohol
3) The polar group and the counter-ion.
4) The solubility of the alcohol in water.

The solubilization sites for the alcohols are probably not fixed in the micelle. An equilibrium between various sites seems likely and there is also a rapid exchange between alcohol molecules in the micelle and in the aqueous medium.

ACKNOWLEDGEMENT

One of us, S.B., thanks NORDISKA FORSKARKURSER for financial support.

REFERENCES

1. J.W. McBain and P.H. Richards, Ind. Eng. Chem., 38, 642 (1946).
2. R.S. Sterns, H. Oppenheimer, E. Simon, and W.D. Harkins, J. Phys. Chem., 15, 496 (1947).
3. W.D. Harkins and H. Oppenheimer, J. Amer. Chem. Soc., 71, 808 (1949).
4. E. Vikingstad, A. Skauge, and H. Høiland, J. Colloid Interface Sci., 66, 240 (1980).
5. E. Vikingstad and H. Sæterdal, J. Colloid Interface Sci., 77, 407 (1981).
6. O. Kvammen, S. Backlund, and H. Høiland, in "Proceedings of the VII Scandinavian Symposium on Surface Chemistry," K.S. Birdi, Editor, Copenhagen, 1981.
7. P. Ekwall and T. Vittasmäki, Acta Chem. Scand., 10, 1177 (1956).
8. L. von Erichsen, Brennstoff Chem., 33, 166 (1952).
9. P. Stilbs, J. Colloid Interface Sci., 87, 385 (1982).
10. A. Wishnia, J. Phys. Chem., 67, 2079 (1963).
11. C. Tanford, "The Hydrophobic Effect: Formation of Micelles and Biological Membranes", Wiley, New York, 1973.
12. P. Molyneux and C.T. Rhodes, Kolloid-Z. u. Z. Polym., 250, 886 (1972).
13. G.H. Bell, Chem. Phys. Lipids, 10, 1 (1973).
14. S. Backlund, K. Rundt, K.S. Birdi, and S. Dalsager, J. Colloid Interface Sci., 79, 578 (1981).
15. J. Gettins, D. Hall, P.L. Jobling, J.E. Rassing, and E. Wyn-Jones, J. Chem. Soc. Faraday Trans. II, 74, 1957 (1978).
16. R. Zana, S.Yiv, C. Strazielle, and P. Lianos, J. Colloid Inter-

face Sci., 80, 208 (1981).
17. A.S.C. Lawrence and J.T. Pearson, Trans, Faraday Soc., 63, 495 (1967).
18. T. Tominaga, T.B. Stem, and D.F. Evans, Bull. Chem. Soc. Japan, 53, 795 (1980).
19. S. Backlund and K. Rundt, Acta Chem. Scand., A34, 433 (1980).
20. S. Candau and R. Zana, J. Colloid Interface Sci., 84, 206 (1981).
21. H. Høiland, J. Solution Chem., 9, 857 (1980).
22. P. Ekwall, L. Mandell, and P. Solyom, J. Colloid Interface Sci., 35, 519 (1971).
23. D.E. Clarke and D.G. Hall, Colloid Polym. Sci., 252, 153 (1974).
24. D. Stigter and K.J. Mysels, J. Phys. Chem., 59, 45 (1955).
25. S. Backlund, H. Høiland, O.J. Kvammen, and E. Ljosland, Acta Chem. Scand., A36, 698 (1982).
26. G. Lindblom, B. Lindman, and L. Mandell, J. Colloid Interface Sci., 42, 400 (1973).
27. K. Shinoda, J. Phys. Chem., 58, 1136 (1954).
28. E. Vikingstad and O.J. Kvammen, J. Colloid Interface Sci. 74, 16 (1980).

SOLUBILIZATION OF PHENOTHIAZINE AND ITS N-ALKYL DERIVATIVES INTO ANIONIC SURFACTANT MICELLES

Y. Moroi, K. Satō, H. Noma, and R. Matuura

Department of Chemistry, Faculty of Science
Kyushu University 33
6-10-1 Hakozaki, Higashi-ku, Fukuoka 812, Japan

Solubilization by surfactant micelles is discussed from a thermodynamic point of view using the stepwise association equilibria between micelles and solubilizates and with the help of probability theory. The maximum additive concentration (MAC) of phenothiazine, N-methylphenothiazine, and N-ethylphenothiazine in aqueous micellar solutions of sodium dodecyl sulfate, manganese(II) dodecyl sulfate, and zinc(II) dodecyl sulfate were determined spectrophotometrically at 20, 25, and 30 °C. The thermodynamic parameters of inter-action between monomeric solubilizate molecule and vacant micelles were evaluated from the stepwise association equilibria. The free energy change per methylene group for solubilization was estimated from differences in the parameters caused by an attachment of alkylgroup to N-atom of phenothiazine molecule, and the solubilizate molecules were found to reside in the palisade layer of micelles. Analysing the data from the viewpoint of gegenions of micelles, it turned out that the aggregation numbers played a more important role than the gegenions for the incorporation of solubilizates into micelles. The Poisson distribution is highly suggested for the distribution of the solubilizates among the micelles considered here.

INTRODUCTION

An enhanced dissolution of otherwise slightly soluble organics
in aqueous solutions is brought about by the presence of surfactant
micelles in the system. This is well known as solubilization.
The solubility increase at concentrations above the cmc is ascribed
to the incorporation of hydrophobic molecules in the micelles present in surfactant solutions. Indeed, many studies have been reported
on solubilization,[1-5] but their interpretations have been rather
qualitative; whereas the papers involving theoretical discussions
are still a few[6-9] and, in addition, include many assumptions.

In studies of kinetics of micelle-catalyzed reaction where
reactant concentrations are comparable to or higher than the micellar
concentrations, a distribution of reactants among micelles must be
considered. Various plots of reaction rates versus micellar concentrations have been reported,[10-13] but their interpretations have
been Michaelis-Menten type in most respects and the distribution of
reactants among micelles has not been considered. In photochemical
processes in micellar solutions, on the other hand, the distribution of photochemical probes among micelles was found to be very
important factor in elucidating the photochemical reactions, and
a generally accepted concept concerning the distribution of probe
molecules among micelles is their Poisson distribution.[14-17] However, the necessity of the Poisson distribution has not been made
clear in this case either.

As a micellar solution is a thermodynamic system even in the
presence of solubilizates, the equilibrium distribution is essentially a thermodynamic problem. However, the association constants
of solubilizates with micelles which determine the distribution of
solubilizate molecules among micelles are too many to be determined.
Therefore, we have analyzed the solubilization process mathematically
and have examined what kind of distribution is preferable to elucidate the present solubilization equilibria.

EQUILIBRIUM DISTRIBUTION OF SOLUBILIZATES
AMONG MICELLES

The monodispersity of micelles in the absence of solubilizate
is assumed to remove the difficulties arising from their polydispersity. However, the following discussion remains essentially
the same for micelles of polydispersity.[9] The association equilibrium between surfactant monomers (S) and micelles (M) is presented
by

$$m\,S \;\; \underset{}{\overset{K_m}{\rightleftharpoons}} \;\; M \tag{1}$$

where K_m is the equilibrium constant of micelle formation and m is
the aggregation number of micelles.   Here, micelles are not a

pseudo-phase but a chemical species, of course. The stepwise association equilibria between micelles and solubilizates (R) can be
represented schematically as follows:

$$M + R \underset{}{\overset{K_1}{\rightleftharpoons}} MR_1, \quad MR_1 + R \underset{}{\overset{K_2}{\rightleftharpoons}} MR_2, \quad \cdots \cdots \quad MR_{n-1} + R \underset{}{\overset{K_n}{\rightleftharpoons}} MR_n \quad (2)$$

where MR_i denotes micelles associated with i molecules of solubilizate, K_i is the stepwise association constant between MR_{i-1} and a
monomer molecule of solubilizate, and n is an arbitrary number.
The total number of components of this system is $n+4$ including water
molecules, and the number of phases is two, micellar solution phase
and a solid or liquid solubilizate phase. A solubilizate phase must
be present in the case where the maximum additive concentration (MAC)
is a point of discussion. The $n+1$ equilibrium equations for the
micellar system reduce the number of degrees of freedom of the phase
rule by $n+1$, resulting in 3 degrees of freedom. Hence, at constant
temperature and pressure, only one other intensive variable can be
selected to specify the thermodynamic system. We selected here the
total surfactant concentration as the variable. From Equations (1)
and (2), we have the following equations for the total micellar concentration ($[M_t]$), the total equivalent concentration of solubilizate
($[R_t]$), and the average number of solubilizate molecules per micelle
($\bar{R}$):

$$[M_t] = K_m[S]^m \{1 + \sum_{i=1}^{n} (\prod_{j=1}^{i} K_j) [R]^i\} \quad (3)$$

$$[R_t] = [R] + K_m[S]^m \sum_{i=1}^{n} i (\prod_{j=1}^{i} K_j) [R]^i \quad (4)$$

$$\bar{R} = ([R_t] - [R])/[M_t] \quad (5)$$

Now let us suppose that the solubilization of $([R_t] - [R])$ molecules
into $[M_t]$ micelles is like placing $([R_t] - [R])$ balls into $[M_t]$
cells randomly, where both balls and cells are independent and indistinguishable. Then the probability $P(i)$ that a specified cell
contains exactly i balls is given as[18]

$$P(i) = \binom{r}{i} \frac{1}{q^i} (1 - 1/q)^{r-i} \quad (6)$$

where $r = ([R_t] - [R])$ and $q = [M_t]$. r and q are of the order of
Avogadro's number and i is very small compared with them, less than
10 at maximum for the present case. This equation is a special case
of the so-called binominal distribution. Using the following approximation, which is reasonable for the present condition,

$$1 - 1/q = \exp(-1/q) \quad (7)$$

and the numerical conditions as to r, q, and i values, we finally
obtain[9]

$$P(i) = \bar{R}^i \exp(-\bar{R})/i! \tag{8}$$

which is the Poisson distribution. The Gaussian distribution can also be obtained from the binominal distribution,[18]

$$G(i) = h/\sqrt{\pi} \exp\{-h^2(i - \bar{R})^2\} \tag{9}$$

$$\sigma = 1/\sqrt{2} \, h \tag{10}$$

where h is the Gaussian distribution constant relating to the standard deviation σ.

The total surfactant concentration which we have selected to specify the present solubility system must determine every intensive variable of the system. Total equilibrium concentration of solubilizate $[R_t]$ must, of course, be determined only by the total surfactant concentration. As the concentration of micelles associated with i solubilizate molecules is

$$[MR_i] = [M_t]P(i) \tag{11}$$

the association constant K_i is given as

$$K_i = \frac{[MR_i]}{[R][MR_{i-1}]} = K_1/i \tag{12}$$

where the Poisson distribution is assumed. Introduction of Equations (8) and (11) into Equation (4) gives

$$[R_t] = [R] + \bar{R}[M_t] \tag{13}$$

or $$([R_t] - [R])/[R] = K_1[M_t] \tag{14}$$

where the summation is over an infinite n. On the other hand, if the Gaussian distribution is assumed, K_i and $[R_t]$ are expressed as

$$K_i = K_1 \exp\{-h^2(2i - 2)\} \tag{15}$$

$$[R_t] = [R][1 + K_1\bar{R} \exp\{h^2(1 - 2\bar{R})\}[M_t]] \tag{16}$$

The above two distributions are complete random distributions free from any restriction.

It should be very enlightening to consider also the distribution where some restriction is applied. In such case, the balls are not independent of each other even though inter-ball interactions are absent. The distribution where balls are indistinguishable and cells are distinguishable is one of the above examples. As for solubilization, if the number of solubilized molecules in a specified micelle has an influence on the succeeding distribution of remaining solubilizate molecules among the other micelles, they cannot be solubilized independently. The random distribution based on the above concept is given by

$$R(i) = \frac{1}{1 + \bar{R}} \left(\frac{\bar{R}}{1 + \bar{R}} \right)^i \tag{17}$$

which is called a geometrical or random distribution.[7] In this case, K_i and $[R_t]$ are given by

$$K_1 = K_2 = \ldots = K_i = \ldots = K_n \tag{18}$$

$$[R_t] = [R]\{1 + K_1(1 + \bar{R})[M_t]\} \tag{19}$$

or $\quad ([R_t] - [R])/[R] = K_1(1 + \bar{R})[M_t] \tag{20}$

However, the difference between the cases with and without restriction becomes small with decreasing $\bar{R}$ value, as is clear from Equations (14) and (20).

EXPERIMENTAL

Materials. Phenothiazine (PTH) of extra pure reagent grade was purified by recrystallization three times from pure benzene and stored in the dark. N-Alkylphenothiazines were synthesized by Bernthsen's method.[19] As for N-methylphenothiazine (MPTH), a mixture of phenothiazine, methyl iodide, and methanol was heated in an autoclave at 95 - 115 °C for 11 hours. As for N-ethylphenothiazine (EPTH), a mixture of phenothiazine, ethyl bromide, and ethanol was heated in an autoclave at 110 - 120 °C for 11 hours. N-Alkylphenothiazines thus prepared were recrystallized from ethanol six times for purification. Their melting points were 184, 102, and 105 °C for PTH, MPTH, and EPTH, respectively, and the elemental analyses agreed with the calculated values within 0.1%.

Sodium dodecyl sulfate (SDS), manganese(II) dodecyl sulfate (Mn(DS)$_2$), and zinc(II) dodecyl sulfate (Zn(DS)$_2$) were prepared and purified by the procedure described elsewhere.[20,21] The water used was distilled twice from alkaline permanganate solution, and all other reagents were of analytical reagent grade.

Solubilization. A suspension of the solubilizate powder in surfactant solution was stirred initially at room temperature by disk rotors in 10-ml injector tubes. The injectors were then dipped into a thermostat until equilibrium was reached, during which any light was shut out from the injectors, and the temperature was kept constant at 20, 25, and 30 °C and controlled within $\pm$ 0.01 °C. The run was timed from this dipping and proceeded for more than 2 hours. At hourly intervals, 2.5 ml of filtrate was withdrawn

by applying pressure upon the injector through the filter of 0.2-μm
pore size (Millipore; FGLP 01300). The phenothiazines have two
absorption bands in the ultraviolet region whose maximum wavelengths are around 250 and 310 nm. The former with higher molar
extinction coefficient was used for MAC determination for surfactant concentrations below the cmc, the latter being used for concentrations above the cmc.

RESULTS AND DISCUSSION

The variation in MAC of PTH with SDS concentration is illustrated in Figure 1. As is evident from the figure, the plots of
MAC against the surfactant concentration can be divided into two
straight lines; one for below the cmc and the other above it.
The former has almost zero slope irrespective of surfactant concentration, whereas above the cmc MAC increases linearly with
increasing surfactant concentration. The intersection and slope

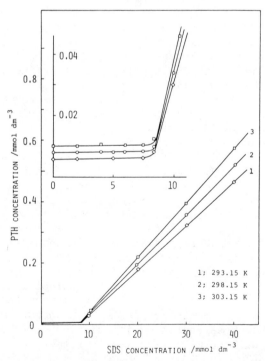

Figure 1. Solubility change of PTH with SDS concentration. The
insert in this figure is an expanded version of the behavior at
low SDS concentration.

of the two lines giving the best fit with the experimental points
were obtained by linear regression analysis. From the above linear
relationship and from an inspection of expressions of Equations (14),
(16), and (20), we can see that (1) where $[M_t]$ is zero below cmc,
$[R_t]$ remains constant as the saturated concentration ($[R]$) of PTH
at the specified temperature and (2) where $[M_t]$ increases with
surfactant concentrations above cmc, the solubilization increases
keeping $\bar{R}$ and $[R]$ constant up to 40 mmol dm^{-3} of SDS, since otherwise a linear relationship between $[R_t]$ and SDS concentrations
would not result. In addition, we can say that the present surfactant concentration range is such that the micelles are too far
apart to influence one another. In Figure 2 are illustrated the
differences in MAC brought about by an attachment of alkyl group
to N-atom of PTH which gives rise to higher hydrophobicity in the
solubilizate molecule.[22] It can be seen from the figure that the
MAC's of N-alkylphenothiazine both below and above the cmc become
smaller as the attached alkyl-chains become more hydrophobic. The
MAC of MPTH below the cmc is about 40% of that of PTH and that
above the cmc it is 75% of PTH at the same concentration of SDS.
The MAC of EPTH below the cmc is 9% of that of PTH and that above

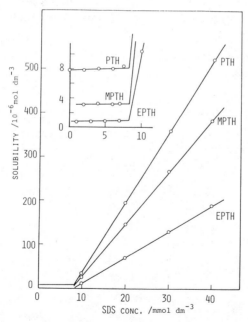

Figure 2. Solubility change of PTH, MPTH, and EPTH with SDS concentration at 298.15 K. The insert in this figure is an expanded
version of behavior at low SDS concentration.

the cmc, it is 35% of PTH. These results indicate that N-alkylphenothiazine with a longer alkyl-chain is more stabilized in a
micellar phase than in an aqueous phase. What has just been said
applies to the other surfactant systems also.

The surfactant concentration and the MAC at an intersection
of the two lines are considered to be the cmc of surfactant[23] and
the monomer concentration of the solubilizate at the cmc, respectively. The cmc values thus obtained are in good agreement with
cmc's determined by other conventional methods (Table I).

We will now try to estimate the thermodynamic parameters for
the solubilization. From Equations (14) and (20) we can determine
K_1 value by plotting $([R_t] - [R])/[R]$ against $C -$ cmc, where $[M_t] =$
$(C -$ cmc$)/N$ and N is the aggregation number of micelles which is
assumed to remain constant. The slope of the line obtained by these
plots gives K_1/N and $K_1(1 + \bar{R})/N$ for the Poisson and the geometrical
distributions, respectively. Here, the Gaussian distribution is
omitted, because it is clear that an application of the distribution to the present case where $\bar{R}$ is less 1.5 is not appropriate.
Figure 3 shows good linear relationships at 20, 25, and 30°C for
PTH – SDS system, which confirm that the analysis is valid within
the experimental accuracy. The point is that K_1/N values decrease
with increasing temperature, notwithstanding that the solubilized
PTH amount increases as the temperature increases. This is due to
the fact that higher solubility at higher temperature is largely

Table I. Cmc Values Determined by Solubilization and Other Methods.

surfactant	temp, K	cmc/(mmol dm^{-3})		
		solubili-zation	conduc-tivity	calcu-lation[a]
SDS	293.15	8.27	8.24	
	298.15	8.34	8.20	
	303.15	8.46	8.25	
SDS/0.15 M NaCl	293.15	1.38		1.22
	298.15	1.29		1.28
	303.15	1.45		1.34
Mn(DS)$_2$	293.15	1.18	1.18	
	298.15	1.05	1.16	
	303.15	1.16	1.18	
Zn(DS)$_2$	293.15	1.29	1.21	
	298.15	1.14	1.20	
	303.15	1.14	1.22	

a) By the method given in the ref. 24.

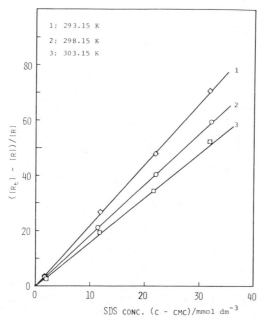

Figure 3. Plots of $([R_t] - [R])/[R]$ of PTH against micellar concentration $(C - \text{cmc})$ of SDS for determination of K_1 value.

counterbalanced by the increased concentration of monomeric R at higher temperature as shown in Figure 1. In Figure 4 is shown the difference in K_1/N between N-alkylphenothiazines. The point is that the order of slope PTH > MPTH > EPTH for MAC above cmc in Figure 2 becomes completely reversed for $([R_t] - [R])/[R]$ in Figure 4. This is also due to the fact that a steeper slope of PTH in Figure 2 is counterbalanced to much extent by a higher concentration of monomeric PTH in an aqueous phase. Now attempts will be made to probe further into the physicochemical meaning of the association constant K_1. As is shown in Equation (2), K_1 is a parameter which relates to the free energy change caused by incorporation of one monomeric solubilizate molecule into micelles free from any solubilizate. In terms of K_1, the free energy change can be expressed as:

$$\Delta G^\circ = - RT \ln K_1 \qquad\qquad (21)$$

Furthermore, ΔG° is made up of three contributions as follows:

$$\Delta G^\circ = \mu^\ominus_{MR_1} - \mu^\ominus_M - \mu^\ominus_R \qquad\qquad (22)$$

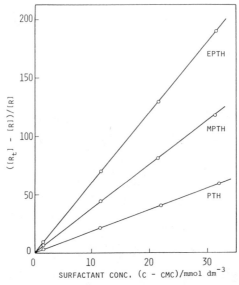

Figure 4. Plots of $([R_t] - [R])/[R]$ of PTH, MPTH, and EPTH against micellar concentration $(C - cmc)$ of SDS for determination of K_1 values at 298.15 K.

where $\mu_{MR_1}^{\ominus}$ is the standard chemical potential of MR_1 at infinite dilution, and $\mu_M^{\ominus}$ and $\mu_R^{\ominus}$ are the corresponding potentials. The difference $\mu_{MR_1}^{\ominus} - \mu_M^{\ominus}$ would depend on the properties of surfactant micelles into which a solubilizate molecule incorporates, in particular on the environment around the site where the solubilizate molecule resides. It was found in the previous paper[25] that PTH molecules reside in the palisade layer of micelles and that a complex formation takes place between PTH and divalent meatl ions. In view of these findings, we can suggest that the aggregation number and the nature of the gegenion of micelles must have a strong influence on K_1 values. In Figure 5 are shown the plots of solubility of PTH against surfactant concentration for four surfactant system; SDS, SDS/0.15 M NaCl, $Mn(DS)_2$, and $Zn(DS)_2$. That the inflection point of plots of the SDS/0.15 M NaCl system is almost the same as those of $Mn(DS)_2$ and $Zn(DS)_2$ suggests an approximately equal aggregation number for the three surfactant systems. The surfactant concentrations at the intersections are given in Table I. Now it is observed that the cmc determination by the solubility method is valid when the solubilizate monomer concentration is lower than 10^{-5} mol dm^{-3}. The ΔG° values thus obtained are given in Table II, where we used aggregation numbers of micelles as

Table II. The Standard Free Energy Change of Association $\Delta G°$ and Incremental Free Energy Changes per CH_2 Group $\Delta(\Delta G°)$ between PTH and MPTH, and between MPTH and EPTH in kJ mol^{-1} Units.

surfactant	solubili-zate	293.15 K $\Delta G°$	$\Delta(\Delta G°)$	298.15 K $\Delta G°$	$\Delta(\Delta G°)$	303.15 K $\Delta G°$	$\Delta(\Delta G°)$
SDS (N = 64)	PTH	-28.9	-1.4	-29.0	-1.7	-29.2	-1.7
	MPTH	-30.3	-1.4	-30.7	-1.4	-30.9	-1.7
	EPTH	-31.7		-32.1		-32.6	
SDS/0.15 M NaCl (N = 95)	PTH	-30.2	-1.6	-30.2	-2.0	-30.3	-2.0
	MPTH	-31.8	-1.8	-32.2	-1.6	-32.3	-1.8
	EPTH	-33.6		-33.8		-34.1	
Mn(DS)$_2$ (N = 95)	PTH	-29.8	-1.9	-29.9	-2.1	-30.0	-2.3
	MPTH	-31.7	-1.9	-32.0	-2.0	-32.3	-2.2
	EPTH	-33.6		-34.0		-34.5	
Zn(DS)$_2$ (N = 95)	PTH	-29.9	-1.7	-30.0	-2.0	-30.1	-2.0
	MPTH	-31.6	-2.0	-32.0	-2.0	-32.1	-2.3
	EPTH	-33.6		-34.0		-34.4	

Table III. The Enthalpy and Entropy Changes, $\Delta H°$ and $\Delta S°$, of the Association Obtained from Equations (23) and (24).

surfactant	solubili-zate	$\Delta H°$/ kJ mol^{-1}		$\Delta S°$/ J $K^{-1} mol^{-1}$	
SDS (N = 64)	PTH	-21.9	10.8	24	22
	MPTH	-11.1	4.5	66	20
	EPTH	- 6.6		86	
SDS/0.15 M NaCl (N = 95)	PTH	-28.2	13.1	7	50
	MPTH	-15.1	- 4.4	57	-9
	EPTH	-19.5		48	
Mn(DS)$_2$ (N = 95)	PTH	-22.4	8.8	25	37
	MPTH	-13.6	5.3	62	25
	EPTH	- 8.2		87	
Zn(DS)$_2$ (N = 95)	PTH	-24.4	5.4	25	19
	MPTH	-18.8	9.6	44	39
	EPTH	- 9.2		83	

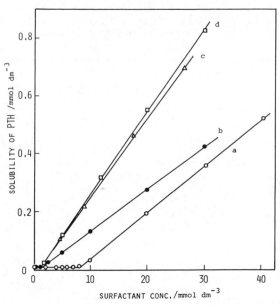

Figure 5. Solubility change of PTH with surfactant concentration at 298.15 K: (a) SDS, (b) SDS/0.15 M NaCl, (c) $Mn(DS)_2$, (d) $Zn(DS)_2$.

$N = 64$ for SDS[26,27] and $N = 95$ for SDS/0.15 M NaCl,[28-30] $Mn(DS)_2$, and $Zn(DS)_2$[31,32] in units of dodecyl sulfate ion, and the Poisson distribution was employed. The reason for using Poisson distribution will be discussed later.

On the other hand, from the variation of $\Delta G°$ with temperature the enthalpy and entropy changes, $\Delta H°$ and $\Delta S°$, of association can be determined.

$$\Delta H° = - R[d \ln K_1/d(1/T)] \qquad (23)$$

$$\Delta S° = (\Delta H° - \Delta G°)/T \qquad (24)$$

The thermodynamic parameters derived from Equations (23) and (24) are summarized in Table III. Of course, $\Delta H°$ and $\Delta S°$ values change with temperature because of effects of temperature on both micellar structure and chemical potentials of other components. However, attention here is paid to the values at 25 °C which can be considered as the mean values for the temperature range 20 - 30 °C. We may here make a few remarks about the behavior of the association constant K_1 with changing gegenions and the aggregation number of surfactant micelles. The following can be deduced from the

values in Tables II and III; (1) the association of solubilizates
with the micelles is an exothermic reaction; (2) an increase in
the aggregation number is favorable for the present association;
and (3) a difference in the gegenions does not play an important
role in the incorporation of the present solubilizates into the
micelles considered here. The third conclusion is contrary to our
expectation, because PTH forms a complex with divalent ions when
solubilized.[25] However, the result comes mainly from such weak
complex formation that the half-life of the reaction is more than
1 day. Thus, the increase of K_1 values for SDS/0.15 M NaCl solution is mainly due to an increase of hydrophobicity of the palisade
layer with an increase of the aggregation number.

It is very useful to consider the effect of alkyl-chain
attached to N-atom of phenothiazine on the thermodynamic parameters so that we may derive some conclusion concerning the present
solubilization process from these values. The most striking feature of $\Delta(\Delta G°)$ in Table II is almost constant differences between
PTH and MPTH and between MPTH and EPTH for each solubilization
system, which clearly suggests that one CH_2 group in the alkyl-
chain makes a constant contribution to the free energy change of
solubilization. In comparison with the values of $- 1.55 \pm 0.15$
kJ mol^{-1} for SDS, those for SDS/0.15 M NaCl, $Mn(DS)_2$, and $Zn(DS)_2$
are somewhat higher, $i.e.$ $- 1.96 \pm 0.20$ kJ mol^{-1}. Those values
are much smaller than many literature values of about $- 3.4$ kJ mol^{-1},
which is an incremental free energy change per CH_2 group on transfer
from dilute aqueous solution to liquid hydrocarbon.[33,34] On the
other hand, its free energy change on transfer into hydrophobic
micellar interior is $- 2.7$ to $- 3.0$ kJ mol^{-1},[33] which is still
higher than the present values. In view of the above results, we
can conclude that the site of the solubilized molecules is rather
hydrophilic part of micelles which is not the micellar interior
but the so-called palisade layer of micelles. In addition, from
smaller values of $\Delta(\Delta G°)$ for SDS micelles the palisade layer of
SDS micelles is looser compared with the other three surfactant
micelles. This conclusion is really consistent with our previous
result.[25] On the other hand, $\Delta H°$ and $\Delta S°$ values increase with
increasing alkyl-chain length, except SDS/0.15 M NaCl system where
properties of aqueous solution are much influenced by the presence
of high NaCl content. The increases in $\Delta H°$ and $\Delta S°$ values are
clear, even if the error from differentiation of $\Delta G°$ with temperature is taken into account. In view of these, with increasing
alkyl-chain length the solubilization reaction becomes less endothermic, and the entropy part, $T\Delta S°$, becomes more important than
the enthalpy part, $\Delta H°$, in the free energy change. In any event,
the latter fact indicates that there exists a strong relationship
between the number of water molecules around solubilitize molecules
in an intermicellar bulk phase and the alkyl-chain length. The
plausible explanation is that the increase in $T\Delta S°$ term results
mainly from the melting of increasing structured water molecules

around the solubilizates due to an increase of alkyl-chain length
in the bulk phase on moving from the bulk phase to the micellar
phase.

SELECTION OF DISTRIBUTION PATTERN

Consider now the distribution patterns of the solubilizates
among micelles. From Equations (14) and (20), the free energy
difference between the Poisson ($\Delta G^{p,o}$) and random ($\Delta G^{r,o}$) distributions can be expressed as

$$\Delta G^{p,o} - \Delta G^{r,o} = RT \ln (1 + \bar{R}) \qquad (25)$$

From the aggregation number (N), 64 for SDS and 95 for the other
surfactant systems, we can determine the $\bar{R}$ values from Equation (5),
using the relation $[M_t] = (C - cmc)/N$. As mentioned above concerning SDS micelles, the value of N changes about 2 % for a
temperature change of 5 °C. In addition, as is seen from Table I,
the cmc change is very little for each present system over the
temperature range 20 - 30 °C, which strongly suggests almost constant
aggregation number over the temperature range. Consequently we
used the above-mentioned numbers for N. The ΔG^o differences from
this term are given in Table IV for PTH - surfactant systems. The
reasons that the Poisson distribution is more suitable than the
other one are the following: (1) If we adopt the random distribution, $\Delta G^{r,o}$ values become almost equal irrespective of the surfactant

Table IV. Average Number of PTH Molecules per Micelle and the
Difference in Standard Free Energy Change between Poisson and
Random Distributions.

surfactant	temp, K	PTH/ micelle	$\Delta G^{r,o} - \Delta G^{p,o}$ kJ mol^{-1}
SDS	293.15	0.93	1.60
	298.15	1.03	1.75
	303.15	1.14	1.92
SDS/0.15 M NaCl	293.15	1.28	2.01
	298.15	1.38	2.15
	303.15	1.50	2.31
Mn(DS)$_2$	293.15	1.21	1.93
	298.15	1.30	2.06
	303.15	1.42	2.22
Zn(DS)$_2$	293.15	1.27	2.00
	298.15	1.35	2.12
	303.15	1.48	2.29

systems, in spite of the fact that the palisade layer where the PTH molecule resides should change its physicochemical properties with the aggregation number. (2) The analysis of many data for the micelle-catalyzed reactions strongly suggests that the Poisson distribution is most preferable.[17] Besides, PTH molecules obey the Poisson distribution among micelles whose aggregation number is 95.[15] (3) The incremental free energy change per CH_2 group, $\Delta(\Delta G°)$, obtained by using the Poisson distribution is very reasonable judging from many literature values. (4) The condition for the association constant $K_i = K_1/i$ which was derived by the Poisson distribution is really acceptable when solubilized molecules are much less in number than the micellar aggregation number, just as in the present case. In view of these facts, it is highly likely that the Poisson distribution is much more suitable for the present systems.

Finally, we may now make a few remarks about distribution patterns from a mathematical point of view. The distributions of phenothiazine and N-alkylphenothiazines for the present micellar systems turned out to obey the Poisson distribution to a good approximation, which results from the fact that both solubilizate molecules and micelles are not only indistinguishable but also independent. This result can be expected from small $\bar{R}$ values less than 2, just as in the present case. Thus, it can generally be said that a discussion of solubilization in terms of the Poisson or binominal distribution is reasonable for small $\bar{R}$ values and for micelles and microemulsions of larger volume, because both micelles and solubilizate molecules can apparently act independently and indistinguishably. On the other hand, the random or geometrical distribution can be one choice for consideration when the $\bar{R}$ value increases, and as a result some restrictions become relevant. At this point, it should be emphasized that the distribution of solubilizate molecules among micelles is determined only by their association constants with micelles, not by probability, as mentioned in the preceding section. As a matter of fact, it is too difficult to determine each association constant. Therefore, we have analyzed the solubilization process with the help of mathematics and here examined which distribution pattern can be approximated by the present solubilization model.

REFERENCES

1. P. H. Elworthy, A. T. Florence, and C. B. Macfarlane, "Solubilization by Surface-Active Agents and its Application in Chemistry and the Biological Sciences", Chapman and Hall, London, 1968.
2. P. Mukerjee and J. R. Cardinal, J. Phys. Chem., 82, 1620 (1978).
3. K. S. Birdi, H. M. Singh, and S. U. Dalsager, J. Phys. Chem. 83, 2733 (1979).

4. A. Goto, M. Nihei, and F. Endo, J. Phys. Chem., $\underline{84}$, 2268 (1980).
5. S. D. Christian, E. E. Tucker, and E. H. Lane, J. Colloid Interface Sci., $\underline{84}$, 423 (1981).
6. E. Ruckenstein and R. Krishnan, J. Colloid Interface Sci., $\underline{71}$, 321 (1979).
7. Y. Moroi, J. Phys. Chem., $\underline{84}$, 2186 (1980).
8. R. Mallikarjun and D. B. Dadyburjor, J. Colloid Interface Sci., $\underline{84}$, 73 (1981).
9. Y. Moroi. K. Sato, and R. Matuura, J. Phys. Chem., $\underline{86}$, 2463 (1982).
10. E. H. Cordes and R. B. Dunlap, Acc. Chem. Res., $\underline{2}$, 329 (1969).
11. E. H. Cordes, "Reaction Kinetics in Micelles", Plenum Press, New York, 1973.
12. E. J. Fendler and J. H. Fendler, Adv. Phys. Org. Chem., $\underline{8}$, 271 (1970).
13. J. H. Fendler and E. J. Fendler, "Catalysis in Micellar and Macromolecular Systems", Academic Press, New York, 1975.
14. K. Kalyanasundaram, M. Gratzel, and J. K. Thomas, J. Am. Chem. Soc., $\underline{97}$, 3915 (1975).
15. M. Maestri, P. P. Infelta, and M. Gratzel, J. Chem. Phys., $\underline{69}$, 1522 (1978).
16. M. Almgren, F. Grieser, and J. K. Thomas, J. Am. Chem. Soc., $\underline{101}$, 279 (1979).
17. T. Harada, N. Nishikido, Y. Moroi, and R. Matuura, Bull. Chem. Soc. Jpn., $\underline{54}$, 2592 (1981).
18. W. Feller, "An Introduction to Probability Theory and Its Applications", 3rd ed. Vol. 1, Wiley, New York, 1967.
19. A. Bernthsen, Lieb. Ann. Chem., $\underline{230}$, 73 (1885).
20. Y. Moroi, K. Motomura, and R. Matuura, Bull. Chem. Soc. Jpn., $\underline{44}$, 2078 (1971).
21. Y. Moroi, T. Oyama, and R. Matuura, J. Colloid Interface Sci., $\underline{60}$, 103 (1977).
22. Y. Moroi, H. Noma, and R. Matuura, submitted for publication in J. Phys. Chem..
23. P. Mukerjee and K. J. Mysels, Natl. Stand. Ref. Data Ser. (U.S. Natl. Bur. Stand.), No.36 (1971).
24. Y. Moroi, N. Nishikido, H. Uehara, and R. Matuura, J. Colloid Interface Sci., $\underline{50}$, 254 (1975).
25. Y. Moroi, M. Saito, and R. Matuura, Nippon Kagaku Kaishi, 482 (1980).
26. J. E. Leiber and J. J. Jacobus, J. Phys. Chem., $\underline{81}$, 130 (1977).
27. N. Muller, in "Micellization, Solubilization, and Microemulsions", K. L. Mittal, Editor, Vol.1, p.229, Plenum Press, New York, 1977.
28. K. J. Mysels and L. H. Princen, J. Phys. Chem., $\underline{63}$, 1696 (1959).
29. M. Emerson and A. Holtzer, J. Phys. Chem., $\underline{71}$, 1898 (1967).
30. S. Hayashi and S. Ikeda, J. Phys. Chem., $\underline{84}$, 744 (1980).
31. I. Satake, I. Iwamatsu, S. Hosokawa, and R. Matuura, Bull. Chem. Soc. Jpn., $\underline{36}$, 205 (1963).
32. Y. Moroi, K. Motomura, and R. Matuura, J. Colloid Interface

Sci., 46, 111 (1974).

33. P. Mukerjee, in "Micellization, Solubilization, and Microemul-
    sions", K. L. Mittal, Editer, Vol.1, p.171, Plenum Press,
    New York, 1977.
34. C. Tanford, "The Hydrophobic Effect: Formation of Micelles and
    Biological Membranes", Wiley, New York, 1973.

CHOLESTEROL SOLUBILIZATION AND SUPERSATURATION IN BILE: DEPENDENCE

ON TOTAL LIPID CONCENTRATION AND FORMATION OF METASTABLE DISPERSIONS

Dov Lichtenberg, Ilana Tamir, Ron Cohen and
  Yochanan Peled
Sackler School of Medicine and Ichilov Medical Center
Tel Aviv Univeristy, Tel Aviv; and School of Pharmacy
Hebrew University, Jerusalem, Israel

Cholesterol solubilization in bile salt-lecithin mixed dispersions depends not only on the relative concentration of these three components but also on several other factors. One important parameter is the total lipid concentration (Carey and Small, J. Clin. Invest., 61, 998 (1978)). Previously, it had been pointed out (Helenius and Simons, Biochim. Biophys. Acta, 415, 29 (1975)) that results of detergent solubilization experiments, carried out at different lipid concentrations, can be compared only after the concentration of aqueous detergent monomers is subtracted from the total detergent concentration. This "normalization procedure" does not appear to be widely appreciated and has not been used for the description of the solubilization of cholesterol and lecithin by bile salts.

In the present work, we have re-plotted Carey and Small's data for various total lipid concentrations on a ternary phase diagram after correcting the mole fractions of lecithin, cholesterol and sodium taurocholate by subtracting the bile salt's critical micellar concentration (CMC) from the bile salt concentration. The micellar-liquid crystalline phase boundary lines in these corrected phase diagrams appear to be independent of the total lipid concentration, indicating that the concentration dependence of cholesterol solubilization is merely a reflection of the fact that monomeric bile salts are not involved in lecithin and cholesterol solubilization. Accordingly, dilution of gall bladder biles resulted in a mixed micelles-mixed bilayers intercon-

981

version as is evident from NMR spectra of the dispersions.

Solubilization of lecithin-cholesterol sonicated vesicles (2:1 molar ratio) by sodium taurocholate occurs when the ratio between the concentration of the non-monomeric bile salt and the sum of lecithin and cholesterol concentrations approaches unity. This value is independent of the total lipid concentration, supporting our interpretation of the dependence of the solubilization on the lipid concentration. The low value of this solubilization ratio suggests that upon solubilization of lecithin-cholesterol membranes by a minimal bile salt concentration, a metastable state of aggregation can be obtained, which can be regarded as a model for supersaturated bile.

INTRODUCTION

Cholesterol - gallstones formation is a rather frequent disease[1-2]. Consequently, much work has been devoted to gain understanding of the mechanism by which these stones are formed, the risk factors in their development and the possibility of solubilizing these cholesterol deposits. Theoretically, the first stage in gallstone formation must involve formation of saturated bile[3]. Accordingly, a general agreement exists that the formation of these deposits is related to the capacity of bile to solubilize cholesterol, in the form of cholesterol - lecithin - bile salts mixed micelles, which, in turn, is a rather complex function of the bile's composition[4]. Much of our understanding of various physico-chemical aspects of bile is due to the pioneering work of Isaksson[5] and Nakayama[6], followed by the systematic phase equilibria studies of Small, Bourges and Dervichian[7-9]. In these investigations, model systems, composed of bile salts, lecithin and cholesterol have been used to determine the maximal equilibrium cholesterol solubility in bile and its dependence on temperature, ionic strength and total lipid concentration. These extensive studies led to the conclusion that the predominant driving force for cholesterol precipitation is the absolute degree of cholesterol supersaturation[4].

However, three findings seemingly undermined this conclusion: (1) cholesterol supersaturation is frequently found in the gall-bladder bile of healthy persons[10]; (2) the hepatic bile of fasting normal patients is regularly supersaturated[11]; (3) unsaturated bile is found in many gallstone patients[12].

These apparent discrepancies were attributed to "deficiencies in the data chosen for the limits of cholesterol solubility in model bile systems"[4]. More specifically, the two major factors that determine the limit of cholesterol solubility are the bile salts: lecithin ratio and the total lipid concentration[4]. In their work[4], Carey and Small emphasized the importance of the latter factor and systematically investigated the effects of variation in total lipid concentration on cholesterol solubility. In these studies, a semilogarithmic relationship was found between cholesterol solubility and total lipid concentration. This dependence has not been explained. However, employing the "appropriate" values for the cholesterol solubility limit, by taking the total lipid concentration into account, Carey and Small concluded that cholesterol stone patients always have supersaturated gallbladder biles; whereas in control and "pigment stone" patients the cholesterol content of the bile is, on the average, 95-98% of the saturation value of cholesterol[4]. Nevertheless, even with respect to the "appropriate limit", 50% of the gallbladder biles of normal patients were supersaturated and fasting hepatic biles were even more so. This led Carey and Small[4] to conclude that "metastable supersaturation is frequent in both normal and abnormal biles".

This conclusion is in fact consistent with that of several authors that "separation of cholesterol-stone patients from controls by means of analysis of biliary lipid composition is impossible"[10,11]. It also underlines the importance of metastable states of aggregation in bile salt-lecithin-cholesterol dispersions. In particular, the "limits of cholesterol saturation" in such non-equilibrium supersaturated model bile systems may be of special importance to the question of the degree of cholesterol saturation of native bile.

The goal of our present study is twofold:

(1)   To simplify the description of the dependence of the cholesterol solubility limit on the total lipid concentration.

(2)   To study the possibility of forming non-equilibrium "metastable mixed dispersions" of bile salt, lecithin and cholesterol. The preliminary results of this study, reported in this communication , emphasize the importance of the bile salt maximal monomer concentration (cmc) in determining the dependence of cholesterol solubilization on the lipid concentration. It also indicates that cholesterol-rich "soluble" mixed micellar structures can exist in supersaturated model bile for long periods of time.

EXPERIMENTAL PART

Sodium taurocholate (Sigma) and egg yolk phosphatidylcholine (lecithin grade I; Makor Chemicals, Jerusalem) were chromatographacally pure and were used without further purification. Cholesterol (Merck) was recrystallized from ethanol. D_2O (99.7%), used for ^{1}H-NMR measurements, was a product of Merck.

Lecithin-cholesterol vesicles were prepared as follows: solutions of lecithin and cholesterol in chloroform were mixed at appropriate molar ratios, the mixed organic solution was evaporated to dryness under nitrogen and then lyphilized. The dried mixture was then dispersed in 150 mM NaCl solutions in water or D_2O. Small sonicated vesicles were obtained from these dispersions by sonication[13] (Heat Systems 350 W sonicator) until clear dispersions were obtained.

The turbidity of the lipid dispersions, before and after the addition of increasing concentrations of sodium taurocholate, was measured using a Payton aggregation module (Model 300 B-5) at 37°C. ^{1}H-NMR spectra were recorded on a Bruker WH 300 spectrometer operating at 300 MHz and equipped with an FT accessory. [1]H linewidths ($\delta v^{1/2}$) were measured using spectral width of 10Hz/cm and sweep rate of 1Hz/cm.

RESULTS AND DISCUSSION

1. The Effect of Total Lipid Concentration on Lipid Solubilization in Bile, Significance and Theoretical Analysis of Previous Data

The phase boundary between purely micellar and other phases in mixed dispersions of cholesterol, lecithin (PC) and sodium taurocholate (NaTC), as described in phase diagrams of these three components, depends on the total lipid concentration in the dispersions. (Figure 1, after Carey and Small[4]). As evident from these phase diagrams, dilution of cholesterol-lecithin-bile salt mixed micelles might result in transformation of a micellar into a lamellar structure. As an example, a mixture of 60 mole percent NaTC, 35 mole percent PC and 5 mole percent cholesterol is micellar when the total lipid concentration is 200 g/ℓ but lamellar for a total lipid concentration of 3g/ℓ.

The importance of the phase dependence on the total lipid concentration is manifested by the ^{1}H-NMR spectra of the native hepatic bile samples presented in Figure 2. While at the present time the different resonances in these spectra cannot be assigned to protons of the various components in the bile in a detailed fashion, it is clear that the biles of spectra A and C are different in

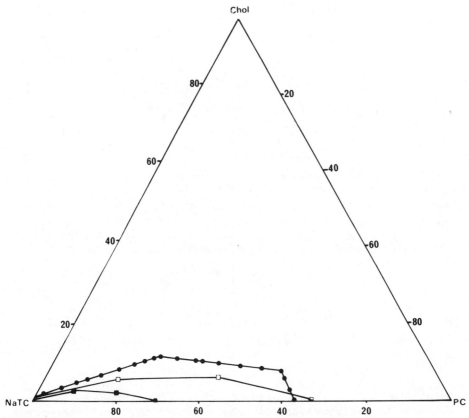

Figure 1. Triangular phase diagram showing the physical state of
aggregation of all combinations of sodium taurocholate, lecithin
and cholesterol in an aqueous solution of 150 mM NaCl at 37°C.
The phase boundary separates compositions at which the system is
mixed micellar (lower part) from compositions at which at least
one other phase (crystals and/or liquid crystals) is present. The
circles represent systems in which the total lipid concentration is
equal to 200 g/ℓ whereas the solid squares represent much more
dilute dispersions (total lipid - 3 g/ℓ). (Both these phase boundaries are taken from the work of Small et al.[9]). The empty
squares represent the non-monomeric composition as defined in the
text.

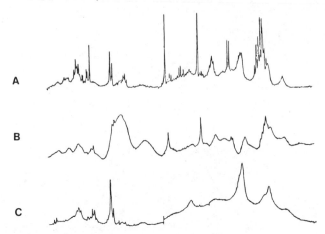

Figure 2.   300 MHz ^{1}H-NMR spectra of hepatic biles of two patients.
The spectra were recorded in D_2O after lyphilization of the native
bile and re-dispersing the residue in D_2O.   Spectra A and C were
of two patients (L.A. and L.L., respectively) and were recorded
after redispersing in D_2O at the original volume.   Spectrum B was
obtained after diluting the bile of spectrum A in D_2O by a factor
of 2.

terms of the state of aggregation of their components.   All the
resonances in spectrum A are clearly narrower than the corresponding peaks of spectrum C.   On the basis of comparison with spectra
of lamellar and micellar dispersions of model bile systems[14,15],
it can in fact be concluded that the hepatic bile of spectrum A is
essentially micellar whereas that of spectrum C is lamellar.   The
mere difference between the two hepatic bile samples is very interesting and deserves further investigation.   However, the most
interesting finding presented in Figure 2 is that dilution of the
bile of spectrum A by a factor of 2 (Figure 2B) is sufficient to
cause changes that make the general spectral features change drastically, such that there is a closer resemblance between spectra B
and C than between spectra B and that of the non-diluted bile A.

One possibility of explaining the dependence of lipid solubilization on the total lipid concentration is that monomeric bile salt
is not involved in the solubilization.   This implies that the

effective concentration of the bile salt is not equal to the bile-salt total concentration and therefore "before experimental results at different lipid concentrations can be compared, it is necessary to subtract the concentration of the aqueous detergent monomers". This view has been pointed out previously[16-19] but it does not yet appear to be widely appreciated. Assuming that the monomer concentration is approximately equal to the detergent's CMC, we have replotted the data of Carey and Small[4] after manipulating them as follows: (1) the actual concentrations of the three components (in mM) have been calculated for each data point on the basis of each component's mole fraction and the total lipid concentration; (2) the CMC of NaTC (2.8 mM) has been subtracted from the total NaTC concentration; (3) the "effective mole fraction" has been calculated from the total concentrations of cholesterol and PC and the corrected concentration of NaTC.

For high total lipid concentrations (e.g. 200 g/ℓ), the above procedure does not change significantly the phase boundary since the CMC of the NaTC is much smaller than the total NaTC concentration. On the other hand, subtracting the CMC from the NaTC concentration in dispersions of low total lipid concentration (e.g. 3 g/ℓ) has a large effect on the calculated molar fraction of the three components. In fact, the phase boundary in the corrected ternary phase diagram of the dilute dispersion is quite similar to the boundary obtained in the much more concentrated solution.

2.   Effect of Total Lipid Concentration on the Solubilization of PC-Cholesterol Vesicles by NaTC

Addition of low concentrations of NaTC to pre-formed sonicated vesicles (liposomes), made of PC and cholesterol (at a 2:1 molar ratio) resulted in a time dependent increased turbidity (Figure 3A). This phenomenon is probably due to NaTC-induced vesicle-vesicle fusion, similar to that described for PC vesicles by Enoch and Stritmatter as leading to the formation of much larger (∿1000 Å diameter) unilamellar vesicles[20]. Upon addition of more NaTC, the turbidity decreased sharply (Figure 3A). Immediately after each addition of NaTC, the turbidity reached a lower value, which did not seem to vary with time. In a vesicle dispersion of 22 mM PC + 11 mM cholesterol, 36 mM NaTC were sufficient to reduce the turbidity to a very low value, which did not change significantly upon further addition of NaTC. To study the solubilization process without the complications due to the initial fusion, we have pre-incubated the sonicated vesicles in the presence of a low concentration of NaTC (12 mM for the dispersion described above) for 48 hours and then studied the effect of additional NaTC on the turbidity of the dispersion of the fused vesicles (Figure 3B). Again, a decrease in turbidity was observed following each addition of NaTC, with no further spontaneous change in turbidity, until a total concentration of 36 mM had been added.

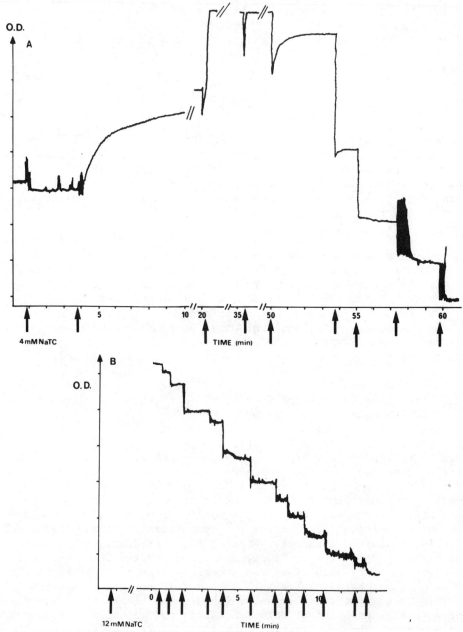

Figure 3. Turbidity measurements (O.D in arbitrary units) of PC (22 mM) cholesterol (11 mM) vesicle dispersions before and after the addition of NaTC. In A, each arrow represents addition of NaTC which corresponds to an increase of 4 mM in its concentration. In B, the sonicated vesicles were incubated in 12 mM NaTC for 48 hours and then titrated with aliquots of 2 mM NaTC.

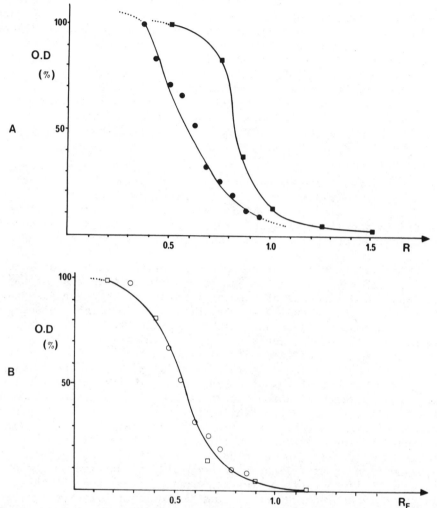

Figure 4. Dependence of the turbidity of fused PC-cholesterol-
NaTC vesicles (as in Figure 3B) on the molar ratio of NaTC to that
of PC+cholesterol. The solid symbols (A) represent the ratio of
total concentrations (R) whereas the empty symbols (B) represent
the effective ratio (R_E see text). In both A and B the circles
are for dispersions in which the sum of concentrations of PC and
cholesterol is equal to 32 mM and the squares are for dispersions
of PC + cholesterol = 8mM. In all cases the PC: cholesterol molar
ratio is equal to 2:1.

Figure 4 describes the turbidity of the dispersion of Figure
3B as a function of the ratio between the concentration of NaTC
and the sum of concentrations of PC and cholesterol. It also

describes the same dependence for a dispersion in which the sum of
PC + cholesterol concentrations is only a fourth of that of Figure
3B. As could have been expected from the phase diagrams of Figure
1, the transformation of the (lamellar) vesicles into mixed micelles occurs at a different ratio (R) of NaTC to PC + cholesterol.
However, the effect of added bile salt on the relative turbidity
of the dispersions appears to be independent of the total lipid
concentration if the added detergent is expressed in terms of the
ratio (R_E) between the non-monomeric bile salt (assumed to be
equal to 2.8 mM) and the sum of concentrations of the other lipids
($R_E = \frac{(NaTC)-2.8}{(PC)+(cholesterol)}$; Figure 4B).

This, of course, strongly supports the assumption that monomeric
NaTC is not involved in the solubilization, which probably occurs
only after a critical ratio of detergent to other lipids (R_E^C) is
exceeded in the lipid bilayer. This critical ratio may not be
reached unless the solution is saturated with detergent monomers.

Since the turbidity of the mixed micelles, formed upon solubilization of the lipid vesicles, is much lower than that of the
vesicles, the dependence of turbidity on R_E can be regarded as a
reflection of the dependence of the percent solubilization on R_E.
Thus, for example, a decrease in the turbidity to a half of its
original value probably means that 50% of the vesicles were trans-
formed into mixed micelles, 25% decrease in turbidity would mean
that 25% of the vesicles were solubilized, and so on (see below
for further support of this conclusion). This interpretation implies that no solubilization occurs at $R_E < 0.2$, all the vesicles
are solubilized at $R_E > 0.8$, while over the range of $0.2 < R_E < 0.8$,
mixtures of mixed micelles and vesicles co-exist. Similar results
have been obtained for the solubilization of bilayers by other
detergents[15,19,21,22] and have been interpreted[19] on the basis of
the assumptions that (1) any given vesicle is ruptured when the
detergent to lipid ratio approaches a value of R_E^C and (2) that
this ratio is approached for various vesicles at different values
of R_E, due to an unequal distribution of detergent molecules between
lipid vesicles. If this distribution is normal (Gaussian) then the
dependence of percent solubilization on R_E (derived from Figure 4B)
should fit a cumulative distribution function of a standardized
normal distribution (of bile salt molecules between lipid vesicles).
For a standardized normal distribution curve [$y = f(u)$], the
standardized cumulative normal distribution curve [$y = \Phi(u)$] has
a well defined value for any given value of u^{23}, where $u = \frac{X-\xi}{\sigma}$
(ξ being the mean and σ^2 the variance of the distribution). Thus,
($X-\xi$) should depend linearly on u and the slope of this function
would then give the value of σ. u values for each value of $\phi(u)$
(percent solubilization) are available from statistical theory[23].
If the value of $R_E = 0.53$, where 50% solubulization occurs, is
taken to be equal to the mean (ξ) and the difference between the

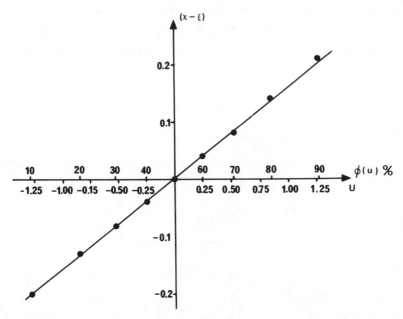

Figure 5.   Dependence of the solubilization of PC-cholesterol
vesicles by NaTC on the effective molar ratio R_E.   Percent solubilization [ϕ (u) %] is taken from Figure 4B (Percent solubilization =
100 % OD) and also expressed in terms of the normalized value U,
derived from the assumption that the percent solubilization curve
is the standardized cumulative normal distribution $\phi(u)$.   (Taken
from Reference 23).   The value of R_E is given in terms of the
difference between the actual R_E and the "mean" R_E (X - ξ).   The
solubilization is thus described by the dependence of the difference
(X-ξ) on the value of U.   The "mean" R_E is that value of R_E at
which 50% solubilization occurs (also denoted in the text R_E^c).

value of R_E at various solubilization percentages (X) and ξ is
then plotted against the value of u^{21}, a linear dependence is
obtained (Figure 5).   This, no doubt, supports the assumption that
Figure 4B is a reflection of a normal distribution and gives a
value of 0.023 for the variance (σ^2) of this normal distribution.

   Our interpretation of the turbidity of the PC-cholesterol-
NaTC ternary systems is consistent with the ^{1}H-NMR spectra of these
mixtures (Figure 6A and B).   In the absence of NaTC, most of the
intensity of the PC methylene groups protons is broadened beyond

detection due to the tight packing of the hydrophobic core constituents, whereas the complete intensity of the choline methyls
protons is observed as a resonance of a linewidth of ca 8.5 Hz.
Thus, the ratio of integrated intensities (methylene/choline) has

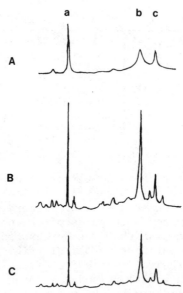

Figure 6. ^{1}H-NMR spectra of: A. sonicated vesicles made of 60 mM
PC + 30 mM cholesterol; B. the same vesicles after solubilization
with NaTC (82.5 mM); C. a dispersion made by lyphilizing sample B,
dissolving the residue in chloroform, evaporating the solution to
dryness and re-dispersing the residue in D_2O. All the dispersions
had 150 mM NaCl in D_2O. Resonance a is due to choline head group
methyls; b, due to methylene groups; c, due to methyl groups of
NaTC and cholesterol and terminal methyl groups of the PC paraffinic chains.

a low value of about 1.7. Addition of NaTC to an R_E of 0.25 had
no significant effect on the latter ratio, (Figure 7B) while the
apparent linewidth of the choline signal was markedly reduced
(Figure 7A). In fact, the latter signal was narrowed at R_E = 0.3
to ca. 3Hz and further addition of NaTC did not cause any signifi-

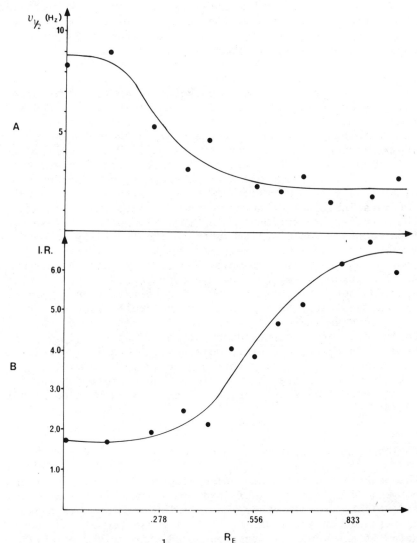

Figure 7. Dependence of ^{1}H-NMR spectral parameters on the effective ratio of NaTC to PC+cholesterol (R_E). A describes the dependence of the linewidths of the choline head group signal (a); B describes the dependence of the ratio of integrated intensities of resonance b to that of resonance a in Figure 6 (I.R.=integration ratio). Linewidths are expressed as the difference between their measured values and the width of the HOD signal in the same spectrum. Every result is an average of at least four consecutive scannings under the same conditions of RF and modulation. Care was taken to ensure reproducibility in the base line determination. This was achieved by the use of the base line of the corresponding spectrum determined at spectral width of 40 Hz/cm.

cant further narrowing. On the other hand, the methylene/choline integrated intensity ratio increases with increasing R_E in a sigmoidal fashion, so that for $R_E > 0.8$, an increase in R_E is not accompanied by changes in the intensity ratio.

These dependencies of the choline head group linewidth and methylene/choline intensity ratio on R_E could in fact have been expected from our interpretation of the turbidity data. The NMR spectrum of a mixture of vesicles and mixed micelles, between which the exchange of PC molecules is slow on the NMR time scale[15], is a superposition of the spectra of vesicles (Figure 6A) and mixed micelles (Figure 6B). The sigmoidal function of the intensities ratio (Figure 7B) is therefore a reflection of the dependence of the solubilized (mixed micelles) PC on R_E. Within experimental error, this solubilization curve is identical to that obtained from the turbidity measurements. As the NMR spectra are especially sensitive to the mixed micelles, whereas the turbidity is predominantly, if not exclusively, due to the much larger lipid vesicles, these two techniques are of complementary nature and the agreement between them strongly supports our interpretation of the results obtained from both these techniques. The co-existence of mixed micelles and vesicles, between which PC molecules exchange only slowly, is further supported by the finding that for $R_E > 0.3$, the choline head group signal does not depend on R_E. Only in the absence of a fast exchange of PC molecules, the narrow signal due to micellar PC, will be the governing factor in determining the apparent linewidth of the choline signal, which is a superposition of the narrow resonance of micellar PC and a much broader signal of lamellar PC.

3.   Cholesterol Supersaturation in Model Bile

The phase boundary in the phase diagram presented in Figure 1 is between "completely solubilized" (mixed micellar) phase and mixtures of this and all other phases. The data of this curve can be presented in terms of the minimal ratio between the concentration of non-monomeric detergent and that of the solubilized lipids that is just sufficient for complete solubilization (R_E^S)[25]. The latter parameter is, no doubt, a function of the lipid composition (Figure 8). For systems in which the cholesterol/PC ratio is larger than 0.2, R_E^S can be expected to rise linearly with increasing cholesterol/PC ratio. It is then expected to be solubilized only at $R_E^S = 2.0$ at cholesterol/PC ratio lower than 0.5, while for concentration ratios higher than 0.5, a much more drastic effect of increasing the concentration ratio is expected. Our data suggest that lamellar aggregates of lecithin and cholesterol at a ratio of 2:1 are completely solubilized when $R_E \simeq 1.0$ (Figures 4B and 7B). This value is much lower than that expected for the same ternary system at equilibrium.

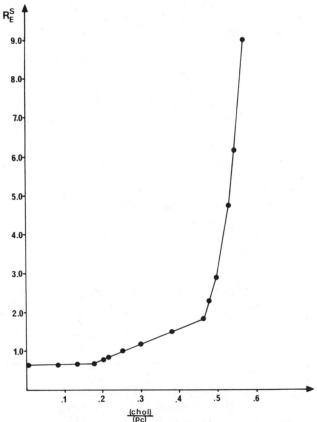

Figure 8. Calculated dependence of R_E^S (see text) on the ratio of cholesterol to PC (data taken from Carey and Small[4], Figure 1 for 200 g/ℓ total lipids).

In view of the large difference between the experimental R_E^S and the value expected for this parameter from Carey and Small's data on the equilibrium systems[4], we can conclude that the mixed micellar system obtained by completely solubilizing PC-cholesterol vesicles with minimal amount of NaTC is not "at equilibrium", i.e., it is supersaturated with cholesterol.

This conclusion is supported by the results presented in Figure 6C. In this experiment the transparent mixed micellar dispersion of Figure 6B was lyophilized to dryness and the residue was dissolved in chloroform, to ensure appropriate mixing of the three components. The solution was then evaporated to dryness and the residue was re-dispersed with the original volume of D_2O. The resultant dispersion was turbid and gave rise to a much less intense NMR spectrum (Figure 6C).

This result clearly demonstrates that while the dispersions studied by Carey and Small[4] and other investigators were probably at equilibrium, since they were made by co-dissolving the three components in a common solvent before being evaporated and dispersed in water, solubilization of lecithin–cholesterol bilayers by bile salts does not necessarily lead to an equilibrium state. Metastable states, which are supersaturated in cholesterol, can in fact be formed and be relatively stable. These and other metastable states are currently under investigation in our laboratories.

ACKNOWLEDGEMENTS

The authors wish to thank the Ministry of Health of the State of Israel for the financial support of this research and Dr. Giora Somjen for helpful discussions.

REFERENCES

1. M. M. Fisher, in "Gallstones", M. M. Fisher, C. A. Goresky, E. A. Shaffer and S. M. Strasberg, Editors, p. 1, Plenum Press, New York, 1978.
2. B. J. Pearlman and L. J. Schoenfield, Med. Clin. North Amer., 62, 87 (1978).
3. L. J. Bennion and S. M. Grundy, New England J. Medicine, 23, 1161 (1978).
4. M. C. Carey and D. M. Small, J. Clin. Invest., 61, 998 (1978).
5. B. Isaksson, Acta Soc. Med. Upsal., 59, 296 (1953-4).
6. F. Nakayama, Clin. Chim. Acta, 14, 171 (1966).
7. M. Bourges, D. M. Small and D. G. Dervichian, Biochim. Biophys. Acta, 144, 189 (1967).
8. D. M. Small, M. Bourges and D. G. Dervichian, Nature, 211, 816 (1966).
9. D. M. Small, in "The Bile Acids", P. P. Nair and D. Kritchevsky, Editors, Volume 1, Chapter 8, Plenum Press, New York, 1971.
10. R. T. Holzbach, M. Marsh, M. Olszewski and K. Holan, J. Clin. Invest., 52, 1467 (1973) and reference cited therein.
11. T. C. Northfield and A. F. Hofmann, Grut, 16, 1 (1975).
12. R. A. Smallwood, P. Jablonski and J. McK. Watts, Br. Med. J., 4, 263 (1972).
13. C. Huang, Biochemistry, 8, 344 (1969).
14. D. M. Small, S. A. Penkett and D. Chapman, Biochim. Biophys. Acta, 176, 178 (1969).
15. D. Lichtenberg, Y. Zilberman, P. Greenzaid and S. Zamir, Biochemistry, 18, 3517 (1979).
16. W. Shankland, Chem. Phys. Lipids, 4, 109 (1970).
17. A. Helenius and K. Simons, Biochim. Biophys. Acta, 415, 29 (1975).

18. G. W. Stubbs and B. J. Litman, Biochemistry, 17, 215 and 220 (1978).
19. M. L. Jackson, C. F. Schmidt, D. Lichtenberg, B. J. Litman and A. D. Albert, Biochemistry, in press (1982).
20. H. G. Enoch and P. Strittmatter, Proc. Nat. Acad. Sci. U.S.A., 76, 145 (1979).
21. E. A. Dennis and J. M. Owens, J. Supramol. Struct., 1, 165 (1973).
22. S. Yedgar, Y. Barenholz and V. G. Cooper, Biochim. Biophys. Acta, 363, 98 (1974).
23. A. Hald, in "Statistical Theory with Engineering Applications", John Wiley and Sons, Inc., New York, 1952.
24. M. P. N. Gent and J. H. Prestegard, Biochemistry, 13, 4027 (1974).
25. D. Lichtenberg, R. J. Robson and E. A. Dennis, Biochim. Biophys. Acta, Reviews in Biomembranes, in press, 1983.

FURTHER INVESTIGATIONS ON THE MICELLAR SOLUBILIZATION OF BIO-

POLYMERS IN APOLAR SOLVENTS

P. Meier, V.E. Imre, M. Fleschar and P.L. Luisi

Technisch-Chemisches Laboratorium, ETH-Zentrum
Universitätsstrasse 6
CH-8092 Zurich, Switzerland

This paper is a follow-up to our previous publications about the solubilization of biopolymers in hydrocarbons via reverse micelles formed by di(2-ethyl-hexyl)
sodium sulfosuccinate (AOT). We show first data pertaining to the solubilization of RNA and DNA in isooc-
tane-AOT-reverse micelles containing 1 - 1.5 % water.
The extent of RNA solubilization does not show marked
influence on pH change, which appears interesting in
view of the fact that both RNA, and AOT micelles are
negatively charged polyelectrolytes. High molecular
weight DNA assumes in reverse micelles a structure
which can be defined as "condensed" - as judged from
circular dichroic spectra, showing the typical appearance of a ψ form. The intensity of the dichroic bands
increases by adding Mg^{2+} ions. Apparently positively
charged ions, by shielding the negative charges of the
anionic guest biopolymers, permit an even closer packing of DNA in the water pools of reverse micelles.
We also report preliminary data on different aspect of micellar chemistry, i.e., the transport properties of protein-containing reverse micelles from one
liquid phase to another. In particular we have studied
the phase transfer of proteins from a water phase to a
supernatant hydrocarbon phase (forward transfer), their
transfer from a hydrocarbon phase (backward transfer),
and the transfer of proteins from a first water phase
to a second water phase across a hydrocarbon phase,
which contains the AOT micelles (double transfer). This

last configuration simulates the transfer of macromolecules across membranes. Various parameters affect the
yields and the rate of such transport, in particular
pH and ionic strength.

INTRODUCTION

In the last couple of years we have shown that enzymes such
as α-chymotrypsin, ribonuclease, alcohol dehydrogenase, alkaline
phosphatase, lipoxygenase, peroxidase, lysozyme, etc. can be solubilized in an AOT/Isooctane solution [AOT = di(2-ethyl-hexyl) sodium sulfosuccinate] while maintaining their catalytic activity[1-4].
Similar studies have been carried out in other laboratories, for
example, Martinek[5], Menger[6] and Douzou[7]. The original notion of
enzymes in reverse micelles can be traced back to the work on phospholipase by Wells and coworkers[8]. A model which provides the dimensions and other structural parameters of protein-containing micelles has been recently presented[9].

The "water-shell model", according to which a hydrophilic enzyme is in the water pool of the reverse micelles and is separated
from the inner surface by a layer of water, is thought to best represent an enzyme in reverse micelle situation. This model, in addition to being the most appealing on the basis of common sense, is
indirectly supported by several observations: i) the fluorescence
properties of α-chymotrypsin[4] and horse liver alcohol dehydrogenase[2]
in reverse micelles are similar to those obtained in bulk water;
this suggests that enzymes in reverse micelles experience basically
an aqueous environment. ii) enzymes in reverse micelles exhibit kinetic behavior and turnover numbers which are rather close to those
found in bulk water[3,4]; this is also pointing to an aqueous environment. iii) the initial velocity is markedly influenced by the percentage of water in the hydrocarbon micellar solutions (that is, by
the W_o value, where $W_o = \frac{H2O}{AOT}$) and iv) the protein conformation, as
investigated by circular dichroism[3,4] (CD), is also often markedly
affected by W_o.

With regard to the last two points, which indicate a direct
interaction between water and the biopolymer, it should be recalled
that the dimensions of the reverse micelles increase with an increase in the water content. In other words, it appears that the
enzyme velocity, as well as the enzymes' conformation can be modu-
lated by simply adjusting the amount of water in the AOT/hydrocar-
bon solution.

The situation outlined above lends itself to several independent lines of investigation. At present, the following appear to us among the most interesting: i) the investigation of the nature of the water surrounding the biopolymer in the water pool, and how this and, possibly, other factors affect the activity and conformation of the guest biopolymer; ii) whether and to what extent enzyme-containing micelles can be utilized technologically, e.g. for the catalytic transformation of lipophilic, water insoluble substrates; iii) whether and to what extent such micellar structures exist in vivo, for example, as a protective or transport device for proteins in membranes. In this regard, a few interesting papers have appeared recently[10],[11],[12]; iv) whether functional biopolymers other than proteins (e.g. nucleic acids) can be solubilized in reverse micelles, and what are the consequences of this; and v) whether and to what extent the reverse micelles can function as transport-vehicles for biopolymers.

In this paper, we will present some experimental observations relative to the last two lines of investigations. In particular, we will discuss some experiments describing the properties of DNA solubilized in AOT-reverse micelles, and some experiments describing phase transfers of protein-containing reverse micelles. These are rather preliminary experimental observations, and results have not yet been incorporated into a solid theoretical framework. Nevertheless, we believe it is important to present these data as they represent quite novel findings which can lead to novel exploitations of reverse micellar systems.

RESULTS AND DISCUSSION

a) The Case of Nucleic Acids

RNA and DNA can be solubilized without denaturation in isooctane with the help of reversed AOT micelles. Optically clear micellar solutions ($OD_{340} < 0.008$) can be obtained over a narrow range of ionic strength, biopolymer concentration and bulk solvent type only. It appears that the possibility of solubilization reflects an interplay of all the physical properties of the micellar system. A full scale investigation of all these factors has not yet been undertaken. In addition to our previous studies, we describe here some more data of general nature. Figure 1 shows the dependance of the transfer of RNA (MW = 20,000 - 30,000 Daltons) on the pH of the starting buffer solution.

These experiments have been carried out by extracting crystalline RNA, with mild stirring in the AOT-isooctane solution. Conditions are given in the figure legend. The influence of pH is very

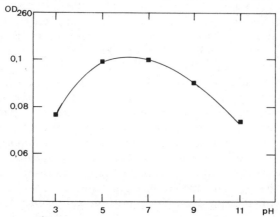

Figure 1.   Solubilization of RNA in isooctane-AOT(50mM)-water reverse micelles. The isooctane-AOT solution, containing also the buffer solution to a W_O = 14.8, is added to RNA in powder form in a
reaction vial, and the percentage of solubilized material is measured
by reading the optical density of the clear solution at 260 nm after
12 hrs. An universal buffer was used in the pH range investigated,
e.g. a mixture of boric acid, KH_2PO_4, citric acid, and sodium 5,5-
diethyl-barbiturate, each 10 mM. The pH was adjusted in the buffered
stock solution, which was then added to the AOT-isooctane solution
with a microsyringe up to the desired W_O. Attempts to extract RNA
in the same way, but without water in the isooctane phase, failed.

slight in the range investigated, with a maximum at about pH 6. It
should be noted that the pH reported in the abscissa of Figure 1 is
the pH of the stock aqueous solution injected into the micellar·sol-
utions in order to obtain the desired W_O value. It is generally
known, and we have reported this also for our systems[14], that the
pH of the water pool may differ somewhat (it is generally lower)
from that of the initial bulk water solution. The precise evaluation
of such pH changes is a difficult problem both theoretically and
experimentally. However, this correction would not affect the main
conclusion to be deduced from Figure 1, i.e., there is no marked
influence of the pH on solubilization.

One may recall in this regard that the pK of the second phosphate dissociation is 7.2. Therefore, we conclude that the density
of negative charges on a guest macromolecule does not affect its
uptake into the reverse micelles. Notice that AOT is a negatively
charged surfactant, and one might expect some inhibition with the
guest polyion. Perhaps under our conditions, this possible repulsion
effect is superseded by the improved solubility of the guest biopolymer in the water pool.

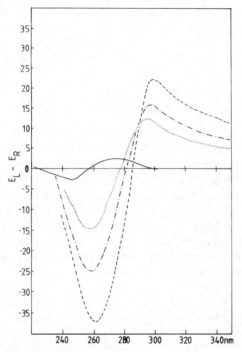

Figure 2. Circular dichroism spectra of DNA (MW = 250,000 Daltons) in aqueous solution (10 mM borate buffer, pH 9.0) in the absence of added magnesium (——); and in 50 mM AOT-isooctane solutions (W_O = 18.5): no added magnesium (···); $[Mg^{2+}]$ = 6 mM (-·-·-); $[Mg^{2+}]$ = 12 mM (---).

We would like to examine some of the circular dichroism properties and their relation to the conformational properties of the
guest biopolymer. Figure 2 gives the CD-spectrum of DNA (Mol. weight
250,000 Daltons) in the AOT-isooctane system with Mg^{2+} present. As
we have already postulated on the basis of experiments carried out
without added ions, such a CD-spectrum is reminiscent of the so-
called ψ-form[15,16]. This should correspond to a situation where DNA
is tightly packed (condensed) into a restricted space.

Note that the positive and negative Cotton effects increase
with an increase in the Mg^{2+} concentration of the DNA stock solution.
This could be interpreted to mean that the positively charged Mg^{2+}
ions neutralize the negative charges of the DNA, thus contributing
to a more tightly packed DNA structure within the reverse micelles.

b) Phase Transport of Biopolymers using Reverse Micelles

Let us now consider another rather unrelated field, where some
progress has been made in our laboratory. This concerns the use of
reverse micelles as vehicles to transport proteins from one phase
to another. These experiments can be carried out with simple glass
vials. The three experimental systems, schematically shown in Figure 3 were used: the forward transfer (Figure 3a) in which the protein is transferred from an aqueous medium, A, to a supernatant
hydrocarbon medium, B, containing the "empty" (no biopolymer) micelles of AOT; the backward transfer (Figure 3b), where the biopolymer contained in the hydrocarbon micellar solution is transferred
to an aqueous solution, which is particularly useful if one wants
to recover an enzyme previously solubilized in hydrocarbon reverse
micelles; and a double transfer, as illustrated in Figure 3c, in
which a protein is transferred frome one aqueous phase, A_1, into
another aqueous phase A_2, through a hydrocarbon solution, B, containing "empty" micelles. This last situation resembles the situation in which a lipophilic liquid membrane separates two aqueous
phases - a model for biological membranes. Preliminary experiments
of this third type have already been described by us[17] and by
Lehn et al.[18]. The experimental conditions used in our experiments
are briefly described in the figure legends.

First let us consider the forward transfer experiments, for
which trypsin, peroxidase and ribonuclease were used. Figure 4 illustrates the results, in particular the influence of pH and the
tpye and concentration of the ions present in the pH water phase on
the percentage of transfer. In several cases experimental difficulties were encountered. Most notably, there was turbidity at the
phase boundary. Generally, a higher turbidity was brought about by

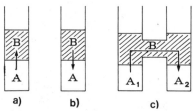

Figure 3. The three experimental systems used for transport of proteins via reverse micelles. A represents an aqueous solution, B the
AOT-hydrocarbon solution (isooctane in our case). Generally, the
vials, which are tightly stoppered, were mildly shaken in a IKA VIBRAX
VXR shaking machine at room temperature. The stirring was interrupted
to permit the removal of small aliquots for measuring the optical
density at 280 nm as a measure of the protein concentration.

a lower ionic strength and a higher AOT concentration. In the figure, only the most representative transfer experiments are reported.

    Under optimum conditions, which for the forward transfer were
found to be 0.1 M NaCl at pH 10, both trypsin and ribonuclease were
transferred nearly quantitatively into the hydrocarbon phase. The
data obtained for trypsin demonstrate the very different effects of
the $CaCl_2$ solution in comparison to that of the optimal NaCl and
NaAc solutions. It is possible that this behavior is due, to some
extent, to the larger charge to radius ratio of Ca^{2+}. Note also that
at pH 10 the transfer of trypsin and ribonuclease is more efficient.

    Peroxidase, which is structurally different from the other two
proteins (it is a glycoprotein with mol-weight 40,000 and contains
a heme group) could be transferred to a substantially less extent.
The highest amount of transfer occurred at pH 4 using either NaCl
or $CaCl_2$, with transfer values of 55 % and 40 %, respectively.

    The backward transfer data are shown in Figure 5. In this case,
the equilibrium values were reached much faster than in the forward
transfer experiments, because the solution could be shaken much more
vigorously without formation of turbidity at the boundary. Some of
the systems could, in fact, be shaken to such an extent that the
equilibrium was reached after about one hour. The most efficient

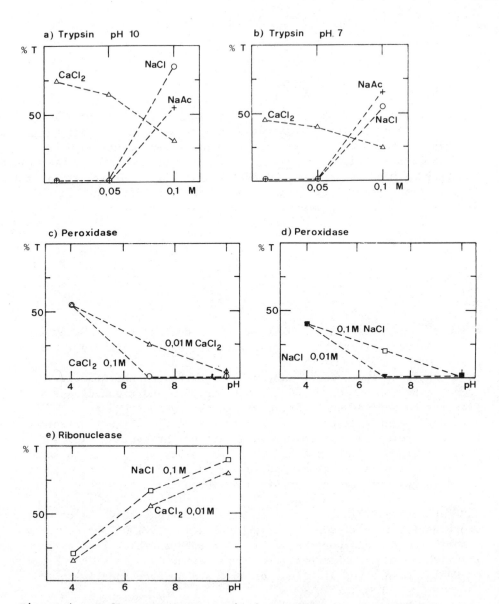

Figure 4. Influence of pH and ionic strength on the efficiency of forward transfer. Readings after ca. 2 days, when the optical densities were constant. The experimental uncertainty of each data point in this and the next figure is about ±5 %, and |AOT| is 50 mM unless otherwise specified. (NaAc = sodium acetate; NaP = sodium phosphate)

ions in the water phase A were found to be sodium phosphate and sodium acetate at high concentrations for both ribonuclease and trypsin.
Furthermore, throughout better results were obtained at pH 10 than
at lower pH values, and this was even more pronounced for trypsin
than for ribonuclease (compare a, b vs d in Figure 5). In the case
of trypsin, the dependence of transfer on the AOT concentration was
also investigated. No pronounced dependence was detected (Figure 5b).

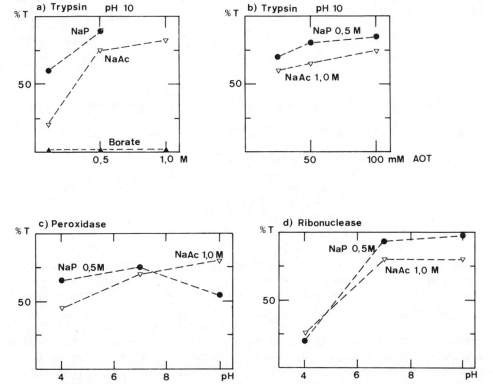

Figure 5.   Influence of various factors upon the efficiency of backward transfer.

The dependence of trypsin transfer upon enzyme concentration was also investigated, but there was no significant effect in the concentration range between 3 and 20 μM.

The behavior of peroxidase is (Figure 5c) different from the above mentioned enzymes. Between pH 4 and pH 10 there is almost no difference in the transfer yield, and the optimum is pH 7 when 0.5 M sodium phosphate is used.

Let us consider double transfer experiments (Figure 3c). From the standpoint of maximal transfer, 0.1 M NaCl and 0.01 M $CaCl_2$ were found to be best suited for the forward step $A_1 \rightarrow B$ (the enzyme is contained in A_1); and 0.1 M NaAc as well as 0.5 M sodium phosphate for the backward step ($B \rightarrow A_2$). The combination of $CaCl_2$ and sodium phosphate could not be used since calcium phosphate is only sparingly soluble and thus precipitated from solution. It was found that the 0.1 M NaCl, pH 10/AOT-isooctane; 25 mM/NaAc 1.0 M; pH 10 combination was best to provide a clear three phase system with an efficient rate of enzyme transfer. Thus, using this system, ca. 45 % of the trypsin could be transferred within 25 hours. No additional enzyme transfer was found after a further 24 hour period.

Table I.   Double Transfer of Ribonuclease using AOT-Reverse Micelles.

Phase A_1[a)	Phase B[b)	Phase A_2[c)	% Transfer[d)	% Activity[e)
$CaCl_2$	25	Acetate pH10	15	65
$CaCl_2$	25	Acetate pH7	15	75
NaCl	25	Acetate pH10	30	85
$CaCl_2$	50	Acetate pH7	30	80
NaCl	50	Acetate pH7	50	80
NaCl	50	Phosphate pH7	35	25

Experiments were carried out with glass apparatus as in Figure 3c and with initial volumes of 5 ml for the phase A_1 and A_2, and 8 ml for B. Initial concentration of the enzyme in A_1 was in the range 10 - 20 μM.
a) $CaCl_2$ 0.01 M, NaCl 0.1M; b) containing AOT in isooctane at the given concentration (mM); c) Acetate 1.0 M, Phosphate 0.5 M; d) data are obtained by making spectroscopic measurements at 280 nm on the protein concentration in A_2 and then corrected for the volume increase of the phase A_2 due to osmotic effects (see text); e) Activity of ribonuclease in A_2, measured under standard conditons in water solution at pH 7.0[19] and with respect to an equally aged control in water solution.

Substitution of sodium acetate by sodium phosphate in the third
phase resulted in 5 - 10 % less transfer; varying the AOT-concentra-
tion produced no significant difference. Table I summarizes the most
successful combinations of phase components for the double transfer
or ribonuclease. Note that a large ionic strength gradient between
A_1 and A_2 is necessary. A decrease in the concentration gradient of
both water phases leads to a decrease in enzyme transport. When this
high ion gradient was maintained, with the same initial concentration
of the enzyme in both A_1 and A_2, a transport of enzyme $A_1 \rightarrow A_2$ was
still observed. For example, at equilibrium (after ca. 3 days) ribonuclease concentration in the second water phase was 65 % of the total enzyme present. This means that during the time necessary to
reach the equilibrium, the enzyme is transported against its own concentration gradient. Conditions for this experiment were: A_1 (NaCl
0.1M pH 10), B (AOT 25 mM), A_2 (NaAc 10M pH 10), $[E]$ 20 μM.

As shown in Table I, the activity of ribonuclease which was
back-transferred in the acetate solution was rather good in contrast
to the back-transfer in the phosphate solution. The activity was
measured in water solution and expressed with respect to an equally
aged control. Trypsin was not well suited for this study, since the
water reference solution inactivates too rapidly.

For ribonuclease, the activity of the double transferred enzyme
was not different from that of the simple back-transfer. The example
of ribonuclease thus shows that under the right condition, an enzyme
can be solubilized in the hydrocarbon micellar solutions, and transferred back from this into an aqueous solution without undergoing
appreciable denaturation. Note that the conditions for the maximal
activity should not be considered to be optimized, on the basis of
the preliminary experiments reported here. Optimization of temperature, apparatus configuration, and concentrations could certainly
yield further improvement in the activity retained.

An interesting phenomenon was observed during the double transfer experiments carried out when NaCl was present in A_1 and NaAc in
A_2 (see Table I). This effect was interesting enough to warrant independent investigations and the water transport was then investigated
in the absence of enzymes. Some data are summarized in Table II.
Experiments were carried out with the same glass apparatus used for
the experiments in Table I. The equilibrium was not achieved before
3 - 5 days, a time which was greater (by a factor of ca. 1.5) than
the time necessary to collect the equilibrium data for enzyme transfer of Table I. In this time interval, as much as 4 ml of water of
A_1 (out of total 5 ml) were transferred as shown in Table II.

Together with water, ions must migrate to varying extents from

Table II. Osmotic Water Transfer and Ion Transport in Double Transfer Experiments.

Phase A$_1$ (M)	Phase A$_2$ (M)	Experimental Observations
NaCl 0.1	NaAc 1.0	ΔVolume 80 %
NaCl 0.1	NaCl 1.0	ΔVolume 70 %
NaCl 0.1	NaAc 0.25	ΔVolume 40 %
Na I 0.1	PbAc$_2$ 0.7	yellow color (PbI) only at A$_1$B phase boundary after 15 min.
NaNCS 0.1	NaAc 1.0 FeNH$_4$(SO$_4$)$_2$	(FeNCS) only at A$_1$B phase boundary (after several hours)

one phase to the other. We have designed some experiments in order to demonstrate this transfer by using ions which would form coloured salts. Some of the results are summarized in Table II. Note that the colour could only be observed at the A$_1$B boundary, indicating that the cations are readily transported from A$_1$ to A$_2$.

It was also shown that anions such as phosphate, chloride, bromide and iodide also migrate from one water phase to the other through the hydrocarbon phase but could analytically be detected in the other water phase only after a number of days.

These experiments show that the ions examined migrate at different rates. In turn, this indicates a certain selectivity of the AOT-isooctane micelles with respect to various ions.

CONCLUDING REMARKS

The solubilization of nucleic acids in reverse micelles complements in an interesting and useful way the studies on enzymes reported earlier. The problem is the same as for proteins (e.g. influence of W_o, dimensions and size of micelles, flexibility of the guest polymers, etc.) but the biological and (perhaps) technological significance is different. For example, it would be interesting to examine to what extent the nucleic acids in reverse micelles are susceptible to nuclease action, and whether such systems can be useful in biotechnology. As already mentioned[13], the novel DNA/micellar aggregate can be considered a model for tightly condensed (super-coiled?) DNA, such as that found in chromatin or phage heads.

The transport data may also offer prospects which are interesting both from the biological as well as from the technological points of view. The phase transfer can be roughly visualized in these terms: the AOT-reverse micelles, localized in the hydrocarbon phase, are able to extract the protein from the water phase. It is surprising that a hydrophilic protein can be solubilized to such an extent in the hydrocarbon phase, albeit surrounded by a layer of water (if one accepts our water-shell model[20]). The energetics accompanying this transport must still be understood. Particularly the first steps of the transfer, which must take place at the boundary between the two phases, are especially intriguing. A few proposals can be made, but at this stage they are all bound to be very speculative.

The backward transfer, and the double transfer have a practical interest inasmuch as they permit the recovery of hydrophilic enzymes from the hydrocarbon micellar solutions. Our data show that this recovery can be made with only moderate losses in activity. The double transfer, as well as the forward transfer, could in principle be used as chromatographic methods, as the transport rates of different enzymes are different from each other. We have shown this before by using a positively charged surfactant[17]. Biopolymers might be separated from one another by adjusting the pH and/or the ionic strength of the water phases, and the structure and size of the micelles in B. An ionic strength gradient from A_1 to A_2 is always necessary, of course. An interesting question about the double transfer experiments is, to what extent they simulate the transport of macromolecules across phase boundaries occurring in vivo. There are not many systems described in the literature which are germane to the transport of proteins across a lipophilic barrier separating two aqueous phases. Also in view of the growing interest in the biological relevance of reverse micellar systems[11,12], the biological aspects of phase transfer of macromolecules contained in reverse micelles are certainly worthy of further investigation.

REFERENCES

1. R. Wolf and P.L. Luisi, Biochem.Biophys.Res.Commun., 89(1), 209 (1979)
2. P. Meier and P.L. Luisi, J. Solid-Phase Biochem., 5(4), 269 (1980)
3. C. Grandi, R.E. Smith and P.L. Luisi, J.Biol.Chem., 256(2), 837 (1981)
4. S. Barbarić and P.L. Luisi, J.Am.Chem.Soc., 103, 4239 (1981)

5. K. Martinek, A.V. Levashov, N.L. Klyachko and I.V. Berezin, Dokl.Akad. Nauk SSR, 236, 951 (1978)
6. F.M. Menger and K. Yamada, J.Am.Chem.Soc., 101, 6731 (1979)
7. P. Douzou, C. Balny and F. Franks, Biochemie, 60, 151 (1978)
8. R.L. Misiorowsky and M.A. Wells, Biochemistry, 13, 4921 (1974)
9. F.J. Bonner, R. Wolf and P.L. Luisi, J. Solid-Phase Biochem., 5(4), 255 (1980)
10. B. de Kruijff, P.R. Cullis and A.J. Verkleij, Trends Biochem. Sci., 5, 79 (1980)
11. B. de Kruijff and P.R. Cullis, Biochem.Biophys. Acta, 602, 477 (1980)
12. A. Sen, W.P. Williams, A.P.R. Brain, M.J. Dickens and P.J. Quinn, Nature, 293, 488 (1981)
13. V.E. Imre and P.L. Luisi, Biochem.Biophys.Res.Commun., 107, 538 (1982)
14. R.E. Smith and P.L. Luisi, Helv.Chim. Acta, 63, 2302 (1980)
15. C.F. Jordan, L.S. Lerman and J.H. Venable Jr., Nature New Biol., 236, 67 (1972)
16. L.S. Lerman, in "Physico-Chemical Properties of Nucleic Acids", J. Duchesne, Editor, Vol. 3, p. 59, Academic Press, London, New York (1973)
17. P.L. Luisi, F.J. Bonner, A. Pellegrini, P. Wiget and R. Wolf, Helv.Chim. Acta, 62, 740 (1979)
18. J.P. Behr and J.M. Lehn, J.Am.Chem.Soc., 95, 6108 (1973)
19. M. Litt, J.Biol.Chem., 236, 1786 (1961)
20. R. Wolf and P.L. Luisi, in "Solution Behavior of Surfactants. Theoretical and Applied Aspects", K.L. Mittal and E.J. Fendler, Editors, Vol. 2, p. 887, Plenum Press, New York (1982)

Part V
Micellar Catalysis and Reactions
in Micelles

MICELLAR EFFECTS ON REACTION RATES AND EQUILIBRIA

L. S. Romsted

Department of Chemistry

Rutgers The State University of New Jersey
New Brunswick, New Jersey 08903

The pseudophase ion exchange model provides, within limits, a unified quantitative interpretation for micellar effects of ionic micelles over a wide range of experimental conditions. The model successfully describes the effect of surfactant charge and chain length, buffers, pH and specific salt effects. Micelles are assumed to act as a separate phase, a pseudophase, of constant properties independent of the composition of the medium. Changes in observed rate constants and apparent ionization constants are described in terms of independently verifiable distribution constants for reactants between the aqueous and micellar pseudophases. A mathematical treatment is outlined and compared briefly with other published approaches. Expressions for observed rate and apparent ionization constants predict the effect of added surfactant and salt for a number of reactions over a wide range of experimental conditions. Virtually all the results published to date, which test the assumptions of the pseudophase model, are collected in Tables. The model also explains the effect of cationic surfactants on indicator equilibria in acidic solution and the effect of anionic surfactants on reaction in basic solution. The pseudophase model fails near the cmc because of specific interactions between surfactant and reactants but also, for reasons not currently understood, for reactive counterion surfactants having very hydrophilic counterions such as OH^- or F^-. Micellar effects on bimolecular reactions between organic substrates and nucleophilic

anions of weak acids are attributed the the product of
micellar induced shift in the apparent ionization constant of the acid and the concentration of the reactants
within the micellar pseudophase. An approach to a
general quantitative treatment of functional, mixed and
comicelles on reaction rates and equilibria is outlined.

INTRODUCTION

In 1934, G.S. Hartley developed a simple set of rules for
interpreting the effects of surfactant solutions on the apparent
acidity constants of visual indicators.[1] Much later, in 1959,
Duynstee and Grunwald showed that anionic surfactants inhibit while
cationic surfactants enhance the rate of alkaline fading of some
triphenylmethane dyes and sulfophthalein indicators.[2] This seminal
paper marked the beginning of the still rapidly growing field of micellar catalyzed reactions. Initial work focused on micelle catalyzed reactions as models for enzyme catalyzed reactions[3], but the
field expanded rapidly to include micellar induced regio and stereoselectivity, and photochemistry, focused in part on energy storage
and conversion and artificial photosynthesis.[4-11] Early experiments
were carried out primarily in dilute aqueous solutions of ionic micelles but recent work includes the effects of microemulsions[12], reversed micelles[13] monolayers[14] and vesicles.[15] Progressively more
sophisticated quantitative treatments of micellar effects on reaction rates and equilibria developed over the same time period. This
literature was reviewed by Bunton and Romsted in 1979[16], by E.J.R.
Sudholter and coworkers in 1980[17] and again by Fendler in 1982.[4]
Chaimovich and coworkers showed recently that the pseudophase ion
exchange model provides a unified approach for interpreting many of
these effects.[18] This model assumes micelles act as a separate medium whose properties are insensitive to composition solution and
that observed rate changes and pK_a shifts depend primarily on the
distribution of reactants, transitions states and products between
the aqueous and micellar pseudophases.

The purpose of this overview is three fold. First, to outline
the mathematical treatment of the pseudophase ion exchange model for
the effect of micellar solutions of ionic surfactants on rates and
equilibria of heterolytic reactions. Second, to show that calculated values of rate and equilibrium constants for reactions within
the micellar phase are single valued over a wide range of experimental conditions and that their values can be understood in terms
of the currently accepted properties of micelles as a separate phase
and its expected effect on the stabilities of reactants, transitions
states and products. Third, to define the models known limits. The
model fails for surfactants with very hydrophilic counterions and,
near the cmc, when the reactants and/or counterions are hydrophobic.
Within these limits the pseudophase ion exchange model accounts for

all the factors contributing to micellar effects on reaction rates
and equilibria. The model aids in the interpretation of the catalytic activity of functionalized micelles used as models for enzyme
active sites[16] and may be applicable to the effects of inversed micelles, microemulsions and vesicles on reaction rates and equilibria.

ASSUMPTIONS OF THE PSEUDOPHASE ION EXCHANGE MODEL

The fundamental assumption of the model embraces a contradiction: micellar solutions are macroscopically homogeneous, but the
total volume of the uniformly distributed dynamic aggregates of surfactant monomers is assumed to act as a separate phase, the micellar
pseudophase, of constant properties.[19-23] Pseudophase formation begins at the critical micelle concentration, the cmc, and all additional surfactant forms micelles with the monomer concentration remaining constant and equal to the cmc. However, because micelles are
not a true phase, the pseudophase assumption is thermodynamically in-
valid. Quantitative treatments of micellar effects on reaction rates
and equilibria generally fit the data very well at high surfactant
concentrations but sometimes fail near the cmc; no doubt because they
are based on the pseudophase assumption. The alternative mass action
model is thermodynamically correct, but more difficult to apply.[17,18]

A corollary of the pseudophase assumption is that micellar
effects on reaction rates and equilibria should be insensitive to
change in micelle size and shape. To date, no exception to this
assumption is known for thermal reactions.

Micelles are assumed to have "hydrocarbon like" interiors surrounded by a Stern layer containing hydrated head groups neutralized
by a fraction of their hydrated counterions and free water.[20,23-25]
The current debate on the detailed structure of micelles, including
the degree of water penetration into the core,[26-29] does not significantly affect the interpretation of micellar effects on rates and
equilibria; probably because the medium properties of micelles are
not very sensitive to solution composition. The outer surface of the
Stern layer is taken as the phase boundary, but its exact location
is not critical. The remaining counterions are in the Gouy-Chapman
layer extending radially into the aqueous phase.

The reaction site within the micellar pseudophase is assumed to
be the Stern layer. A number of studies indicate that micellar bound
substrates, usually polar organic molecules, are in an environment of
moderate polarity similar to that of ethanol and not the non polar
environment expected for the hydrocarbon core.[3,4,30] Finally, the
distribution of all reactants between the two-phases is always at
equilibrium because their diffusion rate is orders of magnitude
faster than the rate of reaction.[4]

The fundamental test of the pseudophase ion exchange model is the isolation of a rate or equilibrium constant for the reaction within the micellar pseudophase that is independent of the distribution of reactants between the two phases.[31,32] To determine micellar rate or equilibrium constants we must know the concentration of the reactants within the micellar pseudophase. Two complimentary approaches are used. The concentration of a reactant can be measured directly, for example spectrophotometrically. Alternatively, the distribution must be described in terms of independently verifiable constants which are part of general expressions for micellar effects on observed rate constants or apparent equilibrium constants. When independently determined values for these constants are unavailable or unreliable, values are selected which give the best fit to the data.

The binding of neutral organic molecules is dominated by the hydrophobic effect mediated to varying degrees by dipole and hydrogen bonding interactions and binding occurs regardless of the surfactant charge type.[4] Binding of hydrophilic ions is controlled primarily by electrostatic interactions.[23] Ions of opposite charge, counterions, to the surfactant head group are attracted to the micelle surface while reverse is true for coions. Binding of amphiphilic ions is governed by hydrophobic and electrostatic effects: additive for counterions, but opposed for coions and the hydrophobic effect must dominate, as it does for the surfactant monomer, for significant binding to occur. Strong binding in dilute surfactant solutions at concentrations just above the cmc produces high concentrations of the bound species, in moles per liter of micellar volume, while their concentration within the total solution volume remains unchanged. For example, in 0.01 M sodium lauryl sulfate (NaLS) with a cmc of 0.008 M, the concentration of sodium ion in the aqueous phase will be only slightly greater than cmc, but on the order of 3-5 M/L within the volume of the Stern layer; a 300-500 fold increase in sodium ion concentration.[31,32] The following sections will develop the equations for the distribution of charged and neutral reactants.

DISTRIBUTIONS OF COUNTERIONS

A large number of micellar catalyzed reactions and all indicator equilibria involve ions, often hydroxide or hydronium ions, that also act as counterions to micelles. Two approaches are used to describe their distribution. The surface potential model which successfully predicts the direction and extent of the shift in apparent pK_a of smaller bound indicators[33-36] and accounts for the large effect of added nonreactive counterions on the rates of bimolecular reactions and apparent pK_a's.[34,39] However, because the model treats all ions as point charges,[40] it fails to account for specific salt effects.[16,18,31,32,41-44]

The ion exchange model views the micelle surface as a loosely cross-linked ion exchange resin and assumes that the micelles surface binds counterions selectively.[31,32] This approach is based, in part, on the Stern layer model for counterion binding to micelles, first developed by Stigter, who introduced specific absorption potentials for counterions to overcome the limitations of the Gouy-Chapman model.[40,41] Stigter's physical picture is very similar to the counterion condensation model developed by Manning for interpreting the distribution of counterions around polyelectrolytes.[45] Both models assume that monovalent counterions are associated with the surface, but not bound to a site, and free to move across the surface.

Ion Exchange

Two additional assumptions are needed to formulate ion exchange equations:

(a)   that the micelle surface is saturated with counterions or equivalently, the experimentally measured degree of counterion binding, β, is constant; and

(b)   that the selectivity of the surface toward different counterions can be described by a simple ion exchange constant.

A review of the literature up to 1975 showed that, with few exceptions, the degree of counterion binding, β, as measured by different surfactants and counterions is essentially independent of surfactant concentration and added counterion within the range $\beta \cong 0.6 - 0.9$.[31] Recent theoretical work supports this conclusion.[23,42,46] Using any one experimental method β shows definite trends. For example, β decreases with increasing temperature[47], surfactant head group size[49] hydrated counterion radius[50], pressure[51], increasing nonelectrolyte concentrations[52], and decreasing surfactant chain length of anionic[53] and cationic surfactants.[54] These trends in β values are generally within the range of β values measured by different experimenters for the same surfactant using the same or different techniques.[31] The exceptions are N-alkylpyridinium iodides, presumably because the iodide ion undergoes a charge transfer interaction with the pyridinium head group[55], and possibly very hydrophilic counterions.[56] Significantly, in Manning's counterion condensation model the fraction of surface charge of a linear polyelectrolyte neutralized of bound counterions is about 0.8.[45]

The concept of ion exchange in micellar solutions has solid experimental support. Sepulveda and coworkers determined ion exchange constants for a number of anions, X, in micellar solutions of cetyltrimethylammonium ions by absorption spectroscopy[57] and ultrafiltration.[58] Representative values of ion exchange constants for monovalent anions relative to hydroxide ion, K_{OH}^X, are listed in Table I. Values for some ions estimated from kinetic data in CTAX[57], indicator equilibria[60], monolayers[61] and loosely crosslinked ion exchange resins[62] are included for comparison. Later Tables list the ion exchange constants used to estimate micellar rate and equilibrium constants.

No information is available on ion exchange constants for cations at anionic micelles surfaces except for H_3O^+ in NaLS estimated by conductance ($K_H^{Na} \cong 1$).[64] Divalent cations are known to bind tightly to anionic micelles[3,65,66], but nothing is known about their ion exchange constants with other mono or divalent cations.

Table I. Ion Exchange Constants for Anions with Hydroxide Ion.

Anion K_{OH}^X

reference	57	59[a]	60[b]	61[c]	62[d]
OH^-	1.0	1.0	1.0	1.0	1.0
F^-	0.98			1.2	1.0
Cl^-	4.2	7	4	3.2	11
Br^-	21	13	12	7.0	31
NO_3^-	23		14	49	42
$CH_3CO_2^-$	2.0				
$C_6H_5SO_3^-$	230				
$CH_3C_6H_4SO_3^-$	480				

a. obtained by fitting kinetic data. b. obtained by fitting apparent basicity constants. c. for polysoap monolayers of N-octylpolyvinylpyridinium bromide. d. Dowex-1 resin has a trimethylbenzylammonium head group.

Derivation of Ion Exchange Equations

The concept of ion exchange has been tested and refined periodically since it was first introduced in 1976 to explain micellar effects on bimolecular reactions between organic substrates and small ions. Chaimovich and Quina extended it to buffered systems, indicator equilibria and reactions with nucleophilic anions of weak acids[18],[67] and Berezin[39], Funasaki[36] and Rydholm and Almgren[41] showed that the ion exchange and surface potential models make virtually identical predictions when only two counterions are present and one is in large excess. In a very different approach Srivastava and Katiyar[68] added the concept of ion exchange to a cooperative binding model, originally developed by Piszkiewicz.[69] The mathematical formalisms used by each group are all different, but based on the pseudophase model, except those of Srivastava and Katiyar and Piszkiewicz. The approach and symbols used here were developed by Bunton and Romsted.[16]

The concentration of micellar bound reactive counterion $m_N{}^s$, in the Stern layer, s, is expressed as the mole ratio of micellar bound reactive ion, N_m, to micellized surfactant, D_n; with aggregation number n:

$$m_N{}^s = [N_m]/[D_n] \qquad (1)$$

$$[D_n] = [D] - cmc \qquad (2)$$

where D is the stoichiometric amount of surfactant, cmc is the critical micelle concentration and square brackets indicate concentration in moles per liter of solution volume, here and in all following equations.

In solutions containing two counterions the sum of micellar bound ions is constant:

$$m_N{}^s + m_X{}^s = \beta \qquad (3)$$

where $\qquad m_N{}^s = [N_m]/[D_n]; \quad m_X{}^s = [X_m]/[D_n] \qquad (4)$

The nonreactive counterion, X, is contributed by the surfactant and any added salt [MX], including buffers. Additional ions, Y, etc., add additional terms to Equation (3), $m_Y{}^s$; etc. These definitions assume that the activity of counterions within the micellar pseudophase are constant and their concentrations are proportional to the total micellar volume or a fraction thereof, i.e. the Stern layer, but not to the total solution volume.

One limitation on this approach is that β values for different ions are not the same, so that Equation (3) is only an approximation. However, it is reasonable approximation given the uncertainty in published values of β. The approximation works, in part, because in many experiments the concentration of reactive counterion, hydrophilic or hydrophobic, is kept low so $m_N{}^s \ll m_X{}^s$ and $m_X{}^s \cong \beta$ and β values are determined by the surfactant's counterion.

The exchange of reactive and nonreactive counterions between the micellar, m, and aqueous, w, pseudophases is described by an ion exchange constant, K_N^X:

$$m_N^s + [X_w] \xrightleftharpoons{K_N^X} m_X^s + [N_w]$$

$$K_N^X = \frac{[N_w]m_X^s}{[X_w]m_N^s} = \frac{[N_w][X_m]}{[X_w][N_m]} \qquad (5)$$

The assumption, β = constant, also works if the hydrophilic reactive counterions, e.g. OH^- to cationic surfactants, bind much less tightly than the other counterion, $K_N^X \gg 1$, Table I. However, at surfactant concentrations just above the cmc or high concentrations of reactive counterion the approximation breaks down unless the β values for two counterions are similar (see below).

Two approaches are used derive expressions for the concentration of micellar bound reactive counterion. Equations (3) and (5) are combined with materials balance equations for N and X and expressed in terms of stoichiometric concentrations.[16,32] Alternatively, following the formalism used by Chaimovich and Quina the equations are expressed in terms of the independently measured concentrations of N and X in the aqueous phase.[67] In either case the final expressions are quadratics which must be solved for the positive root. Chaimovich and Quina's approach can be applied directly to buffered solutions and to solutions containing a third counterion, but the final equation is cubic.[63] Similar expression's using the formalism of Bunton and Romsted can be derived for buffered solutions assuming that the $[OH_w]$ is measured directly and that $[OH_w]$ = antilog (−pOH).[70]

The effects of added surfactant and salt on the concentration of micellar bound counterions are easier to visualize under limiting conditions. If $K_N^X \gg 1$ or if $[X] \gg [N]$ and most of the reactive counterion is in the aqueous phase, then in unbuffered solutions:

$$m_N^s = \frac{\beta[N_T]}{[N_T] + K_N^X[X_T]} \qquad (6)$$

and in buffered solutions where one or both both forms of the buffer is a coion:

$$m_N^s = \frac{\beta[N_w]}{[N_w] + K_N^X[X_T]} \qquad (7)$$

where $\qquad\qquad [X_T] = [D] + [MX] \qquad (8)$

and the subscript, T, stands for stoichiometric concentration.

Equations (6) and (7) are not very useful in themselves because they apply to a very limited set of conditions, but they illustrate the properties of the full quadratic equations.

First. At the cmc the concentration of the reactive ion in moles per liter of Stern layer volume, m_N^S, is at a maximum and decreases with increasing surfactant concentration. Nevertheless, the total quantity of bound reactive counterion always increases with added surfactant.[71] It is this continuous shift in reactant from water to the initially small volume of the micellar phase opposed by the continuous dilution of the reactant within the micellar phase that is responsible for the maxima in the rate-surfactant profiles of bimolecular reactions and in apparent equilibria-surfactant profiles.

Second. The concentration of micellar bound reactive counterion is diminished by added nonreactive counterion, X, and the more tightly the counterion binds, i.e. the larger K_N^X, the more it reduces m_N^S. Additional nonreactive counterions, Y, have a parallel effect and simply add additional terms to the denominator, $K_N^Y[Y_T]$.

Third. The value of m_N^S is linearly dependent upon $[N_T]$ or $[N_w]$ only when its concentration is small. At high concentrations of N, m_N^S approaches a constant value. At this point the micellar surface is completely saturated with N, $m_N^S = \beta$ and is independent of changes in surfactant and salt concentration. This prediction led to the preparation of reactive counterion surfactants and to the first major failure of the pseudophase ion exchange model (see below).

Fourth. The addition of a buffer to a micellar solutions controls only the aqueous pH. The pH at the micelle surface always depends on surfactant and salt concentration, Equation (7). Added surfactant or counterion lowers the hydroxide ion concentration at the surface of cationic micelles or conversely the proton concentration at the surface of anionic micelles whether buffer is present or not; compare Equations (6) and (7). The surface pH will be constant only under the limiting conditions of high surfactant or salt concentration or when K_N^X is large (i.e. $K_N^X[X_T] \gg [N_T]$) and $m_N^S \cong 0$ or, conversely, when the reactive counterion is in large excess or K_N^X is small (i.e. $[N_T] \gg K_N^X[X_T]$ and $m_N^S = \beta$).

A second limiting condition is reached when the reactive counterion is hydrophobic and only a large quantity of hydrophobic nonreactive counterion can displace it.[67] The reactive counterion is usually present in indicator amounts, $m_X^S \gg m_N^S$ and $m_X^S \cong \beta$. At high surfactant concentration, $[D] \gg$ cmc and in excess salt $[X_T] \gg [D]$, the quadratic equation for the concentration of micellized reactive counterion concentration in the micellar pseudophase is given by:

$$m_N{}^s = \frac{\beta[N_T]}{K_N{}^X[X_T]} \qquad (9)$$

Generalizations 1-3, also apply to Equation (9).

BINDING OF NEUTRAL ORGANIC MOLECULES

Uncharged organic molecules may act as nucleophilies, N, or substrates, S, and their distribution between water and the micellar pseudophase is described by a simple binding constant, K_S or K_N.

$$S_w + D_n \underset{}{\overset{K_S}{\rightleftharpoons}} S_m; \qquad N_w + D_n \underset{}{\overset{K_N}{\rightleftharpoons}} N_m$$

$$K_s = \frac{m_S}{[S_w]} = \frac{[S_m]}{[S_w][D_n]} \qquad (10)$$

$$K_N = \frac{m_N}{[N_w]} = \frac{[N_m]}{[N_w][D_n]} \qquad (11)$$

To solve for the concentration of micellar bound nucleophilic or substrate Equations (10) and (11) are combined with the appropriate mass balance equation to give:

$$m_S{}^s = \frac{[S_T]K_S}{1 + K_S[D_n]} \qquad (12)$$

$$m_N{}^s = \frac{[N_T]K_N}{1 + K_N[D_n]} \qquad (13)$$

This derivation assumes that the micellized surfactant is in excess over substrate, $[D_n] \gg [S_m]$, and that the micellar pseudophase occupies only a small fraction of the total solution volume. The former assumption fails only near the cmc and binding constants are usually determined well above the cmc. The later assumption holds because the surfactant concentration seldom exceeds 0.1 M/L or 2-3% of the solution volume. Equations (12) and (13) can be substituted directly into the appropriate rate or equilibrium expression. Binding constants are measured by a variety of methods including: solubility[72], gel filtration[73], absorbance spectroscopy[74], fluorescence spectroscopy[75], and ultrafiltration[74,76,77], or by selecting the value that gives the best fit to the data.[78]

Binding constants are also used to interpret the distribution of organic ions,[76] but their use is questionable because they are written in terms of a single ion being transfered across a phase, albeit pseudophase, boundary. Indeed, in some cases, binding increases with added salt.[79] Nevertheless, the binding for organic ions of opposite charge can often be approximated by simple binding constants because they bind strongly and are not easily displaced except by hydrophobic counterions (see below). Chaimovich and Quina derived an expression for relating binding constants to ion exchange constants.[67]

UNIMOLECULAR REACTIONS

Micellar effects on unimolecular reactions have been reviewed several times.[16,80,81] The results are only briefly summarized here, including a reinterpretation of the effect of hydrophobic salts.

Scheme I illustrates the pseudophase model for unimolecular reactions in micellar solution assuming that the distribution of substrate is described by a simple binding constant.

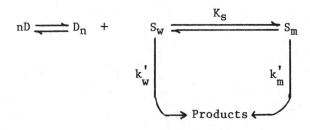

Scheme I

Reaction occurs in either the micellar or aqueous pseudophase and the observed rate is given by:

$$k_\psi [S_T] = k_w' [S_w] + k_m' m_S{}^s [D_n]$$

(14)

where k_w and k_m are first order rate constants in the aqueous and micellar pseudophases respectively. Combining equation (14) with equation (12), the materials balance equation for the substrate and assuming that the micellar pseudophase occupies a small fraction of the total solution volume we obtain:

$$k_\psi = \frac{k_w + k_m K_S [D_n]}{1 + K_S [D_n]}$$

(15)

Equation (15) is similar in form to the Michaelis–Menten equation of enzyme kinetics and successfully fits the sigmoidal rate–surfactant profiles of micellar catalyzed unimolecular reactions; k_ψ increases rapidly at the cmc and then plateaus once all of the substrate is bound. Rearrangement of Equation (15) to the linear double reciprocal form, similar to the Lineweaver–Burke equation, allows both K_S and k_m to be estimated from the kinetic data. Rate enhancements of 3–700 fold are observed for a number of spontaneous hydrolyses and decarboxylations.[80,81] Values of K_S cannot be measured independently for these substrates because the decompose spontaneously, but the kinetically determined values are reasonable.

Salt Effects on Unimolecular Reactions

While equation (15) adaquately describes the rate–surfactant profiles for unimolecular reactions it cannot account for the effect of added salts. At high CTABr concentrations, sufficient to insure complete substrate binding, added hydrophilic counterions inhibit only slightly the spontaneous hydrolysis of 2,4- and 2,6-dinitrophenylphosphate but enhance (about 2 fold at 1.0 M added salt) the spontaneous decarboxylation of 6-nitrobenzisoazole-3-carboxylate ion 1.[82]

However, added hydrophobic anions such as tosylate and benzoate strongly inhibit the hydrolysis of phosphate esters at all salt concentrations, but the decarboxylation rate of 1 initially increases slightly, about 2 fold, then decreases sharply at high concentrations for many of these hydrophobic anions.

The simplest interpretation of the inhibition by added salts is ion exchange. Equation (9) is expanded to include ion exchange by a second anion, Y, and combined with the appropriate materials balance equation and Equation (14) to give:

$$k_\psi = k'_w + \frac{(k_m - k_w)\beta[D_n]}{K_S{}^X[X_T] + K_x{}^Y[Y_T]} \qquad (16)$$

where

$$[X_T] = [D] + [MX]$$

Equation (16) is strictly valid only at high surfactant concentrations, but it does predict the rate-surfactant profile for added surfactant, $Y_T = 0$, and the inhibition produced by added salt. Table II shows the ion exchange constants for the effect of added salts on the spontaneous hydrolyses of 2,4- and 2,6-dinitrophenylphosphates.[82] The original data was analyzed in terms of a competitive inhibition model; the K_s^Y values were calculated from the available constants. However, Equation (16) can not account for the rate increases in the rate-salt profiles for CTABr catalyzed decarboxylation of $\underset{\sim}{1}$.[83]

Comparison of Rate Constants for Unimolecular Reactions in Aqueous and Micellar Pseudophases

Rate constants of unimolecular reactions for fully micellar bound substrates reflect the medium properties of micellar pseudophase. The size of the rate enhancement for spontaneous sulfate and phosphate esters hydrolyses and decarboxylations by cationic surfactants are consistent with the effect of nonaqueous solvents on these reactions.[4,80,81] Anionic micelles have no effect on these reactions, but this is probably because the negatively charged substrates remain in the aqueous phase and not because of the medium properties of anionic micelles are different from the medium properties of cationic micelles. This ambiguity can be resolved by using a very hydrophobic substrate to insure binding to the anionic micelles.

Catalysis appears to occur whenever charge is more dispersed in the transition states than the ground state. For example, the rate of decarboxylation of 1, which increases dramatically with decreasing solvent polarity and is also strongly catalyzed by cationic micelles,[4] is consistent with greater stabilization of the transition state than the ground state by the lower polarity of the micelle surface compared to water.[81] This extreme sensitivity of 1 to solvent polarity indicates that the small rate increase caused by added hydrophilic counterions is probably due to a small decrease in the polarity of the micelle surface. Whereas the rate maxima observed with added hydrophobic counterions is produced by the opposing effects of a decrease in the polarity of the micelle surface opposed by displacement of the substrate by ion exchange.

Catalysis by cationic surfactants is also consistent with a special interaction between aromatic rings and organic anions with dispersed charge and quaternary ammonium head groups.[44,81] However, this interaction cannot be the sole explanation because the decarboxylation of 1 is also promoted by nonionic surfactants.[38] Finally, Bunton showed that the rate enhancements in unimolecular reactions can be rationalized by Pearson's hard and soft classification scheme.[44] The low charge density, soft, quaternary ammonium head group will interact more strongly, with the charge delocalized, soft,

Table II.   Ion Exchange Constants between Substrate and Counterions
for Spontaneous Hydrolyses in CTABr.

Salt	K_N^{Ya}	
	2,4-Dinitrophenylphosphate	2,6-Dinitrophenylphosphate
NaCl	0.002	0.01
CH_3SO_3Na	.....	0.04
$C_6H_5OPO_3Na$	.....	0.4
$o-C_6H_4(CO_2K)_2$	0.2	0.6
$p-C_6H_4(CO_2K)_2$	.....	0.7
$C_6H_5CO_2Na$	0.5	1.3
$p-C_7H_7SO_3Na$	.....	4.4

a. $K_N^Y = \beta K_I N/K$, $\beta = 1.0$, $N = 61$, $K_S = 1.1 \times 10^5$ and 3.9×10^4 for
2,4- and 2,6-dinitrophenylphosphate respectively, and K_I values from
Table VI in reference 83.

transition state than the charge localized, hard, ground state.

Bimolecular Reactions

The simple distribution model applied to unimolecular reactions
fails for higher order reactions. Bimolecular reactions, for example,
show the same increase in observed rate above the cmc, but with increasing surfactant concentration the rate passes through a maximum
and then gradually decreases instead of remaining constant.[32]   The
results for the addition of CN^- to N-alkyl-3-carbamoylpyridinium ions
are typical, Figure 1.[76] This consistent pattern, except for certain
predictable limiting cases[31], was surprising at first because experimental conditions were selected to mimic those of enzyme catalyzed
reactions.[73,84] The concentration of the second reactant was either
buffered, if H^+ or OH^-, or in large excess over the substrate, with
salt added to control ionic strength. The observed rate was expected
to plateau once all the substrate was bound. However, unlike enzyme
kinetics experiments, the surfactant concentration in micellar
catalyzed reactions is usually in large excess over both reactants.

This difference is crucial because, unlike enzymes, increasing the micelle concentration can significantly alter the concentrations of both reactants in both pseudophases.

The maxima in rate-surfactant profiles is produced by two opposing effects. Binding of reactants begins at the cmc, and transfers them into the small volume of the micellar pseudophase. If the binding constant K_S, K_N are large or the ion exchange constant, K_N^X, is small the reactant's concentration within the micellar pseudophase in moles per liter of micellar volume can be 100-1000 times greater than their stoichiometric concentrations. This concentration effect is opposed by continuous dilution of the reactants within the micellar pseudophase with increasing surfactant concentration. Thus the shape of rate-surfactant profiles is primarily a phase transfer phenomena, but the extent of the change depends on both the size of the binding and ion exchange constants and the difference in rate constants for reaction in the micellar and aqueous pseudophases.

Scheme 2 shows reaction between the substrate, S, and nucleophile, N (or any second reactant). The second reactant is generally

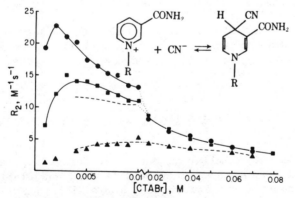

Figure 1. Variation of the observed second order rate constants, k_2, with CTABr in 0.005 M NaCN for addition of CN^- to N-alkyl-3-carbamoylpyridinium ions: ◆, ■, ● alkyl = $C_{12}H_{25}$, $C_{14}H_{29}$ and $C_{16}H_{33}$, respectively. Broken lines are calculated, reference 76.

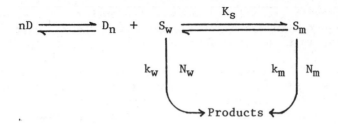

Scheme 2

in large excess over the substrate establishing pseudo first order conditions, so that:

$$k'_w = k_w[N_w] \qquad (17)$$

and

$$k'_m = k_M m^s_N \qquad (18)$$

where k_w and k_M are second order rate constants for reaction in aqueous and micellar pseudophases respectively. Substitution of Equations (17) and (18) into equation (15) gives:

$$k_\psi = k_2[N_T] = \frac{k_w[N_w] + k_M K_S m_N [D_n]}{1 + K'_S[D_n]} \qquad (19)$$

The final form of the kinetic equation will depend upon the properties of the second reactant: whether it is a neutral molecule, a hydrophilic or hydrophobic coion, a counterion to the micelle, or in complex systems, an anion of a weak organic acid NH.

If, N, is a neutral organic molecule, then equation (13) describes its distribution and combined with the materials balance equation for N and Equation (19), and $K_S k_M \gg k_w$ gives:

$$k_2 = \frac{k_\psi}{[N_T]} = \frac{k_w}{1 + K_S[D_n]} + \frac{k_M K_S K_N[D_n]}{(1 + K_S[D_n])(1 + K_N[D_n])} \qquad (20)$$

Equation (20) is very similar to one first derived by Berezin and coworkers for interpreting the effects of NaLS and CTABr on reactions between substituted imidazoles, Table IV.[85] This work sparked the application of the pseudophase model to bimolecular reactions.

If N is a hydrophobic counterion, then it should be treated as an exchangeable ion,[67] but ion exchange constants for most hydrophobic ions have not been measured. However, like unimolecular reactions of charged substrates, these ions bind strongly and their rate-surfactant profiles can often be fit using a simple binding constants. For example, Cuccovia and coworkers successfully fit the rate-surfactant profiles for reaction of completely ionized para substituted thiolphenols with esters in CTABr, Table IV.[86]

If N is a hydrophilic counterion then its concentration within the micellar pseudophase is sensitive to the concentration and type of other nonreactive counterions present in solution, including buffers. The most common approach is to solve the complete expressions for $m_N{}^s$ separately and then substitute into Equation (19). For example, Bunton and coworkers used a computer to simulate rate-surfactant profiles for reaction of some betaine esters with OH$^-$ by solving simultaneously the quadratic expression for mixtures of reactive and nonreactive counterion and Equation (19), Table III.[87] The substrate binding constant, K_s, the rate constant in water k_w and the cmc were determined independently and values for β, $K_N{}^X$ and k_M were selected to give the best fit of the data; β and $K_N{}^X$ were kept close to independently determined values. Chaimovich and coworkers, using their own formalism, analyzed the base promoted ester hydrolysis in borate buffered CTABr solutions,[63] and Bonilha and coworkers applied the same analysis to alkaline ester hydrolysis in several other cationic surfactants Table, III.[88] In a somewhat different approach Bunton and coworkers used an empirical equation for the proton concentration at the surface of NaLS micelles, $m_H{}^s$, as a function of [H$^+$] and [Na$^+$] from conductance data. This equation was used to analyze acid catalyzed hydrolysis of p-nitrophenyl diethyl acetal and the benzidine rearrangement (a 3rd order reaction, 2nd order in H$^+$) in NaLS, Table V.[89] A related approach was used to fit reactions between N-alkyl-3-carbamoylpyridium ions and CN$^-$ in CTABr solutions.

To illustrate the relationship between the parameters and the concentration of the components in solution and their effect on the rate profiles, Equation (7) is modified to include a second nonreactive ion, Y, and the materials balance equation for, N, and substituted into Equation (19) to give:

$$k_2 = \frac{k_w}{(1 + K_s[D_n])} + \frac{k_M K_s \beta [D_n]}{(1+K_s[D_n])([N_w]+K_N{}^X[X_T]+K_N{}^Y[Y_T])} \qquad (21)$$

where $[X_T] = [D] + [MX]$, $[Y_T] = [MY]$, $[N_w]$ = antilog($-$pH) or antilog ($-$pOH) and $k_2 = k_\psi/[N_w]$. The equivalent expression for unbuffered solutions has $[N_T]$ in place of $[N_w]$.

The qualitative relationship between equations like (20) and (21) and rate profiles have been reviewed in detail several times,[16,32,38,39,67,81] and are only summarized here. Quantitative results are in Tables III-V. Equations (20) and (21) are consistent with the following observations:

1. that $k_2 = k_w$ below the cmc and that catalysis begins at the "kinetic" cmc;

2. the maximum in k_2 (or k_ψ) with increasing D_n;[32]

3. that the extent of catalysis k_{rel} $(= k_2^{max}/k_\psi = k_\psi^{max}/k_w')$ increases with increasing surfactant chain length;[32] and

4. that k_{rel} increases with the size of the reactant binding constant K_s or K_N. For example, compare the k_{rel} values for the alkaline the hydrolysis of p-nitrophenylate esters in Table III.

Equation (21) correlates additional micellar effects on reactions between organic substrates and small ions:[32]

1. the sharp inhibition produced by added nonreactive counterions, X and Y;[16,32,43]

2. the dependence on counterion type, i.e. K_N^X and K_N^Y. In buffered and unbuffered solutions of cationic or anionic surfactants, the effectiveness of a series of counterions in inhibiting the reaction follows a Hofmeister series, i.e. increases with the size of the ion.[16] For organic ions, the more hydrophobic the ion, the more effectively it inhibits reaction;[44]

3. that the observed second order rate constant, $k_2 = k_\psi/[N_T]$ or $k_\psi/[N_w]$, depends on [N] provided $[N] \cong [X]$ or $[Y]$; and

4. buffering the aqueous phase, $[N_w]$ = constant, does <u>not</u> buffer the micellar pseudophase, m_N^s = constant,[63] and rate-surfactant and rate-salt profiles have the same form in buffered and unbuffered solutions.[67]

The pseudophase ion exchange model also account for micellar induced inhibition when N is a coion and its concentration is diminished at the micelle surface relative to the aqueous phase. Menger and Portnoy[90], who first applied the pseudophase model to micellar systems, fit the sodium laurate inhibition of alkaline ester hydrolysis by assuming the rate in the micellar pseudophase is zero; i.e. either k_M or m_N^s in Equation (19) is zero and

$$k_2 = \frac{k_\psi}{[N_T]} = \frac{k_w}{1 + K_S[D_n]} \qquad (22)$$

However, rate constants for bimolecular reactions are generally close to their values in water and the hydroxide ion concentration

Table III. Bimolecular Reactions with Hydrophilic Nucleophiles in Cationic Surfactants.

Surfactant	Substrate	N	K_s, M^{-1}	k_X^Y	k_{rel}	k_2^m/k_w	Ref.
CTABr[a]	DNCB[c]	N_3^-	67	2	67,000	52	95
CTACl[b]	DNCB	N_3^-	82	1.3	76,000	52	95
CTABr	DNCN[d]	N_3^-	600	2		400	95
CTACl	DNCN	N_3^-	600	1.3		400	95
CTABr	$PhCO_2C_6H_3(NO_2)_3$	N_3^-	650	2		1	95
CTAMes[e]	$PhSO_3Me$	N_3^-	55	1.1		0.71	95
CTABr	benzoic anhydride	OH^-	650	10		0.06	102
CTABr	benzoic anhydride	HCO_2^-	650	10		0.21	102
CTABr	$(O_2NC_6H_4O)_2CO$	OH^-	1000	10		0.70	102

a. cetyltrimethylammonium bromide, b. cetyltrimethylammonium chloride, c. 2,4-dinitro-chlorobenzene, d. 2,4-dinitrochloronaphthalene, e. cetyltrimethylammonium mesylate,

(continued)

Table III. (continued)

Surfactant	Substrate	N	K_S,M^{-1}	K_X^Y	K_{rel}	k_2^m/k_w	Ref.
CTACl	$\overset{+}{C_{12}H_{25}NMe_2CH_2CO_2Me}$	OH^-	26	4	5	0.23	87
TDTACl[f]	$\overset{+}{C_{12}H_{25}NMe_2CH_2CO_2Me}$	OH^-	22	4	5	0.28	87
CTABr	$N-C_{12}H_{25}-3-CPg$	CN^-	20	1	360	2.3	76
CTABr	$N-C_{14}H_{25}-3-CPg$	CN^-	390	1	1000	2.4	76
CTABr	$N-C_{16}H_{33}-3-CPg$	CN^-	3500	1	1640	2.5	76
CTABr	Malachite green	BH_4^-	29		274	2.7	79
CTABr	Malachite green	OH^-	10	5	13	0.65	79
CTABr	PNPA[h]	OH^-	54	12.5		0.14	63
CTABr	PNPA	OH^-	50	40		0.6	41
CTABr	PNPB[i]	OH^-	530	10.0		0.13	36
CTABr	PNPO[j]	OH^-	15,000	12.5		0.11	63

f. tetradecyltrimethylammonium chloride, g. N-alkyl-3-carbamoylpyridinium ion,
h. p-nitrophenylacetate, i. p-nitrophenylbutyrate, j. p-nitrophenyloctanoate.

Surfactant	Substrate	N	K_S, M^{-1}	K_X^Y	k_{rel}	k_2^m/k_w	Ref.
CPyCl[k]	PNPA	OH^-		5.6		0.17	88
CPyCl	PNPO	OH^-	15,000	5.6		0.09	88
CTABr	ethyl benzoate	OH^-	220	10		0.012	112
CTABr	ethyl p-aminobenzoate	OH^-	250	10		0.0043	112
CTABr	ethyl p-nitrobenzoate	OH^-	240	10		0.034	112
CPyCl	N-dodecylPyCN[l]	OH^-		5.6		3.8	88
CPyCl	N-decylPyCN	OH^-		5.6		3.4	88
CTABr	N-decylPyCN	OH^-		12.5		3.4	88
CTABr	N-dodecylPyCN	OH^-		12.5		4.2	88
CTABr	DTNB[m]	OH^-	20,000	12.5	15	0.096	108
DDOAC[n]	DTNB	OH^-	40,000	7	1500	0.89	108

k. cetylpyridinium chloride, l. 4-cyano-N-alkylpyridinium ion, m. 5,5'-dithiobis(2-nitrobenzoic acid), n. dioctadecyldimethylammonium chloride.

Table IV. Bimolecular Reactions with Organic Nucleophiles.[a]

Surfactant	Substrate	Nucleophile	K_S, M^{-1}	K_N, M^{-1}	k_{rel}	k_2^m/k_w	Ref.
CTACl	PNPDPP[b]	$C_6H_{13}OC_6H_4C(NH_2)=NOH$	16,000			0.15	113
CTABr	PNPDPP	2-quinolinecarbaldoxamate	16,000			0.30	113,114
CTABr	PNPDPP	p-nitrobenzaldoximate	16,000			0.33	113,114
CTABr	PNPDPP	benzimidazole	16,000			1	100
CTABr	PNPDPP	naphthimidazole	16,000			1	100
CTABr	PNPDPP	PhO⁻	16,000		3,000	0.53	114,115
CTABr	PNPDPP	p-MePhO⁻	16,000		4,000	0.50	114,115
CTABr	PNPDPP	p-EtPhO⁻	16,000			0.88	114,115
CTABr	PNPDPP	p-PrPhO⁻	16,000			0.82	114,115
CTABr	PNPDPP	n-BuPhO⁻	16,000			0.82	114,115
CTABr	PNPDPP	t-AmPhO⁻	16,000			0.82	114,115
CTABrr	DNF[c]	PhO⁻		54	750	1.5	115
CTABr	DNF	PhNH₂		54	37	0.17	115

a. See footnotes in Table III for definitions of most abbreviations, b. p-nitrophenyldiphenylphosphate, c. 2,4-dinitrofluorobenzene.

Surfactant	Substrate	Nucleophile	K_S, M^{-1}	K_N, M^{-1}	k_{rel}	k_2^M/k_w	Ref.
CTABr	Malachite green	BDHNA[d]	10	400	2	0.003	79
CTABr	PNPA	p-ClPhS[-]	27	15,000	42	0.24	86
CTABr	PNPA	PhS[-]	27	3,900	69	0.42	86
CTABr	PNPA	p-MePhS[-]	27	7,700	55	0.32	86
CTABr	PNPA	p-MeOPhS[-]	27	2,300	57	0.36	86
CTABr	Mn(cydta)[-][e]	Co(edta)[2-][f]	325		600	0.048	116
CTABr	Mn(cydta)[-]	Co(cydta)[2-]	325		160	0.029	116
CTABr	PNPHpg	N-heptylimidazole	4,500	100	1.4	0.0021	85
CTABr	PNPB	N-heptylimidazole	530	130	1.6	0.0028	85
CTABr	PNPA	N-heptylimidazole	27	120	1.9	0.012-0.024	85
DODAC	PNPO	$C_7H_{15}S^-$	4,000		2.7×10^6[h]	5	109
NaLS[i]	DNF	PhNH2	14	285	15	0.12	115
NaLS	malachite green	BDHNA	8,000			0.01	79

d. 1-benzyldihydronicotinamide, e. cydta[4-]= trans-1,2-diaminocyclohexane N,N,N',N'-tetra-
cetate, f. edta[4-]= 1,2-diaminoethane-N,N,N',N'-tetraacetate, g. p-nitrophenylheptanoate,
h. pH = 5, i. sodium lauryl sulfate.

Table V. Bimolecular Reactions with Electrophiles in Sodium Lauryl Sulfate.[a]

Substrate	N	K_s, M^{-1}	k_{rel}	k_2^M/k_w	Ref.
N-trifluoroacetylindole	H^+	420	3.6[b]	0.027	117
p-$O_2NC_6H_4CH(OEt)_2$	H^+	100	20[c]	0.052	89
1-benzyldihydronicotinamide	H^+	285	6.6[d]	0.044	118
1-benzyl-3-acetyl-1,4-dihydropyridine	H^+	405	14.3[d]	0.066	118
benzidine	H^+[e]	220	2,000	0.013	89
ferrocene	Fe^{3+}	340		0.43	119
n-butylferrocene	Fe^{3+}	ca. 10^4		~0.55	119
PADA[f]	Ni^{2+}[g]	2800		~1	110
2,2'-bipyridyl	Ni^{2+}[g]	40	130	~1	110
4,4'-dimethyl-2,2bipypyridyl	Ni^{2+}[g]	250		~1	110

a. For acid catalyzed reactions $K_{Na}^H \simeq 1$, b. at 0.03 M HCl, c. at 0.00316 M HCl, d. at 0.001 M HCl, e. 3rd order reaction, 2nd order in H^+ at 0.001 M HCl, f. Pyridine-2-azo-p-dimethylaniline, g. estimated binding constant for Ni^{2+}, $K_S^+ \simeq 10^3$.

at the surface of anionic micelles is no doubt much lower than in the aqueous phase, but not zero. Micelles generally shift the pK_a's of indicators 1-2 log units indicating a large change in local pH, but not $[H^+] = 0$ at cationic or $[OH^-] = 0$ at anionic interfaces.[35]

Recently, Quina and coworkers measured the extent of inhibition of the rate of alkaline hydrolysis of a series of long chain N-alkyl-4-cyanopyridinium ions and p-nitrophenylalkanoates in NaLS.[91] As the chain length increased the rate at high NaLS approached a small limiting value but not zero, indicating that hydroxide ion reacts with the substrate within the micellar pseudophase and not just with unbound substrate in the aqueous phase. Significantly, at high NaLS concentration, added NaCl increased the observed rate. The authors explained this increase in terms of ion exchange; displacement of H^+ from the surface by added Na^+ produces a reciprocal increase in OH^- at the micelle surface ($K_w = [H^+][OH^-]$). A complete quantitative treatment was not attempted because the autoprotolysis constant for water at the surface of NaLS micelles is not known.

Srivastava and Katiyar obtained similar results for reaction of Setoglaucin, a triphenylmethane dye, with CN^- in NaLS at pH = 10.7.[92] Added NaLS strongly inhibited the reaction, but at high NaLS concentration added salts increased the rate a small but noticeable amount. The effectiveness of the cation depends on its radius: $Cs^+ > K^+ > Na^+ > Li^+$. Following the interpretation of Quina et al., the proton concentration should be substantially higher at the NaLS micelle surface than in the aqueous phase. Because HCN is a weak acid, $pK_a = 9.4$, substantial amounts of HCN may be present at the micelle surface. Added cations will displace protons, shifting the position of equilibrium at the micelle surface,

$$HCN_m \rightleftharpoons H_m^+ + CN_m^-$$

increase CN_m^- and therefore the observed reaction rate.

Reactive Counterion Surfactants:
A Partial Failure of the Pseudophase Model

Quantitative interpretation of bimolecular reaction is complicated by counterion effects and buffer effects, uncertainties in the cmc and specific interactions with the substrate and the large number of assumptions required to fit the data. To test the assumption that the micelle surface is saturated with counterion, Bunton and coworkers synthesized several different reactive counterion surfactants, Table VI. This work was recently reviewed.[93] In these systems the counterion is the second reactant, e.g. CN^-, OH^-, and Br^- with cationic surfactants and H^+ for anionic surfactants. Complications caused by buffers and nonreactive counterions are absent and:

$$m_N{}^s = \beta \qquad (23)$$

Combining Equation (23) with the materials balance equation for N and substituting into Equation (19) gives:

$$k_\psi = \frac{k_w[N_T] + \beta(k_M K_s - k_w)[D_n]}{1 + K_s[D_n]} \qquad (24)$$

Equation (24) predicts a plateau in k_ψ at high [D] and that added counterion only affects the reaction in water, i.e. k_ψ is independent of N when the substrate is fully micellar bound.

Equation (24) fits the long chain sulfonic acid catalyzed hydrolysis of acetals including the effect of added HCl,[94] the addition of CN^- to N-alkyl-3-carbamoylpyridinium ions in CTACN, Figure 2 (compare to Figure 1),[76] and the reaction of azide ion with 2,4-dinitrochlorobenzene (DNCB) and 2,4-dinitrochloronaphthalene (DNCN) in CTAN.[95]

However, Equation (24) fails for reactions of hydrophilic nucelophiles in cationic surfactants, Table VII. For example, Figure 3 shows the reaction of CTAOH with DNCN.[56] Instead of rising rapidly to a plateau, k_ψ increases steadily to very high [CTAOH] far above the concentration of CTABr required to completely bind DNCN ($K_s = 1,610$).[94] In addition, added KOH produces a large increase in rate at all [CTAOH], but eventually k_ψ reaches a limiting value near [KOH] = 1.0 M. This sensitivity to counterion concentration is completely different from the small predictable effects of added HCl in sulfonic acid catalyzed acetal hydrolysis. This failure of the model is repeated for other small hydrophilic counteranions to cationic micelles, Table VII. Several explanations are possible; all require substantial modification of the pseudophase model.

First. Ordinary micelles are present, but their properties as a medium are very sensitive to ionic strength so k_M increases rapidly with added surfactant and counterion. This possibility is inconsistent with other reactive counterion surfactants and the large numbers of surfactant systems containing two counterions where single values of k_M can be used used to fit the data.

Second. The volume element of reaction shrinks toward a limiting value with increasing surfactant and salt concentration. This explanation was recently used by Hicks and Reinsborough to interpret the effect of increasing chain length of sodium alkylsulfonates on the rate-surfactant profiles for the formation of Ni^{2+} complexes with pyridine-2-azo-p-dimethylaniline.[96] This is an interesting possibility especially if the entire double layer around the micelle is actually part of the reaction volume, but there is currently no other physical or kinetic data consistent with this assumption.

Table VI. Reactive Counterion Surfactants, Pseudophase Model Succeeds[a]

Surfactant	Substrate	$K_s M^{-1}$	k_2^m/k_w	Ref.
CTAN[b]	DNCB	115	28	95
CTAN	DNCN	>600	200	95
CTAN	$PhSO_3Me$	70	0.43	95
CTACN[c]	$N-C_{12}H_{25}-3-CP$	70	2.3	75
CTACN	$N-C_{14}H_{25}-3-CP$	390	2.4	75
CTACN	$N-C_{16}H_{33}-3-CP$	3500	2.6	75
$C_{14}H_{29}SO_3H$	$ArCH(OEt)_2$	73	0.02	93
p-octylOPhSO$_3$H	$ArCH(OMe)_2$	27	0.055	93
p-octylOPhSO$_3$H	$ArCH(OEt)_2$	36	0.049	93
p-dodecylOPhOSO$_3$H	$ArCH(OMe)_2$	37	0.076	93
p-dodecylOPhOSO$_3$H	$ArCH(OEt)_2$	91	0.056	93

a. See footnotes in Table I for definitions of most abreviations.
b. Cetyltrimethylammonium azide, c. cetyltrimethylammonium cyanide d. Ar = $p-O_2NC_6H_4$.

Third. Reaction occurs across the interfacial boundary. For example, in CTAOH solutions, reaction of hydroxide ion in the water with micellar bound substrate is an additional reaction path beside reaction within the two pseudophases. This explanation was originally proposed by Bunton et al,[94] and was also used by Nome and coworkers to explain the failure of the pseudophase model at high [OH⁻] in CTABr catalyzed dehydrochlorination of DDT and some of its derivatives.[97] However, this explanation cannot account for saturation of k_ψ at high counterion concentrations.[56,98]

    Fourth. Micellization in these surfactant systems is not a
highly cooperative phenomena unlike other surfactants and large
amounts of small aggregates, n-mers, which bind organic substrates
but not counterions are present even at very high surfactant concen-

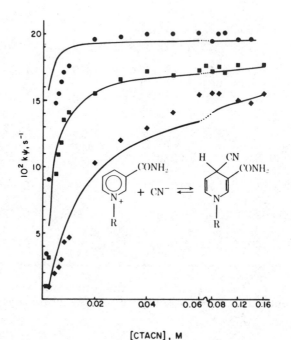

[CTACN], M

Figure 2. Variation of the first-order rate constants with CTACN
for additions of CN⁻ to N-alkyl-3-carbomoylpyridinium ions: ◆,
■,   ● , alkyl = $C_{12}H_{25}$, $C_{14}H_{28}$ and $C_{16}H_{23}$, respectively. Solid
lines are calculated, reference 76.

trations. As the surfactant concentration increases full micelles
appear which bind both substrate and counterions (β = constant).
Added counterions shift size distribution in favor of full micelles.
There is some circumstantial evidence in support of this model.

Phase transfer catalysts such as tri-n-octylethylammonium ions form
non-micellar aggregates which catalyze bimolecular reactions between
two organic reactants but not between bound organic substrates and

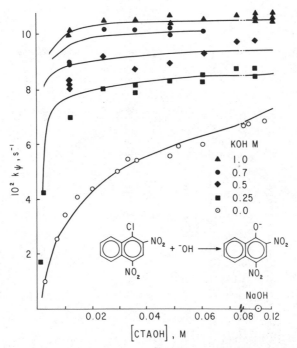

Figure 3. Variation of k_ψ for reaction of 2,4-dinitrochloronaphtha-
lene with [CTAOH] in the presence and absence of added KOH. Solid
lines are theoretical; see text. In 0.1 M NaOH, $k_\psi = 6.4 \times 10^{-1} \text{ s}^{-1}$.

hydroxide ion.[99] CTAOH solutions do not scatter light except at
high concentrations of added NaOH.[56] The equivalent conductance
versus [CTAOH]$^{0.5}$ plot has a maximum around 0.001 M CTAOH, consis-

Table VII. Reactive Counterion Surfactants, Pseudophase Model Fails.[a]

Surfactant	Substrate	K_S, M^{-1}	K'_X, M^{-1}	k_2^m/k_w	Ref.
CTAOH[b]	benzoic anhydride	650	55	0.06	102
CTAOH	bis(4-nitrophenyl)carbonate	1000	55	0.70	102
CTAFor[c]	benzoic anhydride	650	70-80	0.21	102
CTAFor	bis(4-nitrophenyl)carbonate	1000	70-80	d	102
CTAAc[e]	bis(4-nitrophenyl)carbonate	1000	75	f	102
CTAOH	PNPDPP	10,000	55	0.23	56
CTAOH	DNCB	100-250[g]	55	5-7	56
CTAOH	DNCN	1,600	55	2.2	56
CTAF[h]	PNPDPP	10,000	40	0.6	56
Octyl0PhCH$_2$NMe$_3^+$OH$^-$	DNCB	100	8	4.3	56
Octyl0PhCH$_2$NMe$_3^+$OH$^-$	DNCN	1,000	8	2	56

a. See footnotes in Table III for definitions of some abbreviations, b. cetyltrimethyl-ammonium hydroxide, c. cetyltrimethylammonium formate, d. k_w to small to measure, $k_m = 0.017s^{-1}$, e. cetyltrimethylammonium acetate, f. k_w to slow to measure, $k_m \simeq 0.014$-$0.016s^{-1}$, g. K_S depends upon the quantity of added NaOH, h. cetyltrimethylammonium fluoride.

tent the formation of small highly conductive aggregates.[100]

Fifth. The micelle surface is not saturated with counterions (β = constant) and counterion binding is described by a mass-action model:

$$N_w + D_n \underset{}{\overset{K_N'}{\rightleftharpoons}} N_m$$

$$K_N' = \frac{[N_m]}{[N_w]([D_n] - [N_m])} \tag{25}$$

Bunton and coworkers successfully applied this approach to all rate-surfactant profiles of reactions catalyzed by reactive counterion surfactants including the effect added counterion, Table VII,[56,101] that could not be fit by the assumption of β = constant and Equation (24). The complete solution for the concentration of micellar bound counterion is again a quadratic equation and a computer was used combined this expression with Equation (19). The values of K_N, k_M and K_S were selected to give the best fit to the data. No constraints were put on the degree of counterion binding and β was allowed to vary from 0 to 1. All the curves in Figure 3 were produced from the set of parameters in Table VII for reaction of DNCN in CTAOH. The authors noted that for counterions which bind strongly to micelles, $K_N > 2000$, that β rises rapidly to a plateau and then remains essentially constant with increasing surfactant concentration, thus approximating the assumption of constant β.[56]

Like approaches 1-4, the mass-action model has several draw-backs. Currently there is no independent evidence for β being a sensitive function of surfactant concentration and the recently published cell model used to derive the distribution of ions around charged micelles predicts that β decreases gradually with added surfactant.[27] Also, when K_N is small, about 50, the mass-action model predicts a very high surface charge ($\beta < 0.2$) at low micelle concentrations which seems physically unreasonable; and the mass-action model appears to be difficult to apply to solutions containing mixtures of counterions. However, because the mass action model makes no assumptions about micelle size, it is consistent with the physical picture of the n-mer model (approach 4) provided the substrate binding constant is insensitive to aggregate size.

Comparison of Rate Constants for Bimolecular Reactions in Aqueous and Micellar Pseudophases

The units for k_M are in reciprocal seconds because the concentration of the second reactant is expressed as the ratio of bound reactant to micellized surfactant, $m_N{}^s$. To convert these rate constants to conventional units for comparison with rate constants for reaction in water, a volume element for reaction within the micellar pseudophase must be chosen; usually the Stern layer because the micelle surface is assumed reaction site for most micelle catalyzed reactions. To estimate the Stern layer volume of NaLS micelles, for example, Bunton assumed that the micelles were spherical with a density of one, a Stern layer thickness equal to the diameter of the hydrated head group and the thickness of the core equal to the length of the extended hydrocarbon chain.[16,79] The Stern layer volume of the equals roughly half the total micelle volume, so the Stern layer volume of one mole of NaLS, molecular weight of 288, is about 0.14L. The conventional second order rate constant for reaction the the Stern layer, $k_2{}^m$ $M^{-1}s^{-1}$ is given by:

$$k_2{}^m = 0.14k_M \qquad (26)$$

The volume of one mole of CTABr Stern layer is close to that of NaLS,[79] and to simplify comparisons Equation (26) is used for all reactions listed in Tables III - VII.

The numerical value of this conversion factor clearly depends upon the assumed structure of micelles and several factors may make the actual volume of reaction somewhat larger: uncharged reactants distribute themselves between the Stern layer and the core;[102,103] water and ionic reactant may partially penetrate the core;[26] micellized monomer oscillates around an average radius greater than the length of the extended monomer;[23,102] and the boundary surface defining the volume available to bound counterions may extend beyond the hydrated head group.[96] However, the effect of these factors on the value of k^m is probably no greater than the two fold difference between the Stern layer and total micelle volume and errors in the conversion factor should not dramatically affect the comparison of rate constants.

Tables III - VII summarize virtually all the published results for micellar catalyzed bimolecular reactions analyzed by pseudophase kinetic models using the assumptions outlined in the previous sections, except for results missing relevant constants or analyzed using different models. Beside $k_2{}^m/k_w$ ratios the tables include values for K_S, K_N and k_{rel} (= $k_\psi{}^{max}/k'_w$ or $k_2{}^{max}/k_w$) when available. Values for other parameters, β, cmc, k_M and k_w are in the references. Several conclusions can be drawn from the results in these tables.

First. Catalysis, k_{rel}, usually increases with the hydrophobicity of the substrate, K_S, or the second reactant, K_N. This generalization holds for the addition of CN^- to N-alkyl-pyridinium ions, Table III, and the N-heptylimidazole catalyzed hydrolysis of p-nitrophenylesters, Table IV, but not the reaction of PNPA with p-substituted phenylthiolates, Table IV.

Second. Although micelles increase the observed reaction rate 100-1000 fold, most rate constants for reaction within the micellar pseudophase are close to their rate constants in water, $k_2^m/k_w < 1$. Many of these reactions are alkaline hydrolyses of carboxylate or phosphate esters in cationic micelles whose rates of reaction should not be very sensitive to the polarity of the medium.[106] Micellar rate constants for acid catalyzed acetal hydrolysis and the benzidine rearrangement, which is second order in H^+, are 10 to 20 times smaller than their rate constants in water, Tables V and VI. This inhibition might be caused by enhanced stabilization of the proton by hydrogen bonding or because the sulfonic acid is not strong at the micelle surface. Alternatively, the acetal may be partially solubilized by the micellar core reducing its concentration surface.[81]

Hinze and Fendler made the first application of the pseudophase ion exchange model to a vesicle catalyzed reaction using the formalism of Quina and Chaimovich.[107] Vesicles of dioctadecyldimethyl-ammonium chloride (DDOAc) speed the alkaline hydrolysis of 5,5'-dithiobis(2-nitrobenzoic acid) (DTNB) $k_{rel} = 1500$, but $k_2^m/k_w \cong$ 0.9, so the observed rate increase was attributed entirely to the concentration effect, Table III. The rate constant in CTABr micelles was about 10 times slower, consistent with the difference in surface polarities for the vesicles and micelles estimated by the authors. Using the same formalism, Cuccovia and coworkers reported a better than million fold acceleration of the thiolysis of p-nitrophenyloctanoate by n-heptyl mercaptan also in DODAc vesicles, but the micellar rate constant was only slightly more than an order of magnitude greater than the rate constant in water, Table IV.[108]

Using a somewhat different mathematical treatment Reinsborough and Robinson used the pseudophase model to analyzed the effect of NaLS micelles on the rate of Ni^{2+} complex formation with 3 neutral bidentate ligands, Table V.[109] To estimate the micellar rate constants they measured the binding constants of the ligands by solubility and assumed that Ni^{2+} was completely micellar bound. All the calculated micellar rate constants were close to their values in water indicating an aqueous reaction environment for reaction in the micellar pseudophase. Frahm and Diekmann obtained similar results for Ni^{2+} complex formation in sodium decyl sulfate, but the Ni^{2+} concentration at the interface was estimated from their model calculations.[110]

Third. Micellar rate constants for nucleophilic aromatic substitutions are all greater than their values in water, Tables III and VI. The most dramatic increase is the attack of azide ion on DNCN with $k_2^m/k_w \cong 400$; the first example of substantial "true" micellar catalysis.[95] These rate enhancements can be attributed to the same factors which catalyze unimolecular reactions: transition state stabilization by a medium of lower polarity and/or by interaction with a cationic head group.

Fourth. For the same reaction, micellar rate constants, k_2^m, decrease steadily with increasing substrate hydrophobicity. This decrease implies that as substrate hydrophobicity increases a greater fraction of substrate is located in the micelle core away from the reaction site, consistent with the two site model for substrate distribution within the micellar pseudophase.[101,102] This explanation was first proposed by Berezin and coworkers for benzimidazole catalyzed ester hydrolysis in CTABr.[39] The same trend occurs in base promoted ester hydrolysis, Table III and N-heptyl-imidazole catalyzed hydrolysis of p-nitrophenylbenzaldehyde diethyl acetals, Table VI.

Conversely, charged substrates and reactants are not expected to penetrate the micellar core but remain fixed at the surface and the volume available for reactions and the micellar rate constant should be independent of substrate hydrophobicity. The k_2^m/k_w ratios are independent of substrate chain length for the alkaline hydrolysis of N-alkyl-4-cyanopyridinium ions and addition of CN⁻ to N-alkyl-3-carbamoylpyridinium ions, Tables III and VI, and the reaction of thiolates with PNPA and phenolates with p-nitrophenyl diphenylphosphate, Table IV.

It is not possible at this time to attribute differences in k_2^m/k_w ratios for all reactions unambiguously to medium effects or to distribution differences within the micelle or to substrate orientation within the micelle. For example, low k_2^m/k_w ratios for reaction of N-heptylimidazole with esters and the much lower ratios of benzoate compared to p-nitrophenyl esters could be either medium or distribution effects, Table IV. The 10 fold smaller ratio for ethyl p-aminobenzoate compared to the other two benzoate esters might be a difference in orientation; the amino group at the interface forces the ester group into the interior, Table III.

ACID-BASE EQUILIBRIA

Shortly after Hartley published his rules predicting the apparent pK_a shifts of indicators produced by aqueous surfactant solutions[1], Hartley and Roe showed that these shifts could be related quantitatively to the effect of the surface potential on the hydrogen ion concentration at the micelle surface.[33] Since then spectrophotometric indicators have become attractive probes for estimating surface pH because they have high molar absorbtivities and can be used at very low concentrations, minimizing the perturbation of the interface. A wide variety of indicators have been studied[4] and although the shift is observed pK_a is generally 1-2 pK_a units, it is greater than 4 for some carbon acids.[124] The surface potential model has been applied to a variety of charged aqueous

interfaces in addition to micelles[34],[35],[119], including: functional
micelles[120],[121] monolayers[122], microemulsions[12], vesicles and
biological membrances.[123]

In 1964 Mukerjee and Banerjee showed that acidity constants of
micellar bound indicators depend on the polarity of the interface as
well as the hydrogen ion concentration.[34] Recently, Fernandez and
Fromhertz demonstrated that shifts in the apparent pK_a's of hydrophobic micellar bound coumarin dyes could be separated quantitatively into two parts.[35] First a shift in the intrinsic pK_a of the indicator produced by the lower polarity of the micelle surface which
the authors assumed was the same for all surfactants, ionic and nonionic. The contribution of this effect was estimated from the shift
in apparent pK_a produced by a nonionic surfactant. Second, this
shift was either enhance or diminished in ionic surfactants because
of changes in local pH; increased by anionic micelles and decreased
by cationic micelles. However, the utility of this approach may be
limited. Although surface polarities of different types of micelles
are certainly similar[4],[30],[107], they need not be identical; and
intrinsic acidity constants like micellar rate constants may be
affected by specific interactions between the indicator and the
micellized surfactant. And, as noted earlier, surface potential
models do not account for specific salt effects.[18],[41]

Application of the pseudophase ion exchange model to indicator
equilibria overcomes both these problems because it makes no
assumptions about intrinsic pK_a's of micellar bound indicators and
specific salt effects are treated directly in terms of ion exchange.
To date the model has only been applied to a few indicator
equilibria in basic solutions of cationic surfactants, but the
results are encouraging. The same assumptions used for treating
micellar effects reaction rates are used for quantitatively interpreting micellar effects on indicator equilibria. The size and
direction of the shift will depend on a combination of factors
including the charge type and concentration of the surfactant, the
charge on each form of the indicator and their hydrophobicity, and
the bulk pH and concentration of electrolyte. The purpose here is
to outline the elements required for a quantitative treatment and to
summarize the published results.

Scheme III illustrates the pseudophase ion exchange model for
the apparent basicity constant, K_B, of a uncharged weak acid, NH, in
an aqueous alkaline solution of cationic micelles. The constants
K_N^X, K_B^W and K_B^m are respectively, the ion exhange constant for the
charged form of the indicator, the basicity constant of the indicator in water and the intrinsic basicity constant of the indicator in
the micellar pseudophase. The equilibria are expressed in terms of
basicity constants instead of the more conventional acidity constants because K_B^m can be estimated directly without assuming a
value for the unknown dissociation constant of water at the micelle

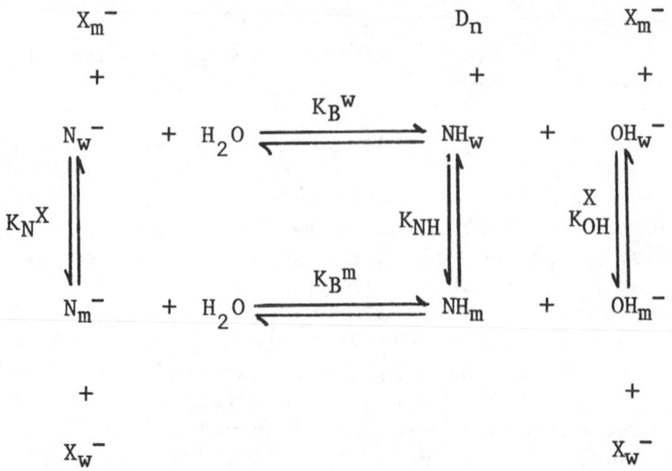

$$X_m^-$$ $$\hspace{4cm}$$ $$D_n$$ $$\hspace{3cm}$$ $$X_m^-$$

$$+$$ $$\hspace{5.5cm}$$ $$+$$ $$\hspace{4cm}$$ $$+$$

$$N_w^- \; + \; H_2O \; \underset{}{\overset{K_B^w}{\rightleftharpoons}} \; NH_w \; + \; OH_w^-$$

$$K_N^X \Big\updownarrow \hspace{3cm} K_{NH} \Big\updownarrow \hspace{1cm} K_{OH}^X \Big\updownarrow$$

$$N_m^- \; + \; H_2O \; \underset{}{\overset{K_B^m}{\rightleftharpoons}} \; NH_m \; + \; OH_m^-$$

$$+$$ $$\hspace{7cm}$$ $$+$$

$$X_w^-$$ $$\hspace{7cm}$$ $$X_w^-$$

Scheme III

surface and also because OH⁻and not the H⁺ is displaced by added
counteranions, X⁻. In anionic micelles Scheme III should be rewritten in terms of acidity constants and exchange of the proton by
countercations. Further modifications would be required for indicators of different charge types or for other types of equilibria.

The intrinsic basicity constant is defined by:

$$K_B^m = \frac{m_{NH}^s m_{OH}^s}{m_N^s} = \frac{[NH_m] \, m_{OH}^s}{[N_m^-]} \qquad (27)$$

Like micellar rate constants for bimolecular reactions, K_B^m depends
on concentrations within the micellar pseudophase, not the total
solution volume. Other research groups have developed different
formalisms for expressing apparent basicity (or acidity) constants
of weak organic acids based on the distribution of species illustrated by Scheme III;[36,67] but regardless of the formalism used the
final expressions correctly predict the effect of added surfactant
and salt. Also, just as with bimolecular reactions, the concentration of hydroxide ion is controlled by ion exhange and added buffers
will not control surface pH.

Bunton and Romsted used a computer to simulated the biphasic
apparent basicity constant-CTAX (X = Cl⁻, Br⁻, and NO⁻) profile for
benzimidazole in aqueous NaOH and the effect of added Cl₃⁻ at several
NaOH concentrations at constant CTACl for benzimidazole and naphth-
imidazole, Table VIII.[60] Equation (27) was solved for the apparent
basicity constant, K_B, using the full quadratic expression for m_{OH}^s,
Equation (13) for m_{NH}^s and the benzimidolide anion was assumed to be
completely bound under all conditions. The fit was excellent except

for benzimidazole at high concentrations of added Cl^- (> 0.2 M), but
the fit improved markedly when the salt induced binding of
benzimidazole was taken into account.

In a separate treatment valid under the limiting conditions $m_X{}^s$
>> $m_{OH}{}^s$ and $m_X{}^s \cong \beta$, Bunton and coworkers fit the apparent basicity
constants of benzimidazole in CTABr and CTANO$_3$ (but not CTACl) and a
series of substituted phenols and oxime in CTABr using single
values of the intrinsic basicity constants for each compound Table
VIII.[125] The apparent basicity constants of the phenols and oximes
were calculated from apparent acidity constants, assuming the pK_w of
water is 14, because the measurements were made in borate buffered
solutions.

· In both treatments reasonable values of β and $K_X{}^{OH}$ were used to
calculate $K_B{}^m$. Just as with micellar catalyzed bimolecular
reactions, fitting apparent basicity constant data will not give
unique values of $K_B{}^m$ because more than one set of reasonable values
of these constants will fit the data.[60] An unambiguous estimate of
$K_B{}^m$ requires that all other constants be measured independently.

Using their formalism, Chaimovich and coworkers simulated the
apparent acidity constant-CTABr concentration profiles for phenol,
n-heptylmercaptan and an oxime.[18] Their treatment explicity
accounts for the effect of buffers and ion exchange of the basic
form of the organic acid.

Funasaki used a somewhat different form of the pseudophase ion
exchange model to simulate the effect of CTABr and added NaBr on the
apparent acidity constant of Thymol Blue.[36] Both forms of this indicator are negatively charged and were assumed to be completely
micellar bound so that all shifts in the apparent acidity constant
were related to changes in surface pH. As with the alkaline
hydrolysis of p-nitrophenyl butyrate in CTABr, the author found that
the surface potential model fit the data equally well.

The pseudophase ion exchange model also provides a plausible
physical explanation for surfactant and salt effects on apparent
ionization constants that may be obscured if analyzed in terms of
the surface potential effects on interfacial pH. Scheme III is valid for any neutral acid indicator which binds to cationic micelles
even if its intrinsic basicity constant is so small that measurements must be carried out in acidic solution. For example, Mukerjee
and Banerjee found that the measured ratio, R, of the acidic to basic forms of completely micellar bound Bromothymol Blue in acidic
buffer, increased with added CTABr and that added Br^- increased R
more than added Cl^-.[34] They attributed the increase in R to negative adsorption of protons from the surface of the cationic micelle.
These observations are completely consistent with Scheme III
($R = [NH_m]/[N_m{}^-]$, $[NH_w] = [N_w{}^-] = 0$). As with all micellar solu-

tions, increasing the surfactant concentration dilutes, NH_m, N_m^- and OH_m^- within the micellar pseudophase shifting the position of equilibrium to the right and increasing R. Added counterion displaces OH^- from the micelle surface with Br^- being more effective than Cl^- because $K_{OH}^{Br} > K_{OH}^{Cl}$ (Table VIII), again increasing R. Analogous arguments hold for the effect of anionic micelles on strongly basic indicators in basic solution. This analysis is identical to the one used by Chaimovich and coworkers to analyze alkaline hydrolysis of esters and 4-cyanopyridinium derivatives in NaLS (see above).[91]

Comparison of Basicity Constants

The intrinsic basicity constant, K_B^m, is a unitless number and to compare it with basicity constants in water it must first be converted to ordinary basicity constants with units of M/L. Assuming the equilibrium occurs only in the Stern layer, the micellar basicity constant K_b^V is:

$$K_B^V = K_B^m/0.14 \qquad (28)$$

The conversion factor is the same as the one used for converting rate constants.

The K_B^V/K_B^W ratios in Table VIII show that micellar basicity constants, like micellar bimolecular rate constants are not too different from their values in water. The ratio of roughly 10 for phenols and oximes may reflect the assumption about the ion product of water used in converting the measured acidity constants into basicity constants. However, the basicity constants of the phenols do not depend on their hydrophobicity, but do parallel their aqueous basicity constants.

BIMOLECULAR REACTIONS IN FUNCTIONAL, MIXED AND COMICELLES

Micelles are models for enzymes in the sense that they are similar in size and shape and both have polar surfaces and hydrophobic interiors. But reactions catalyzed by unfunctionalized micelles generally fail to mimic the enormous catalytic activity and regio and stereoselectivity of the modeled enzyme. However, micelles functionalized with groups that model the amino acid side chains responsible for enzyme activity are much more impressive catalysts.[4,8,126] A variety of functional groups have been used including alcohols, thiols, imidazoles, and oximes and in most cases the active forms are their anions. Some systems have catalytic activities similar to that of the native enzyme[127,129] and recently a number of micellar induced stereoselective[130-132] and regioselective[134], reactions have been studied.

Table VIII. Indicator Equilibria[a]

Surfactant	Indicator	K_S, M^{-1}	K_X^Y	K_m^v/K_b^w	Ref.
CTACl	benzimidazole	43	4	0.6	60
CTABr	benzimidazole	36	12	0.6	60
CTANO$_3$[b]	benzimidazole	36	14	0.6	60
CTABr	naphthimidazole	1,100	12	0.6	60
CTABr	benzimidazole	36	21	0.47[c]	125
CTANO$_3$	benzimidazole	36	23	0.33[c]	125
CTABr	p-MePhOH	485	21(12.5)[d]	8(14)[d]	125
CTABr	p-n-PrPhOH	1,350	21(12.5)	8(14)	125
CTABr	p-t-BuPhOH	1,700	21(12.5)	11(19)	125
CTABr	p-t-AmPhOH	4,330	21(12.5)	9(15)	125
CTABr	2-napthol	1,390	21(12.5)	6.7(10)	125
CTABr	p-t-ButoxyPhOH	1,630	21(12.5)	13(22)	125
CTABr	p-nitrobenzaldoxime	240	21(12.5)	8.6(14)	125
CTABr	acetophenoneoxime	150	21(12.5)	3.9(6.5)	125

a. See footnotes in Table III for definitions of some abbreviations. b. cetyltrimethylammonium nitrate, c. average values, d. values in parentheses show basicity constant ratios
based on ion exchange constant in parentheses.

Functional groups are introduced into micelles several different ways, see Figure 4. The substrate, S, and the nucleophile, [NH] + [N$^-$] = [N$_T$], such as phenol are simply added to the solution and the surfactant is introduced in progressively larger amounts to study its effect (System 1). Alternatively a hydrophobic tail is attached to the nucleophile to insure substantial binding at low surfactant concentrations and is comicellized with a nonfunctional surfactants (System 2). The nucleophile is covalently attached to a surfactant molecule and mixed with a nonfunctional surfactant (System 3), or the functionalized surfactant molecule is water soluble and self-micellizes, (System 4).

The outer equilibria in Figure 4 show the binding of the neutral (left) and anionic (right side) forms of the nucleophile. The inner equilibria illustrate the ionization or titration of the nucleophile. In cationic micelles, the neutralization of the acidic group, NH, must be accompanied by dissociation of a counterion from the micelle surface. In studies that fit System 1 the nucleophile is usually present in indicator amounts, but in Systems 2 and 3 the nucleophile may occupy a substantial fraction of the total micellar volume and in System 4, the nucleophile covers the micelle surface. Thus, the usual assumption that micelle structure is not perturbed by the reactants may not hold in Systems 2-4 because changing the nucleophile concentration or the pH will significantly alter both the surface charge density and the extent of counterion binding. However, in System 4 for example, if the operational pH is substantially below the apparent pK_a of the nucleophile, the concentration of its anionic form will be in indicator amounts and micelle structure should remain unperturbed.

The experimental conditions for most studies are: [D] ⩾ [N$_T$] ≫ [S] with buffers added to control pH and salt added to control ionic strength. The rate-surfactant profiles are, like other bimolecular reactions, biphasic, rising rapidly to a maximum followed by a gradual decrease in rate at higher surfactant concentrations. The position of the rate maximum and the extent of catalysis, k_{rel}, depend on the same factors: surfactant structure, substrate and nucleophile (neutral and anionic forms) binding constants, pH, ionic strength and the rate constant within the micellar pseudophase. For these reasons, comparison of rate enhancements in different systems is fraught with uncertainty and comparison of micellar rate constants for reaction within the micellar pseudophase free from concentration effects should be more meaningful.

The pseudophase ion exchange model treats micellar effects on bimolecular reaction between an organic substrate and the nucleophilic anion of a weak acid as a product of two effects: concentration of the reactants into the small volume of the micelles pseudophase and micellar induced shifts in the extent of ionization of the nucleophile. The final rate equation will be a composite of

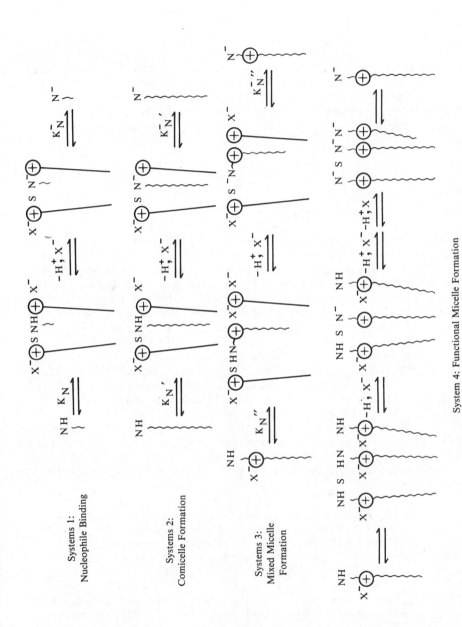

Figure 4. Distribution and dissociation of the nucleophile in micelles, and functional, mixed comicelles, reference 16.

the expressions used for micellar effects on bimolecular reactions
and on acid-base equilibria. Complete rate expressions using the
formalism presented here were recently derived but remain
untested.[135]

To avoid complications caused by buffers and added salts and
uncertainties in treating the apparent pK_a of the nucleophile, Bunton and Sepulveda measured the concentration of micellar bound anion
of the nucleophile, N_m^-, spectrophotometrically for reaction of a
series of oximates, imidazoles, hydroxamic acids and substituted
phenols with p-nitrophenyl diphenylphosphate, Table IV.[112-114]
Reaction in water is very slow and neglected, and equation (19)
reduces to:

$$k_\psi = \frac{k_M K_s [N_m^-]}{1 + K_s [D_n]} \qquad (29)$$

where
$$[N_m^-] = m_N{}^s [D_n]$$

Chaimovich and coworkers derived a complete set of equations
based on their own formalism to fit the CTABr catalyzed thiolysis
and oximolysis[18] and vesicle catalyzed thiolysis of p-nitrophenyloctanoate,[108] Table IV. All parameters were measured independently
except the micellar rate constant which was selected to give the
best fit to the data. The observed rate for the oximate reaction
was corrected for the competing reaction with hydroxide ion.

Fornasier and Tonellato, using a somewhat different approach,
derived kinetic equations based on the apparent pK_a of the functional group which was measured independently for the reaction of thiol,
hydroxyl and imidazole functionalized surfactants reacting with PNPA
and PNPH.[105] Despite the dependence of the apparent pK_a on
surfactant and salt concentration, their use of the apparent instead
of intrinsic acidity constant worked because the high concentration
of added salt, $\mu = 0.1$ (KCl), suppressed shifts in the apparent pK_a
with added surfactant; i.e. $[X_T] = [D] + [M_X] \gg [OH_w]$, Equation
(7).

Bunton used estimates of the apparent pK_a's from kinetic data
to calculate micellar rate constants for reaction of a β-hydroxy-
ethyl functionalized cationic surfactant with several different
substrates Table IX.[136] Note that a substantial fraction of the
surfactant was in zwitterionic form, $X = [N_m]/[D_n]$.

Recently Bunton and coworkers also tested the pseudophase model
for reaction of p-NPDPP in mixed and comicelles of CTABr with an
oxime, hydroxamic acid and an imidazole, Table IX.[136,137] The surfactant concentration was kept high enough to insure complete binding of the p-NPDPP and the nucleophile, and a high pH was maintained
to insure complete ionization of the nucleophile. Rate constants

were corrected for contribution of the reaction with hydroxide ion. A single value of the micellar rate constant fit the data reasonably well at several different ratios of nucleophile to surfactant and with the nucleophile occupying a substantial fraction of the total micellar volume.

Comparison of Micellar and Aqueous Rate Constants

As with most bimolecular reactions, micellar rate constants for reactions of nucleophiles of weak organic acids expressed in terms of the concentration of its active form (usually its anion) are close to their values in water; $k_2^m/k_w \simeq 1$. As before, for reactions which pass through charged transition states that are more delocalized than their ground states, $k_2^m/k_w > 1$; for example, reaction of 2,4-dinitrochlorobenzene (DCNB) with the β-hydroxyethyl surfactant, Table IX. For the neutral hydrophobic substrate p-NPDPP, $k_2^m/k_w < 1$, which implies that some of this substrate is located in the micellar core as well as the surface; Tables IV and IX.

Toward a General Treatment of Counterion Binding in Functional, Mixed and Comicelles

As noted above, the nucleophile in Systems 2-4 of Figure 4 may cover a substantial fraction of the micelle surface. Thus, titration of these systems with base deprotonates the weak acid and counterions dissociate, transforming the cationic micelles into a mixed zwitterionic/cationic micelles, or in the limit, completely zwitterionic micelles. This process reduces the fraction of the surface which will bind counterions and participate in ion exchange, diminishes the concentration of micellar bound hydroxide and shifts the apparent basicity (or acidity) constant of the bound nucelophile.

Quantitative interpretation of titration curves of weak acids at aqueous interfaces is a major theoretical problem in surface chemistry. Most current approaches are based on the surface potential model, which has been applied to micelles[119-121], monolayers[122], and phospholipid vesicles.[123] The intrinsic acidity constant is usually, but not always[138], equated with the measured value when the net surface charge is zero. The surface potential model predicts the shape of the titration curve but cannot account for specific salt effects.[119-121]

Several new assumptions are required to apply the pseudophase ion exchange model to titration curves or reactions in functional, mixed or comicelles at high ratios of nucleophile to surfactant. For example, for a functionalized cationic surfactant, System 4 (Figure 4), if counterions are assumed to associate only with the

Table IX. Bimolecular Reactions with Functional, Mixed and Comicelles.[a]

Surfactant	Substrate	N	k_2^m/k_w	X^b	Ref.
R^+-OH^c	DNCB	R^+-O^-	3.36	0.29,0.67	136
RY-OH	$DNFB^d$	R^+-O^-	1.05	0.29,0.67	136
R^+-OH	malachite green	R^+-O^-	0.49	0.63	136
R^+-OH	$O_2NC_6H_4CO_2PO_3^{2-}$	R^+-O^-	0.38	0.29,0.55	136
R^+-OH	PNPDPP	R^+-O^-	0.036	0.29,0.67	136
CTABr	PNPDPP	$C_{12}H_{25}\overset{+}{N}Me_2CH_2C(Ph)=NO^-$	0.42	0.06-0.17	136
CTABr	PNPDPP	$C_{13}H_{27}CON(Ph)O^-$ Na^+	0.42-0.48	0.06-0.17	136
CTABr	PNPDPP	$C_{16}H_{33}\overset{+}{N}Me_2CH_2Im$ Cl^{-e}	0.5	0.3-0.5	137

a. See footnotes in Table III for definitions of some abbreviations. b. $X = [N_m]/[D_n]$,
c. β-hydroxyethylcetyldimethylammonium bromide, d. 2,4-dinitrofluorobenzene,
e. I_m = imidazole.

fraction, n, of the surface covered by untitrated positively charged head groups, then equation (3) becomes:

$$m_X^s + m_{OH}^s = \beta n \tag{30}$$

$$n = \frac{[NH_m^+]}{[N_m^\pm] + [NH_m^+]} \tag{31}$$

and where $[NH_m^+]$ and $[N_m^\pm]$ are the concentrations of micellized protonated and deprotonated functional surfactant. For mixed and comicelles the concentration of the nonfunctional surfactant, [D], must be added to the numerator and denominator and the derivation modified accordingly. Combining Equation (30) with Equation (27) for the intrinsic basicity constant and the mass balance Equation, $[N_T] = [NH_m^+] + [N_m^\pm]$, with Equation (30) gives:

$$m_X^s + m_{OH}^s = \frac{\beta K_B^m}{K_B^m + m_{OH}^s} \tag{32}$$

This derivation assumes that ideal mixing exists between the two forms of the micellized surfactant and that β remains constant regardless of the surface charge, that is the value of n.

Equation (32) substitutes for Equation (3) when deriving equations for rates of bimolecular reactions or apparent basicity (or acidity) constants in functional, mixed and comicelles. The complete solution is complex and implicitly assume that rate and equilibrium constants for reaction within the micellar pseudophase are also independent of surface charge. The partial success of the surface potential model supports this assumption.[35,119,121] In the limit, $m_{OH}^s \ll m_X^s$, when only a small fraction of the functional group is ionized, Equation (32) reduces to Equation (3), and the usual assumptions hold because the active form of the nucleophile is present in indicator amounts. Conversely, when, $m_{OH}^s \gg K_B^s$ and n → 0, the micelle should become completely zwitterionic with zero counterion binding. Equation (32) explicitly includes the concentration of other counterions, m_X^s, and the final expression for the apparent basicity (or acidity) constant should account for the specific salt effects on titration curves of micelles and other types of aqueous interfaces.

CATALYSIS NEAR THE CMC: A SECOND PARTIAL FAILURE OF THE PSEUDOPHASE MODEL

The pseudophase ion exchange model often fails to fit rate-surfactant profiles near the cmc. At these concentrations the amount of micellized surfactant is not in large excess over reactant concentration and the assumption that reactants do not perturb micelle structure may fail. Also, near the cmc the concentration of monomer is sensitive to specific interactions with the reactants and other reagents in solution and is probably not constant, and small shifts in the cmc will have large effects on the concentration of micellized surfactant, Equation (2). This uncertainty is particularly serious for bimolecular reactions and acid-base equilibria because of the second order dependence on the surfactant concentration, Equations (20), (21) and (27).

Unfortunately, reliable estimates of the cmc are difficult to obtain.[139] The cmc is not a single surfactant concentration marking a sharp phase boundary but an experimentally defined quantity assigned to a narrow range of surfactant concentrations. The cmc is also very sensitive to the presence of additives. Added salts and nonelectrolytes and small amounts of hydrophobic organic molecules, e.g. reactants, induce micelle formation at lower surfactant concentrations. While the concentration dependence of the cmc on some individual additives is well known, the composite effect of added salts, buffers and organic reactants used in rate and equilibrium studies is not.

To solve this problem the cmc is usually defined operationally as the surfactant concentration marking the onset of catalysis or shift in the apparent ionization constant.[16] For example, the rate-surfactant profiles a variety of unimolecular and bimolecular reactions can be fit over the entire concentration range using the "kinetic" cmc, including: acid catalyzed hydrolysis of acetals in NaLS and long chain sulfonic acids[94], addition of CN^- to N-alkyl-3-carbamoylpyridinium ions in CTACN[76], thiol catalyzed ester hydrolysis in CTABr[86], and alkaline ester hydrolysis in CTABr.[36] In these cases the binding or ion exchange constants are not extremely large so that reactant concentrations, within the micellar pseudophase $m_S{}^S$, $m_N{}^S$, are probably low even near the cmc.

However, if the substrate is very hydrophobic and especially if it is of opposite charge to the surfactant, the model usually fails near the cmc. For example the reactions of malachite green with borohydride ion, hydroxide ion and 1-benzyl-dihydronicotinamide (BDHNA) are all catalyzed by hydroxide ion while NaLS inhibits the first two but catalyzes the later.[79] The pseudophase ion exchange model fits all these reactions at high surfactant but fails near the cmc for reactions in CTABr and with BDHNA in NaLS. The interaction between malachite green and NaLS is very strong ($K_S \simeq 8,000$) and catalysis begins at about 1×10^{-4} M/L NaLS, almost two orders of magnitude below the literature cmc of 0.008 M/L.[139] Even using this kinetic cmc the predicted rate maximum at 1×10^{-3} M/L NaLS does not match the observed maximum at 3×10^{-3} M/L.

Failures of the pseudophase model near the cmc is sometimes attributed to catalysis by premicellar aggregates formed by specific

interactions between reactants and small preformed or induced
aggregates of surfactant.[79,89,140] Reeves found strong evidence for
the formation of catalytically active hydrophobic ion pairs between
CTABr monomers and oppositely charged hydrophobic substrates[141] and
existence of premicellar aggregates has been proposed several
times.[20,142]

One possible approach to interpreting catalysis near the cmc is
to use a cooperativity model like the one developed by Piszkiewicz
for micellar catalyzed reactions between organic substrates and
hydrophilic ions. His model is based on the Hill equations used to
estimate the cooperative binding of substrates to enzymes.[69] The
estimated size of the catalytic unit is general small, 1 to 4 surfactant molecules per substrate. However, this model interprets
micellar effects solely in terms of the distribution of the
substrate and ignores the distribution of second reactant.

Another possibility is the formation of hydrophobic reactive
ion-pairs. Kunitake explained the large catalytic activity of the
phase transfer catalyst trioctylmethylammonium chloride on reaction
between long-chain, but not short-chain, hydroxamates and imidazoles
with PNPA by the phase transfer catalyst induced increase in the
concentration and reactivity of the nucleophile by ion-pair formation with its anion.[143] However, the latter assumption contradicts
substantial evidence that ion pair formation reduces nucleophili-
city.[144] Also, recently Bunton and coworkers successfully interpreted the large rate enhancement of benzimidazole and napthimida-
zole catalyzed hydrolysis of p-NPDPP by phase transfer catalysts in
terms of the pseudophase model.[145] The estimated rate constants for
reaction within the aggregate were only slightly larger than those
in water. Nevertheless, it is clear that the pseudophase model
fails near the cmc and new approaches are needed to account for
specific interactions.

CONCLUSIONS, PROSPECTS AND POSSIBILITIES

Results from a number of studies support the contention of
Chaimovich, Quina and coworkers that, within definable limits,
the pseudophase ion exhange model provides a unified quantitative
interpretation for the effect of ionic micelles on reaction rates
and equilibria of heterolytic reactions.[18] The model correctly
interprets the effect of surfactant charge and chain length, buffers, pH and specific salt effects. In most cases the distribution
of reactants between the aqueous and micellar pseudophases are
described by a set of independently verifiable constants; the cmc,
the degree of counterion binding and binding and ion exchange
constants for reactants. The calculated rate and equilibrium
constants for reaction within the micelles pseudophase are generally
constant over a wide range of experimental conditions.

Micellar rate and equilibrium constants are usually close to their values in water supporting the assumption that reaction occurs in the Stern layer, a medium with a polarity similar to ethanol. Reactions with rate constants larger in micelles than in water are more charge dispersed in the transition state than in the ground state and represent "true" micellar catalysis. Whereas reactions with smaller rate constants in micelles than water are examples of micellar stabilization of the ground state, such as reduced H^+ activity at the surface of anionic micelles in micelle catalyzed acetal hydrolysis, or solibilization of neutral hydrophobic reactants in the micelle core reducing their concentration at the reactive site at the micelle surface.

The model fails twice and in both cases the failure is related to the pseudophase assumption. The model fails near the cmc when specific interactions between reactants(s) and surfactant molecules become more important than noncooperative binding of reactants to micelles. A new approach is needed to describe the properties of these systems. The model also fails to describe the catalytic properties of cationic surfactants with hydrophilic counterions such as OH^- and F^-. The reason for this failure is not known, but one or more of the pseudophase assumptions, that micelle formation is highly cooperative or that the degree of counterion binding is constant, are probably breaking down. Careful measurement of the physical properties of these novel surfactants should provide new information on the energetics of micelle formation and counterion binding at aqueous interfaces.

The pseudophase ion exchange model has yet to be extensively tested on other types of aqueous interfaces such as microemulsions, inverse micelles, monolayers and vesicles. Comparison of aqueous and micellar rate constants may aid in the interpretation of micellar induced regio and stereoselectivity. Finally, extension of the pseudophase ion exchange model to titration curves of weak acids bound to aqueous interfaces may provide a quantitative interpretation of specific salt effects and a more precise estimate of surface pH.

ACKNOWLEDGEMENTS

I would like to thank Madeline Mendoza for patiently typing the manuscript and K. L. Mittal for waiting patiently to receive it; the American Chemical Society for permission to use Figures 1-3; all my colleagues and friends who contributed so much to this work; and especially C. A. Bunton whose continuing support and guidance over the past 7 years made this work possible.

REFERENCES

1. G.S. Hartley, Trans. Faraday Soc., 30, 44 (1934).
2. E.F.J. Duynstee and E. Grunwald, J. Am. Chem. Soc., 81, 4540, 4542 (1959).
3. J.H. Fendler and E.J. Fendler, "Catalysis in Micellar and Macromolecular Systems," Academic Press, New York, 1975.
4. J.H. Fendler, "Membrane Mimetic Chemistry," Wiley-Interscience, New York, 1982.
5. F.M. Menger, Pure Appl. Chem., 51, 999 (1979).
6. N.J. Turro, M. Gratzel and A.M. Braun, Angew. Chem. Int. Ed., Engl., 19, 675 (1980).
7. T. Kunitake and S. Shinkai, Adv. Phys. Org. Chem., 17, 435 (1980).
8. C.J. O'Connor, R.E. Ramage and A.J. Porter, Adv. Coll. Interface Sci., 15, 25 (1981).
9. K.L. Mittal, Editor, "Micellization, Solibilization and Microemulsions," Vols. 1 and 2. Pleum Press, New York, 1977.
10. K.L. Mittal, Editor, "Solution Chemistry of Surfactants," Vols. 1 and 2, Plenum Press, New York, 1979.
11. K.L. Mittal and E.J. Fendler, Editors, "Solution Behavior of Surfactants: Theoretical and Applied Aspects" Vols 1 and 2, Plenum Press, New York, 1982.
12. R.A. Mackay, Adv. Coll. Interface Sci., 15, 131 (1981).
13. A. Kitahara, Adv. Coll. Interface Sci., 12, 109 (1980).
14. S.J. Valenty, J. Am. Chem. Soc., 101, 1 (1979).
15. J.H. Fendler, Acc. Chem. Res., 13, 7 (1980).
16. C.A. Bunton and L.S. Romsted, in "The Chemistry of the Functional Groups. Supplement B: The Chemistry of Acid Derivatives," S. Patai, Editor, Part 2, p. 945, Wiley-Interscience, London, 1979.
17. E.J.R. Sudholter, G.B. van de Langkruis and J.B.F.N. Engberts, Recl. Trav. Chim. Pays-Bas Belg., 99, 73 (1980).
18. H. Chaimovich, R.M.V. Aleixo, I.M. Cuccovia, D. Zanette and F.M. Quina, in "Solution Behavior of Surfactants: Theoretical and Applied Aspects" K.L. Mittal and E.J. Fendler, Editors, Vol. 2, p. 949 Plenum Press, New York, 1982.
19. K. Shinoda, T. Nakagawa, B-I. Tamamushi, T. Isemura, "Colloidal Surfactants," Academic Press, New York, 1963.
20. P. Mukerjee, Adv. Coll. Interface Sci., 1, 241 (1967).
21. G.C. Kresheck, in "Water: A Comprehensive Treatise", F. Franks, Editor, Vol. 4, p. 95, Plenum Press, 1975.
22. D.G. Hall, B.A. Pethica, "Nonionic Surfactants," M.J. Shick, Editor, p. 634, Marcel Dekker, New York, 1967.
23. B. Lindman and H. Wennerstrom, Topics Curr. Chem., 87, 1 (1980).
24. L.R. Fisher and D.G. Oakenfull, Quart. Rev. Chem. Soc., 6, 25 (1977).
25. H. Wennerstrom and B. Lindman, Phys. Rep., 52, 1 (1979).
26. F.M. Menger, Acc. Chem. Res., 12, 111 (1979).

27. H. Wennerstrom and B. Lindman, J. Phys. Chem., 83, 2931 (1979).
28. P. Fromherz, Chem. Phys. Lett., 77, 460 (1981).
29. P. Fromherz, Ber. Bunsenges. Phys. Chem., 85, 891 (1981).
30. E.H. Cordes and C. Gitler, Prog. Bioorg. Chem., 2, 1 (1973).
31. L.S. Romsted, Ph.D. Thesis, Indiana University, (1975).
32. L.S. Romsted, in "Micellization, Solubilization and
     Microemulsions", K.L. Mittal, Editor, Vol. 2, p. 509, Plenum
     Press, New York 1977.
33. G.S. Hartley and J.W. Roe, Trans. Faraday Soc., 36, 101 (1940).
34. P. Mukerjee and K. Banerjee, J. Phys. Chem., 68, 3567 (1964).
35. M.S. Fernandez and P. Fromherz, J. Phys. Chem., 81, 1755
     (1977).
36. N. Funasaki, J. Phys. Chem. 83, 1998 (1979).
37. K. Shirahama, Bull Chem. Soc. Jap., 48, 2673 (1975).
38. I.V. Berezin, K. Martinek and A.K. Yatsimirski, Russ. Chem.
     Rev., (Engl.), 42, 487 (1973).
39. K. Martinek, A.K. Yatsimirski, A.V. Levashov and I.V. Berezin,
     in "Micellization, Solubilization and Microemulsions," K.L.
     Mittal, Editor, Vol. 2, p. 489, Plenum Press, New York, 1977.
40. D. Stigter, J. Phys. Chem., 68, 3603 (1964).
41. M. Almgren and R. Rydholm, J. Phys. Chem., 83, 360 (1979).
42. D. Stigter, Prog. Colloid and Polymer Sci., 65, 45 (1978).
43. E.H. Cordes and R.B. Dunlap, Acc. Chem. Res., 2, 329 (1969).
44. C.A. Bunton, in "Reaction Kinetics in Micelles", E.H. Cordes,
     Editor, p. 73, Plenum Press, New York, 1973.
45. G.S. Manning, Acc. Chem. Res., 12, 443 (1979).
46. G. Weisbuch and M. Gueron, J. Phys. Chem., 85, 517 (1981).
47. B.W. Barry and G.F.J. Russell, J. Colloid Interface Sci.,
     40, 174 (1972). Recently β was determined over a very large
     temperature range, from 25° to 166°, ref. 48.
48. D.F. Evans and P.J. Wightman, J. Colloid Interface Sci. 86, 515
     (1982).
49. E.W. Anacker, in "Solution Chemistry of Surfactants," K.L.
     Mittal, Editor, Vol. 1, p. 247, Plenum Press, New York, 1979.
50. E.W. Anacker and H.M. Ghose, J. Phys. Chem., 67, 1713 (1963).
51. G. Sugihara and P. Mukerjee, J. Phys. Chem., 85, 1612 (1981).
52. G.D. Parfitt and J.A. Wood, Kolloid -Z. Z. Polym., 229, 55
     (1969).
53. H.C. Evans, J. Chem. Soc., Pt. 1, 579 (1956).
54. J.J. Pearson and K.J. Humphreys, J. Pharm. Pharmacol., 22,
     Suppl. 126S (1970).
55. K.D. Heckmann and R.F. Woodbridge, Proc. IVth Intern. Congr.
     Sur. Act. Sub., Brussels, 2, 519 (1964).
56. C.A. Bunton, L-H. Gan, J.R. Moffatt, L.S. Romsted, and G.
     Savelli, J. Phys. Chem., 85, 4118 (1981).
57. D. Bartet, C. Gamboa and L. Sepulveda, J. Phys. Chem., 84, 272
     (1980).
58. C. Gamboa, L. Sepulveda and R. Soto, J. Phys. Chem., 85, 1429
     (1981).

59. H. Chaimovich, J.B.S. Bonilha, M.J. Politi and F.H. Quina, J. Phys. Chem., 83, 1851 (1979).
60. C.A. Bunton, Y-S. Hong and L.S. Romsted, in "Solution Behavior of Surfactants: Theoretical and Applied Aspects" K.L. Mittal and E.J. Fendler, Editors, Vol. 2, p. 1137, Plenum Press, New York 1982.
61. M. Plaisance and L. Ter-Minassian-Saraga, J. Colloid Interface Sci., 56, 33 (1976).
62. O. Samuelson, "Ion Exchange Separations in Analytical Chemistry," p. 71, John Wiley & Sons, New York, 1963.
63. F.H. Quina, M.J. Politi, I.M. Cuccovia, E. Baumgarten, S.M. Martins-Franchetti and H. Chaimovich, J. Phys. Chem., 84, 361 (1980).
64. C.A. Bunton, K. Ohmenzetter and L. Sepulveda, J. Phys. Chem., 81, 2000 (1977).
65. J. Holzwarth, W. Knoche and B.H. Robinson, Ber. Bunsenges, Phys. Chem., 82, 1001 (1978).
66. S. Diekmann and J. Frahm, J.C.S. Faraday I. 76, 446 (1980).
67. F.H. Quina and H. Chaimovich, J. Phys. Chem., 83, 1844 (1979).
68. S.K. Srivastava and S.S. Katiyar, Ber. Bunsenges. Phys. Chem., 84, 1214 (1980).
69. D. Piszkiewicz, J. Am. Chem. Soc., 99, 1550, 7695 (1977).
70. L.S. Romsted, unpublished results.
71. For sample calculations, see reference 60.
72. C.A. Bunton and L. Robinson, J. Org. Chem., 34, 773, 780 (1969).
73. R.B. Dunlap and E.H. Cordes, J. Am. Chem. Soc., 90, 4395 (1968).
74. C. Hirose and L. Sepulveda, J. Phys. Chem., 85, 3689 (1981).
75. E. Lissi, E. Abuin and A.M. Rocha, J. Phys. Chem., 84, 2406 (1980).
76. C.A. Bunton, L.S. Romsted and C. Thamavit, J. Am. Chem. Soc., 102, 3900 (1980).
77. G.B. van de Langkruis and J.B.F.N. Engberts, Tet. Lett., 3991 (1979).
78. A.K. Yatsimirski, A.P. Osipov, K. Martinek and I.V. Berezin, Kolloid Z. (Engl.), 37, 470 (1975).
79. C.A. Bunton, N. Carrasco, S.K. Huang, C.H. Paik and L.S. Romsted, J. Am. Chem. Soc., 100, 5420 (1978).
80. C.A. Bunton, Catal. Rev.-Sci. Eng., 20, 1 (1979).
81. C.A. Bunton, in "Solution Chemistry of Surfactants," K.L. Mittal, Editor, Vol. 2, 519, Plennum Press, New York, 1979.
82. C.A. Bunton, E.J. Fendler, L. Sepulveda and K-U. Yang, J. Am. Chem. Soc., 90, 5512 (1968).
83. C.A. Bunton, J.J. Minch, J. Hidalgo and L. Sepulveda, J. Am. Chem. Soc., 95, 3262 (1973).
84. L.R. Romsted and E.H. Cordes, J. Am. Chem. Soc., 99, 4404 (1968).
85. A.P. Osipov, K. Martinek, A.K. Yatsimirski, and I.V. Berezin, Izv. Akad. Nauk. SSSR Ser. Kim, 1984 (1974).

86. I.M. Cuccovia, E.H. Schroter, P.M. Monteiro and H. Chaimovich, J. Org. Chem. 43, 2248 (1978).

87. H. Al-Lohedan, C.A. Bunton and L.S. Romsted, J. Phys. Chem., 85, 2123 (1981).

88. J.B.S. Bonilha, G. Chiericato, Jr., S.M. Martins-Franchetti, E.J. Ribaldo and F.M. Quina, private communitation.

89. C.A. Bunton, L.S. Romsted and H.J. Smith, J. Org. Chem., 43,. 4299 (1978).

90. F.M. Menger and C.E. Portnoy, J. Am. Chem. Soc., 89, 4698 (1967).

91. F.H. Quina, M.J. Politi, I.M. Cuccovia, S.M. Martins-Franchetti and H. Chaimovich, in "Solution Behavior of Surfactants: Theoretical and Applied Aspects," K.L. Mittal and E.J. Fendler, Editors, Vol. 2, 1125, Plenum Press, New York 1982.

92. S.K. Srivastava and S.S. Katiyar, Ber. Bunsenges. Phys. Chem., 84, 1214 (1980).

93. C.A. Bunton and L.S. Romsted, in "Solution Behavior of Surfactants: Theoretical and Applied Aspects," K.L. Mittal and E.J. Fendler, Editors, Vol. 2, 975, Plenum Press, New York (1982).

94. C.A. Bunton, L.S. Romsted and G. Savelli, J. Am. Chem. Soc., 101, 1253 (1979).

95. C.A. Bunton, J.R. Moffatt and E. Rodenas, J. Am. Chem. Soc., 104, 2653 (1982).

96. J.R. Hicks and V.C. Reinsborough, Aust. J. Chem., 35, 15 (1982).

97. F. Nome, A.F. Rubira, C. Franco and L.G. Ionescu, J. Phys. Chem., 86, 1881 (1982).

98. C.A. Bunton, J. Frankson and L.S. Romsted, J. Phys. Chem., 84, 2607 (1980).

99. C.A. Bunton, Y-S. Hong, L.S. Romsted and C. Quan, J. Am. Chem. Soc., 103, 5788 (1981).

100. L.S. Romsted, Unpublished data.

101. H. Al-Lohedan and C.A. Bunton, J. Org. Chem., 47, 1160 (1982).

102. M. Almgren, F. Grieser and J.K. Thomas, J. Am. Chem. Soc., 101, 279 (1979).

103. P. Mukerjee, Ber. Bunsenges. Phys. Chem. 82; 931 (1978).

104. E.A.G. Aniansson, J. Phys. Chem. 82, 2805 (1978).

105. R. Fornasier and U. Tonellato, J.C.S., Faraday I, 76, 1301 (1980).

106. T.H. Lowry and K.S. Richardson, "Mechanism and Theory in Organic Chemistry," 2nd Edition, p. 324 ff., Harper and Row, New York, 1981.

107. J.H. Fendler and W.L. Hinze, J. Am. Chem. Soc., 103, 5439 (1981).

108. I.M. Cuccovia, F.H. Quina and H. Chaimovich, Tetrahedron, 917 (1982).

109. V.C. Reinsborough and B.H. Robinson, J.C.S. Faraday I, 75, 2395 (1979).

110. S. Diekmann and J. Frahm, J.C.S. Faraday I, 75, 2199 (1979).
111. N. Funasaki and A. Murata, Chem. Pharm. Bull., 28, 805 (1980).
112. C.A. Bunton, S.E. Nelson and C. Quan, J. Org. Chem., 47, 1157 (1982).
113. C.A. Bunton and L. Sepulveda, Israel J. Chem., 18, 298 (1975).
114. C.A. Bunton, G. Cerichelli, Y. Ihara and L. Sepulveda J. Am. Chem. Soc., 101, 2429 (1979).
115. A.A. Balekar and J.B.F.N. Engberts, J. Am. Chem. Soc., 100, 5914 (1978).
116. A. Cipiciani, P. Linda, G. Savelli and C.A. Bunton, J. Org. Chem., 46, 911 (1981).
117. C.A. Bunton, F. Rivera and L. Sepulveda, J. Org. Chem., 43, 1166 (1973).
118. C.A. Bunton and G. Cerichelli, Intern. J. Chem. Kinetics, 12, 519 (1980).
119. F. Tokiwa, Adv. Colloid Interface Sci., 3, 389 (1972).
120. S.H. Yalkowsky and G. Zografi, J. Pharm. Sci., 59, 798 (1980).
121. S.H. Yalkowsky and G. Zografi, J. Colloid Interface Sci., 34, 525 (1970).
122. E.D. Goddard, Adv. Collid Interface Sci., 4, 45 (1974).
123. A.S. Waggoner, Ann. Rev. Biophys. Bioeng., 8, 47 (1979).
124. M.J. Minch, M. Giaccio and R. Wolff, J. Am. Chem. Soc., 97, 3766 (1975).
125. C.A. Bunton, L.S. Romsted and L. Sepulveda, J. Phys. Chem., 84, 2611 (1980).
126. U. Tonellato, in "Solution Chemistry of Surfactants," K.L. Mittal, Editor, Vol. 2, 541, Plenum New York, 1979.
127. T. Kunitake, Y. Okahata and T. Sakamoto, J. Am. Chem. Soc., 98, 7799 (1966).
128. R.A. Moss, G.O. Bizzigotti and C.W. Huang., J. Am. Chem. Soc., 102, 754 (1980).
129. L. Anoardi, R. Fornasier and U. Tonellato, J.C.S. Perkin II, 260 (1981).
130. R.A. Moss, Y-S. Lee, K.W. Alwis, J. Am. Chem. Soc., 102, 6648 (1980).
131. Y. Ihara, M. Nango and N. Kuroki, J. Org. Chem., 45, 5011 (1980).
132. Y. Ihara, R. Hosako, M. Nango and N. Kuroki, J.C.S. Chem. Comm., 393 (1981).
133. D.A. Jaeger and R.E. Robertson, J. Org. Chem., 42, 3298 (1977).
134. C.M. Link, D.K. Jansen and C.N. Sukenik, 102, 7798 (1980).
135. L.S. Romsted, unpublished results.
136. C.A. Bunton, F.H. Hamed and L.S. Romsted, J. Phys. Chem., 86, 2103 (1982).
137. J.M. Brown, C.A. Bunton, S. Diaz and Y. Ihara, J. Org. Chem., 45, 4169 (1980).
138. N. Funasaki, J. Colloid Interface Sci., 62, 189 (1977).

139. P. Mukerjee and K.J. Mysels, "Critical Micelle Concentrations of Aqueous Surfactant Systems", National Bureau of Standards, Washington, D.C., 1971.

140. R. Shiffman, Ch. Rav-Acha, M. Chevion, J. Katzhendler and S. Sarel, J. Org. Chem., 42, 3279 (1977).

141. R.L. Reeves, J. Am. Chem. Soc., 97, 6019, 6025 (1975).

142. R.S. Farinato and R.L. Rowell, in "Solution Chemistry of Surfactants," K.L. Mittal, Editor, Vol., 1, 311, Plenum Press, New York, 1979.

143. Y. Okahata, R. Ando, and T. Kunitake, J. Am. Chem. Soc., 99, 3067 (1977).

144. R.W. Alder, R. Baker and J.M. Brown, "Mechanism in Organic Chemistry," Chapter 3, Wiley-Interscience, New York, 1971.

145. C.A. Bunton, Y.S. Hong, L.S. Romsted and C. Quan, J. Am. Chem. Soc., 103, 5788 (1981).

REVERSED MICELLAR ENZYMOLOGY

A.V. Levashov, Yu.L. Khmelnitsky, N.L. Klyachko
and Karel Martinek

Department of Chemistry
Lomonosov State University of Moscow
Moscow 117234, USSR

Micellar enzymology utilizes organic solvent-sur-
factant-water systems as media for enzyme-catalyzed
reactions. Such systems allow one to vary purposefully
the nature of the microenvironment of enzyme molecules
by proper surfactant selection and by variation of the
hydration degree of the latter. By using oil-surfac-
tant-water colloid systems and in particular reversed
micellar systems as the reaction media for enzyme catalyzed processes, it is possible to achieve higher catalytic efficiency of an enzyme as compared with that
in aqueous solution due to concentrating effect and/or
increase of true reactivity of the enzyme. Moreover,
it is possible to influence the substrate specificity
of the enzyme and to shift the equilibrium of enzyme
catalyzed reactions. The suggested structural similarity between reversed micelles and biomembranes provides close parallelism between reversed micellar
enzymology and enzyme function in vivo.

INTRODUCTION

Most enzymologists are aware of the fact that the most significant experiments aiming at the elucidation of physico-chemical mechanisms of biocatalysis could be performed only when it
became possible to isolate enzymes (or enzyme complexes) in
quite a pure form[1]. On the other hand, such "pure" experiments
naturally raise (as was noted time and again) the question, whether the enzyme properties observed in vitro can be correlated
adequately with the conditions of its functioning in the living
cell. The fact is that enzymes are usually dissolved in water for
in vitro studies. The enzymatic reactions in the living cell most
often proceed in close proximity to (or at) the interface; more
precisely, enzymes are either adsorbed on biological membranes,
or embedded in the inner part of membranes, or located inside
closed membrane formations. Even plasmic (diffusionally free)
enzymes do function in a particular medium which, by its physical
parameters (dielectric permeability, polarity, viscosity, etc.)
and chemical composition, much differs from that of aqueous solutions, used in most enzymologic studies. Furthermore, there are
good reasons to believe[2] that many plasmic enzymes function in
adsorbed (on subcellular structures) state. Finally, the properties of water in proximity to the interface differ significantly
from those of the bulk water[3]. That is why many researchers are
inclined to believe that the traditional enzymology (studying the
properties of enzymes in water solutions) is quite an artificial
science[4].

An evident way out of the situation is the elaboration and
study of model systems. For this purpose enzymes are entrapped
into artificial membranes of various types including liposomes
and monolayers of surfactants, etc.[5-7], or else enzymes are immobilized on (or in) carriers of various kinds (see for instance[8-10].
Unfortunately, the model systems known so far have at least one
of the following drawbacks: (i) their creation (in particular,
with enzyme entrapment) calls for special techniques (sometimes
quite complicated compared with the simple dissolution of an enzyme in traditional enzymological studies); (ii) the listed model
systems may usually be created only under certain (often quite
specific) conditions and it is difficult to predetermine (or to
vary in a sufficiently wide range) the properties of enzyme microenvironment in them, for instance the degree of hydration, etc.,
and (iii) as these systems are usually macroheterogenous, it is
difficult to observe the enzyme directly, for instance by spectral
methods.

We found[11] and other authors then confirmed (their works are
discussed below) that water-soluble enzymes may be solubilized in
organic solvents with the aid of surfactants, the catalytic acti-

vity and specificity of the enzymes being retained. On solubilization, the enzyme is entrapped into the inner cavities of reversed
micelles and is thus protected from the denaturing effect of the
organic solvent.

This approach (the dissolution of an enzyme in the colloidal
solution of water in an organic solvent) offers a more "natural"
model for enzymological studies in vitro, than a usual water solution. In fact, in the colloidal system the enzyme molecule is
located near the interface (see Figure 1 discussed below). In addition, the water inside reversed micelles differs notably in the
properties from the aqueous macrophase[12-16]. Moreover, the enzyme-
containing colloid system has the following important properties
distinguishing it from the membrane models developed before: (i)
it can be prepared very easily (see below); (ii) the water content
in such a system and hence the amount of water in the enzyme microenvironment (inside a reversed micelle) can be varied over a
wide range, from a practically "dry" enzyme to a system containing
approximately equal amounts of aqueous and organic components;
(iii) such a colloid system is optically transparent, which allows
the use of conventional spectral methods to observe the structure
of enzyme and the course of enzymatic reaction.

In this review we have tried to discuss briefly the principle
aspects and progress of "micellar enzymology", the novel methods
for studying the behaviour and properties of enzymes in the systems of reversed micelles.

TECHNIQUE OF ENTRAPMENT OF PROTEIN(ENZYME)INTO REVERSED MICELLES

Homogeneous (optically transparent, i.e. non turbid) solutions of proteins (enzymes) in organic solvents may be obtained by
one of the three following procedures: (a) a complete solubilization of an aqueous protein solution in a surfactant – organic solvent system, i.e. a complete dissolution of the introduced aqueous component; (b) a complete or partial solubilization of a dry
(lyophilized) protein in a solution of surfactant in organic solvent containing a predetermined amount of water, and (c) a partial
capture of aqueous protein solution by surfactant solution in
organic solvent.

The first procedure has been suggested by us[11] and at present
is most widely used[17-24]. In this case a small amount (about a
few % v/v) of an aqueous protein solution is injected into the
surfactant solution in an organic solvent (a particular volume ratio of aqueous and organic components is determined depending on
the purpose of the experiment, i.e. by the required degree of
surfactant hydration). The resultant mixture is vigorously shaken

(for dozens of second) to produce an optically transparent solution. Instead of mechanical agitation, the sonication of the mixture is also used, as described by Douzou et al.[25,26]. It is to be
emphasized that the choice of the agitation technique is of great
importance in the study of enzymes and especially of enzymatic kinetics because it influences the equilibration time in the colloidal system. The fact is that in some cases the equilibration time
in the system of reversed micelles (in the absence of solubilized
protein) may be as long as several hours and even tens of hours
(see, for instance,[27,28]). The equilibration time in surfactant
solutions is strongly dependent on the nature of surfactant and/or
solvent. It was shown for solutions of one of the most widely used
surfactants, sodium diisooctylsulfosuccinate (Aerosol OT or AOT),
in aliphatic hydrocarbons[29,30] that the equilibrium (stable sizes
of micelles) was achieved quickly and sonication was not necessary. We arrived at the same conclusion when using this surfactant
to solubilize enzymes[31]. The necessity of ultrasonic treatment may
appear when using other surfactants. It should be, however, taken
into account that sonication may inactivate enzymes[32]. The solubilization of aqueous enzyme solutions without sonication does not
inactivate enzyme[11,21,33,34]. To prepare concentrated enzyme solutions in systems of reversed micelles, the second procedure seems
to be preferable.

In the second procedure suggested by Menger[35] (see also refs.
21,31) the necessary amount of water is first added to the surfactant solution (to achieve a desired degree of surfactant hydration, and then a dry (lyophilized) protein is dissolved in the resultant transparent solution with vigorous shaking. The time required to dissolve the dry protein is usually longer than that required when using aqueous protein solutions. It shortens with increasing the degree of surfactant hydration[31]. When it is necessary to obtain highly concentrated protein solutions, the dry protein is added in excess and the portion remained undissolved after a few hours is removed by centrifugation[31]. A repeated procedure of protein dissolution in the resultant supernatant increases
the protein content in the micellar solution. Attention should,
however, be given to the fact that this method of producing concentrated protein solutions requires a strict control of the surfactant and water contents in the supernatant layer after separation of the undissolved portion.

In the third procedure used by Luisi and his coworkers[36,37]
in their early works, the solubilization was achieved by a spontaneous transfer (partition) of water and protein in a two-phase
system of about equal volumes of an aqueous protein solution and
organic solvent with surfactant. After the solubilization at mild
stirring or without stirring has been completed, the organic phase
containing the dissolved protein was separated. A drawback of this

procedure is long duration for the solubilization process. Besides, some problem arises in determining the amount of water solubilized by the micelles (the degree of surfactant hydration).

MOLECULAR MECHANISMS OF ENTRAPMENT OF PROTEIN INTO REVERSED MICELLES

A microheterogeneous structure of systems of reversed micelles in organic solvents offers a methodologically unique possibility to dissolve proteins of various nature under such standard conditions when a molecule of the dissolved protein can choose itself an optimal environment corresponding to its nature. In fact, the scheme in Figure 1a shows that a molecule of a true hydrophilic protein can avoid a direct contact both with the organic solvent and surface of the reversed micelle inner cavity formed by polar "heads" of surfactant molecules and become localized in the aqueous core of the hydrated reversed micelle. The surface-active enzyme, such as lipases, can interact with the surface layer of a reversed micelle, or even be partially buried in it (Figure 1b). Finally, the typical membrane enzymes can come in contact with the organic solvent (Figure 1c).

Literature provides examples of solubilization of various enzymes according to "a"[11,17-25,33-40], "b"[41] or "c"[26,42] types (Figure 1). In all cases, solubilization results in the formation of an optically transparent solution while the solubilized enzymes retain catalytic activity.

A sufficiently detailed quantitative study was made of the molecular mechanism of solubilization of water-soluble proteins (enzymes) (Figure 1a). In particular, using sedimentation analysis it was shown[43] that molecules of typical water-soluble proteins (lysozyme, trypsin, α-chymotrypsin, ovalbumin, alcoholdehydrogenase, γ-globulin), under a wide range of experimental conditions, can be entrapped by reversed AOT micelles in octane without changing the initial size of the micelles. This result agrees with the scheme in Figure 1a and attests to the absence of specific protein surfactant interactions. It may be expected that the sedimentation properties of the protein-containing reversed micelles of the types "a", "b" and "c" (Figure 1) will specifically differ from each other. It is quite possible that this will lead to the elaboration of stringent criteria for classification of proteins by the "a", "b" and "c" types.

In the sedimentation studies[31,43] conditions were also found when the protein-containing micelles of various stoichiometry (containing one, two and more enzyme molecules) are present in so-

lution. This appears to be very promising in the study of the
functioning of subunit regulator enzymes.

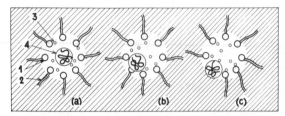

Figure 1. Schematic representation of a reversed micelle containing (a) hydrophilic (b) surface-active and (c) membrane hydrophobic protein; 1, polar head and 2, hydrocarbon tail of surfactant
molecule; 3, counter ion and/or water molecule, 4, protein molecule.

LIMITS OF PROTEIN (ENZYME) SOLUBILIZATION
IN THE SYSTEMS OF REVERSED MICELLES

     A large amount of protein comparable with that in aqueous so-
lutions can be solubilized in organic solvents with the aid of reversed surfactant micelles. Exact limits of solubilization are
unpredictable and depend on many factors-temperature, pH of aqueous solution, nature of organic solvent and surfactant, concentration of water and surfactant in the system, and nature of protein.
As an example, Figure 2 shows the solubilization curves of lysozyme and α-chymotrypsin in the system of reversed AOT micelles in
octane[31], demonstrating the maximal solubility of lyophilized enzymes as a function of degree of hydration of reversed micelles,
the latter value being expressed in terms of ratio of molar water
and surfactant concentrations, w_0. The lysozyme solubilization
curve similar to that in Figure 2 was obtained earlier for the
system of reversed AOT micelles in isooctane[21]. These results was
discussed in detail elsewhere[31]. Here we should emphasize that at

very low w_0 values the proteins studied are insoluble in the micellar system (Figure 2). Judging by the experimental results at
our disposal, this phenomenon is of general character. The protein solubilization starts only at some critical w_0 value after
which the protein solubility over a certain range of surfactant
hydration is almost a linear function of w_0, as is shown in Figure 2. The number of water molecules necessary to solubilize a protein molecule can be determined from the slope of this linear dependence. The values for lysozyme and α-chymotrypsin are about
1250 and 2000 H_2O molecules, respectively, i.e. approx. 1.5 g
water per 1 g protein.

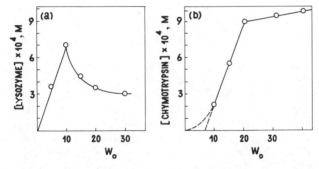

Figure 2.   Solubilization curves of (a) lysozyme and (b) α-chymo-
trypsin in the system of reversed AOT micelles in n-octane (0.1 M
AOT, 0.05 M Tris·HCl buffer, pH 8.0, 25°C). $w_0 = [H_2O]/[AOT]$.

     It should be noted that the solubilization curves present a
mean pattern of the process and do not provide information on
protein distribution between micelles. The assay of lysozyme system by the sedimentation method attests to the presence of a wide
set of protein associates, the fraction of the largest ones increasing with rising w_0[31].

KINETIC THEORY OF REACTIONS CATALYZED BY ENZYMES SOLUBILIZED
BY REVERSED SURFACTANT MICELLES IN ORGANIC SOLVENTS

The rate of any chemical reaction, including enzymatic ones, is a function of concentrations of reagents. The kinetic assay is aimed at elucidating the type of this functional dependence and calculating the kinetic parameters independent of concentrations of reactants. The essential point to ascertain the kinetic regularities of chemical reactions in the colloidal systems is the monitoring of partition of reactants between the micelles and the bulk phase. This problem was adequately solved for the case of micellar catalysis in aqueous solutions using the pseudophase model[44]. The pseudophase approach was also successfully applied in the kinetic assay of organic (nonenzymatic) reactions in the system of reversed micelles[33]. It will be shown below that the pseudophase model is reliable for the description of regularities of catalysis by enzymes solubilized by reversed surfactant micelles in organic solvents.

We chose, as an example, the enzymatic reaction obeying the Michaelis-Menten equation. This, however, does not restrict the possible application of the suggested theory for the explanation of other kinetic mechanisms. The Michaelis-Menten equation describes most enzymatic reactions in water solutions. In follows from the experimental data available to date[11,19-21,25,33-35,38,39] that it also holds for the system of reversed micelles. These data evidence that the apparent kinetic parameters of enzymatic reactions in the micellar systems often considerably differ from those in aqueous solution. The principal reasons for observed discrepancies may be accounted for in terms of the kinetic theory we suggest.

Let us consider that the kinetics of the reaction between an enzyme, E, and substrate, S, taking place in the organic solvent-surfactant system follow the Michaelis scheme:

$$E + S \rightleftharpoons ES \longrightarrow E + \text{products} \tag{1}$$

Let (i) the surfactant solution consists of the bulk "phase" of the organic solvent and of the "phase" of hydrated micelles[45]. Let us then assume that (ii) the equilibrium distribution of the substrate has been established between the phases[46]:

$$(S)_o \underset{}{\overset{P_S}{\rightleftharpoons}} (S)_{mic} \tag{2}$$

The distribution coefficient may be presented as follows:

$$P_S = [S]_{mic} / [S]_b . \tag{3}$$

Here and below, the subscripts "mic" and "b" denote micellar and bulk phases, respectively.

We shall not take into account the enzyme distribution because (iii) proteins are actually insoluble in hydrophobic solvents[47], besides, enzymes in nonaqueous media usually denaturate
(see review[48]). So, we shall assume that the catalytic activity
is located only in the micellar phase.

Let (iv) the reaction rate in the micelles obey the Michaelis-Menten equation. Then at the initial moment, when the concentration of products is low compared with the initial concentration
of the substrate and the concentration of the substrate is much
greater than that of the enzyme, the steady-state rate referred
to the total volume of the system may be expressed as follows:

$$V = \frac{k_{cat,mic} [E]_{o,mic} [S]_{o,mic}}{K_{M,mic} + [S]_{o,mic}} \cdot \theta , \qquad (4)$$

where θ is the volume fraction of the micellar phase and "o" denotes the initial concentrations.

Let (v) the exchange of substrate molecules between the phases proceeds fairly quickly, i.e. the course of the enzymatic
reaction (1) does not distort the equilibrium (2). The validity
of this assumption is supported by the results obtained in the
study of the rate of transfer of low-molecular weight compounds
from the micellar phase into the organic solvent phase and backed
by studies using techniques of NMR[49], stopped-flow[50,51], etc.[52].

The concentrations of reactants may then be determined from
Equation (3) and the equations of material balance:

$$[S]_{o,total} = [S]_{o,mic} \cdot \theta + [S]_{o,b}(1-\theta),$$
$$[E]_{o,total} = [E]_{o,mic} \cdot \theta. \qquad (5)$$

It should be noted, however, that (vi) Equation (3) holds only
for sufficiently diluted solutions and, therefore, the concentrations of reactants should be much lower than the surfactant concentration.

The substitution of Equations (3) and (5) into Equation (4)
gives

$$V = \frac{k_{cat,app} [E]_{o,total} \cdot [S]_{o,total}}{K_{M,app} + [S]_{o,total}} , \qquad (6)$$

where

$$k_{cat,app} = k_{cat,mic} \qquad (7)$$

and

$$K_{M,app} = K_{M,mic} \frac{1+\Theta(P_S-1)}{P_S} . \qquad (8)$$

In the case of the reaction involving a charged substrate, Equation (8) can be simplified by assuming that (vii) the substrate molecules are present only in the water-micellar phase and hence, both $P_S \gg 1$ and $P_S\Theta \gg 1$. Then

$$K_{M,app} = K_{M,mic} \cdot \Theta . \qquad (9)$$

We have discussed here only the simplest pseudophase model that assumes a uniform distribution of reactants over the entire volume of hydrated micelles. In a general case the model would actually be more involved and should allow for the microheterogeneity of the inner cavity of reversed micelles; in particular, it would be possible to take into account the distribution of substrate between the inner surface layer and water core of the reversed micelle. Nevertheless, even the above-discussed simplest model allows a successful kinetic analysis of enzymatic reactions occurring in the systems of reversed micelles.

To determine the "true" parameters of the Michaelis equation characterizing the enzymatic reaction in the water-micellar phase it is necessary to study the observed rate of the enzymatic reaction as a function of the volume ratio of the bulk (organic solvent) and micellar phases. In this case the volume fraction of the water-micellar phase (Θ) should be varied in such a way that the w_O remains constant so as to prevent the alteration of the properties of the micellar phase proper. Figure 3 gives an example of the Θ-dependence of the apparent parameters of the Michaelis equation for the trypsin-catalyzed hydrolysis of p-nitroanilide N-benzoyl-D, L-arginine in the system of cetyltrimethylammonium bromide reversed micelles in the chloroform - octane (1:1) mixture[33]. According to Equation (9) the apparent Michaelis constant should be a linear function of Θ; this is actually the case (Figure 3). The slope of this linear dependence is the true Michaelis constant for the reaction in the water-micellar phase, $K_{M,mic}$. The example in Figure 3 shows that the $K_{M,mic}$ is somewhat worse than K_M for the same reaction in aqueous solution (Figure 3, broken line). It follows from Equation (9), however, that conditions may be realized when the $K_{M,app}$ will be much better than the K_M in aqueous solution (Figure 3, low Θ values) and because of this much higher observed rates of reaction may be attained in the system of reversed micelles than in aqueous solution. This phenomenon, caused by the concentrating of reactants in micelles, is well known for organic (nonenzymatic) reactions[44]. Similar concentration effects occur in reactions catalysed by enzymes immobilized in/on carriers[10].

The concentration effect does not influence the first-order rate-constants measured as "time^{-1}". In fact, Figure 3 shows that the observed k_{cat} value is independent of the volume ratio of the micellar and bulk phases. According to Equation (7) the observed k_{cat} is equal to the true value of this constant for the reaction

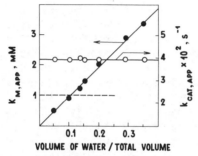

VOLUME OF WATER / TOTAL VOLUME

Figure 3. Hydrolysis of N-benzoyl-D,L-arginine p-nitroanilide catalyzed by trypsin solubilized in a system of CTAB+aqueous buffer (0,02 M acetate/phosphate/borate, pH 8.0)+1:1 chloroform/octane mixture. The Michaelis parameters $K_{M,app}$ and $k_{cat,app}$ are calculated depending on the volume proportion of water in the system (at H_2O/CTAB=25, concentration of the surfactant being varied from 0.1 to 0.8 M); $[E]_{o,total}$=5·10^{-6}M, 26°C. Broken line shows the value of K_M in aqueous solution.

in the micellar phase. Thus it characterizes the true reactivity of the enzyme and should be dependent on the state and environment of the enzyme molecule entrapped into the reversed micelle. Some aspects of such analysis are discussed in the following sections.

CATALYTIC ACTIVITY OF ENZYMES SOLUBILIZED
BY REVERSED SURFACTANT MICELLES IN ORGANIC SOLVENTS

The effect of w_0. One of the most striking effects observed
in the study of enzymes in the reversed micelle systems is the dependence of the catalytic activity of solubilized enzymes on the
degree of hydration of reversed micelles, w_0, i.e. on the parameter defining the size and properties of the inner cavity of micelles. The w_0-dependence of the catalytic activity has been shown
for enzymes such as α-chymotrypsin[20,33,39], trypsin[53], lysozyme[21],
phospholipase A_2[41], alcoholdehydrogenase[19,34], lactatedehydrogena-
se[33], pyrophosphatase[33,38]. This dependence is usually described
by a bell-shaped curve, suggesting the existence of some optimal
w_0 values at which the catalytic activity of solubilized enzyme is
maximal.

The optimal w_0 value is different for each enzyme. We shall
discuss, as an example, the w_0-dependence (Figure 4) of the rate
constant, k_{cat}, for deacylation of N-trans-cinnamoyl-α-chymotryp-
sin solubilized by reversed AOT micelles in octane. The possible
reasons for appearance of an optimum on the k_{cat}-w_0 dependence
curve have been analysed in detail elsewhere[39,54]. Attention
should be drawn to the fact that the optimum k_{cat} exceeds that in
aqueous solution (Figure 4, broken line). This phenomenon of "superactivity" of α-chymotrypsin solubilized by reversed AOT micelles was also observed by other authors[20,35]. It seems that the
"superactivity" phenomenon partially arises from the effects of
the microenvironment of enzyme molecule in reversed micelle. In
fact, it is seen in Figure 4 that the enzyme shows the optimal activity at medium degrees of surfactant hydration, i.e. when the
microenvironment of the enzyme molecule inside the micelle greatly
differs from that in aqueous solution. In particular, Figure 4
shows that the hyperfine splitting constant $a(-)$ of the ESR signal
of the spin label introduced into the active site of α-chymotryp-
sin is lower at the optimum catalytic activity than in aqueous solution[54]. This implies that the polarity of the medium in which
the active site of the solubilized enzyme is located is much lower
than that of aqueous solution (see ref. 55). The "superactivity"
may also stem from conformational changes in the enzyme molecule
during its entrapment into reversed micelle. So, it is seen from
Figure 4 that at the optimum of catalytic activity, the minimum
is observed in the w_0-dependence of the rotation frequency, ν $(-)$,
characterizing the spin label situated in the active site of α-
chymotrypsin, i.e. the minimum of conformational mobility of the
enzyme molecule. Barbaric and Luisi[20] observed, in the study of
circular dichroism, an increase in rigidity (helix content) of α-
chymotrypsin solubilized by reversed AOT micelles in isooctane in
the region $w_0 \simeq 10$. The increased conformational rigidity of the
enzyme molecule at $w_0 \simeq 10$ is likely accounted for by the fact that

under these conditions the diameter of the inner cavity of an empty reversed AOT micelle[56] is comparable with the size of α-chymo-
trypsin molecule (40x40x50 Å, see ref. 57).

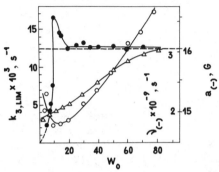

Figure 4. First-order rate constant, $k_{3,lim}$(●), for deacylation
of N-trans-cynnamoyl-α-chymotrypsin, rotation frequency, ν (-)(○),
and hyperfine splitting constant, $a_{(-)}$(△), for spin-labelled α-
chymotrypsin versus w_0 in the system AOT (0.1 M)-n-octane-water
(0.05 M Tris. HCl buffer, pH 8.0) at 25°C. Spin label: 2,2,5,5-te-
tramethyl-4-iodoacetamidepyrrolidine-1-oxyl. Broken line shows the
value of k_{cat} and $a_{(-)}$ in aqueous solution.

        The conformation is an important point in the study of enzy-
matic catalysis including that in micellar systems. The conformation of proteins solubilized by reversed surfactant micelles in
organic solvents was explored by many researchers, Luisi[17,19-21,
36,37], Douzou[25,26,53] and Fendler[24] who studied the spectral characteristics (absorption spectra, fluorescence, circular dichroism) of α-chymotrypsin, trypsin, pepsin, ribonuclease, lysozyme,
horse liver alcoholdehydrogenase, horse radish peroxidase, cyto-
chromes c, c_3 and P_{450}. According to the data obtained, the transition from aqueous solution to the reversed micellar system does
not usually cause an appreciable change in spectral properties of
the enzymes. In other words, judging from the spectral data, the

conformation of enzymes solubilized by reversed micelles is usually similar in main features to that in aqueous solution. This inference is in line with the experimental data on catalytic activity, according to which the enzyme activity in micellar systems under optimal conditions is usually close to that observed in aqueous solution, as is the case for horse liver alcoholdehydrogenase[19,34], trypsin[25] and pyrophosphatase[38].

It should be emphasized, however, that the relationship between conformational (by spectroscopic data) and catalytic characteristics of enzymes is not always simple. For instance, the catalytic activity of lysozyme solubilized by reversed AOT micelles in isooctane is close, under optimal conditions, to the value observed in aqueous solution, though the conformation of the enzyme, according to the data of circular dichroism, undergoes a considerable range on entrapment into the reversed micelle[21].

On the contrary, the spectral characteristics of another enzyme, horse radish peroxidase, solubilized by reversed AOT micelles do not actually alter[26,40]. Nevertheless, the optimal catalytic activity of solubilized peroxidase, as shown in Figure 5, is about 20 times that exhibited by the enzyme in aqueous solution (Figure 5, broken line). Furthermore, it follows from the detailed kinetic assay of this system[40] that the optimal k_{cat} value in Figure 5 is reduced by more than 10-fold because of the inhibition of the enzyme by the surfactant. In other words, the true reactivity of solubilized peroxidase exceeds hundreds times the catalytic activity of the enzyme in aqueous solution.

The effect of pH. The rate of enzymatic reactions and the kinetic parameters describing them are largely pH-dependent. On transfer of an enzymatic reaction from aqueous solution into reversed micellar system, marked shifts in pH-profiles are noted[17, 20,21,25,33,35]. This phenomenon may be due to various reasons. First, the use of an ionogenic surfactant forming a charged (double electric) layer around the enzyme may cause a local pH shift (depending on the charge sign and degree of surfactant ionization), as is the case for the enzymes immobilized in polyelectrolyte matrices[58-60]. In this instance, the value of the apparent shift in pH and, accordingly, in pK_a is usually about 1-2 units. Second, the acid-base properties of the ionogenic groups of the solubilized enzyme may alter as a result of a change of the nature of the enzyme microenvironment, which may cause, for instance, its partial dehydration. It is well-known that the dehydration of ionic groups occurring on their transfer from aqueous to water-organic or organic solution leads to a change in pK_a up to a few pH units[61]. Finally, it is necessary to allow for possible conformational changes of enzyme on solubilization and the related shift

in the pK_a of its ionogenic groups, including those controlling
the enzymatic reaction.

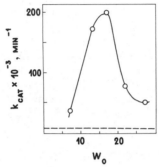

Figure 5. First-order rate constant, k_{cat}, for peroxidase oxidation of pyrogallol <u>versus</u> w_0 in the system AOT (0.1M)-n-octane-
water (0.025 M phosphate buffer, pH 7.0) at 26°C. Broken line is
the value of k_{cat} in aqueous solution.

SUBSTRATE SPECIFICITY OF ENZYMES ENTRAPPED INTO REVERSED
SURFACTANT MICELLES IN ORGANIC SOLVENTS

It has been shown in the preceding section that the conformation and catalytic properties of enzymes solubilized by reversed micelles may markedly differ from those observed in aqueous
solution. It is possible that in these cases a change in substrate
specificity will occur as well. The latter, however, may also be
observed in reactions catalyzed by solubilized enzymes not undergoing any marked alterations on their transfer from aqueous solution to the system of reversed surfactant micelles. This effect
may be due to the shifts of the observed Michaelis constant values varying with substrate partition in the reversed micellar
system, see Equation (8).

A striking example of the enzyme substrate specificity alteration resulting from the effects of substrate partition was obtained in our laboratory[34] on studying the reaction of aliphatic alcohols oxidation catalyzed by horse liver alcoholdehydrogenase in the system of reversed AOT micelles in octane. Figure 6a shows that the transition from aqueous solution to the system of reversed micelles causes a shift in the maximum on the plot of the second-order rate constant of the reaction, $k_{cat}/K_{M,app}$ versus the length of hydrocarbon chain in an alcohol substrate molecule (n) from octanol (n=8) to butanol (n=4). In other words, the substrate specificity, a criterion for which is $k_{cat}/K_{M,app}$[62], markedly alters on transition from aqueous to micellar solution. Yet, the conformation of solubilized alcoholdehydrogenase does not actually differ from that in aqueous solution[19]. This observation is in agreement with the constant pattern of the "k_{cat}-n" dependence both in aqueous solution and in the micellar system (Figure 6b). The observed alteration in the substrate specificity arises from a change in the pattern of the "$K_{M,app}$-n" dependence (Figure 6c), caused by the effects of alcoholic substrate partition between the micellar and organic phases.

THE MODELLING OF STRUCTURE AND FUNCTION OF BIOLOGICAL MEMBRANES

It is quite possible that the structure and properties of the microenvironment around an enzyme molecule entrapped into a reversed micelle may serve as a good model of its functional state in vivo. A formal ground for such an optimistic prognosis is the physico-chemical similarity between surfactants forming reversed micelles and natural lipids forming biological membranes. The possibility of using the reversed micelles as a realistic model of biomembrane structures, frequently supposed by various authors[15,63-65], has found an experimental support owing to the works of Dutch[66-68], British[69] and Soviet[70] researchers. These authors have revealed that bilayer membranes often contain non-bilayer structures, so called lipidic particles, distinctly discerned by electron microscopy photographs. These particles are shown to be aggregates of lipid molecules structurally similar to reversed micelles and sandwiched between the monolayers of the bilayer membrane (Figure 7). Attempts were made in the literature[71,72] to give another treatment to the nature of the lipidic particles. Their origin was explained as due to some curvatures and deformation of planar bilayer membranes. However, this explanation does not provide a full picture. In particular, it does not explain the NMR data unequivocally indicative of the presence of the micellar-type structures[73].

Of principal importance are the data[68] pointing to the fact that some proteins, for instance cytochrome c, are capable of inducing the formation of reversed micelles in a bilayer membrane by

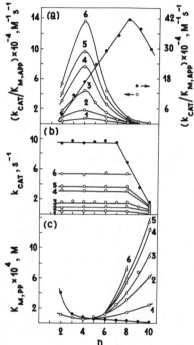

Figure 6. The Michaelis equation parameters for the oxidation of
aliphatic alcohols by horse liver alcoholdehydrogenase versus the
hydrocarbon chain length in an alcohol substrate molecule, n, as
determined in the system AOT (0.1 M)-n-octane-water (0.05 M phosphate buffer, pH 8.8) (o) and in aqueous solution (•) at 25°C (a)
Second-order rate constant, $k_{cat}/K_{M,app}$; (b) first-order rate constant, k_{cat}; (c) Michaelis constant, $K_{M,app}$. Degree of surfactant
hydration, w_o: (1) 12.2; (2) 23.0; (3) 29.0; (4) 37.7; (5) 42.3;
(6) 48.9.

embedding in the inner cavity of such a micelle, as the superposition of Figures 1a and 7 implies. This mechanism may play an important role in the interaction between cytochrome c and the membrane enzyme, cytochrome c oxidase[68].

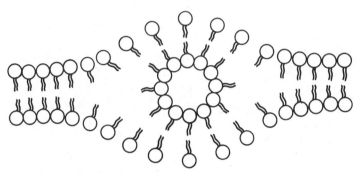

Figure 7. Schematic representation of a lipidic particle as proposed by de Kruijff et al.[67].

Research on the role of lipidic particles in membrane processes (the functioning of membrane enzymes, the transmembrane transport, etc.) has only just been started. Nevertheless, the first results support the validity of using the reversed micelles to model the biomembrane systems and stimulate the development of micellar enzymology as a science capable of describing a real picture of the enzyme functioning in vivo [74].

CONCLUSION

At present reversed micellar enzymology is in a state of infancy and a wide range of problems are awaiting their solution. However, the accumulated experimental material attests to the successful development of this trend for providing a deeper insight into the fundamental knowledge of enzymes and the wider possibilities of the applications of enzymatic catalysis [74].

Theoretically, the most important questions are the structure-function relationship of enzymes entrapped into reversed micelles and the role of the microenvironment and micellar template (matrix) in the catalytic activity of solubilized enzyme. The discovery of "superactivity" of some enzymes in reversed micellar systems compared to aqueous solution points to the possibility of revealing the biocatalysts which conceal their catalytic potential in traditional experiments in vitro. It is possible that the most essential discrepancies in catalytic activities will be revealed for membrane enzymes. The use of reversed micellar systems as models of biomembranes also provides information on the mechanism of regulation of enzyme catalytic activity and substrate specificity in vivo. The fact that the enzyme substrate specificity in vivo may markedly differ from that observed in vitro in aqueous solutions is evidenced by the data on oxidation of aliphatic alcohols, cacalyzed by alcoholdehydrogenase solubilized in the system of reversed AOT micelles in octane[33,75]. Besides, the results on equilibrium of alcohol oxidation in reversed micellar systems point to the fact that the equilibrium constant of the reaction is highly sensitive to the change in the degree of surfactant hydration[75]. It is possible that the content of some metabolites in the cell is controlled by a similar mechanism, i.e. by a change in hydration of biomembranes.

At the present time it is difficult to assess all the aspects of the functioning of the enzymic systems in vivo which might be simulated by means of reversed micellar enzymology. One may only surmise that such systems will prove highly useful in the study of mass transfer across the interface, in the elaboration of the models of membrane transport, and in the study of the role of protein-protein and protein-lipid interactions in regulating the enzymatic activity of subunit enzymes and enzymic complexes. Actually reversed micellar systems have found application in cryoenzymology[76].

The reversed micellar systems containing solubilized enzyme are also highly promising for applied biochemistry[9,77]. Such systems may be used in organic synthesis, primarily for enzymatic conversion of water-insoluble compounds, such as steroids[78,79], petroleum products[34,75], lipids[80], etc.; and for carrying out the reactions, whose equilibrium in aqueous solution, by thermodynamic reasons, is shifted in favour of initial reactants[77]. As for analytical chemistry, the systems of reversed micelles offer wide possibilities to utilize enzymes for determination of water-insoluble compounds (parallel to water-soluble compounds), which is particularly important for the monitoring of environmental pollution, as many pollutants are hydrophobic. These systems may also be applied to solve other analytical problems, for instance the determination of molecular mass and effective size of protein globules[81] and in thin-layer chromatography, as is shown by the example of amino acids[82].

ACKNOWLEDGEMENT

We are very greateful to Professor I.V. Berezin for support of our work and helpful discussion.

REFERENCES

1. M. Dixon and E. Webb, "Enzymes", 2nd Edition, Longmans, 1964.
2. C.J. Masters, CRC Crit. Rev. Biochem., 11, 105 (1981).
3. F. Franks, Editor, "Water. A comprehensive treatise". Vol. 5, Plenum Press, New York - London, 1975.
4. H. Brockerhoff and R.G. Jensen, "Lipolytic Enzymes", Academic Press, New York, 1974.
5. C. Tanford, "The Hydrophobic Effect. Formation of Micelles and Membranes", Wiley, New York, 1980.
6. D. Thomas and G. Broun, in "Methods in Enzymology", K. Mosbach, Editor, Vol. XLIV, p. 901, Academic Press, New York, 1976.
7. G. Gregoriadis, in "Methods in Enzymology", K. Mosbach, Editor, Vol. XLIV, p. 218, Academic Press, New York, 1976.
8. "Methods in Enzymology", K. Mosbach, Editor, Vol. XLIV, Academic Press, New York, 1976.
9. I.V. Berezin and K. Martinek, Editors, "Introduction to Applied Enzymology", Moscow State University Press, Moscow, 1982.
10. R. Goldman, L. Goldstein and E. Katchalski, in "Biochemical Aspects of Reactions on Solid Supports", G.R. Stark, Editor, p. 1, Academic Press, New York, 1971.
11. K. Martinek, A.V. Levashov, N.L. Klyachko and I.V. Berezin, Dokl. Acad. Nauk SSSR, 236, 920(1977); Engl. Ed., 236, 951 (1978).
12. J.H. Fendler, Accounts Chem. Res., 9, 153 (1976).
13. H.F. Eicke, in "Microemulsions", I.D. Robb, Editor, p. 17, Plenum Press, New York, 1982.
14. M.A. Wells, Biochemistry, 13, 4937(1974).
15. K. Martinek, A.V. Levashov, V.I. Pantin and I.V. Berezin, Dokl. Acad. Nauk SSSR, 238, 626(1978).
16. A.V. Levashov, V.I. Pantin and K. Martinek, Kolloid Zh., 41, 453 (1979).
17. R. Wolf and P.L. Luisi, Biochem. Biophys. Res. Commun., 89, 209(1979).
18. E.J. Bonner, R. Wolf and P.L. Luisi, J. Solid-Phase Biochem., 5, 255(1980).
19. P. Meier and P.L. Luisi, J. Solid-Phase Biochem., 5, 269(1980).
20. S. Barbaric and P.L. Luisi, J. Am. Chem. Soc., 103, 4239(1981).
21. C. Grandi, R.E. Smith and P.L. Luisi, J. Biol. Chem., 256, 837(1981).
22. M.P. Pileni, Chem. Phys. Lett., 81, 603(1981).
23. D. Balasubramanian, J. Indian Chem. Soc., 58, 633(1981).
24. A.J.W.G. Visser and J.H. Fendler, J. Phys. Chem., 86, 947 (1982).
25. P. Douzou, E. Keh and C. Balny, Proc. Natl. Acad. Sci. USA, 76, 681(1979).
26. C. Balny and P. Douzou, Biochimie, 61, 445(1979).
27. P. Becher and N.K. Clifton, J. Colloid Sci., 14, 519(1959).

28. J.M. Corkill, J.F. Goodman and T. Walker, Trans. Faraday Soc., 61, 589(1965).
29. M. Zulauf and H.F. Eicke, J. Phys. Chem., 83, 480(1979).
30. E. Keh and B. Valeur, J. Colloid Interface Sci., 79, 465(1981).
31. N.L. Klyachko, A.V. Levashov and K. Martinek, (1983), Biokhimia (Russ.), submitted for publication.
32. M. Joly, "A Physico-Chemical Approach to the Denaturation of Proteins", Academic Press, New York, 1965.
33. K. Martinek, A.V. Levashov, N.L. Klyachko, V.I. Pantin and I.V. Berezin, Biochim. Biophys. Acta, 657, 277(1981)
34. K. Martinek, Yu.L. Khmelnitsky, A.V. Levashov and I.V. Berezin, Dokl. Acad. Nauk SSSR, 263, 737(1982).
35. F.M. Menger and K. Yamada, J. Am. Chem. Soc., 101, 6731(1979).
36. P.L. Luisi, F. Henninger and M. Joppich, Biochem. Biophys. Res. Commun., 74, 1384(1977).
37. P.L. Luisi, F.J. Bonner, A. Pellegrini, P. Wiget and R. Wolf, Helv. Chim. Acta, 62, 740(1979).
38. N.L. Klyachko, A.A. Baykov, A.V. Levashov, K. Martinek and S.M. Avaeva, Bioorgan. Khimiya, 6, 1707(1980).
39. A.V. Levashov, N.L. Klyachko and K. Martinek, Bioorgan. Khimiya, 7, 670(1981).
40. N.L. Klyachko, A.V. Levashov and K. Martinek, (1983), Dokl. Acad. Nauk SSSR, submitted for publication.
41. R.L. Misiorowski and M.A. Wells, Biochemistry, 13, 4921(1974).
42. A.N. Eremin and D.I. Metelitsa, Dokl. Acad. Nauk SSSR, 267, 221(1982).
43. A.V. Levashov, Yu.L. Khmelnitsky, N.L. Klyachko, V.Ya. Chernyak and K. Martinek, J. Colloid Interface Sci., 88, 444 (1982).
44. K. Martinek, A.K. Yatsimirski, A.V. Levashov and I.V. Berezin, in "Micellization, Solubilization and Microemulsions", K.L. Mittal, Editor, Vol. 2, p. 489, Plenum Press, New York, 1977.
45. K. Shinoda, in "Proc. IVth International Congress on Surface Active Substances", Vol. 2, p. 527, Gordon and Breach, New York, 1967.
46. D.G. Herries, W. Bishop and E.M. Richards, J. Phys. Chem., 68, 1842(1964).
47. S.J. Singer, Adv. Protein Chem., 17, 1(1962).
48. K. Martinek and I.V. Berezin. J. Solid-Phase Biochem., 2, 343(1978).
49. O.A. El-Seoud, E.J. Fendler and J.H. Fendler, J. Chem. Soc., Faraday Trans. I, 3, 459(1970).
50. K. Tamura and Z.A. Schelly, J. Am. Chem. Soc., 103, 1018(1981).
51. A. Yamagishi, T. Masui and F. Watanabe, J. Phys. Chem., 85, 281(1981).
52. M. Wong and J.K. Thomas, in "Micellization, Solubilization, Microemulsions", K.L. Mittal, Editor, Vol. 2, p. 647, Plenum Press, New York, 1977.

53. C. Balny, G. Hui Bon Hoa and P. Douzou, Jerusalem Symp. Quantum Chem. Biochem., 12, 37(1979).
54. G.I. Likhtenstein, O.V. Belonogova, A.V. Levashov, Yu.L. Khmelnitsky, N.L. Klyachko and K. Martinek, Biokhimiya, (1983) in press.
55. G.I. Likhtenstein, "Spin-labelling in Molecular Biology", "Nauka" Press, Moscow, 1974.
56. H.F. Eicke and J. Rehak, Helv. Chim. Acta, 59, 2883(1976).
57. D.M. Blow, in "The Enzymes", 3rd ed., P.D. Boyer, Editor, Vol. III, p. 185, Academic Press, New York, 1971.
58. L. Goldstein, Y. Levin and E. Katchalski, Biochemistry, 3, 1913(1964).
59. L. Goldstein, in "Methods in Enzymology", K. Mosbach, Editor, Vol. XLIV, p. 397, Academic Press, New York, 1976.
60. I.V. Berezin, A.M. Klibanov and K. Martinek, Russ. Chem. Revs. (Usp. Khim.), 44, 9(1975).
61. C. Reichardt, "Losungsmittel-Effekte in der Organischen Chemie", Verlag Chemie, 1969.
62. A. Fersht, "Enzyme Structure and Mechanism", W.H. Freeman, San Francisco, 1977.
63. P.H. Elworthy, A.T. Florence and C.B. Macfarlane, "Solubilization by Surface Active Agents", Chapman and Hall, London, 1968.
64. J. Sunamoto, T. Hamada, T. Seto and S. Yamamoto, Bull. Chem. Soc. Japan, 53, 583(1980).
65. J.H. Fendler, J. Phys. Chem., 84, 1485(1980).
66. P.R. Cullis and B. de Kruijff, Biochim. Biophys. Acta, 559, 399(1979).
67. B. de Kruijff, P.R. Cullis and A.J. Verkleij, Trends Biochem. Sci., 5, 79(1980).
68. B. de Kruijff, A.J. Verkleij, C.J.A. van Echteld, W.J. Gerritsen, P.C. Noordam, C. Mombers, A. Rietveld, J. de Gier, P.R. Cullis, M.J. Hope and R. Nayar in "International Cell BIology 1980-1981", H.G. Schweiger, Editor, p. 599, Springer-Verlag, Berlin Heidelberg, 1981.
69. A. Sen, W.P. Williams, A.P.R. Brain, M.J. Dickens and P.J. Quinn, Nature, 293, 488(1981).
70. V.V. Chupin, I.P. Ushakova, S.V. Bondarenko, I.A. Vasilenko, G.A. Serebrennikova, R.P. Evstigneeva, G.J. Rozenberg and G.N. Koltsova, Bioorgan. Khimiya, 8, 1275(1982).
71. R.G. Miller, Nature, 287, 166(1980).
72. S.W. Hui and T.P. Stewart, Nature, 290, 427(1981).
73. A.J. Verkleij and B. de Kruijff, Nature, 290, 427(1981).
74. K. Martinek, A.V. Levashov, Yu.L. Khmelnitsky, N.L. Klyachko and I.V. Berezin, Science, 218, 889(1982).
75. K. Martinek, Yu.L. Khmelnitsky, A.V. Levashov, N.L. Klyachko, A.N. Semenov and I.V. Berezin, Dokl. Acad. Nauk SSSR, 256, 1423(1981).

76. P. Douzou, Adv. Enzymol. Relat. Areas Mol. Biol., 51, 1(1980).
77. K. Martinek and A.N. Semenov, J. Appl. Biochem., 3, 93(1981).
78. E. Antonini, G. Carrea and P. Cremonesi, Enzyme Microb. Technol., 3, 291(1981).
79. M.D. Lilly, J. Chem. Technol. Biotechnol., 32, 162(1982).
80. E.A. Malakhova, B.I. Kurganov, A.V. Levashov, I.V. Berezin and K. Martinek, (1983) Dokl. Acad. Nauk SSSR, submitted for publication.
81. A.V. Levashov, Yu.L. Khmelnitsky, N.L. Klyachko, V.Ya. Chernyak and K. Martinek, Anal. Biochemistry, 118, 42(1981).
82. D.W. Armstrong and M. McNeely, Anal. Lett., 12, 1285(1979).

COMPARISON OF RATE ENHANCEMENTS IN MICELLAR AND

NONMICELLAR AGGREGATES

C.A. Bunton

Department of Chemistry
University of California
Santa Barbara, California 93106

Hydrophobic quaternary ammonium ions such as tri-n-octylethylammonium bromide and mesylate (TEABr and TEAMS, respectively) effectively catalyze reactions between hydrophobic substrates and hydrophobic nucleophilic anions. The quaternary ammonium ions can be functionalized and at high pH the hydroxyethyl derivatives are very effective nucleophiles in dephosphorylation and aromatic nucleophilic substitution. Analogies can be drawn between this mode of catalysis and that by aqueous micelles or surfactants at submicellar concentration.

INTRODUCTION

Rate enhancements of bimolecular reactions by normal micelles in water are generally ascribed to equilibrium incorporation of both reactants into the micelles, which are treated as a pseudophase, distinct from bulk solvent.[1-4] (For thermal reactions it is reasonable to assume that equilibrium is maintained between monomeric and micellized surfactant and between aqueous and micellar-bound solutes, but this assumption may not be justified in other systems). This increased concentration of reactants should, of itself, speed the reaction, and the variations of rate constant with surfactant concentration have been analyzed by a number of workers in terms of reactant concentrations in the aqueous and micellar pseudophases and the second order rate constants in each pseudophase.[5-11] For most of the reactions examined to date the second order rate constants in the micellar pseudophase are not very different from those in water. This generalization also applies to reactions with functional micelles and comicelles.[12-14] Typically for reactions between nonionic reactants rate constants are lower than in water, because the micellar surface appears to be less polar than water,[15] and these reactions are inhibited by a decrease in the polarity of the reaction medium. Reactions between very hydrophobic substrates and hydrophilic anions also seem to have lower second order rate constants in the micellar pseudophase than in water, probably because the anions are located in the Stern layer at the micelle-water interface whereas the substrate may be, on the average, more deeply in the micelle.[13] Spontaneous, water-catalyzed, reactions are generally micellar inhibited, but, except for S_N1 reactions, the inhibition is not large.[16,17]

This ability of submicroscopic aggregates to influence reactivity in bimolecular reactions is also shown by vesicles in water,[18] and the rate effects are similar to those in aqueous micelles. Bimolecular reactions in o/w microemulsion droplets also have second order rate constants similar to those in water or aqueous micelles.[19,20] The only exception to these generalizations of which I am aware is aromatic nucleophilic substitution by azide ion.[21]

Hydrophobic quaternary ammonium ions which are phase transfer catalysts can also give large rate enhancements of nucleophilic deacylation [22] and dephosphorylation in water.[23] Typical salts are tri-n-octylmethylammonium chloride (TMAC) and tri-n-octylethylammonium bromide and mesylate (TEABr and TEAMs). These salts do not form micelles in water, e.g., their solutions do not exhibit a critical micelle concentration,[22] but they can affect indicator equilibria involving relatively hydrophobic species, and markedly speed some nucleophilic displacements. One requirement for rate enhancement is that the nucleophilic

anion should be hydrophobic, for example, TMAC and TEABr do not
speed deacylations or dephosphorylations by OH⁻.

These observations strongly suggest that both the substrate
and the hydrophobic nucleophile have to associate with the
quaternary ammonium ion, or with a cluster of them, for reaction
to be speeded.

Dephosphorylations of p-nitrophenyl diphenyl phosphate
(pNPDPP) by benzimidazolide and naphthimidazolide ion, acting as
nucleophiles, are strongly accelerated by TEAMs, and the first
order rate constants, k_ψ, go through maxima with increasing
[TEAMs].[23] This kinetic form is very similar to that observed
for bimolecular reactions in aqueous micelles, where these rate
maxima can be explained in terms of extensive incorporation of
both reactants into the micelles,[5-11] suggesting that there is
extensive binding of both pNPDPP and imidazolide ion to the hydrophobic ammonium ions or more probably to clusters of them.
The deprotonation of benzimidazole is increased by TEAMs,
probably because it effectively binds the imidazolide ions, but
not the imidazole, and the solubility of pNPDPP is increased by
TEAMs, suggesting that substrate is bound to the ammonium ions.[23]

These changes in deprotonation of benzimidazole, and solubilization of pNPDPP, can be used to estimate the fractions of
reactants bound to TEAMs and these data can be treated quantitatively using equations similar to those applied to rate
enhancements by micelles.[23]

The overall first order rate constants are given by:

$$k_\psi = k_M \ f \ [B\bar{I}_M]/[R_4\overset{+}{N}] \tag{1}$$

In Equation 1 $R_4\overset{+}{N}$ is a hydrophobic ammonium ion, either
$(C_8H_{17})_3\overset{+}{N}$ Et (TEA) or a cationic surfactant in a micelle, e.g.,
$C_{16} H_{33} \overset{+}{N} Me_3$ in a micelle of $C_{16} H_{33} NMe_3$ Br (CTABr), f is the
fraction of substrate bound to the ammonium ion and BI_M^- is bound
benzimidazolide ion. The second order rate constant, k_M, is
written in terms of the mole ratio of BI^- to quaternary ammonium
ion. (It is important to note that k_M has units of s^{-1}, and cannot be compared directly with second order rate constants in
water, for example, where the units are typically $M^{-1} s^{-1}$).

Equation 1 can also be applied to reaction of benzimidazolide
ion with pNPDPP in micelles of CTABr and $k_M = 7 \ s^{-1}$. [24] The
dimensions of this rate constant can be converted into $M^{-1} s^{-1}$
by allowing for the molar volume of the micellar Stern layer,
giving a second order rate constant which is very similar to
that in water,[24] as is often found for bimolecular reactions
in aqueous micelles. For reaction of benzimidazolide ion with

pNPDPP in TEAMs, k_M is in the range of 11-31 s^{-1}.[23] The similarity
in the values of k_M for reactions in micelles of CTABr and in non-
micellizing TEAMs suggests that the same rate enhancing factors
are at work in both systems. But there are differences. For
example, in micellar systems values of k_M are generally constant,
over a range of surfactant concentration, suggesting that micellar
structure is essentially independent of added reactants, at least
when they are in low concentration. The variation of k_M with
[TEAMs] suggests that here the solution contains clusters or
aggregates which are probably small and whose structure depends
not only upon the quaternary ammonium ion, but also upon the
(variable) concentrations of the reactants.

The inability of salts such as TMAC and TEABr to speed reactions of hydrophilic anions such as OH⁻ suggests that the
ammonium ions form only small clusters which cannot bind small
anions, whereas normal micelles are large enough, with ca. 10^2
monomers, to bind both hydrophilic and hydrophobic counterions.[1-4]

The key point is that both cationic micelles and ions such
as TMA^+ and TEA^+ or their clusters speed bimolecular reactions by
bringing the two reactants together in a small volume element, by
coulombic and hydrophobic interactions. This concentration or
proximity effect is present in other systems, including some in
which no "catalyst" is present. For example, Diels Alder
reactions of cyclopentadiene with butenone or acrylonitrile are
much faster in water than in organic solvents, and addition of
LiCl speeds reaction in water.[25] These observations suggest
that hydrophobic interactions bring the apolar reactants together
in water, and this association is strengthened by addition of
LiCl which "salts out" the reactants, although in these reactions
we do not know whether there is 1:1 association or formation of
small clusters. Rate enhancements due to 1:1 association have
been postulated for reactions in water,[26] and the evidence has
been critically analyzed by Guthrie and coworkers.[27]

FUNCTIONALIZED QUATERNARY AMMONIUM IONS

Functionalized micelles have been studied extensively in
aqueous solution and are effective reagents in deacylation,
dephosphorylation and nucleophilic addition and substitution.
The functional groups can be amino, imidazole, oximate, hydrox-
amate, thiolate or hydroxyl,[12-14,28-36] and in some systems the
functional group has to be deprotonated to be an effective
reagent. Most of the groups act as nucleophiles, but base
catalysis is sometimes observed.[32,33] In many systems, e.g.,
in deacylation, a covalent intermediate is formed and the final
product is observed as the result of a second deacylation step.
Phosphorylation and subsequent hydrolysis of a phosphorylated
intermediate has also been observed.[13,24]

A quantitative analysis of rate enhancements by functional micelles can be complex because of micellar effects upon deprotonation, but, if they are allowed for, it appears that the rate enhancements are due largely to the high concentration of nucleophile or general base at the micellar surface. When this concentration is taken into account the second order rate constants at the surface of functional micelles are similar to those for reactions of a similar, but monomeric, reagent in water. [12,14,33]

We have directed our attention to functionalized quaternary ammonium ions (TEA$^+$) containing the 2-hydroxyethyl group (<u>1</u>):

$$(C_8H_{17})_3\overset{+}{N}CH_2CH_2OH + O\overline{H} \rightleftarrows (C_8H_{17})_3\overset{+}{N}CH_2CH_2O^-$$

<u>1</u>   <u>1a</u>

Micelles of the analogous hydroxyethyl surfactants, <u>2</u>, are effective reagents for deacylation, dephosphorylation and nucleophilic addition and substitution, and reaction involves attack of the alkoxide moiety.[12,36]

$$R\overset{+}{N}Me_2CH_2CH_2OH \overset{OH^-}{\underset{\leftarrow}{\rightarrow}} R\overset{+}{N}Me_2CH_2CH_2O^-$$

<u>2</u>   <u>3</u>

a, R = $C_{12}H_{25}$

b, R = $C_{16}H_{33}$

The apparent pKa of micellized <u>2b</u> is ca. 12.4.

Reactions of <u>1</u> and <u>2b</u> at high pH with p-nitrophenyl diphenyl phosphate (pNPDPP) and 2,4-dinitrochlorobenzene (DNCB) involve nucleophilic attack by the alkoxide zwitterion (<u>1a</u>, <u>3</u>), and are very much faster than the corresponding reactions of the cholinate zwitterion or of OH$^-$.

A problem in examining reactions mediated by <u>1</u> (and by TMAC and TEAMs) is that salts of these hydrophobic ammonium ions are only sparingly soluble in water, and the covalent intermediates formed by nucleophilic attack (Scheme 1) are even less soluble. Reactions were carried out in water containing acetonitrile. [37]

Scheme 1

$$(C_8H_{17})_3\overset{+}{N}CH_2CH_2\overline{O} + (PhO)_2PO.OC_6H_4NO_2$$

$$\rightarrow (C_8H_{17})_3\overset{+}{N}CH_2CH_2OPO(OPh)_2 + \overline{O}_6H_4NO_2$$

$$(C_8H_{17})_3\overset{+}{N}CH_2CH_2\overset{-}{O} \;+\; \text{[2-Cl-4,6-dinitrobenzene]} \;\longrightarrow\; \text{[2-(O-CH_2CH_2\overset{+}{N}(C_8H_{17})_3)-4,6-dinitrobenzene]} \;+\; Cl^-$$

$$\downarrow OH^- \text{ (slow)}$$

$$\text{[2,4-dinitrophenoxide ion]}$$

Formation of the phosphorylated intermediate from 1, Scheme 1, was demonstrated by precipitating it as the perchlorate salt, dissolving the precipitate in ethanol and showing that the uv spectrum was characteristic of an aryl phosphate. Formation of the ether (Scheme 1) was demonstrated spectrophotometrically, and its formation was much faster than its conversion into 2,4-dinitrophenoxide ion, cf., ref. 38.

Medium Effects

The first order rate constants increase sharply with increasing concentration of 1 for reactions of both DNCB and pNPDPP (Figures 1 and 2), but rate enhancements decrease markedly as the acetonitrile content of the solvent is increased. This solvent effect is as expected, because addition of acetonitrile disrupts the three dimensional structure of water, and therefore the hydrophobic binding of the solutes, and at the same time increases solubility of the apolar substrates and therefore decreases their interaction with the hydrophobic ammonium ions. (Similar solvent effects are observed with reactions in solutions of normal micelles.) [1,15]

Effect of pH

The rate constants increase with increasing [OH$^-$] (Figure 2), but with 10% and 20% acetonitrile the increase is linear with [OH$^-$] showing that there is no extensive deprotonation of 1 (Scheme 2). However, in 30% acetonitrile relatively high concentrations of OH$^-$ could be used and plots of k_ψ against [OH$^-$] are curved. This curvature allows estimation of an apparent pK_b for the alkoxide zwitterion (1), assuming that curvature is due wholly to changes in the protonation equilibrium. (Scheme 2)

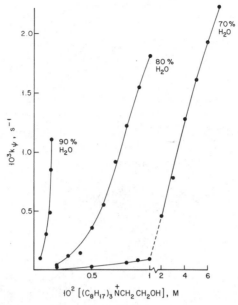

Figure 1. Reaction of 2,4-dinitrochlorobenzene with $\underline{1}$ in 0.01 M NaOH in H_2O-MeCN.

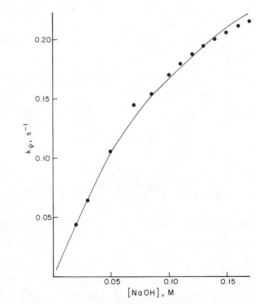

Figure 2.   Reaction of p-nitrophenyl diphenyl phosphate with $\underline{1}$ in H_2O:MeCN 70:30 w/w.

Scheme 2

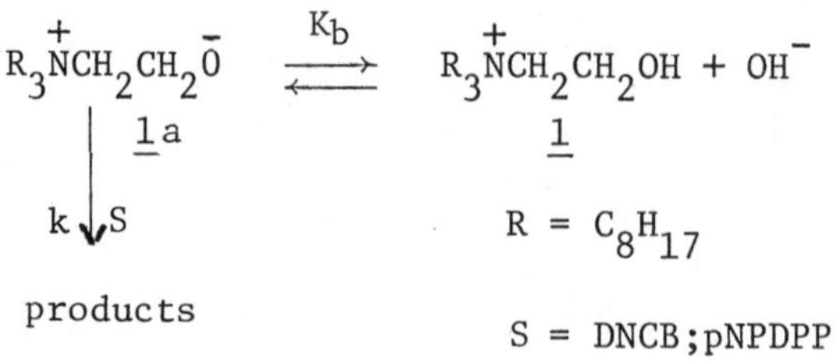

$$R_3\overset{+}{N}CH_2CH_2\overset{-}{O} \quad \overset{K_b}{\underset{\longleftarrow}{\longrightarrow}} \quad R_3\overset{+}{N}CH_2CH_2OH + OH^-$$

$$\underset{\mathbf{\underline{1}a}}{\Big|}$$

$$k\Big\downarrow S \qquad\qquad\qquad R = C_8H_{17}$$

$$products \qquad\qquad\qquad S = DNCB; pNPDPP$$

For reaction in 0.05M $\underline{1}$, Scheme 2 gives:

$$k_\psi = 0.05k[\overline{OH}]/(K_b + [\overline{OH}]) \qquad\qquad (2)$$

or: $\qquad\qquad 0.05/k_\psi = K_b/k[\overline{OH}] + 1/k \qquad\qquad (3)$

This treatment gives for reaction of DNCB, $pK_b = 0.7$ (app) and $k = 0.8$ $M^{-1}s^{-1}$, and for dephosphorylation of pNPDPP, $pK_b = 0.8$ (app) and $k = 8$ $M^{-1}s^{-1}$, in H_2O:MeCN 70:30 w/w at 25.0°C, and the curve in Figure 2 is predicted.

It is important to recognize that these values of pK_b do not take into account that deprotonation of the hydroxyethyl derivative ($\underline{1}$) may change its state of aggregation and its ability to bind substrate. This is a general problem in interpreting acid dissociation constants in micellar and similarly associated systems. Nonetheless, the values of pK_b and k are reasonable in comparison with those in similar systems.

In water, choline has $pK_b = 0.1$,[39] and micellized hydroxyethyl surfactant ($\underline{2}a$) has an apparent $pK_b \sim 1.6$, so that discounting any solvent effect, $\underline{1}$ is a slightly stronger acid than choline, but weaker than micellized $\underline{2}$, which is reasonable because micellization of $\underline{2}$ should increase its deprotonation.

The rate constants, k, for reactions of the alkoxide zwitterion are much larger than the corresponding second order rate constants for reactions of the cholinate zwitterion in water, suggesting that direct association between substrate and the functionalized ammonium ions plays a major role in the enhanced rate of reaction.

It is more difficult to relate the rate constants for reactions of the alkoxide zwitterion ($\underline{1}a$) with those for reactions in micellized $\underline{2}$ because in H_2O:MeCN 70:30w/w there is probably only limited binding of the substrates to $\underline{1}$ whereas substrate binding is essentially complete in the micellar systems.[34-36,38]

Rate Enhancements by Cationic Micelles and
by Hydrophobic Cations.

Overall rate enhancements of bimolecular reactions by
micelles depends upon several factors: (i) the distribution of
both reactants between micellar and aqueous pseudophases, (ii)
the second order rate constant in the micellar pseudophase, and
(iii) micellar effects upon formation of anionic nucleophiles
by deprotonation of weakly acidic precursors, and these factors
can often be separated.

This situation also applies to comparisons between rate enhancements by cationic micelles and hydrophobic, but non-micellizing,
quaternary ammonium ions. It appears that second order rate constants, based on nucleophile concentration written in terms of the
mole ratio of nucleophile to monomeric cation, are similar for
cationic micelles and hydrophobic ammonium ions, suggesting that
differences in size and shape of micelles and of clusters are
relatively unimportant. This generalization seems to apply to a
wide variety of submicroscopic aggregates, including o/w microemulsions, which is understandable because the surfaces of normal
micelles and o/w microemulsion droplets are water-rich. [40,41]

There are however, marked differences between cationic micelles
and cations such as 1 or TMAc or TEAMs, especially as regards
interactions with hydrophilic anions. The charge density at the
surface of ionic micelles is such that even hydrophilic counterions
can be strongly micellar-bound, [6,42,43] but non-micellizing
cations such as $(C_8H_{17})_3NEt$ do not effectively bind hydrophilic
anions.

The different apparent dissociation constants of the
functionalized quaternary ammonium ions, 1, and the corresponding
micelle-forming ions, 2, can similarly be related to the binding
of hydroxide ion.

Overall rate enhancements by cationic micelles and by
hydrophobic quaternary ammonium ions do not provide evidence
regarding the relative efficiencies of these systems as "catalysts".
The hydrophobic cations form aggregates which are probably much
smaller than micelles, so that reactants bound to them could be
much closer than they would be in a micelle. This proximity effect
would make the hydrophobic cations the more effective in rate
enhancement, but, on the other hand, much less effective in binding solutes, so the overall rate effects depend upon the hydrophobicities of the reactants.

There is considerable discussion as to the significance of
cooperativity in these submicroscopic systems. One model of

micellar rate effects assumes that micellar structure is relatively unaffected by the reactants, at least if they are dilute and the surfactant concentration is well above the critical micelle concentration (cmc). [1-4] However there are suggestions that micellar rate effects depend upon cooperative interactions of reactants and surfactants, [44] and these interactions are almost certainly of considerable importance at surfactant concentrations close to, or below, the cmc. For example, micellar rate effects tend to be small at submicellar concentrations of surfactant when reactants are not very hydrophobic, but large rate enhancements under these conditions are often observed when the reactants are hydrophobic. [45-47]

Cooperativity is apparently of major importance in rate enhancements by hydrophobic quaternary ammonium ions, such as TEAMs and 1.

Implicit in all these comparisons of rate constants in micelles, or similar clusters, with those in water, is the assumption that micelles can be regarded as a reaction medium distinct from that provided by the bulk solvent. In other words we assume that reaction does not occur across the shear surface at the micelle-water interface. [1-4] This assumption seems to be justified for most reactions, but the kinetic form of reactions involving micelles having a reactive hydroxide counterion was initially ascribed to a contribution of reaction between micellar bound substrate and OH^- in water, [48] (a different explanation was used in subsequent discussions.[49]) Recently micellar effects upon reaction of chlorohydrocarbons with relatively concentrated OH^- have been ascribed, in part, to reaction across the micellar shear surface. [50]

This question of the validity of the pseudophase model of catalysis by micelles and similar aggregates is linked to another, which is the already noted, question of the volume element of reaction, and the appropriate measure of reactant concentration in the micellar pseudophase. There is no general agreement on this issue, and perhaps there is no unique volume element of reaction for any given aggregate. In addition there seems to be no obvious way of defining the volume element of reaction in clusters of such quaternary ammonium ions as TEAMs, so that rate constants have been defined in terms of mole ratios as concentration units. [19,23] This measure is unambigous, but does not allow comparison of second order rate constants in the clusters with those in water.

It may well be that for reactions in normal micelles, for example, the volume of the Stern layer, or of the whole micelle, is an appropriate volume element for reactions of relatively hydrophobic reactants, but that for very hydrophilic anions, e.g., OH^-, we cannot make such a clear distinction between micellar bound and free counterions. In that event one loses the clear distinction between reactions in the Stern layer and in the aqueous pseudophase. Then micellar rate effects upon reactions of OH^-, for example, would be interpreted in terms of a binding of the hydrophobic reactant at the micelle-water interface and a probability of OH^- being located close to that interface. (The mass-action treatment of counterion binding with its assumed variation of α is an approximation of such a distinction). [49]

As the amphiphilic cluster becomes smaller, e.g., at submicellar surfactant concentrations, or with hydrophobic ammonium ions such as TEAMs or 1, it could hydrophobically bind apolar reactants, or bulky, low charge density, counterions, but there would be no concentration of hydrophilic counterions. Thus, rate enhancements by cations such as TEAMs will be very similar to those by cationic micelles, or similar aggregates, only when the nucleophilic reagents are very hydrophobic.

ACKNOWLEDGEMENTS

Support of this work by the National Science Foundation (Chemical Dynamics Program) and the Army Office of Research is gratefully acknowledged.

REFERENCES

1. J. H. Fendler and E. J. Fendler, "Catalysis in Micellar and Macromolecular Systems", Academic Press, New York, 1975.
2. E. H. Cordes, Pure Appl. Chem., 50, 617 (1978).
3. F. M. Menger, in "Bioorganic Chemistry, III. Macro-and Multicomponent Systems," E. E. van Tamelen, Editor, p. 137, Academic Press, New York, 1977.
4. C. A. Bunton, Pure Appl. Chem., 50, 617 (1978).
5. K. Martinek, A.K. Yatsimirski, A. V. Levashov, and I.V. Berezin, in "Micellization, Solubilization and Microemulsions", K. L. Mittal, Editor, Vol. 2, p. 489, Plenum Press, New York, 1977.
6. L. S. Romsted, in "Micellization, Solubilization and Microemulsions", K. L. Mittal, Editor, Vol. 2, p. 509, Plenum Press, New York, 1977.
7. C. A. Bunton, Catl. Rev. Sci.-Eng.,20, 1 (1979).
8. I. M. Cuccovia, E. M. Schroter, P. M. Monteiro, and H. Chaimovich, J. Org. Chem., 43, 2248 (1978).
9. A. A. Bhalekar and J. B. F. N. Engberts, J. Am. Chem. Soc., 100, 5914 (1978).
10. M. Almgren and R. Rydholm, J. Phys. Chem., 83, 360 (1979).
11. N. Funasaki and A. Murata, Chem. Pharm. Bull., 28, 805 (1980).

12. C. A. Bunton, in "Solution Chemistry of Surfactants", K. L. Mittal, Editor, Vol. 2, p. 530, Plenum Press, New York, 1979.

13. C. A. Bunton, F. H. Hamed, and L. S. Romsted, J. Phys. Chem., 86, 2103 (1982).

14. R. Fornasier and U. Tonellato, J. Chem. Soc., Faraday Trans. 1, 76, 1301 (1980).

15. E. H. Cordes and C. Gitler, Prog. Bioorg. Chem. 2, 1 (1973).

16. F. M. Menger, H. Yoshinaga, K. S. Venkatasubban, and A. R. Das, J. Org. Chem., 46, 415 (1981).

17. H. Al-Lohedan, C. A. Bunton, and M. M. Mhala, J. Am. Chem. Soc., 104, 6654 (1982).

18. J. H. Fendler and W. L. Hinze, J. Am. Chem. Soc., 103, 5439 (1981).

19. C. Hermansky and R. A. Mackay, in "Solution Chemistry of Surfactants", K. L. Mittal, Editor, Vol. 2, p. 723, Plenum Press, New York, 1979.

20. C. A. Bunton and F. de Buzzaccarini, J. Phys. Chem., 85, 3142 (1981).

21. C. A. Bunton, J. R. Moffatt, and E. Rodenas, J. Am. Chem. Soc., 104, 2653 (1982).

22. Y. Okahata, R. Ando, and T. Kunitake, J. Am. Chem. Soc., 99, 3067 (1977).

23. C. A. Bunton, Y. S. Hong, L. S. Romsted, and C. Quan, J. Am. Chem. Soc., 103, 5788 (1981).

24. C. A. Bunton, Y. S. Hong, L. S. Romsted, and C. Quan, J. Am. Chem. Soc., 103, 5784 (1981).

25. D. C. Rideout and R. Breslow, J. Am. Chem. Soc., 102, 7816 (1980).

26. C. A. Blyth and J. R. Knowles, J. Am. Chem. Soc., 93, 3017, 3021 (1971).

27. J. P. Guthrie and Y. Ueda, Can. J. Chem., 54, 2745 (1976).

28. U. Tonellato, in "Solution Chemistry of Surfactants", K. L. Mittal, Editor, Vol. 2, p. 541, Plenum Press, New York, 1979.

29. R. A. Moss, R. C. Nahas, and S. Ramaswami, J. Am. Chem. Soc., 99, 627 (1977).

30. T. Kunitake, Y. Okahata, and T. Sakamoto, J. Am. Chem. Soc., 98, 7799 (1976).

31. L. Anoardi, R. Fornasier, and U. Tonellato, J. Chem. Soc. Perkin Trans. 2, 260 (1981).

32. J. M. Brown, C. A. Bunton, and S. Diaz, J. Chem. Soc., Chem. Commun., 971 (1974).

33. J. M. Brown, C. A. Bunton, S. Diaz, and Y. Ihara, J. Org. Chem., 45, 4169 (1980).

34. C. A. Bunton, L. Robinson, and M. Stam, J. Am. Chem. Soc., 92, 7393 (1970).

35. C.A. Bunton and L. G. Ionescu, J. Am. Chem. Soc., 95, 2912 (1973).

36. A. Pillersdorf and J. Katzhendler, Isr. J. Chem., 18, 330 (1979).

37. C. Quan and Z. Yang, (1982) unpublished data.
38. C. A. Bunton and S. Diaz, J. Am. Chem. Soc., 98, 5663 (1976).
39. R. M. C. Dawson, D. C. Elliott, W. H. Elliott, and K. M. Jones, "Data for Biochemical Research", Clarendon Press, Oxford, 1959.
40. F. M. Menger, Acc. Chem. Res., 12, 111 (1979).
41. C. A. Bunton and F. de Buzzaccarini, J. Phys. Chem., 85, 3139 (1981).
42. D. Stigter, J. Phys. Chem., 68, 3603 (1964).
43. G. Gunnarsson, B. Jonsson, and H. Wennerstrom, J. Phys. Chem., 84, 3114 (1980).
44. D. Piskiewicz, J. Am. Chem. Soc., 99, 7695 (1977).
45. R. Shiffman, Ch. Rav-Acha, M. Chevion, J. Katzhendler, and S. Sarel, J. Org. Chem., 42, 3279 (1977).
46. C. A. Bunton, L. S. Romsted, and H. J. Smith, J. Org. Chem., 43, 4299 (1978).
47. J. M. Brown and J. R. Darwent, J. Chem. Soc. Chem. Comm., 169, 171 (1979).
48. C. A. Bunton, L. S. Romsted, and G. Savelli, J. Am. Chem. Soc., 101, 1253 (1979).
49. C. A. Bunton, L-H. Gan, J. R. Moffatt, L. S. Romsted and G. Savelli, J. Phys. Chem., 85, 4118 (1981).
50. F. Nome, A. F. Rubira, C. Franco, and L. G. Ionescu, J. Phys. Chem., 86, 1881 (1982).

ON THE VALIDITY OF THE PSEUDO-PHASE MODEL FOR MICELLAR CATALYSIS

Lavinel G. Ionescu[*] and Faruk Nome
Laboratório de Química de Superfícies
Universidade Federal de Santa Catarina
Florianópolis, S.C., Brasil
                        and
Sarmisegetuza Research Group[*]
Las Cruces and Santa Fe, N. Mexico, U.S.A.

As part of our study of the decomposition of
various pesticides and insecticides, we have analyzed
in detail the dehydrochlorination of DDT and some of
its derivatives and the hydrolysis of di- and tri-
substituted phosphate esters. The dehydrochlorination
of 1,1,1-trichloro-2,2-bis(p-chlorophenyl)ethane (DDT);
1,1-dichloro-2,2-bis(p-chlorophenyl)ethane (DDD); and
1-chloro-2,2-bis(p-chlorophenyl)ethane (DDM) with
hydroxide ion was studied in the presence of cetyltrimethylammonium bromide (CTAB) micelles at 25° C. The
experimental results indicate that there is a good
agreement between experimental values and theoretical
models of micellar catalysis for OH^- in the range of
0.001 to 0.01 M. For higher values of OH^-, the pseudophase model fails and there is significant deviation
between the theoretically calculated and observed values
for the pseudo-first order rate constant.

The hydrolysis of lithium p-nitrophenyl ethyl
phospahte was studied at 25°, 35° and 45° C in aqueous
solutions containing OH^- and CTAB in the presence and
absence of NaCl. The concentration of NaOH varied
between 0.050 M and 5.00 M and NaCl ranged between
0.005 and 0.030 M. A detailed analysis of the experimental results for high concentration of OH^- indicates
that conventional models of micellar catalysis, considering partition coefficients for the substrate and
ion exchange between water and the micellar phase, are
not applicable. This conclusion is supported by recent

1107

tensiometric, quasi-elastic light scattering and viscosity studies of the $NaCl-CTAB-H_2O$ and $NaOH-CTAB-H_2O$ systems, suggesting significant changes in micellar structure and liquid crystalline growth. A model analogous to phase transfer catalysis may be more suitable for the reactions described above.

INTRODUCTION

Conventional models of micellar catalysis generally involve partition of the substrate and ion exchange between the aqueous and micellar phases.[1-6] All the quantitative treatments of the kinetic data for micellar catalyzed bimolecular reactions of the type A + B ⟶ Products are based on Equation (1).

$$\text{Rate} = k_{2w}(A)_w(B)_w + k_{2m}(A)_m(B)_m \qquad (1)$$

The terms k_{2w} and k_{2m} are the second order rate constants in the aqueous and micellar phases and the subscripts w and m refer to incorporation of the reactants in water and micelles, respectively. Equation (1) is consistent with a phase separation model, where the reaction occurs in the aqueous phase or in both phases. It is important to note that this approach assumes a closed thermodynamic system, since it involves partitioning of the substrate and the reactant between the aqueous and micellar phases and corresponds, in principle, to ion exchange phenomena that are in the final analysis dependent on a limited number of sites.

The quantitative treatment outlined above appeared to be satisfactory until recently when Bunton, Romsted and Savelli[7] reported a partial failure of the pseudo-phase model and proposed an additional reaction pathway across the interfacial boundary between micellar bound organic substrate and ionic reactant in the aqueous phase for the reaction of 2,4-dinitrochlorobenzene and 2,4-dinitrochloronaphthalene with hydroxide ion in the presence of the surfactant p-octyloxybenzyltrimethylammonium hydroxide. They explained the failure of the pseudo-phase model of micellar catalysis by introducing an additional region in the structure of the micelle, i.e., a shear surface between the Stern and Gouy-Chapman layers.[7]

This paper reviews some recent studies that we have undertaken in an attempt to analyze the existence of this additional reaction pathway in reactions catalyzed by cetiltrimethylammonium bromide (CTAB). The reactions investigated included the dehydrochlorination of 1,1,1-trichloro-2,2-bis(p-chlorophenyl)ethane (DDT) and some of its derivatives[8-11] given by Equation (2) and the

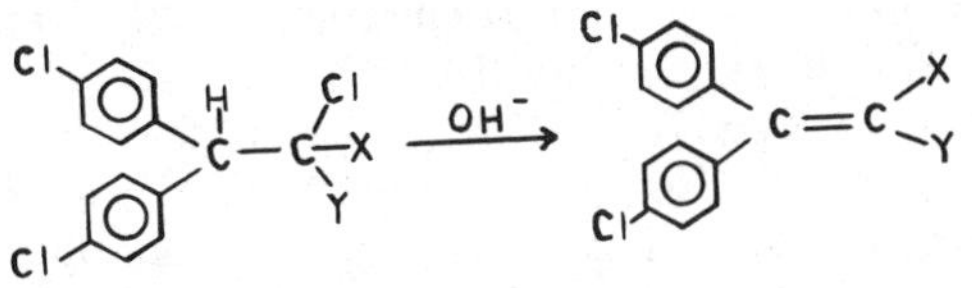

(2)

DDT,   X=Y=Cl   DDE,   X=Y=Cl
DDD,   X=Cl, Y=H   DDMU,   X=Cl, Y=H
DDM,   X=Y=H   DDNU,   X=Y=H

hydrolysis of lithium ethyl p-nitrophenyl phosphate and other
phosphate esters exemplified by Equation (3).[12-20] In contrast to
previous studies of micellar catalyzed reactions, these systems
were studied at high hydroxide ion concentration.

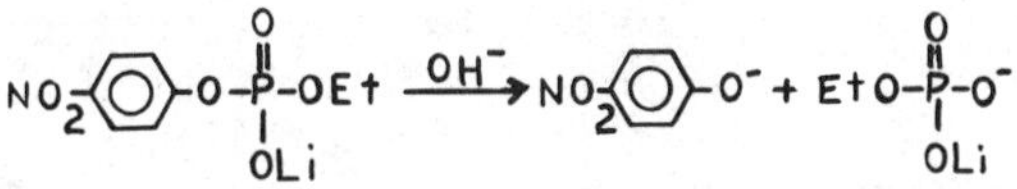

(3)

EXPERIMENTAL PROCEDURE

Materials

   The compounds 1,1,1-trichloro-bis(p-chlorophenyl)ethane (DDT)
and 1,1-dichloro-2,2-bis(p-chlorophenyl)ethane (DDD) were purchased
from Aldrich Chemical Company, Milwaukee, Wisconsin, U.S.A. . The
compound 1-chloro-2,2-bis(p-chlorophenyl)ethane (DDM) was prepared
by a slight modification of a previously described procedure and
crystallized from ethanol.[21] Lithium p-nitrophenyl ethyl phosphate
was prepared from diethyl phosphate and sodium p-nitrophenoxide
using LiCl and dry acetone followed by precipitation with diethyl
ether.[22,23] The surfactant cetyltrimethylammonium bromide (CTAB)
was obtained from Aldrich Chemical Company. It was purified by
recrystallization from ethanol and dried under vacuum in the presence of P_2O_5.

Kinetics

   The reactions were followed spectrophotometrically using
Varian 634 or Shimadzu UV-210-A spectrophotometers equipped with
water-jacketed cell compartments. Rates of dehydrochlorination of
DDT, DDD and DDM were determined by measuring the appearance of

1,1-dichloro-2,2-bis(p-chlorophenyl)ethylene (DDE) at 260 nm;
1-chloro-2,2-bis(p-chlorophenyl)ethylene (DDMU) at 257 nm, and
1,1-bis(p-chlorophenyl)ethylene (DDNU) at 252 nm, respectively.[8]
These reactions were studied at 25° C and the OH⁻ varied from
0.001 M to 1.00 M at different concentrations of CTAB. The
hydrolysis of lithium ethyl p-nitrophenyl phosphate was monitored
by determining the absorbance of the p-nitrophenoxide ion formed
at 403 nm. The concentration of NaOH used ranged from 0.05 M to
5.00 M and NaCl from 0.005 M to 0.030 M. The reaction was
studied at 25°, 35° and 45° C at a constant concentration of CTAB
of 0.0088 M.The pseudo-first order rate constant $k_{\psi m}$ in sec⁻¹ was
determined from linear plots of ln $(A_\infty - A_t)$ vs. time. The second
order rate constants k_2 were calculated from $k_{\psi m}$ and the hydroxide
ion concentration. Activation parameters such as E_a, $\Delta H^\ddagger$, $\Delta G^\ddagger$ and
$\Delta S^\ddagger$ were also determined.

RESULTS AND DISCUSSION

Essentially all the present models of micellar catalysis[1-6]
consider the partition coeficient for the substrate between the
micellar and aqueous phases and the distribution of the reagents
between the two phases. The hydrolysis of lithium ethyl p-nitrophenyl phosphate and the dehydrochlorination of DDT, DDD and DDM
with hydroxide ion in aqueous solutions of CTAB can be considered
as bimolecular reactions between hydroxide ion and the substrate.
Since the concentration of hydroxide ion in the micellar phase is
dependent on the concentration of bromide ion and surfactant, a
quantitative treatment of the reaction rate must consider ion
exchange phenomena on or near the micellar surface.

For the present case, the model of Quina and Cahimovich[5]
reduces to Equation (4) that gives the theoretical dependence of
the pseudo-first order rate constant, $k_{\psi m}$, as a function of the
total concentration of hydroxide ion $(OH)_T$.

$$k_{\psi m} = \frac{\left[\dfrac{k_{2m}}{\overline{V}} K_s K_{OH/Br} \dfrac{(Br)_m}{(Br)_w} + k_2^o \right] (OH)_T}{(1 + K_s C_D) (1 + K_{OH/Br}(Br)_m/(Br)_w)} \qquad (4)$$

where

C_D = concentration of micellized detergent

$k_{\psi m}$ = pseudo-first order rate constant

k_{2m} = second-order rate constant in the micellar phase

k_2^o = second-order rate constant in the aqueous phase

$K_{OH/Br}$ = ion-exchange constant

K_s = binding constant for the substrate

$(Br)_m$ = concentration of Br^- in micellar phase

$(Br)_w$ = concentration of Br^- in aqueous phase

$(OH)_T$ = total OH^- concentration

Considering substrates that are very insoluble in water and are solubilized by CTAB, Equation (4) can be reduced to a simpler form. For the particular case of DDT, the solubility in water at 25° C is approximately 0.2 - 1.2 ppb. In the presence of 6.0×10^{-3} M CTAB a concentration of 1.0×10^{-5} M DDT can be easily solubilized and the K_s value can be readily estimated from $(DDT)_m/(DDT)_w = 1 + K_s C_D$ to be higher than 5.0×10^{-5} M^{-1}. Equation (4) can be simplified to a more convenient expression, since $K_s C_D \gg 1$ and the mole fraction of organic substrate in the aqueous phase approaches zero.

$$k_{\psi m} = \frac{k_{2m}}{C_D \bar{V}} (OH)_T \frac{K_{OH/Br}(Br)_m/(Br)_w}{1 + K_{OH/Br}(Br)_m/(Br)_w} \qquad (5)$$

Equation (5) can be applied to the other substrates under consideration, i.e., DDD, DDM and lithium ethyl p-nitrophenyl phosphate that are all solubilized by CTAB and are highly insoluble in water.

We have calculated theoretical values for $k_{\psi m}$ for the substrates mentioned above using various concentrations of OH^-, $\bar{V}$ = 0.37 and $K_{OH/Br}$ = 0.08 . The term $\bar{V}$ corresponds to the effective volume per mole of micellized detergent. The concentration of Br^- in the aqueous and micellar phases was calculated according to Quina and Chaimovich[5] using a value of 0.2 for the degree of ionization α.

Figure 1 illustrates some typical experimental results obtained for the dehydrochlorination of DDT as a function of hydroxide ion concentration compared to calculated results obtained by treating the experimental data with Equation (5) according to the model of Quina and Chaimovich. We have used a wide range of values for k_{2m}, the second-order rate constant in the micellar phase. As can be seen, there is reasonable agree-

ment between experimental data and theoretically calculated
results only at concentrations of OH⁻ between 0.001 and 0.01 M.

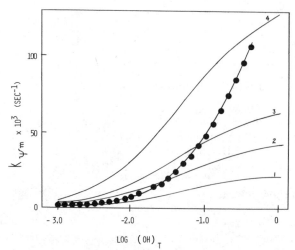

Figure 1. Plot of the experimental pseudo-first-order rate
constant ($k_{\psi m}$) for the dehydrochlorination of DDT as a
function of hydroxide ion concentration at 25° C (•).
Curves 1,2,3 and 4 represent fits of experimental data
with Equation (5) using values of 0.01, 0.02, 0.03 and
0.04 $M^{-1}s^{-1}$, respectively, for k_{2m}.

At higher concentrations of OH⁻ there is a drastic deviation
from theoretically calculated values and $k_{\psi m}$ increases much more
than predicted by the ion-exchange theory. The values of k_{2m} that
were obtained for the best fit of the experimental data have been
reported in the literature[8]. The fact that a small but consistent
increase in k_{2m} is observed as a function of surfactant concentration also indicates a lack of validity of the ion-exchange model
of Quina and Chaimovich in the pH range studied, since k_{2m} should
be independent of the surfactant concentration.

Figure 2 illustrates some representative results obtained
for the dehydrochlorination of DDD as a function of hydroxide
ion concentration. Different values of k_{2m} were used in an
attempt to fit the experimental data to Equation (5). Again,
there is reasonable agreement between the experimental $k_{\psi m}$ values
and theoretically calculated results only at low concentrations
of hydroxide ion. Similar results were obtained for the dehydrochlorination of DDM.[8-11]

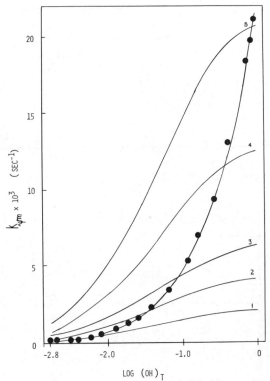

Figure 2. Plot of the experimental pseudo-first-order rate
          constant for the dehydrochlorination of DDD at 25° C
          as a function of hydroxide ion concentration (•).
          Curves 1,2,3,4 and 5 represent fits of the experimental
          data with Equation (5) using values of 0.001, 0.002,
          0.003, 0.006 and 0.01 $M^{-1}s^{-1}$, respectively, for k_{2m}.

       The reaction of lithium p-nitrophenyl ethyl phosphate with
hydroxide ion in aqueous solutions of CTAB was studied both in
the presence and absence of NaCl at various temperatures. Some of
the experimental pseudo-first-order rate constants obtained for
the hydrolysis of the phosphate ester at 25° C as a function of
NaCl for selected values of hydroxide ion concentration are shown
in Figure 3. The values obtained for low concentrations of NaOH
($<$ 0.55 M) decrease as a function of the added salt and agree

well with the results previsously reported for the same system for 0.01 and 0.10 M NaOH.[12] At higher concentrations of NaOH ($>$ 1.0M) the experimental values of $k_{\psi m}$ are almost constant and are not affected by added NaCl. Although the lack of inhibition at high OH$^-$ concentration is consistent with ion exchange phenomena, the values of k_{2m} needed to adjust the data according to Equation (5) increase with increasing (OH$^-$), a result which indicates deviation from the pseudo-phase model of micellar catalysis.

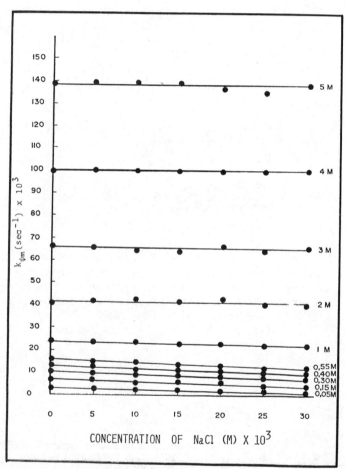

Figure 3. Plot of the pseudo-first-order rate constant($k_{\psi m}$) versus the concentration of NaCl for the hydrolysis of lithium p-nitrophenyl ethyl phosphate at 25° C in the presence of 0.0088 M CTAB and different concentrations of NaOH.

Figure 4 illustrates the dependence of the experimental second order rate constant k_2 ($k_2 = k_{\psi m}/(OH)$) on the hydroxide ion concentration in the absence of salt. The values of k_2 decrease exponentially at low concentration of NaOH (0.01 - 2.0 M), reach a minimum at 2.0 M and increase at higher concentrations of NaOH (2.0 - 5.0 M).

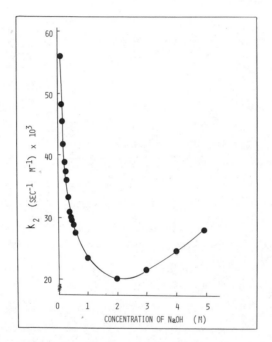

Figure 4. Plot of the observed second order rate constant (k_2) as a function of the NaOH concentration for the hydrolysis of lithium ethyl p-nitrophenyl phosphate at 25° C.

Figure 5 shows some of the experimental results obtained for the pseudo-first-order rate constant $k_{\psi m}$ for the hydrolysis of lithium ethyl p-nitrophenyl phosphate as a function of hydroxide ion compared to theoretical results calculated by means of Equation (5). Again, the agreement between experimental results and the theoretical model is reasonable only for low concentration of OH⁻.

The activation parameters determined from experimental $k_{\psi m}$ of values at 25°, 35° and 45° C, both in the presence and absence of NaCl at different NaOH concentration are comparable to other values reported in the literature for micellar catalyzed reactions.[10] In general, all the activation parameters decrease as a function of NaOH. The addition of NaCl causes a further decrease in the activation parameters. For example, in the absence of salt $\Delta S^{\neq}$ decreases from -25.6 eu (0.10 M NaOH) to -30.4 eu (1.0M NaOH).[18-20] It thus appears that the addition of both NaOH and NaCl lead to a more ordered transition state. The change in the activation parameters could also be a reflection of changes in the reaction medium.

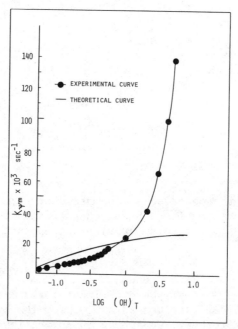

Figure 5. Plot of the pseudo-first-order rate constant as a
          function of the logarithm of the concentration of NaOH
          for the hydrolysis of lithium ethyl p-nitrophenyl ethyl
          phosphate at 25° C.

Attempts to explain the experimental data presented above by means of other theoretical models of micellar catalysis proposed in the literature [2-6] lead to the same result, i.e., failure of the models at higher hydroxide ion concentration. Similar behavior at higher hydroxide ion concentration has been observed by Bunton, Romsted and Savelli[7] and by Lapinte and Viout[24]. This is not surprising, since all models assume a constant micellar structure, involve partitioning of the substrate and reactant between the aqueous and micellar phases and a fixed number of sites. All of them assume a closed thermodynamic system for a case when the kinetics under consideration may be more akin to flow kinetics, particularly when the micelles change in form, size and shape or undergo transitions to liquid crystalline mesophases.

As previously mentioned, Bunton, Romsted and Savelli[7] explained the discrepancy between the theoretical models and the experimental results by considering an additional reaction pathway across the micellar boundary at the shear surface between the Stern and Gouy-Chapman layers of the micelles. We have suggested that conceptually this interfacial boundary or micelle-water interface, as it has been called, should not be very different from the interface present in systems where phase transfer catalysis is taking place[8,16-20].

A more careful analysis of the data presented in Figure 4 for the hydrolysis of lithium ethyl p-nitrophenyl phosphate suggests that the pronounced variation in k_2 may be indicative of a change in the mechanism of catalysis, from micellar to phase transfer.

Recent surface tension and viscosity studies of the NaCl-CTAB-H_2O system show that the apparent critical micellar concentration (CMC) and the shape of CTAB micelles changes markedly as a function of added salt. Experimental parameters of micellization such as ΔH°_{mic} indicate that the NaCl-CTAB-H_2O system undergoes a phase transition at 25° C when the concentration of NaCl is about 0.5 M[25,26].

Recent quasi-elastic light scattering studies of the NaOH-CTAB-H_2O[27] and NaCl-CTAB-H_2O[28] systems also suggest the occurrence of significant structural changes as a function of NaOH and NaCl concentration. The diffusion coefficient (D) of aggregate particles in the NaOH-CTAB-H_2O system at 25° C increases for fixed NaOH concentration as a function of surfactant in the 0.01 - 1.0 M range, is constant at 2.0 M NaOH and subsequently decreases, indicating formation of liquid crystals[27]. For NaCl-CTAB-H_2O at 25° C, D has a positive slope as a function of CTAB in the 0.01 to 0.20 M range of NaCl, D is constant for 0.50 M to 1.0 M NaCl and has a negative slope at higher concentrations of NaCl.

L. G. IONESCU ET AL.

Consideration of the kinetic results presented above in light of the structural changes occurring in the systems appears to exclude the theoretical models presented in the literature[2-6] for the higher concentrations of hydroxide ion. Changes in form, size and shape of the CTAB micelles and subsequent transitions to lamellar and liquid crystalline phases are not considered in the conventional models[2-6]. For solutions of surfactants containing higher concentrations of hydroxide ion or salts, serious consideration should be given to a phase transfer model of catalysis.

ACKNOWLEDGEMENTS

Financial support received from CNPq-National Research Council of Brazil, FINEP and Sarmisegetuza Research Group, Las Cruces and Santa Fe, N. Mexico, U.S.A., is gratefully acknowledged.

REFERENCES

1. J. H. Fendler and E. J. Fendler, "Catalysis in Micellar and Macromolecular Systems", Academic Press, New York, 1975.
2. I. V. Berezin, K. Martinek and A. K. Yatsimirskii, Russ. Chem. Rev., $\underline{42}$, 787 (1973).
3. K. Martinek, A.K. Yatsimirskii, A. V. Levashov and I. V. Berezin, in "Micellization, Solubilization and Microemulsions", K. L. Mittal, Editor, Vol. 2, p.489, Plenum Press, New York, 1977.
4. L. S. Romsted, in "Micellization, Solubilization and Microemulsions", K. L. Mittal, Editor, Vol. 2, p. 509, Plenum Press, New York, 1977.
5. F. Quina and H. Chaimovich, J. Phys. Chem., $\underline{83}$, 1844 (1979).
6. N. Funasaki, J. Phys. Chem., $\underline{83}$, 1998 (1979).
7. C. A. Bunton, L. S. Romsted and G. Savelli, J. Am. Chem. Soc., $\underline{101}$, 1253 (1979).
8. F. Nome, A. F. Rubira, C. Franco and L. G. Ionescu, J. Phys. Chem., $\underline{86}$, 1881 (1982).
9. F. Nome. A. Neves and L. G. Ionescu, in "Solution Behavior of of Surfactants – Theoretical and Applied Aspects", K. L. Mittal, Editor, Vol. 2, p.1157, Plenum Press, New York, 1982.
10. F. Nome, E. Schwingel and L. G. Ionescu, J. Org. Chem., $\underline{45}$, 705 (1980).
11. A. F. Rubira, M.S. Thesis, Universidade Federal de Santa Catarina, Florianópolis, S.C., Brazil, 1981.
12. C. A. Bunton and L. G. Ionescu, J. Am. Chem. Soc., $\underline{95}$, 2912 (1973).
13. C. A. Bunton, S. Diaz, J. M. Hellyer, Y. Ihara and L. G. Ionescu, J. Org. Chem., $\underline{40}$, 2313 (1975).

14. L. G. Ionescu, Bull. N. Mex. Acad. Sci., 14(2), 65 (1973).
15. L. G. Ionescu and D. A. Martinez, J. Colo. Wyo. Acad. Sci., /(5), 13 (1974).
16. L. G. Ionescu and D. A. R. Rubio, Supl. Cienc. Cult., 33(7), 436 (1981).
17. L. G. Ionescu and D.A.R. Rubio, Arq. Biol. Tecnol., 24, 76 (1981).
18. L. G. Ionescu, D. A. R. Rubio and T. M. H. Do Aido, Proc. 1st Conf. Phys. Org. Chem., Florianópolis, S.C., Brazil, p.217, 1982.
19. L. G. Ionescu, D. A. R. Rubio and T. M. H. Do Aido, J. Chem. Soc. Perkin Trans., submitted, 1982.
20. D. A. R. Rubio, M. S. Thesis, Universidade Federal de Santa Catarina, Florianópolis, S.C., Brazil, 1981.
21. S. J. Cristol, N. L. Hause, A. J. Quant, H. W. Miller, K. R. Eilar and J. S. Meek, J. Am. Chem. Soc., 74, 3333 (1952).
22. A. S. Kirby and M. Jounas, J. Chem. Soc., B., 1165 (1970).
23. A. M. Roos and J. Toet, Rec. Trav. Chim., 77, 946 (1958).
24. C. Lapinte and P. Viout, Tetrahedron, 35, 1931 (1979).
25. L. G. Ionescu and T. H. M. Do Aido, Arq. Biol. Tecnol., 24, 77 (1981).
26. T. H. M. Do Aido, M. S. Thesis, Universidade Federal de Santa Catarina, Florianópolis, S.C., Brazil, 1981.
27. L. G. Ionescu, C. A. Bunton, D. F. Nicoli and R. Dorshow, unpublished results, 1982.
28. L. G. Ionescu and T. H. M. Do Aido, Supl. Cienc. Cult., 34(7), 458 (1982).

ANALYSIS OF THE EFFECT OF MICELLES AND VESICLES ON THE REACTIVITY

OF NUCLEOPHILES DERIVED FROM THE DISSOCIATION OF WEAK ACIDS

Hernan Chaimovich, João B.S. Bonilha, Dino Zanette and
Iolanda Midea Cuccovia
Departamento de Bioquímica
Instituto de Química da Universidade da São Paulo
Caixa Postal 20780, Sao Paulo, SP, Brazil

Micellar solutions of hexadecyltrimethylammonium
bromide(CTAB)accelerate the thiolysis of p-nitrophenyl
octanoate(NPO)and acetate(NPA)by n-heptyl mercaptan(HM)
by several orders of magnitude.The rate of thiolysis of
NPO by HM increases two hundred fold in the presence of
the zwitterionic detergent dodecyldimethylammonium pro-
pane sulfonate(DDAPS)while sodium dodecyl sulfate (SDS)
produces a seventy fold rate decrease.The detergent-
induced rate effects were analyzed within the framework
of a pseudophase ion—exchange formalism,the requisite
parameters being determined independently of the kinetic
results. Excellent stimulations were obtained using ratios
of second order rate constants in the aqueous and
micellar phases close to one. CTAB and vesciles derived
from dimethyldioctadecylammonium chloride accelerate
the thiolysis of esters by cysteinylhexadecyl amide
by more than a million fold. This rate enhancement was
also analyzed using pseudophase ion-exchange formalism.
The maximum rate effects, produced by the addition of
amphiphile aggregates on the thiolysis of esters, extend over a range of one hundred million. Although
changes of the interphase affect the reactivity of the
nucleophile,this factor does not determine the extent
of catalysis or inhibition.The main kinetic effect of
micelles and vesicles is due to changes in the local
concentration of the reactants in the restricted
environment of the organized pseudophase.

INTRODUCTION

Pseudophase models, with explicit consideration of ion exchange, successfully describe micellar effects on reaction kinetics and equilibria involving both neutral and ionic reagents.[1-6] Application of such models shows that nucleophilic reactivities, assesed by the values of calculated rate constants, are of comparable magnitude in the micellar pseudophase and in the intermicellar aqueous phase. Similar reactivities have been found for reactions such as hydroxide ion attack on esters,[5,7] ester thiolysis,[7,8] acid catalyzed benzidine rearrangement,[9] acid catalyzed hydration of pyridinium ions,[10] nucleophilic substitution on halobenzenes,[11] cyanide addition on nicotinamides[12] and ester oximolysis.[6,13] The use of pseudophase models has recently been extended to the analysis of reactions in synthetic amphiphile vesicles.[14,15]

Our aim in this work was to examine the limits of the applicability of the ion exchange formalism, and of the concept of similar reactivities in amphiphile aggregates and in the aqueous phase. For that purpose we analysed the effect of positive,negative and zwitterionic detergents on the rate of thiolysis of esters. In addition we compared the effect of micelles and synthetic amphiphile vesicles on the same reaction. The maximum rate effects ranged from seventy fold inhibition by negatively charged micelles to ten million fold acceleration by positively charged vesicles. Pseudophase ion-exchange formalism was shown to apply in all cases. The analysis of such wide range of kinetic effects demonstrated that local reagent concentration in the main contributor to the observed "catalysis" or "inhibition" by these amphiphile aggregates.

EXPERIMENTAL SECTION

Materials. Hexadecyltrimethylammonium bromide (CTAB) (Merck, Darmstadt, Germany) and dioctadecyldimethylammonium chloride (DODAC) (Herga Ind. Quim. Rio de Janeiro, Brasil) were purified and analyzed as previously described.[8,16] Sodium dodecylsulfate (SDS) (Specially pure grade, BDH, Poole, England) gave no mininum in surface tension plots, was used as received. Heptyl mercaptan (HM) (Aldrich) was distilled (37 - 38°C, 1 mmHg) and maintained at -18°C under Ar. p-Nitrophenyl acetate (NPA) was purchased from Sigma Chem. Co.. Dodecyldimethylammoniumpropanesulfonate (DDAPS) and p-nitrophenyl octanoate (NPO) were synthesized by Dr. O.A. El Seoud of this Institute. Cysteinylhexadecylamide (HCys) was synthesized as shown in Scheme 1, while compounds I and II were synthesized according to the literature [17].p-Nitrophenylester of L-4-carboxy-3-formyl-2,2-dimethylthiazolidine (III): 25 g of dicyclohexylcarbodiimide (DCC) was added to a stirred suspension of II (18.9 g) in chloroform (300ml) containing p-nitrophenol (16 g) at

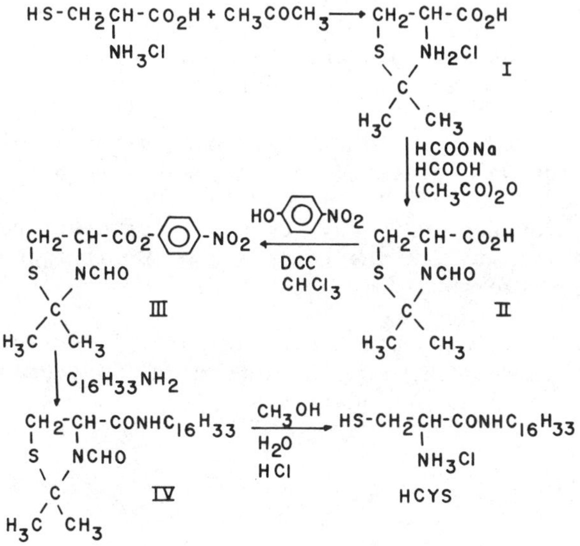

SCHEME I

15°. The mixture was stirred for 4 hrs at room temperature (22°). The copious precipitate of dicyclohexylurea was filtered and the solution evaporated to dryness. The residue was crystallized from absolute ethanol (m.p. 113-114, nmr, ir.).

L-3-Formyl-2,2-dimethylthiazolidine-4-(hexadecyl) carboxamide (IV): 21.7 g of III and 14.5 g of hexadecylamine, dissolved in chloroform, were left overnight. The solvent was removed and residue (orange-red liquid) was added to a dilute solution of sodium carbonate (100 ml, 10%) to remove p-nitrophenol. The aqueous layer was decanted and the residue was washed with ethanol. Cysteinyl hexadecylamide (HCys): Crude IV was dissolved in a mixture of 10 ml of concentrated HCl and 120 ml of methanol and refluxed on a steam bath for 30 min and then left at room temperature for 48 hrs. The solvent was evaporated and the residue recrystallized from MeOH-ether. Elemental analysis: Required for $C_{19}H_{41}ClN_2OS$: C, 59.88; H, 10.84; N, 7.35. Found: C, 58.91; H, 10.43; N, 7.05. NMR, IR. and SH analysis are in agreement with those required by structure.

METHODS

DODAC vesicles were prepared by alcohol injection as described previously.[16] DODAC and CTAB concentrations were determined by halide titration with $Hg(NO_3)_2$.[16] Stock solutions (ca. 10^{-2} M) of HM (in CH_3CN) or HCys (in CH_3OH) were maintained at $-18°C$ under Ar. SH concentration was determined by reaction with 5,5' dithiobis (2-nitrobenzoic acid).[18] Stock solutions of NPA and NPO (ca. 10^{-3} M) were prepared in CH_3CN. Pseudo first order rate constants $k\psi$ (excess mercaptan) for the thiolysis of p-nitrophenyl esters ($30.0 \pm 0.1°C$) were obtained as described previously.[8,15]

The distribution constant for the partitioning of HM between the micellar pseudo-phase and the intermicellar aqueous phase (K_{HM}) can be described by Equation (1):[6]

$$K_{HM} = \frac{(HM_b)}{(HM_f)C_D} \qquad (1)$$

where (HM_b) and (HM_f) are the analytical concentration of HM bound to the micelles and free in the aqueous phase, respectively. The concentration of micellized detergent, C_D, is equal to the total detergent concentration (C_T) minus the critical micelle concentration (CMC). CMC's in the presence of HM were determined by measuring surface tension as a function of detergent concentration. In all other cases the CMC's were corrected for salt and/or the increase in detergent concentration by numerical iteration of Equation 2:[19a]

$$\log CMC = A - b \log (salt + CMC + \alpha\, C_D) \qquad (2)$$

The constants A and b for CTAB and SDS were taken from the literature.[19a,b] The degree of counterion dissociation of the micelle (α) was taken as 0.2 for CTAB and 0.218 for SDS.[20] Equation (1) can be expressed as

$$(HM_o) = (HM_f) + (HM_f)\, K_{HM}\, C_D \qquad (3)$$

where HM_o is the total analytical concentration of HM at a particular detergent concentration. In practice, 0.10 ml of HM were added to Teflon-stoppered Ar-flushed tubes containing 2.00 ml of 0.01 M HCl. The solutions were equilibrated for 2 days at $30°C$ with gentle agitation. After aspiration of the HM layer, HM_o was determined in an aliquot of the aqueous layer. K_{HM} was calculated from the slope of the linear plot relating HM_o with C_D (Equation 3).

The apparent acid dissociation constant (Kap) of a weak acid, such as HM or HCys, in the presence of an amphiphilic aggregate can be defined[6] as:

$$Kap = H_f \frac{(M_b) + (M_f)}{(HM_b) + (HM_f)} = H_f \frac{(M_T)}{(HM_T)} \qquad (4)$$

where H_f represents the intermicellar concentration of hydrogen ions. HM and M are the undissociated and dissociated forms of the weak acid. The subscripts b, f and T represent the bound, free and total analytical concentrations of each species.

The absorptivities of the protonated (E_{HM}) and anionic (E_M) forms of the mercaptans, determined in 0.01 M HCl and 1.0 M NaOH, were found to be insensitive to the presence of detergent. For n-heptylmercaptan E_M and E_{HM} were 5650 and 115 M^{-1} cm^{-1} and for cysteinylhexadecylamide the corresponding values were 5495 and 153 M^{-1} cm^{-1}. The absorbance (240 nm) of a solution of HM (or HCys) can be described by

$$Abs = E_\psi HM_o = E_{HM} HM_T + E_M M_T \qquad (5)$$

where E_ψ is the observed absortivity of the added mercaptan. From Equations (4) and (5) it can be shown that

$$pKap = pH + \log \left| \frac{E_M - E_\psi}{E_\psi - E_{HM}} \right| \qquad (6)$$

The variation of pKap with C_D was determined (Equation 6) at a single, fixed pH using a fixed mercaptan concentration (ca. 5×10^{-5} M).

RESULTS

Cationic detergents, such as CTAB, increase the rate of thiolysis of esters.[6,8,21] The effect of CTAB on the rate of thiolysis of NPA and NPO by HM was determined as a function of detergent concentration at several pH's (Figure 1). The shape of the curves relating k_ψ with CTAB is characterized by a maximum at low detergent concentration, followed by a smooth decrease upon increasing $|CTAB|$. The concentration of detergent at which the maximum rate is attained (k_ψ^{max}), and the values of k_ψ^{max} are pH dependent (Figure 1). For comparison, we calculated the pseudo-first order rate constant (k_ψ^o) that would be observed for the thiolysis of NPO and NPA on the absence of micelles under comparable conditions. The second order rate constant for reaction of NPA with HM in the aqueous phase (k_2^o) was found to be 30 $M^{-1}s^{-1}$.[22] The relationship k_ψ^{max}/k_ψ^o gives the maximum rate acceleration in each case (Table I). The value of k_ψ^{max}/k_ψ^o increased with pH in the case of NPO as a reflection of the decrease in the concentration of bromide needed to maintain the pH with Tris buffer. For NPA the small variations of the values of k_ψ^{max}/k_ψ^o were attributed to a change in the ionic composition of the borate and borate-borax buffers in this pH region and to the corresponding differences in the affinity of these ionic species to the CTAB micelle.

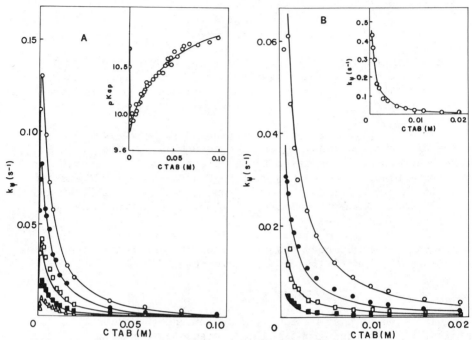

Figure 1. Effect of CTAB on the rate of thiolysis of NPA (A) and
NPO (B) by n-heptyl mercaptan. A. (NPA) = 5 x 10⁻⁶ M,   (HM) =
5 x 10⁻⁵ M, Borate buffer (0.02 M) pH's 8.9 (O), 8.6 (●) and
8.3 (□). Borax/borate buffer (0.02 M) pH's 8.0 (■) and 7.7 (Δ).
Inset shows the variation of pKap a function of (CTAB),   borate
buffer (0.02 M) pH 10.10. B. (NPO) = 5 x 10⁻⁶ M, (HM) = 5 x 10⁻⁵ M
all reactions on 0.02 M Tris/HBr pH's 7.7 (■), 8.0 (□), 8.3 (●),
8.55 (O) and 8.90 (Inset).

Table 1. Effect of CTAB on the rate of thiolysis of NPO and NPA
by n-heptyl mercaptan.

Ester	pH^a	k_ψ^{max}/k_ψ^o $(x10^3)$	k_2^m/k_2^o
NPA	7.7	6.09	3.3
	8.0	6.35	3.6
	8.3	6.94	4.2
	8.6	6.13	4.8
	8.9	5.55	4.6
NPO	7.7	3.05	1.0
	8.0	5.02	1.2
	8.3	5.06	1.0
	8.55	5.80	1.0
	8.9	15.80	1.2

a. Buffer composition is given in the legend of Figure 1. b. CTAB
concentration at k_ψ^{max} was 2 x 10⁻³M for NPA and 1 x 10⁻³M for NPO

The quantitative analysis of a reaction of a nucleophile derived from the dissociation of a weak acid in a micelle solution starts by considering the effect of the detergent on the acid dissociation. The relationship between Kap and C_D for a neutral weak acid such as HM in the presence of a positively charged detergent (CTAB) can be expressed[6] as:

$$Kap = Ka \frac{1 + K_T \frac{(Br_b)}{(Br_f)}}{1 + K_{HM} C_D} \qquad (7)$$

where Ka is the acid dissociation constant of HM. The analytical concentration of bromide bound to the micelle (Br_b) can be described by[4,6]

$$(Br_b) = (1 - \alpha) C_D - OH_b - M_b \qquad (8)$$

where OH_b represents the analytical concentration of hydroxide ion bound to the micelle. The analytical concentration of bromide in the intermicellar aqueous phase (Br_f) can be expressed as

$$(Br_f) = \alpha C_D + OH_b + M_b + Br_{AD} \qquad (9)$$

where Br_{AD} is the concentration of added bromide. In the presence of a buffer which contains ions other than bromide the binding of all ionic species have to be considered in Equations (8) and (9).[7a] The selectivity constant for the mercaptide/bromide ion exchange (K_T) was calculated by fitting the data for the variation of pKap with C_D (inset Figure 1A) with Equation (7) using K_T as a variable parameter and a value of 3700 M^{-1} for K_{HM} (see Methods). In this particular case K_T can also be calculated from the data (inset Figure 1A) at the point where pKap = pKa = 10.7.[23] Under such condition Equation (6) reduces to

$$K_T = K_{HA} C_D \frac{(Br_f)}{(Br_b)} \qquad (10)$$

Both procedures yield a value of 75 for K_T.

The kinetic data for the thiolysis of NPA and NPO by HM (Figure 1) were analyzed quantitatively using an ion-exchange model.[6] This formalism takes into account: a) the partitioning of the neutral form of the mercaptan (HM) and of the neutral esters (NPA and NPO) between the micellar pseudophase and the aqueous phase b) selective ion exchange of all ionic species, including the dissociated form of the mercaptan, with bromide counterions and c) independent reactivities in the micellar and aqueous phases. We have demonstrated[6] that this system can be described by Equation (11).

$$k_\psi = HM_o \ Ka \ \frac{k_2^o + (k_2^m/\overline{V}) \ K_S \ K_T \ (Br_b/Br_f)}{H_f \ (1+K_{HM}C_D) + Ka \ (1+K_T(Br_b/Br_f))(1+K_SC_D)} \quad (11)$$

k_2^m and $\overline{V}$ represent the second order rate constant in the micellar phase and partial molar volume of the micellized detergent, respectively. H_f and maintained constant by the use of an appropriate buffer[7a] and the distribution constants for NPA and NPO (K_S) were taken as 54 and 1.5×10^4 [7a] respectively. The values of K_T and K_{HM} were those determined previously (vide supra). The data in Figure 1 were fitted to Equation (11) by iteration varying excusively the value of $(k_2^m/\overline{V})$. The correspondence between the data and the k_ψ vs. CTAB functions generated at several pH's by the ion exchange model was excellent (Figure 1).

Calculation of second order rate constants in the micelle pseudophase in conventional units is possible after selection of a volume element for the reaction. We[7a,15] and others[1,6] have used the partial molar volume of the micellized detergent as such volume element. The k_2^m's for the thiolysis of NPO and NPA were calculated from the respective best fit values of $(k_2^m/\overline{V})$ taking $\overline{V} = 0.37$ LM[-1] 24 (Table I). For the thiolysis of NPO by HM the values of k_2^m/k_2^o are identical within experimental error. The values of k_2^m/k_2^o for the NPA series exhibit a definite increase upon changing the buffer from borax/borate to borate.

Since the early work of Hartley[25] it is known that negatively charged detergents increase the apparent acid dissociation constant of neutral weak acids. In fact, the addition of SDS caused an increase in the pKap of HM (Inset Figure 2). The dissociated form of the mercaptan (M), bearing a negative charge, will distribute as a pseudo monomer of the detergent[26]. With these consideration and the usual assumptions of ion exchange[4,20] it can be shown that the function relating the variation of Kap of HM with SDS concentration is

$$Kap = Ka \ \frac{1 + (K_b \cdot K_{HM} \cdot C_D/K_{H/Na} \cdot K_a)((Na_f)/(Na_b)_\ell)}{(1 + K_{HM} C_D)} \quad (12)$$

The acid dissociation constant of HM in the micellar phase (K_b) was defined[6] as:

$$K_b = \frac{(M_b)_\ell (H_b)_\ell}{(HM_b)_\ell} \quad (13)$$

where the subscript ℓ represents the local concentration in the micellar pseudophase. The value of the sodium/hydrogen selectivity coefficient ($K_{H/Na}$) for ion exchange was taken as 1.2[7b] (Na_f) and $(Na_b)_\ell$ represent the analytical concentrations of free sodium

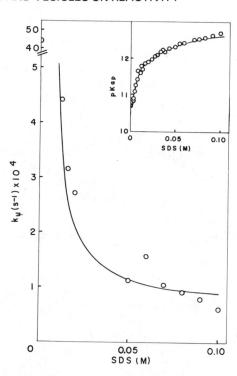

Figure 2. Effect of SDS on the rate of thiolysis of NPO by HM.
$[HM] = 2 \times 10^{-4}$ M, $[NPO] = 4 \times 10^{-6}$ M, $[NaOH] = 2.0 \times 10^{-3}$ M (pH =
11.13, 30°C). Inset shows the variation of pKap with SDS. Sodium
phosphate (0.02 M) buffer pH 11.30, HM = 7×10^{-5} M.

ions and the local concentration of bound sodium ions respectively.
The mathematical expressions relating Na_f and $(Na_b)_\ell$ with detergent
concentration, salt and pH have been published.[7b,12]
Equation (12) is equivalent to Equation (14)

$$Kap = Ka \; \frac{1 + (K_T^M/m_f) \; C_D}{1 + K_{HM} \; C_D} \tag{14}$$

The distribution constant of the pseudo monomer M (K_T^M) can be
described by Equation (15)

$$K_T^M = \frac{M_b \; m_f}{M_f \; C_D} \tag{15}$$

where m_f represents the analytical concentration of free monomers
in the intermicellar aqueous phase. The solid line in the inset of
Figure 2 was obtained by iteration to best fit (Equation 12) using
K_b as the adjustable parameter. The calculated value for K_b
was 1×10^{-12}.

SDS strongly inhibits the thiolysis of NPO by HM (Figure 2), the maximum inhibitory effect, at the plateau region, being ca. 70 fold (Table II). The kinetic data presented in Figure (2) were analyzed using Equation (16).This equation can be derived from a ion exchange formalism[4,6,20] taking into account all the factors described above for the variation of pKap with [SDS] and the distribution of NPO

$$k_{\psi} = HM_o \frac{k_2^o \, Ka \, K_{H/Na} + (k_2^m/\overline{V}) \, K_b \, K_{HM} \, K_S \, C_D \frac{(Na_f)}{(Na_b)_\ell}}{(1+K_S C_D)\left[K_a K_{H/Na} + K_{H/Na} H_f (1+K_{HM} C_D) + K_b K_{HM} C_D \frac{(Na_f)}{(Na_b)_\ell}\right]} \quad (16)$$

The values of Ka, $K_{H/Na}$, K_{HM}, K_b and the calculation of C_D, (Na_f) and $(Na_b)_\ell$ were those described in the analysis of the variation of pKap with [SDS] and K_S for NPO was taken as 1×10^4 M^{-1} .Equation (16) was fitted to the experimental results by varying $(k_2^m/\overline{V})$. Taking $\overline{V}$ as 0.25[24] the best fit value of k_2^m is lower than k_2^o by a factors of ca. 3 (Table II).

The effect of a zwitterionic detergent (DDAPS) on the pKap of HM is similar to that observed for SDS (Insets Figures (2) and (3)). In this case the pKap vs. DDAPS profile can be described by Equation 17[6] employing a (constant) value of 0.002 M for the cmc of DDAPS.

$$Kap = Ka \frac{1 + K_M C_D}{1 + K_{HM} C_D} \quad (17)$$

The partitioning of M(K_M) and of HM(K_{HM}) were represented as distribution constants between the micellar pseudophase and the aqueous phase. The value of the distribution constant for neutral,

Table II. Effect of SDS and DDAPS on the rate of thiolysis of NPO by HM.

Detergent	pH[a]	k_ψ^{max}/k_ψ^o [b]	k_2^m/k_2^o
SDS	11.30	0.09	0.27
DDAPS	10.15	266	1.0

a. The buffer compositions are given in the legends of Figures (2) and (3).b. For SDS this ratio refers to the k_ψ observed at 0.10 M, for DDAPS the concentration of detergent at k_ψ^{max} was 0.004 M.

long chain reagents between the micellar pseudophase and the
aqueous phase depends mostly on the length of the alkyl chain of
the surfactant.[27] Thus K_{HM} in DDAPS was taken as equal to the
value in SDS. Equation (17) was fitted to the experimental data
by varying K_M and the best fit value was found to be 100 M^{-1}.

Although both DDAPS and SDS cause an increase in the pKap of
HM, the zwitterionic detergent produces an increase in the rate
of thiolysis of NPO (Figure 3), the maximum rate enhancement
(k_ψ^{max}/k_ψ^0) being ca. 250 fold (Table II). The effect of the
variation of [DDAPS] on the pseudo first order rate constant for
the thiolysis of NPO by HM can be described by Equation (18).

$$k_\psi = HM_o \frac{k_o + (k_2^m/\overline{V}) K_M K_S C_D}{(1+K_M C_D)(1+K_S C_D)(1+H_f/Kap)} \qquad (18)$$

Equation (18) was fitted to the experimental data (Figure 3) using
$(k_2^m/\overline{V})$ as a variable parameter. Taking $\overline{V}$ as 0.25 LM^{-1}, the second
order rate constant in the micellar pseudophase is equal to that
in the aqueous phase[28] (Table II).

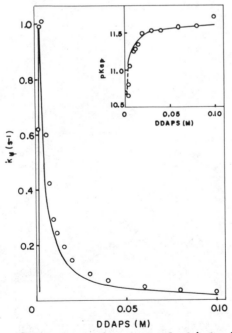

Figure 3. Effect of DDAPS on the rate of thiolysis of NPO by
HM. [HM] = 5.7 x 10^{-5} M, [NPO] = 4 x 10^{-6} M in borate buffer 0.02
pH 10.15. Inset shows the variation of pKap with DDAPS under the
same conditions.

We have also employed a functional detergent derived from cysteine (HCys) as a model to study SH reactivity in micellar and vesicular systems.CTAB produced an effect on the pKap of HCys entirely analogous to that observed with HM (compare Figure 1 with Figure 4).

For a single dissociation step

$$RNHCOCH\begin{smallmatrix} \overset{+}{NH_3} \\ SH \end{smallmatrix} \rightleftharpoons RNHCOCH\begin{smallmatrix} \overset{+}{NH_3} \\ S^- \end{smallmatrix} + H^+$$

the effect of CTAB on the pKap the SH group of HCys can   be described by Equation (19)

$$Kap = \frac{Ka\ (1 + K_{HC}\ C_D)}{1 + \dfrac{K_w'\ Ka\ K_{HC}\ C_D\ ((Br_f)/(Br_b)_\ell)}{K_b\ K_{OH/Br}}} \tag{19}$$

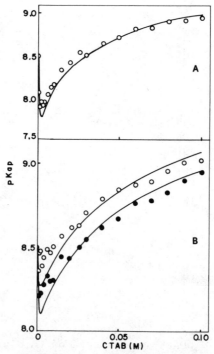

Figure 4. Effect of CTAB on the pKap of cysteinyl hexadecylamide [HCys] = 4 x 10^{-5} M.   A. Borate buffer 0.01 M pH 8.5 (O) B. Tris. HBr 0.01 M pH 8.55 (O), Tris.Acetate 0.01 M pH 8.60 (●).

where K_b is the corresponding dissociation in the micelle, K_{HC} the distribution constant of the zwitterionic form of HCys between the aqueous phase and micellar pseudo phase and K_w' the ratio between the ionic product of water in the micellar pseudophase and in the aqueous phase, respectively. The value of K_{HC} was taken as 1×10^4 M^{-1}, a value which is reasonable in view of the hydrophobicity of HCys. In fact, the pKap vs [CTAB] function can be simulated by Equation (18) with any value of K_{HC} exceeding 1×10^4 M^{-1}. pKa for HCys was taken as 8.50^2. Equation (19) was fitted to the experimental data using K_w'/K_b as a adjustable parameter. In the case of borate and tris.HBr buffers the fit was reasonable when the exchange constants employed were those used previously ($K_{OH/Br} = 0.08$; $K_{OH/Borate} = 2.25$). The best fit value for K_w'/K_b was found to be 8.3×10^9 for borate and 9.1×10^9 for Tris. HBr. Study of the variation of pKap as a function of CTAB in tris.acetate furnished a value of 7.1×10^9 for the parameter K_w'/K_b. In the latter case, the hydroxide/acetate selectivity constant was used as a varying parameter in order to obtain the fit shown in Figure 4B. For this system $K_{OH/Acetate}$ was calculated as 0.1.

The addition of CTAB causes a remarkable increase in the rate of thiolysis of NPO by HCys. The reaction can be conveniently followed down to pH 4.0, the maximum acceleration being ca. 10^6 fold (Figure 5, Table III).

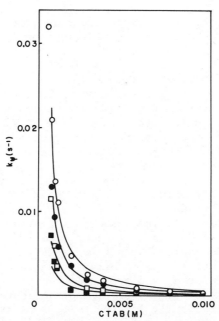

Figure 5. Effect of CTAB on the rate of thiolysis of NPO by HCys. [HCys] = 4×10^{-5}, [NPO] = 4×10^{-6} M. Na acetate buffer 0.01 M pH 5.5 (O), 5.05 (●), 4.55 (□), 4.05 (■).

Table III. Effect of CTAB on the rate of thiolysis of NPO by hexadecylcysteinamide.

pH [a]	k_ψ^{max}/k_ψ^o [b] $(\times 10^{-6})$	k_2^m/k_2^o
4.05	1	13
4.55	0.39	17
5.05	0.14	20
5.50	0.12	23

a. In 0.01 M acetate buffer   b. The value of k_2^o was taken as 6.6 M^{-1} s^{-1}.[30]

The variation of k_ψ with $|CTAB|$ (Figure 5) for the reaction of HCys with NPO was analyzed using Equation (20),

$$k_\psi = HCys \frac{k_2^o + (k_2^m/\overline{V}) K_S K_{HC} C_D}{(1+K_S C_D)\left[(1+K_{HC}C_D) + \frac{H_f}{K_a}(1+ \frac{K_{HC}K_w'K_a C_D}{K_{OH/Br}K_b} \frac{(Br_f)}{(Br_b)_\ell})\right]} \qquad (20)$$

The values of the parameters were those obtained by the analysis of the variation of pKap with $|CTAB|$, except for $(k_2^m/\overline{V})$ that was varied to fit Equation (20) with the data in Figure (5).

The effect of a vesicle prepared with a synthetic amphiphile (DODAC) on the same reaction was also investigated. The rate enhancement produced by DODAC vesicles is almost one order of magnitude larger than that observed in the CTAB, and is dependent upon the nature and concentration of ions present in the solution. (Figure 6, Table IV). The kinetic data were analyzed using Equation (20), and the following parameters $K_{HC} = 8 \times 10^4$ M^{-1}, $K_S = 3 \times 10^4$, $\alpha = 0.15$, $(K_w'/K_b) = 7.1 \times 10^9$ $K_{OH/Cl} = 0.14$, and varying $(k_2^m/\overline{V})$ to best fit. Taking the value of $\overline{V}$ as 0.37 LM^{-1}, the second order rate constant in the vesicle is larger than the value of k_2^o by a factor of 60[28] (Table IV). The maximum rate effects obtained are summarized in Table V. The most remarkable feature of these results lies in the range of the kinetic effect observed for a single reaction type. It is important to note that within this 10^8 fold rate difference, and over all conditions (surfactant type and concentration, thiol, pH, salt), the pseudophase ion exchange model accounts successfully for the effects of amphiphile aggregates on the rate of thiolysis. Although partial failures of the pseudo phase model have been reported,[31a,b] under conditions in which the micelle coverage by the reacting nucleophile is small, the pseudophase model is widely applicable. The comparison between the maximum rate effects and the ratio of second order rate

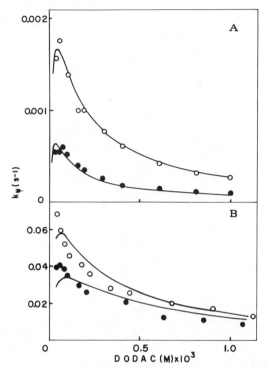

Figure 6. Effect of dioctadecyldimethylammonium vesicles on the rate of thiolysis of NPO by HCys. $[HCys]$ = 4.0 x 10^{-5}, $[NPO]$ = 4 x 10^{-6} M. A. HCl = 2.65 mM, pH 2.65 (O), 5 mM, pH 2.35 (●) B. Na Acetate buffer 0.005 M pH 4.50 (O), 3.75 (●).

Table IV. Effect of DODAC vesicles on the rate of thiolysis of NPO by HCys.

pH	k_Ψ^{max}/k_Ψ^o [c] $(x10^{-6})$	k_2^m/k_2^o
2.35 [a]	3.2	60.6
2.65 [b]	5.2	60.6
3.75 [b]	8.8	53
4.15 [b]	5.8	60.6

a. HCl   b. Acetate buffer 0.005 M c. All k_Ψ^{max} values were obtained at 1 x 10^{-5} M DODAC.

Table V. Maximum rate effects and calculated second order rate
constants for the thiolysis of p-nitrophenol octanoate in the
presence of amphiphile aggregates.

Thiol	Detergent (structure)	k_ψ^{max}/k_ψ^o	k_2^m/k_2^o
HM	SDS (micelle)	0.09	0.27
HM	DDAPS (micelle)	266	1.0
HM	CTAB (micelle)	1.6×10^3	1.2
HCys	CTAB (micelle)	1×10^6	15
HCys	DODAC (vesicle)	9×10^6	50

constants (Table V) indicates unequivocally that the main
contributing factor for the rate enhancements is concentration of
the reagents in the amphiphile pseudophase (s). The quantitative
analysis of the effect of vesicles on the thiolysis (Tables IV and
V, see also Ref. 15) suggests that in this case there may be a
significant difference in the reactivity of the thiol between the
aqueous and vesicular phases. It should be pointed out, however
that there are major problems with the analysis of the data in
the vesticular systems. Firstly it is not clear whether the
reaction is taking place in the outer or inner surfaces of the
vescles[32]. Moreover, although vesivles do exchange ions[15] absolute values for the extent of ion dissociations from the surface (s)
are presently lacking. Thus the quantitative analysis must be
considered tentative at this stage.

CONCLUSIONS

We have presented here the initial part of a systematic study
of the effect of variations of the structure and charge of
amphiphile aggregates on the thiolysis of esters. The main
features of the results are: (i) the kinetic effects range from a
tenfold inhibition by negatively charged micelles to 10^7 fold
rate enhancement by positively charged vesicles, (ii) all these
kinetic effects can be analyzed quantitatively using a pseudophase
model with explicit consideration of ion exchange, and (iii) the

analysis demonstrates that the main factor determining the effect of the amphiphile aggregates on the thiolysis is reagent concentration in the dimensionally restricted environment of the micellar or vesicular pseudophase.

ACKNOWLEDGEMENTS

This work was supported, in part, by grants (to HC) of the Fundação de Amparo a Pesquisa do Estado de São Paulo (FAPESP), FNUD UNESCO (RLA 028/78), CNPq and FINEP.

REFERENCES

1. C.A. Bunton and L.S. Romsted in "The Chemistry of Functional Groups, Suppl. B: The Chemistry of Acid Derivatives", Part 2, p. 945, S. Patai Editor ,J.Wiley & Sons Ltd., New York (1979).
2. N. Funasaki, J. Phys. Chem., 83, 1988 (1979).
3. L.S. Romsted in "Micellization, Solubilization and Microemulsions", Vol. 2, p. 509. K.L. Mittal, Editor, Plenum Press, N.Y., 1977.
4. F.H. Quina and H. Chaimovich, J. Phys. Chem., 83, 1844 (1979).
5. M. Almgren and R. Rydholm, J. Phys. Chem., 83, 360 (1979).
6. H. Chaimovich, R.M.V. Aleixo, I.M. Cuccovia, D. Zanette and F.H. Quina in "Solution Behavior of Surfactants - Theoretical and Applied Aspects" K.L. Mittal and E.J. Fendler, Editors, Vol. 2, pp. 949-973 Plenum Press, New York (1982).
7. a) F.H. Quina, M.J. Politi, I.M. Cuccovia, E. Baumgartem, S.M. Martins-Franchetti and H. Chaimovich, J. Phys. Chem., 84, 361 (1980); b) F.H. Quina, M.J. Politi, I.M. Cuccovia, S.M. Martins-Franchetti and H. Chaimovich in "Solution Behavior of Surfactants-Theoretical and Applied Aspects" K.L. Mittal and E.J. Fendler Editors, Vol. 2 pp. 1125-1136. Plenum Press, New York (1982).
8. I.M. Cuccovia, E. Schroter, P.M. Monteiro and H. Chaimovich, J. Org. Chem., 43, 2248 (1978).
9. C.A. Bunton, L.S. Romsted and H.J. Smith, J. Org. Chem., 43, 4299 (1978).
10. C.A. Bunton, F. Rivera and L. Sepulveda, J. Org. Chem., 43, 1166 (1978).
11. C.A. Bunton, G. Cerichelli, Y. Ihara, and L. Sepulveda, J. Am. Chem. Soc., 101, 2429 (1979).
12. C.A. Bunton, L.S. Romsted and C. Thamavit, J. Am. Chem. Soc., 102, 3900 (1980).
13. J. Epstein, J.J. Kaminski, N. Bodor, R. Enever, J. Sawa and T. Higuchi, J. Org. Chem., 43, 2816 (1978).
14. J.H. Fendler and W.L. Hinze, J. Am. Chem. Soc., 103, 5439 (1981).
15. I.M. Cuccovia, F.H. Quina and H. Chaimovich, Tetrahedron, 38, 917 (1982).
16. I.M. Cuccovia, R.M.V. Aleixo, R.A. Mortara, P. Berci, J.B.S.

Bonilha, F.H. Quina and H. Chaimovich, Tetrahedron Lett. 3065 (1979).

17. J.C. Sheehan and D.H. Yang, J. Am. Chem. Soc., 80, 1158 (1958).
18. G.L. Ellman, Arch. Biochem. Biophys., 82, 70 (1959).
19. a) N. Funasaki J. Phys. Chem., 83, 237 (1979). b) R.J. Williams, J.N. Phillips and K.J. Mysels. Trans. of Faraday Soc., 51, 728 (1955).
20. L.S. Romsted, Ph. D. Thesis, Indiana University, Bloomington, IN (1975).
21. P. Heitman, Europ. J. Biochem., 5, 305 (1968); R.A. Moss, G. Bizzigotti and C.W. Huang, J. Am. Chem. Soc., 102, 754 (1980).
22. This value is comparable to published values for similar mercaptans. D.J. Hupe and W.P. Jencks, J. Am. Chem. Soc., 99, 451 (1977).
23. D.L. Yabroff Ind. Eng. Chem., 32, 257 (1940).
24. K. Martinek., A.P. Osipov, A.K. Yatsimirski and I.V. Berezin, Tetrahedron, 31, 709 (1975).
25. G.S. Hartley, Trans. Faraday Soc., 30, 44 (1934).
26. M.J. Politi, I.M. Cuccovia, D. Zanette and H. Chaimovich Unpublished results; M.J. Politi, Dissertação de Mestrado, Instituto de Química da Universidade de São Paulo, SP, Brasil (1980).
27. J.H. Fendler and E.J. Fendler, "Catalysis in Micellar and Macromolecular Systems", Academic Press N.Y., 1975.
28. A reaction volume element within micelles or vesicles must be chosen to convert the calculated rate constants $(k_2^m/\bar{V})$ to conventional units for comparison with rate constants for the same reaction in water. In this work the values of $\bar{V}$ for micellar DOAPS and vesicular DODAC were taken as 0.25^{24} and 0.37^8 LM^{-1} respectively in order to facilitate comparisons with reactions in SDS and CTAB. The partial molar volume   of aggregated DDAPS and DODAC were 0.31 and 0.81 LM^{-1} respectively (Cuccovia, I.M., unpublished). The choice of a reaction volume element is in fact arbitrary and has been usually taken as either the partial molar volume of the aggregated detergent[8,24] or the calculated volume of the Stern Layer[1]. For a cogent analysis of the importance of the choice and significance of the reaction volume element in the analysis of bimolecular reactions in micelles the reader should consult the article by L.S. Romsted in these proceedings.
29. R.A. Moss, R.C. Nahas and T.J. Lukas, Tetrahedron Lett. 507 (1978).
30. J.R. Whitaker, J. Am. Chem. Soc., 84, 1900 (1962).
31. a) C.A. Bunton, L.S. Romsted and G. Savelli, J. Am. Chem. Soc. 101, 1253 (1979). b) F. Nome, A.F. Rubira, C. Franco, and L.G. Ionescu, J. Phys. Chem. 86, 1881 (1982).
32. R.A. Moss and G.O. Bizzigotti J. Am. Chem. Soc., 103, 6512 (1981).

MICELLE-MEDIATED LUMINESCENCE AND CHROMATOGRAPHY

L. J. Cline Love, Robert Weinberger and Paul Yarmchuk

Seton Hall University
Department of Chemistry
South Orange, New Jersey 07079 U.S.A.

Improved selectivity and competitive sensitivities have been achieved in the analysis of complex mixtures of aromatic species by exploiting micellar solution dynamics. This has been accomplished by coupling micellar chemical effects to the powerful resolving technique of high performance liquid chromatography (HPLC) and the impressive sensitivity of total luminescence measurements. The differentiating power of the total method is greatly enhanced, thanks to the increased dimensionality of the sample data matrix. Using this approach, we have delineated many of the major considerations in using micellar solutions as the mobile phase in HPLC and in their use to stabilize room temperature phosphorescence as the detection mode in the analysis of mixtures of aromatic hydrocarbons, polynuclear aromatics and substituted aromatic compounds. Selectivity is improved because the chromatographic separation provides temporal resolution while the phosphorescence results in spectral differentiation, not all molecules that fluoresce will phosphoresce, and for those that do, the emission is redshifted relative to fluorescence and occurs in a less crowded spectral region. The requisite micellar reagents can be introduced into the system either as the mobile phase or as post-column reagents. Linear dynamic ranges covering three orders of magnitude and limits of detection as low as 5 ng were obtained for several compounds with a precision of <2% relative standard deviation.

1139

INTRODUCTION

There is little doubt of the enormous impact of micelles in areas such as organic reaction kinetics[1-5], catalysis[6,7], and biochemical membrane modeling[8,9]. However, they have had few applications in basic or applied analytical chemistry research[10-13], perhaps reflecting the emphasis on instrumentation and automation to solve chemical problems in recent years. The impressive ability of micelles to manipulate the microenvironment experienced by an analyte[5,8,14-17] augurs well for a reexamination of how solution dynamics can be used to achieve new limits of selectivity and sensitivity in analysis. The heterogeneous chemical microenvironments provided by micelles influence excited-state equilibria and partitioning equilibria by imposing additional constraints and pathways on the molecule's behavior. Using these effects, we have developed a new chromatographic separation/phosphorescence detection system, using favorable micelle-mediated interactions, for use in the analysis of complex mixtures.

The search for better measurement schemes has led to development of techniques combining a separation step with spectroscopic detection[18]. This paper describes specifically the combination of high performance liquid chromatography (HPLC) separation with phosphorescence spectroscopic detection. For micelles to be useful in this type of chemical analysis, the analyte must have an appreciable residence time (inverse of the exit rate, k_-, of the analyte molecule from the micelle) to allow the unique properties of the micelle to have an effect. For a mixture of analyte molecules with differing polarities, we would expect a range of residence times producing selectivity and differentiation. However, these preferential interactions are only useful if they also produce variations in the instrumental observables. In other words, the entrance and exit rates of the analyte from the micelle will overlay the simultaneous spectroscopic rates and chromatographic rates the analyte is experiencing, and the resultant effect may or may not be analytically useful. For enhanced analysis, we must match an analyte with the proper surfactant and bulk solvent to obtain the optimum k_+/k_- ratio. We will see later that if this ratio is too large, chromatographic retention will decrease, and if it is too small, appreciable heavy atom-induced spin-orbit coupling will not occur and phosphorescence intensity will diminish.

Because of recent developments in solution chemistry, it is now possible to conveniently observe phosphorescence from species contained in micellar chromatography effluent.[19-26] Several improvements in the reliability of the phosphorescence measurement are described which are a direct result of coupling the detector to a chromatographic sample introduction device.

EXPERIMENTAL

The HPLC system uses a Constametric III pump and UV III monitor (Laboratory Data Control), and a Model 7120 sample injector with a 20 μL loop (Rheodyne, Inc.). The column was packed with 5 micron Supelcosil LC-1 and was 15 cm x 4.6 nm i.d. (Supelco, Inc.). The cationic and anionic surfactants were restricted to different columns and the void volume of the system calculated using the displacement peak caused by actuation of the sampling valve. The void volumes for sodium dodecyl sulfate (SDS) and dodecyltrimethylammonium bromide (DTAB) were 0.94 mL and 0.96 mL, respectively. Column temperature was held at 298 K by a circulating heater.

The HPLC system used for the phosphorescence detection studies consisted of modular components of a Technicon FAST·LC which was connected as required by the different experimental modes. These included a pneumatic injector with a six-port valve and 20 μL loop, high-pressure LC pumps, a 254 nm UV detector, stripchart recorder, and system controller. Luminescence measurements were performed with a Schoeffel FS 970 HPLC detector. A Laboratory Data Control HPLC pump was added when a three-pump configuration was required. Propyl-cyano and C8 bonded phase columns (15 cm x 4.6 mm i.d., 5 μm particle size) (Supelco, Inc.) were used for all studies. To facilitate studies on surfactant concentration effects. the Na/Tl ratio in the mobile phase and postcolumn reagent was varied using a two- or three- pump configuration, respectively. For example, to vary the Na/Tl ratio of a micellar mobile phase, the outputs of separate pumps delivering 0.15 M SDS and 0.15 M SDS/TlDS (70/30) were joined with a zero dead volume tee (Valco). The effluent from the tee was then hooked directly to the sampling valve. The pumps were calibrated to deliver ±3% of their stated flow rates. For postcolumn studies, one pump provided for mobile phase flow through the column while two other pumps were connected as described above except the output was connected to mobile phase effluent postcolumn with another zero dead volume union. For most studies, the excitation wavelength used was 270 nm, with a 550 nm cutoff filter to select the emission wavelengths.

The SDS was obtained from BDH Chemicals, Ltd., and was recrystallized twice from 50% aqueous methanol. The DTAB (Fisher Scientific) was recrystallized twice from acetone/chloroform. Water was deionized, continuously distilled and batch distilled two times. Benzene, toluene, nitrobenzene, phenol, methanol and 2-naphthol were obtained from Fisher Scientific Co. p-Nitrophenol, p-nitroaniline, biphenyl, and phenanthrene (MC/B) were used as received. Naphthalene (J. T. Baker) was used as received. Benzylamine (Eastman) was redistilled. High-purity nitrogen was further purified by passing the gas through an indicating oxygen trap (Alltech Assoc.).

The micellar chromatography mobile phases were prepared by dissolving the appropriate quantity of surfactant in water and filtering through a 0.5 μm cellulosic membrane filter (Rainin Instru-Co., Inc.). The pH for SDS was 6.0–6.4 and for DTAB 5.4–5.8, depending on surfactant concentration. Stock solutions of the test solutes were prepared in methanol and then diluted to the appropriate working concentration with 0.10 M SDS or DTAB. The working concentrations are: phenol (52 μg/mL), p-nitrophenol (60 μg/mL), p-nitroaniline (17 μg/mL), benzene (380 μg/mL), nitrobenzene (9.0 μg/mL), 2-naphthol (31 μg/mL), toluene (330 μg/mL), and benzylamine (450 μg/mL). Individual solutes were chromatographed using a flow rate of 2.0 mL/min, and their retention times and peak widths measured manually.

For the luminescence/chromatography studies, the glassware used for preparing the mobile phases and postcolumn reagents was soaked in 0.15 M SDS overnight followed by rinses with water, methanol, water, and methanol followed by heating at 105 °C until dry. For the mobile phase and postcolumn reagent preparation, it was first necessary to degas one liter of 3 times distilled water under vacuum for 10 min. To prepare a 0.15 M micellar mobile phase with a 70/30 Na/Tl ratio, 30.28 g of SDS and 21.14 g of TlDS were added and dissolved with stirring and gentle heat. The solution was further degassed under mild vacuum taking care not to induce excessive foaming. The mobile phase was generally not filtered unless particulates were noted. Residual oxygen was scrubbed by bubbling high-purity nitrogen through the solution, and the nitrogen flow was maintained throughout the analyses. The 0.15 M SDS was prepared as described above using the appropriate weight of SDS and omitting the TlDS. Conventional HPLC mobile phases of 50% and 75% methanol were prepared by volume, degassed, and nitrogen purged. The mobile phase flow rates were varied as required by the mode of separation and postcolumn studies. Stock solutions of 10 mg/mL of each solute were prepared in methanol. Serial dilutions were performed with the appropriate mobile phase to obtain working solutions down to 50 ng/mL. Repetitive injections of solutes were made until the rate of increase of signal height leveled or oscillated randomly. This was necessary to ensure the removal of residual oxygen from the mobile phase and postcolumn reagents.

RESULTS AND DISCUSSION

Description of our recent studies using micellar solution dynamics to enhance analytical measurements is most conveniently done by dividing the results into two sections. First, a discussion of the effects of micelle type and concentration on chromatographic characteristics will be given, followed by phosphorescence detector results.

Micellar Chromatography

The use of micellar phases represents the first significant advance in mobile phase research since gradient elution [27] and ion-pair chromatography[28]. We have further characterized the micellar aggregate's ability to mimic mixed solvents[29-32], and investigated triple partitioning of solutes between the micelles-bulk solvent-stationary phases using the theory proposed by Armstrong and Nome.[12] Micellar chromatography owes its uniqueness to this triple equilibrium consisting of three partition coefficients: K_{wm}, bulk water to micelle; K_{ws}, bulk water to column stationary phase; and K_{ms}, micelle to column stationary phase. The relative influence of these separate effects can be used to explain the electrostatic and hydrophobic effects which can produce enhanced selectivity for separations of interactive solutes.

Surfactant Concentration Effects. The anionic surfactant, SDS, and the cationic surfactant, DTAB, both have a C-12 hydrophobic moiety and similar critical micelle concentrations (CMC) of 0.008 M [33] and 0.016 M[34], respectively. However, they do differ in the polar head group, which results in quite different electrostatic interactions with the test solutes to be described in the next section. As the number of micelles in the mobile phase increases, there is a resultant decrease in retention in a manner analogous to what is seen with increasing mobile phase modifier concentration in conventional reversed phase chromatography. However, the data in Figure 1 shows that the linear plots of log capacity factor, k', versus log surfactant concentration are not parallel, but intersect one another for some compounds. This change in separation factor, α, with surfactant concentration can be a very powerful mechanism by which retention times can be adjusted and the desired selectivity achieved. Thus, by varying the SDS concentration, shown in Figure 1, the relative retentions of nitrobenzene, toluene and 2-naphthol can be adjusted by working at SDS levels above or below the crossing points. Similar results, although with different crossings of plots, were observed for the same compounds using varying concentration of DTAB micelles. As will be shown, the partition coefficient, K_{wm}, causes the divergence or convergence of the retention plots for different solutes, and is one controlling parameter of selectivity via surfactant concentration.

Figure 2 contains two chromatograms of a mixture of three test solutes run with SDS mobile phases differing in concentration by a factor of ten. As can be seen, the retention order of the three solutes are reversed by changing the SDS concentration from 0.02 M to 0.2 M. These data also show that improved separation is achieved at the lower SDS concentration, but the time for the analysis is increased from <5 min to >20 min. Similar, but different crossings, were observed for these test compounds when using DTAB micellar mobile phases.

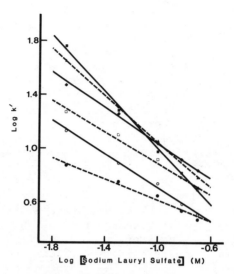

Figure 1. Effect of SDS concentration on the retention of - ● -
phenol, - ○ - p-nitroaniline, - □ - benzene, - ◆ - nitrobenzene,
- ▲ - toluene, - ■ - 2-naphthol.

Unlike conventional reversed phase chromatography mobile phases,
the micellar mobile phase is microscopically heterogeneous being composed of the bulk aqueous solvent and the micellar aggregates. As a
result, additional equilibria involving the solute must be considered
in the interpretation of the results. Both K_{ws} and K_{wm}, describing
the partitioning of the solute from the water to the column stationary
phase and to the micelle, respectively, affect retention. However,
their effects are entirely opposite. As K_{wm} increases, retention
decreases due to increased partitioning into the micelle, but as K_{ws}
increases, the retention increases.

The magnitudes of the K_{ws} and K_{wm} partition coefficients were
estimated using a proposed equilibrium theory[12] and used to explain
trends in the data observed for the six test solutes given in Figure 1. The ratio $V_s/(V_e - V_m)$ was calculated for each compound from
the retention data, where V_s is the volume of the stationary phase,
V_e is the elution volume of the solute, and V_m is the volume of the
mobile phase. V_s was taken to be the total column volume minus V_m,

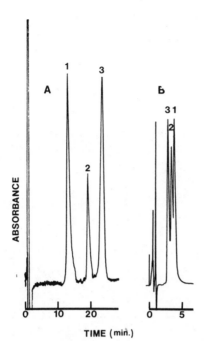

TIME (min.)

Figure 2. Micellar chromatograms of (1) nitrobenzene, (2) toluene, (3) 2-naphthol; mobile phases: (A) 0.02 M SDS, (B) 0.20 M SDS: detector sensitivity: (A) 0.004 AUFS, (B) 0.020 AUFS.

where V_m is the void volume. This ratio was then plotted versus the surfactant concentration in micelles (total surfactant concentration minus CMC). The values of K_{ws} and K_{wm} can then be calculated from the slope and intercept obtained from the plots for each compound. Unfortunately, this approach produces large errors in the two partition coefficients for compounds with long retention times (small intercepts). This occurs because K_{ws} is the reciprocal of the intercept and K_{ws} is used to calculate K_{wm}. The values calculated by this method for the seven test solutes are given in Table I.

The solutes are listed in Table I in their elution order using a 0.02 M SDS mobile phase. Increasing values of K_{ws} produce increased retention, in a manner similar to conventional reversed phase chromatography. One would expect this relationship between K_{ws} and retention to be unaffected by surfactant (micelle) concentration, as long as the column stationary phase has been equilibrated at surfactant concentrations above the CMC, since the concentration of absorbed SDS or DTAB above the CMC is constant[35]. On the other hand, the magnitude of K_{wm} is affected by the concentration of surfactant

Table I. Calculated Equilibrium Constants of Test Compounds in SDS and DTAB Micellar Systems.

Compound	K_{ws}	K_{wm}[a]	K_{ws}	K_{wm}[a]
Phenol	5.0	39	23	111
p-nitrophenol	8.7	77	99	305
p-nitroaniline	9.4	84	39	192
benzene	13.8	77	18	80
nitrobenzene	22.4	90	23	89
toluene	41.3	215	47	188
2-naphthol	59.5	393	235	875

[a] The partial specific volume, v, must be known in order to calculate the values of K_{wm}. A value of v=0.929 mL/g has been determined for DTAB by Guveli, et. al[34] A value of v=0.862 mL/g was determined by Mukerjee for SDS.[36] The K_{wm} values are per surfactant molecule.

due to the increase in the number of micelles in the mobile phase. This results in K_{wm} having a more predominant role in governing the solute equilibria as the surfactant concentration is increased. Indeed, the larger the value of K_{wm}, the greater the effect of increasing concentration of surfactant, as evidenced by the steeper slopes in the log-log plots of k' versus SDS concentration shown in Figure 1. Thus, with compounds with large K_{wm}, a greater decrease in retention time is obtained with increasing surfactant concentration compared to compounds with lower K_{wm}. This effect leads to retention order reversals with changing surfactant concentration. This will occur for compounds with similar K_{ws} values but different K_{wm} values, such as 2-naphthol, nitrobenzene or toluene. In contrast, for compounds with large differences in their K_{ws} values (phenol and 2-naphthol), extremely large concentrations of surfactant are required for reversal of elution order to occur. This would be impractical in some cases, and not truly comparable because the nature of the micelle can change at very high surfactant concentrations. It should be noted that the K_{ws} values given in Table I are in agreement with what would be expected based on their lipophilicity. Similar relationships are seen using the cationic DTAB micellar mobile phases.

Electrostatic Interaction Effects. Selectivity can also be achieved for polar solutes by changing the charge type of the surfactant head group. Since an appreciable amount of surfactant is adsorbed onto the nonpolar column stationary phase, both hydrophobic and electrostatic interactions can occur with the stationary phase and with the micelle, especially for ionizable molecules.

First, consider the case of a solute and surfactant of like charge, such as an anionic surfactant and an acidic solute. Repulsive interactions between the solute and the micelle should not affect retention appreciably since the solute would reside a large portion of the time in the bulk polar mobile phase and, thus, still move down the column. Repulsion between the solute and the stationary phase would result in a decrease in retention, however. This repulsion would be somewhat counterbalanced because a hydrophobic interaction between the stationary phase and the solute is also possible. This can be seen for two acidic solutes (2-naphthol and phenol) using like-charged anionic SDS mobile phase, in Figure 1. They are well separated due to retention on the column stationary phase by hydrophobic interactions. These same effects are also observed for cationic surfactants, such as DTAB, and cationic solutes, such as benzylamine, where the latter is completely unretained on the column due to repulsive effects.

A second case involves the combination of a solute with an oppositely charged surfactant, where considerable electrostatic attraction between the solute and the surfactant is possible. In the micelle, the electrostatic interaction would reinforce the hydrophobic interaction, resulting in the molecules spending more time in the mobile phase and a decrease in retention. However, the situation is more complex than this simple reasoning suggests. Experimentally, an increase in retention is observed for phenol in going from SDS to DTAB (3.1 min versus 8.9 min). A similar increase is seen for p-nitrophenol (4.1 min versus 18.0 min in SDS and DTAB, respectively). The most probable explanation of this trend can be put forth by considering the electrostatic interactions between the surfactant adsorbed on the stationary phase and the solute. If one considers classic ion-interaction chromatography which uses surfactants in the mobile phase, but below the CMC, two types of interactions are possible. A straight forward ion-exchange interaction can take place where a solute in the bulk mobile phase would be attracted to the oppositely charged surfactant adsorbed on the stationary phase, thus increasing retention. A second type of interaction involves formation of an ion-pair between the solute and a monomer in the bulk solution, which then interacts hydrophobically with the stationary phase. Both of these models would increase retention, and both are probably occuring. The net effect of the combined electrostatic attractions and hydrophobic attractions of the solute and the stationary phase is to cause the solute to be retained on the column longer, thus, effectively competing with the opposite charge attraction between the solute and the micelle.

An experiment where both of these cases are operative involves the separation of p-nitroaniline and p-nitrophenol. When this cationic solute and anionic solute were chromatographed with the anionic SDS mobile phase, no separation was achieved at any concentration of SDS. The admixture of electrostatic and hydrophobic effects from the micelle and the modified stationary phase were balanced for each solute. Complete separation was achieved by going to the cationic DTAB system, where p-nitrophenol's retention is greatly increased.

A third case involving nonpolar solutes such as benzene and toluene illustrates the sole effect of hydrophobic interactions and the absence of electrostatic interactions. Both SDS and DTAB have the same hydrocarbon moiety (C-12) and should produce similar retention times for nonpolar species. This was observed experimentally, with only slightly longer retrntions measured in DTAB. Examination of the values of K_{ws} and K_{wm} for compounds chromatographed in SDS and DTAB reveals they are virtually the same between surfactants, and supports the evidence that retention is not affected by the nature of the head groups for nonpolar solutes.

One would expect that for real samples containing both polar and nonpolar compounds, effective separation could be achieved by the correct choice of surfactant since nonpolar compounds are virtually unaffected by electrostatic effects. Figure 3 illustrates this point. Using a 0.05 M SDS micellar mobile phase, 2-naphthol elutes between nitrobenzene and toluene, with extremely poor resolution. Using a 0.05 M DTAB micellar mobile phase, 2-naphthol elutes considerably after the other two compounds, producing excellent resolution.

Efficiency of Micellar Chromatography. Examination of the preceding chromatograms shows increased band broadening over what is normally expected for conventional reverse phase chromatography. The decline in chromatographic efficiency in micellar chromatography is a consequence of increased resistance to mass transfer of the solute within the mobile phase and the stationary phase. Since mass transfer is a function of the solute's exit rate from the micelle, mobile phase viscosity, stationary phase surface area, pore size and other features, steps can be taken to optimize the performance of micellar chromatography. Since the columns employed in these studies represent state-of-the-art technology, little can be expected from alteration of column parameters.

Two areas were explored to determine their effect on chromatographic efficiency. First, small amounts of conventional mobile phase modifiers, e.g. methanol, were added to the

micellar mobile phases. As expected, the efficiency improved somewhat but this approach was abandoned for two reasons. Part of the
elegance of micellar chromatography is that it allows chromatography
without nonaqueous solvents. More importantly, even small amounts of
these solvents affect the micelle integrity. Although this may improve the chromatography by invoking a hybrid system combining features of micellar and reverse phase chromatography, these solvents
will result in a decline in phosphorescence intensity. This feature
is undesirable since our ultimate goal is a sensitive and selective
system employing micellar chromatography with phosphorescence
detection.

The second area investigated was the temperature effects on
chromatographic efficiency. Since mass transfer from the micelle to
the stationary phase depends greatly upon the solute's exit rate from
the micelle, increased temperature would be expected to have a
favorable impact on chromatographic efficiency. The results (Table
II) support that argument. Substantial improvement in band widths is
seen for the compounds that had the worst efficiency. These are the
higher molecular weight multi-ring compounds like anthracene where
the efficiency is improved four-fold at 70°C over that at 25 °C. Most
importantly, the 1700 plate peaks obtained are comparable to that of
reverse phase chromatography. Therefore we can enjoy the advantages
of micellar chromatography without suffering its potential liabili-
ties.

Table II. Effect of Temperature on Micellar Chromatography Efficiency
Number of Theoretical Plates per Column, N^a

Temp, K	Phenol	Benzene	2-Naphthol	Naphthalene	Anthracene
298	2300	3000	750	1200	460
303	2800	3400	1000	1300	530
313	3200	3800	1300	1700	860
323	4100	4400	1900	2300	1200
333	4400	5000	2400	3000	1500
343	4100	4900	2700	3100	1700

[a] Chromatographic conditions: Mobile Phase, 0.10 M DTAB; Column,
LC-1 (15 cm, 4.6 mm I.D.), (Supelco, Inc.); Flow Rate, 2.0 mL/
min.; Detector, UV at 254 nm; $N = 5.54 (t_r/w_{1/2})^2$ where t_r =
retention time of solute and $w_{1/2}$ = peak width at half height.

The relatively new technique of micellar chromatography offers a unique multi-mechanistic approach to the general separation problem. Contrasted to conventional mobile phases which are homogeneous, micellar mobile phases are heterogeneous consisting of the micellar aggregate and the surrounding bulk water. The multiple equilibria arising from this coupled with the wide variety of surfactants available including ionic, nonionic and zwitterionic will allow many unique separations and applications to be developed.

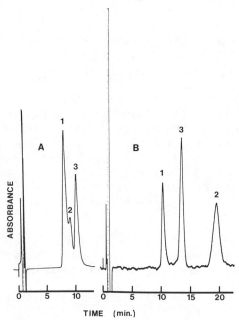

TIME (min.)

Figure 3. Anionic and cationic micellar chromatograms of (1) nitrobenzene, (2) 2-naphthol, and (3) toluene; mobile phases: (A) 0.05 M SDS, (B) 0.05 M DTAB; detector sensitivity: (A) 0.02 AUFS, (B) 0.01 AUFS.

Phosphorescence Detection for HPLC

The ability of micellar aggregates to organize reactants on a molecular basis and to form a somewhat protective environment has been exploited in recent years to study the photophysics of the triplet state of aromatic molecules in fluid solution at room temperature.[19,23-26] These fundamental studies led to the development of a new method of analysis based on micelle stabilized room temperature phosphorescence (MS-RTP).[20-22] Previous studies of phosphorescence at room temperature required immobilizing the sample, making flow systems impossible.[37-41] The unique nature of micellar solutions now

permits observation of phosphorescence using a conventional fluoro-
meter and cuvettes or micro flowcells, since fluid solutions at room
temperature are used. We have taken advantage of this instrumental
simplification to develop a new spectroscopic detector for use in
high performance liquid chromatography. A description of its characteristics, advantages and limitations is given below.

Incorporation of a chromatographic column prior to the phosphorescence detector proved to be quite advantageous. Prior to this
addition, conventional static measurements of MS-RTP were labor intensive, subject to variability ($\sim$6%) from the incomplete and inconsis-
tant removable of oxygen (a triplet state quencher) from solution,
and required extensive sample purification. Imprecision also resulted from variable illumination time when performing measurements. The
intrinsic precision of HPLC can minimize all of these sources of
error. An in situ, on-column purification allows the detection of a
solute, unencumbered by quenchers, including oxygen, and illumination
time is held constant. Using this approach, relative standard deviations of the phosphorescence signal are reduced to <2% for replicate
measurements. Sample throughput is now dependent on chromatographic
time and not on chemical preparation. Thus, it is no longer necessary to deaerate samples for phosphorescence measurements when using
the routinely deaerated, oxygen-free HPLC mobile phases, thereby
reducing sample preparation/measurement time from 45 minutes to <5
minutes.

Characterization of the Phosphorescence Signal. Verification
of the source of the photometric signal (fluorescence, phosphorescence, scattered light, or a combination) requires clever experimental design when using a HPLC flow cell. Normal phosphorimeters
permit scanning of the spectrum of the sample to allow identification
of the emission based on energy/intensity profiles. This is not
possible with most chromatographic fluorescence detectors which use
cutoff filters to pass a broad band of radiation. Additionally, some
sort of temporal discrimination is normally used to select an observation time window when the short-lived fluorescence and scatter has
decayed to zero, but the long-lived phosphorescence is still present.
Since no time delay was available on the detector used for this study,
it was necessary to devise other means of determining the nature of
the observed signal.

Three approaches can be used to discriminate between the various
types of signals. These rely on the fact that the excited triplet
state of a molecule is difficult to populate, especially for those
with large fluorescence quantum yields, and, once achieved, it is
very easy to deactivate it by radiationless processes. The first
approach is elimination of the heavy atom (usually Tl or Ag) as a
counterion in the micelle. The presence of a heavy atom is required
for generation of any appreciable amount of phosphorescence in fluid

solution at room temperature. Without the heavy atom, there is usually insufficient spin-orbit coupling which is necessary for intersystem crossing from the excited singlet state to the triplet state. Thus, the triplet state is not populated and phosphorescence cannot occur. If a signal is still observed in the absence of a heavy atom, it can be ascribed to fluorescence or scatter. An exception would be if the molecule contains a heavy atom substituent.

A second approach involves oxygenating the sample which serves to introduce a quencher of the triplet state phosphorescence. If a signal is present in the presence of oxygen it can be safely designated as being fluorescence or scatter. This is accomplished experimentally by simply oxygenating the chromatographic mobile phase reservoir. These two methods of signal discrimination are illustrated in Figure 4. The micellar chromatogram (Figure 4A) of 2-naphthol, biphenyl and phenanthrene was obtained using a thallium-free, oxygenated SDS mobile phase. The signals observed are due to fluorescence passed by a 418 nm lower cutoff emission filter in the detector. Upon substitution of 30% of the sodium counterions in the SDS micelle with thallium, one observes a similar chromatogram, shown in Figure 4B, but with altered and diminished relative intensities. These signals also arise from fluorescence, and are diminished in intensity compared to Figure 4A because thallium is effectively depopulating the excited singlet state from which fluorescence occurs by intersystem crossing to the triplet state. Thus, Figure 4A describes a sample in which virtually no triplet states are populated and Figure 4B describes a sample in which the triplet state is appreciably populated but is radiationlessly quenched by oxygen. Figure 4C was obtained with an oxygen-free, SDS/TlDS micellar mobile phase, and represents phosphorescence emission from the three analytes. Some fluorescence may be adding to the signal, depending on the wavelength cutoff filter used.

The two approaches to signal identification described above are for use with micellar chromatography with phosphorescence detection. However, if one does not wish to employ micellar chromatography, but is interested in studying phosphorescent molecules in a complex mixture, the requisite micellar reagents can be added via an additional pump post-column. This would allow the use of well-characterized and efficient reverse phase separations (no micelles present) and selective MS-RTP detection. Some results using the post-column reaction (PCR) mode phosphorescence detector (Figure 5) are given for the same test solutes used for the data in Figure 4. A 50% methanol/water mobile phase with a conventional reversed phase chromatographic column was used to obtain the data in Figure 5A. The detector emission filter had a 550 nm cutoff which removed most of the fluorescence which was occurring at shorter wavelengths. The signal is due to fluorescence only because no micelles or heavy atoms were present (PCR was off). Figure 5B illustrates the effect

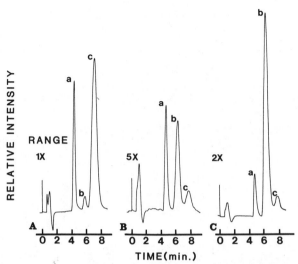

Figure 4. Micellar chromatograms: column, Supelco CN; flow rate, 2 mL/min; excitation wavelength, 250 nm; emission filter, 418 nm. Solutes: (a) 2-naphthol, 800 ng; (b) biphenyl 16 μg; (c) phenanthrene, 400 ng. Mobile phases: (A) oxygenated 0.1 M SDS, (B) oxygenated 0.1 M SDS/TlLS(70/30), (C) oxygen-free version of (B). Full scale range expansions are so noted on each chromatogram.

of adding the heavy atom substituted micelles via turning on the PCR pump. The signals observed for biphenyl and phenanthrene are phosphorescence, while that observed for 2-naphthol is a combination of phosphorescence and fluorescence. The complex nature of the signal requires careful verification to positively identify it as phosphorescence emission. Clearly, this was not done in a recent publication of a purported HPLC phosphorescence detector[32].

Surfactant Concentration Effects. Since the surfactant micelle is an integral part of the chromatography and spectroscopic detection systems, it is important to understand the effect of varying the surfactant concentration. As we saw in Figure 1, as the surfactant concentration decreases, retention times always increase. The effect this will have on the shape of the chromatographic peak is band broadening and reduced peak height. This manifests itself in the spectroscopy as diminished signal. An optimum surfactant concentration is one which gives adequate resolution of peaks,

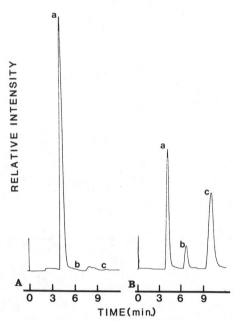

Figure 5. Reversed phase chromatograms without the PCR activated
(curve A) and with the PCR activated (curve B):   column, Supelco CN;
mobile phase, 50% methanol/water; flow rate, 1 mL/min; PCR flow
rate, 3 mL/min; solutes, 10 μg each of (a) 2-naphthol, (b) biphenyl,
and (c) phenanthrene.

minimal band broadening (maximum peak height for photometric response)
and stable micelles for production of phosphorescence. From 0.15 M
to 0.05 M sodium/thallium (70/30) dodecyl sulfate, the phosphorescence signal remains constant in the PCR mode. The import of this
is that there is substantial latitude in micellar chromatography to
optimize the separation without affecting the phosphorescence signal.

    Effect of Heavy Atom Concentration. The spin-orbit coupling
induced by the presence of heavy atoms is essential for observation
of phosphorescence in fluid solution at ambient temperature. Both
the $S_1 \to T_1$ and $S_0 \to T_1$ transitions are spin forbidden and will not
occur to any significant extent for many molecules unless there is
some mixing of states which relaxes selection rule requirements.
Silver and thallium are both effective in promoting these types
of transitions and have been used extensively for generation of
MS-RTP. The effect of the percent thallium counterions in the
micelle is shown in Figure 6. Unfortunately, an experimental

difficulty prevents optimization of heavy atom concentration. In the
case of thallium, one is limited to studies of 0-30% thallium, above
which precipitation of the soap occurs. This has been found true
with several other heavy atom micelles, e.g. Cs, Hg, where no
significant solubility exists in aqueous solution. It does appear
that the integrity of the micellar assembly is not being adversely
affected by the incorporation of the thallium atoms into the counterion layer. Determination of the CMC's of the Ag/NaDS and Tl/NaDS
micelles gave similar values to that obtained for the NaDS micelle.[42]
Interestingly, it is generally true that silver produces more intense
MS-RTP with heterocyclics, and thallium is best for carbocyclics.

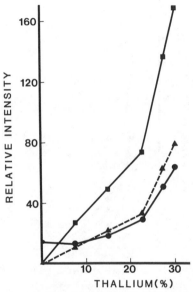

Figure 6. Effect of thallium/sodium ratio in mixed counterion
Na/Tl dodecyl sulfate micelles on MS-RTP intensity for micellar
chromatography: (-●-) 2-naphthol, (-▲-) phenanthrene, and
(-■-) naphthalene; column, Supelco CN; mobile phase, 0.15 M
total surfactant.

Effect of Methanol Concentration. In the post-column reaction
phosphorescence detector, organic mobile phase modifiers are introduced from the chromatographic column. Conventional reversed phase
chromatography uses a blend of organic/aqueous mobile phases, often
methanol/water or acetonitrile/water, to effect separation, and the
PCR pump is used to mix the micelle reagents with the mobile phase/

solute stream. The effect of methanol on the phosphorescence intensity using the PCR phosphorescence mode is shown in Figure 7. The
post-column micellar reagent is acting as a diluent for the methanol
in the mobile phase as well as the solute. It is apparent from the
data that its effect on the methanol is greater than its effect on
the solute, and intense phosphorescence is observed from apparently
stable micelles at methanol concentration less than 10%. The data
in Figure 7 was corrected for this dilution effect, such that the
% methanol axis represents actual concentrations. Other data on the
effect of PCR flow rates show that the greater the PCR flow rate,
the more sensitive the detection technique. The data given in Figure
7 were generated by increasing the PCR flow rate incrementally which
served to dilute the chromatographic effluent by known amounts, thus
allowing calculation of the precise amount of methanol in the
detector flow cell. No significant band broadening was observed.

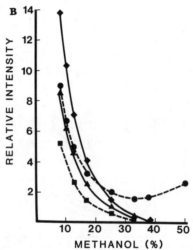

Figure 7. Effect of percent methanol on the MS–RTP intensity using
the post-column reaction mode: mobile phase: 50% methanol; flow
rate, 1 mL/min; (- ● -) 2-naphthol, (- ◆ -) biphenyl, (- ▲ -)
phenanthrene, (- ■ -) naphthalene.

        Analytical Utility. The addition of chromatography phosphor-
escence detection to the arsenal of analysis techniques makes
available a means of obtaining selectivity in the analysis of complex
mixtures. The sensitivity is competitive with fluorescence detection
for many compounds, being generally in the low nanogram range for
limits of detection. Figure 4 is a good example of how selectivity
may be achieved by proper

manipulation of spectroscopic variables. In Figure 4C, the phosphorescence response for biphenyl is enhanced 30 times and 170 times over 2-naphthol and phenanthrene, respectively, as compared to their relative response using fluorescence (Figure 4A). This relative response enhancement for biphenyl could be further increased by using a longer wavelength cutoff filter which would more efficiently discriminate against unwanted signals.

ACKNOWLEDGEMENTS

Support of this work by the National Institutes of Health Grant No. GM-27350 and the National Science Foundation Grant No. PRM-8111335 is gratefully acknowledged. The authors also thank Technicon, Inc. for the loan of some of the chromatographic equipment, and Supelco, Inc., for aid in obtaining some of the columns and Kratos, Inc. for providing the HPLC fluorescence detector. We also acknowledge R. F. Hirsch for input to some of the chromatographic studies performed. Figures 4-7 are reprinted with permission from reference 43. Table I, Figures 1-3 are reprinted from "Selectivity in High Performance Liquid Chromatography Via Micellar Mobile Phases", in press, Analytical Chemistry, 1982, ref. 44. Copyright 1982 American Chemical Society.

REFERENCES

1. E. H. Cordes, Pure Appl. Chem., 50, 617 (1978).
2. F. M. Menger, Pure Appl. Chem., 51, 999 (1979).
3. J. H. Fendler and W. L. Hinze, J. Am. Chem. Soc., 103, 5439 (1981).
4. J. K. Thomas, Chem. Rev., 80, 283 (1980).
5. J. H. Fendler and E. J. Fendler, "Catalysis in Micellar and Macromolecular Systems", Academic Press, New York, 1975.
6. C. A. Bunton, Pure Appl. Chem., 49, 969 (1977).
7. C. A. Bunton, in "Solution Chemistry of Surfactants", K. L. Mittal, Editor, Vol. 2, 519-540, Plenum Press, New York, 1979.
8. C. Tanford, "The Hydrophobic Effect", 2nd ed., John Wiley & Sons, New York, 1980.
9. J. H. Fendler, Acc. Chem. Res., 13, 7 (1980).
10. W. L. Hinze, in Solution Chemistry of Surfactants", K. L. Mittal, Editor, Vol. 1, 79-127, Plenum Press, New York, 1979.
11. M. Kodama and S. Miyagawa, Anal. Chem., 52, 2358 (1980).
12. D. W. Armstrong and F. Nome, Anal. Chem., 53, 1662 (1981).
13. G. L. McIntire and H. N. Blount, in "Solution Behavior of Surfactants--Theoretical and Applied Aspects", K. L. Mittal and E. J. Fendler, Editors, Vol. 2, Plenum Press, New York, 1982.
14. F. M. Menger, Acc. Chem. Res, 12, 111 (1979).
15. J. K. Thomas, Acc. Chem. Res, 10, 133 (1977).

16. L. R. Fisher and D. G. Oakenfull, Chem. Soc. Rev., 6, 25 (1977).
17. N. J. Turro, M. Gratzel, and A. M. Braun, Angew. Chem. Int. Ed. Engl., 19, 675 (1980).
18. T. Hirschfeld, Anal. Chem., 52, 297A (1980).
19. M. Almgren, F. Grieser, and J. K. Thomas, J. Am. Chem. Soc., 101, 279 (1979).
20. L. J. Cline Love, M. Skrilec, and J. G. Habarta, Anal. Chem., 52, 754 (1980).
21. L. J. Cline Love, J. G. Habarta, and M. Skrilec, Anal. Chem., 53, 437 (1981).
22. L. J. Cline Love and M. Skrilec, Anal. Chem., 53, 1872 (1981).
23. L. J. Cline Love and M. Skrilec, Anal. Chem., 52, 1559 (1980).
24. R. Humphry-Baker, Y. Moroi, and M. Gratzel, Chem. Phys. Lett., 58, 207 (1978).
25. N. J. Turro, K. C. Liu, M. F. Chow, and P. Lee, Photochem. Photobiol., 27, 523 (1978).
26. M. Almgren, F. Grieser, and J. K. Thomas, J. Am. Chem. Soc., 101, 2021 (1979).
27. L. R. Snyder and J. J. Kirkland, "Introduction to Modern Liquid Chromatography", 2nd ed., John Wiley & Sons, New York, 1979.
28. C. Horvath, W. Melander, I. Molnar, and P. Molnar, Anal. Chem., 49, 2295 (1977).
29. D. W. Armstrong and R. Q. Terrill, Anal. Chem., 51, 2160 (1979).
30. D. W. Armstrong and M. NcNeely, Anal. Lett., 12, 1285 (1979).
31. D. W. Armstrong and S. J. Henry, J. Liq. Chromatogr., 3, 657 (1980)
32. D. W. Armstrong, W. L. Hinze, K. H. Bui and H. N. Singh, Anal. Lett., 14, 1659 (1981).
33. P. H. Elworthy and K. J. Mysels, J. Colloid Interface Sci., 21, 331 (1966).
34. D. E. Guveli, J. B. Kayes, and S. S. Davis, J. Colloid Interface Sci., 82, 307 (1981).
35. C. T. Hung and R. B. Taylor, J. Chromatogr., 209, 175 (1981).
36. P. Mukerjee, J. Phys. Chem., 66, 1733 (1962).
37. M. Zander, "Phosphorimetry", Academic Press, New York, 1968.
38. C. M. O'Donnell and J. D. Winefordner, Clin. Chem., 21, 285 (1975).
39. T. Vo-Dinh, G. L. Walden, and J. D. Winefordner, Anal. Chem., 49, 1126 (1977).
40. R. M. Von Vandruszka and R. J. Hurtubise, Anal. Chem., 49, 2164 (1977).
41. G. J. Niday and P. G. Seybold, Anal. Chem., 50, 1577 (1978).
42. J. G. Habarta, Ph.D. thesis, "Study of the Enhancement of Luminescence Properties of Various Pyrene Derivatives, Pyrene, Napthalene and Biphenyl in Normal Micellar Media and Analytical Applications", Seton Hall University, South Orange, NJ, 1982.
43. R. Weinberger, P. Yarmchuk and L. J. Cline Love, Anal. Chem., 54, 1552 (1982).
44. P. Yarmchuk, R. Weinberger, R. F. Hirsch, L. J. Cline Love, Anal. Chem., 54, 2233 (1982).

MICELLAR EFFECTS ON KINETICS AND EQUILIBRIA OF ELECTRON TRANSFER REACTIONS

Ezio Pelizzetti and Edmondo Pramauro

Istituto di Chimica Analitica
Universitá di Torino
10125 Torino, Italy

Dan Meisel

Chemistry Division
Argonne National Laboratory
Argonne, Illinois 60439, USA

Enrico Borgarello

Institut de Chimie Physique
EPFL
1015 Lausanne, Switzerland

Five electron transfer equilibria have been measured in an effort to obtain information on the effect of micellar systems on the equilibrium quotients
and on the kinetics of these reactions. The micelles
have been shown to strongly affect the equilibrium
quotients and the kinetics of the reactions. In the
analysis of the kinetic data, a method for the computation of the local ion concentrations in the vicinity of the charged surface has been applied. The contribution of the electrostatic and the hydrophobic
interactions in the micellar system could thus be
separately evaluated. Although for the charged ions
the predominant effect is electrostatically controlled, a strong hydrophobic effect (either catalytic or inhibiting) can be observed for some of the
inorganic complexes.

INTRODUCTION

The relatively simple structure of micellar aggregates[1] provides an opportunity to study the effects of microheterogenous aggregation on kinetic and thermodynamic parameters. These effects are rather important in biological aggregations mimicked by the micellar systems.[2] Furthermore, in recent years such systems have been designed to enhance charge separation in light-driven electron transfer reactions and thus are highly relevant to energy conversion problems.[3]

Due to their importance in biological as well as in technological processes, attention has been recently drawn to the micellar effects on outer-sphere electron transfer reactions of metal ion complexes. Although these effects on kinetics of electron transfer reactions have extensively been studied,[4-10] reports on the equilibria of such reactions are quite rare.[10] The analysis of these kinetic effects is often carried out using models originally developed for organic reactions.[11-12]

In order to obtain a better understanding of these processes both from the thermodynamic and from the kinetic aspects, some electron transfer equilibria have been presently studied. A method for the computation of the local ion concentration in the vicinity of the charged micellar surface has been applied.[13] The effect of anionic micelles on electron transfer equilibria in several systems is compared here with previously studied related systems.[14-15]

EXPERIMENTAL AND RESULTS

Reagents

Tris(4,4'-dimethyl-2,2'-bipyridine)osmium(II) perchlorate (OsL_3^{2+}) was prepared following the literature procedure, starting from $(NH_4)_2OsBr_6$ and the corresponding ligand.[16] The complex was purified and its absorption spectra were in agreement with the literature data.[17] Octacyanomolybdate(IV) was prepared according to the procedure described in the literature.[18] The corresponding oxidized complexes were obtained by electro-oxidation. N-methylphenothiazine (MPTZ) was obtained from Pfaltz and Bauer and purified by recrystallization from ethanol. Sodium dodecyl sulfate (SDS, Fluka) was purified by recrystallization from methanol/water mixtures.

All other inorganic salts were analytical grade quality and their solutions were standardized according to usual titrimetric procedures. Water was doubly distilled.

Instrumentation

Absorption spectra were recorded on a Cary 219 spectrophotometer. Kinetic parameters were measured using a Durrum-Gibson stopped-flow apparatus.

Equilibrium Measurements

Reaction (4) (see Table 1) was followed at $\lambda=511$ nm, $(OsL_3^{3+})=$ $1-3 \times 10^{-5}$ and $(MPTZ)=1-2 \times 10^{-5}M$.

The following equilibrium quotients were obtained: in absence of surfactant $K_4=0.40 \pm 0.05$; $(SDS)=0.02M$, $K_4=9.5 \pm 1.8$; $(SDS)=0.10M$, $K_4=11 \pm 2$.

Reaction (5) was followed at $\lambda=488$ nm, $(OsL_3^{2+})=1.52.5 \times 10^{-5}M$; $(Mo(CN)_8^{3-})=2-5 \times 10^{-5}M$; $(Mo(CN)_8^{4-})=5 \times 10^{-5}-5 \times 10^{-4}M$; The equilibrium quotients were: in the absence of surfactant $K_5=75 \pm 12$; $(SDS)=$ $0.02M$, $K_5=3.1 \pm 0.6$; $(SDS)=0.10M$, $K_5=3.8 \pm 0.8$.

The measurements were carried out in $(HClO_4)=0.02M$, $\mu=0.10M$ $(NaClO_4)$.

DISCUSSION

Equilibria

Among the reactions reported in Table I, equilibria (3) and (5) are those in which one reactant couple $(Mo(CN)_8^{3-/4-})$ is excluded from the micellar subphase. Nevertheless, while for Equation (3) the electron transfer reaction takes place in the bulk, for Equation (5) it occurs at the micellar interface (see kinetic considerations below). Under these conditions, the experimentally observed equilibrium constant of reaction (3), $K_{3,exp}^{SDS}$ can be related to the equilibrium constant in water, K_3^W, through the binding constant of MPTZ and MPTZ$^+$ with SDS, K_B and K_B^+ respectively:

$$K_{3,exp}^{SDS} = K_3^W \frac{1 + K_{B^+}(SDS)}{1 + K_B(SDS)} \qquad (6)$$

Table I. Equilibrium Quotients in the Absence of Surfactant and in 0.10 M SDS[a].

		K^W_{exp} no SDS	K 0.10M SDS	Ref.
(1)	$OsL_3^{2+} + Fe^{3+} \rightleftharpoons OsL_3^{3+} + Fe^{2+}$	3.8^b	0.07^b	14
(2)	$Fe^{3+} + MPTZ \rightleftharpoons Fe^{2+} + MPTZ^+$	$3.2^{b,e}$ $(1.9)^f$	0.6^b $(0.6)^f$	15
(3)	$Mo(CN)_8^{3-} + MPTZ \rightleftharpoons Mo(CN)_8^{4-} + MPTZ^+$	$3.0^{c,e}$	15^c	15
(4)	$OsL_3^{3+} + MPTZ \rightleftharpoons OsL_3^{2+} + MPTZ^+$	$0.4^{d,e}$	11^d	this work
(5)	$Mo(CN)_8^{3-} + OsL_3^{2+} \rightleftharpoons Mo(CN)_8^{4-} + OsL_3^{3+}$	75^d	3.8^d	this work

a 25°, μ = 0.10 M(NaClO₄), spectrophotometrically evaluated

b (HClO₄) = 0.01 M

c (HClO₄) = 0.10 M

d (HClO₄) = 0.02 M

e 10% ethanol

f (HClO₄) = 0.01 M, μ = 0.02 M (NaClO₄)

In Equation (6), (SDS) refers to the micellized surfactant concentration in mole/ℓ. The ratio K_{B^+}/K_B = 5±1 (at (H^+) = μ = 0.10M) and 30±3 (at (H^+) = μ = 0.02M) and a value K_B=1700 M^{-1} has been previously estimated.[15] It should be noted that the ratio K_{B^+}/K_B also represents the decrease in the reduction potential of the $MPTZ^+/MPTZ$ couple on micellization and thus the increase in the equilibrium constant. For Equation (5), since the couple $OsL_3^{3+/2+}$ strongly associates with the micelles, through electrostatic and hydrophobic interactions, the reaction is most likely to occur at the water/micelle interface (indeed, this is substantiated by kinetic results for similar reactions).[19-20] Interestingly, in this system the equilibrium constant is decreased on micellization. The reduction potential of the $OsL_3^{3+/2+}$ couple thus has to increase on micellization in SDS. Indeed, the reversible half-wave potentials for some similar $OsL_3^{3+/2+}$ complexes have been recently reported to increase by ca. 0.1V in the presence of 0.1 M SDS.[21] Contrary to perhaps simplistic intuition, the reduction in the equilibrium constant also indicates a stronger reduction in the rate constant for the forward reaction than for the backward reaction.

For the other three equilibria reported in Table I (Equations (1), (2) and (4)), all the species interact with the micelles. The equilibrium constants can then be expressed in terms of the water equilibrium constant and the binding constants. Thus for example, for equilibrium (2), Equation (7) applies:

$$K_{2,exp}^{SDS} = K_2^W \frac{(1 + K_{B^+}(SDS))(1 + K_{Fe^{2+}}(SDS))}{(1 + K_B(SDS))(1 + K_{Fe^{3+}}(SDS))} \tag{7}$$

Where $K_{Fe^{2+}}$ and $K_{Fe^{3+}}$ represent the binding constants of Fe^{2+} and Fe^{3+} to SDS micelles. The electrostatic contribution to the micellar binding can be estimated from:[11]

$$P_{es} = e^{z\psi/25.69} \qquad (at\ 25.0°C) \tag{8}$$

Where P_{es} is the partition coefficient, Z is the ionic charge of the counter ion and ψ the surface potential of the micelle. Using the values of ψ reported in the literature, (80±5 mV)[22-23] the ratio for the electrostatic contribution of Fe^{2+} and Fe^{3+} to the binding constant $((5±1)x10^{-2})$ are in good agreement with those experimentally derived by applying Equation (7) and the K_{B^+}/K_B ratios reported[15].

On the other hand, it can be shown that if the partition coefficient for $MPTZ^+$ could be expressed as the product of a

hydrophobic (the same as for MPTZ) and an electrostatic contribution, no change in the equilibrium constant should be observed in
the presence of SDS. This means that the hydrophobic contribution
is influenced by the electrostatic part.   In this case the charge
decreases the hydrophobic contribution by a factor of ca. 5 in
the overall binding constant of MPTZ$^+$ with SDS micelles.

Finally the five equilibria in Table I are related to each
other. Thus cross checks for self consistency could be made.   For
example, $K_4=K_2/K_1$ is to be expected and the experimentally obtained values do agree with this expectation in SDS solutions
considering the slight differences in the experimental conditions.
For equilibrium (5), however, the self consistency is not so
satisfactory.   While in the micellar system $K_5=K_3/K_4$ differs by
ca. a factor of 2 from the experimentally determined K_5, there is
an order of magnitude difference for the same comparison in the
absence of SDS.   A possible explanation for this discrepancy is
perhaps ion pairing of the reactants and products of reaction
(5).   Considering the high opposite charges in reaction (5)
($Z_A Z_B=6$; $Z_C Z_D=-12$) ion pairing might be expected in this system.
This effect is expected to be minimized upon micellization.

Kinetics

The simplest case among the reactions in Table I which were
investigated kinetically (the first three reactions in Table I) is
equilibrium (3).   The observed inhibition can be treated according to the simple model:

$$
\begin{array}{ccccc}
(MPTZ)_w & & \overset{k_3^w}{\underset{k_{-3}^w}{\rightleftharpoons}} & (MPTZ^+)_w & \\
K_B \updownarrow & + \ Mo(CN)_{8w}^{3-} & & \updownarrow K_B^+ & + \ Mo(CN)_{8w}^{4-} \\
(MPTZ)_m & & \overset{k_3^{SDS}}{\underset{k_{-3}^{SDS}}{\rightleftharpoons}} & (MPTZ^+)_m &
\end{array}
$$

The dependence on [SDS] indicates that the reaction occurs in the
bulk aqueous phase and thus the reduced equation (9) can be
derived:

$$
\frac{1}{k_{3,\text{exp}}^{SDS}} = \frac{1}{k_3^w} + \frac{K_B}{k_3^w} \ (SDS) \qquad\qquad (9)
$$

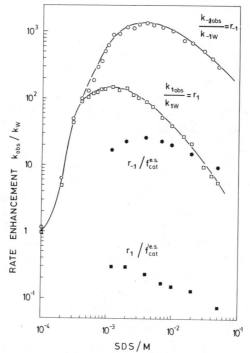

Figure 1. Experimental rate enhancement r_i of the forward and backward reactions in the system $OsL_3^{3+/2+}/Fe^{3+/2+}$ as a function of SDS concentration. The non-electrostatic catalytic factors have been calculated from the total rate enhancement using the calculated values for the electrostatic enrichment factors.

From the experimental data, since $\dfrac{1}{k_3^W}$ cannot be reliably deter-ined, only the value K_B/k_3^W can be evaluated (5.4×10^{-6} s). Since the value of K_B has been independently determined (by gel filtration and comparison with cationic surfactant), an estimate of k_3^W could be made ($3.2 \times 10^8 M^{-1}s^{-1}$). This value was found to be in good agreement with that calculated using the Marcus theory[24] and with the value determined experimentally later by pulse radiolysis ($9.6 \times 10^7 M^{-1}s^{-1}$).

Reactions (1) and (2) allow the comparison of the micellar effect on the kinetics of reactions where both couples interact with the micelles. Furthermore, since one couple ($Fe^{3+/2+}$) is common to both reactions, the difference in the behavior between the other two couples ($OsL_3^{3+/2+}$ and $MPTZ^+/MPTZ$) could be studied. The kinetic analysis has been carried out using the model proposed by Frahm and Diekmann[13] (also reported in these proceedings), which essentially allows the computation of the local ion concentrations in the vicinity of the charged micellar surface. The term f^{es} (electrostatic catalytic factor) represents the ratio between this estimated concentration and the concentration of free species if dissolved in the free volume of the aqueous phase. Figures 1 (see previous page) and 2 show the experimentally observed rate enhancement factors, denoted r, and the ratios of these over the electrostatic catalytic factors, f^{es}. It is evident that in both cases a higher enhancement ratio is observed for the reverse reaction (involving Fe^{2+}) while the contrary should be expected based solely on electrostatic considerations (if one assumes complete micellization of the other pair).

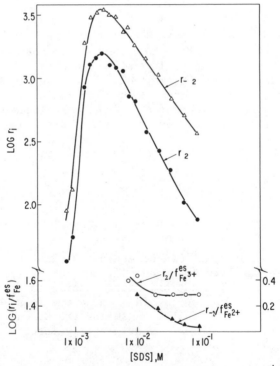

Figure 2. Same as in Figure 1 for the system $MPTZ^{+/o}/Fe^{3+/2+}$. Right hand scale refers to log ($r_2/f_{Fe^{3+}}^{es}$).

The ratios r/f^{es} differ significantly from unity, and can be attributed to non-electrostatic contributions. It is noteworthy that in the case of the $OsL_3^{3+/2+}$, $r_{-1}/fes > 1$ and $r_1/f^{\cdot es} < 1$, whereas for $MPTZ^+/MPTZ$ both are greater than one.

Since the rate constant for an electron transfer reaction is controlled by the free energy of the reaction and by the rate constant for electron self exchange between each of the two reactanting couples, we can analyze which is the one that affects the reaction rate at the micellar interface. For reaction (1) we hypothesize that the r/f^{es} ratios for $OsL_3^{3+/2+}$ are mainly derived from the changes in the free energy of the reaction. On the other hand, for the $MPTZ^+/MPTZ$ couple in reaction (2), the r/f^{es} ratios seem to result primarily from the changes in the electron exchange rate on performing the reaction at the micellar interface. These two explanations are consistent with the presently measured trends as well as with the observation that the self exchange rate for $MPTZ^+/MPTZ$ increases by an order of magnitude in changing the medium from water to less polar solvents.[25-26]

CONCLUSIONS

The reported electron transfer equilibria have been found to be very useful probes for the investigation of the effect of organized structures (in the present case, micelles) on the equilibrium quotients and on the kinetics of such reactions. Furthermore, the computational model provides an estimate both of the electrostatic and the non-electrostatic contribution to the reaction rates and equilibria.

Although the "catalytic" effect observed in the presence of anionic micelles is largely due to enrichment in the concentration of the reactants at the micellar interface, other effects can and do exhibit appreciable changes in the rate constants and the equilibrium quotients of the electron transfer reactions.

ACKNOWLEDGEMENT

E. Pelizzetti and D. Meisel gratefully acknowledge the support of NATO Grant Number 1780. Work at ANL is done under the auspices of the Office of Basic Energy Sciences, US-DOE.

REFERENCES

1.   J. H. Fendler and E. J. Fendler, "Catalysis in Micellar and Macromolecular Systems," Academic Press, New York, 1975.

2. S. Wherland and H. B. Gray, in "Biological Aspects of Inorganic Chemistry," A. W. Addison, W. R. Cullen, D. Dolphin, and B. R. James, Editors, Wiley, New York, 1977.

3. M. Grätzel, in "Micellization, Solubilization, and Microemulsions", K. L. Mittal, Editor, Vol. 2, p. 531, Plennum Press, New York, 1977.

4. A. A. Bhalekar and J. B. F. N. Engberts, J. Am. Chem. Soc., 100, 5914 (1978).

5. H. Bruhn and J. Holzwarth, Ber. Bunsenges. Phys. Chem., 82, 1006 (1978).

6. C. A. Bunton and G. Cerichelli, Int. J. Chem. Kin., 12, 519 (1980).

7. K. V. Ponganis, M. A. De Arujo, and H. L. Hodges, Inorg. Chem., 19, 2704 (1980).

8. D. Meisel, M. Matheson, and J. Rabani, J. Am. Chem. Soc., 100, 117 (1978).

9. E. Pelizzetti and E. Pramauro, Inorg. Chem., 19, 1307 (1980).

10. M. Almgren, F. Grieser, and J. K. Thomas, J. Phys. Chem., 83, 3232 (1979).

11. I. V. Berezin, K. Martinek, and Y. A. Yatsimirskii, Russ. Chem. Rev. (Engl. Transl.), 42, 787 (1973).

12. L. S. Romsted, in "Micellization, Solubilization, and Microemulsions," K. L. Mittal, Editor, Vol. 2, p. 509, Plenum Press, New York, 1977.

13. J. Frahm and S. Diekmann, J. Colloid Interface Sci., 70, 440 (1979).

14. E. Pramauro, E. Pelizzetti, S. Diekmann, and J. Frahm, Inorg. Chem., 21, 2432 (1982).

15. C. Minero, E. Pramauro, E. Pelizzetti, and D. Meisel, J. Phys. Chem., 87, 399 (1983).

16. F. P. Dwyer, F. H. Brustall, and E. C. Gyarfas, J. Chem. Soc., 953 (1950).

17. B. J. Pankuch, D. E. Lacky, and G. A. Crosby, J. Phys. Chem., 84, 2061 (1980).

18. N. H. Furman and C. O. Miller, Inorg. Synth., 3, 160 (1950).

19. E. Pelizzetti and E. Pramauro, Inorg. Chem., 18, 882 (1979).

20. E. Pelizzetti and E. Pramauro, Inorg. Chim. Acta, 46, L29 (1980).

21. Y. Oshawa, Y. Shimazaki and S. Aoyagui, J. Electroanal. Chem., 114, 235 (1980).

22. M. S. Fernandez and P. Fromherz, J. Phys. Chem., 81, 1755 (1977).

23. D. Stigter, J. Colloid Interface Sci., 23, 379 (1967).

24. R. A. Marcus, Am. Rev. Phys. Chem., 15, 155 (1964).

25. B. A. Kowert, L. Marcoux, and A. J. Bard, J. Am. Chem. Soc., 94, 5538 (1972).

26. S. P. Sorensen and W. H. Bruning, J. Am. Chem. Soc., 95, 2445 (1973).

CATALYSIS OF ESTER HYDROLYSIS BY FUNCTIONALIZED COUNTERION

SURFACTANTS

M. Gobbo, R. Fornasier, and U. Tonellato

Centro "Meccanismi di Reazioni Organiche" del C.N.R.
Istituto di Chimica Organica, Università di Padova
35100 Padova, Italy

Micelles of cetyltrimethylammonium ions and ω-
-mercaptoalkanoates 1 or ω-(imidazol-4-ylcarbamoyl)al-
kanoates 2 are effective catalysts for the esterolysis
of p-nitrophenyl esters at pH 8.7. The kinetic parameters obtained from the analysis of rate-surfactant profiles indicate that the reactivity of the functionalized counterions depends only on the extent of their
binding to the micellar surface, i.e., on the hydrophobic and electrostatic interactions between counterions and micelles. The electrostatic factor, however,
may be strongly masked in the presence of other electrolytes. The results indicate that functionalized
counterion surfactants may provide a simple but
effective micellar catalytic system whose kinetic
features compare well with those of functional comicelles.

INTRODUCTION

Most of the studies[1,2] of esterolytic processes in neutral
or moderately alkaline micellar solutions have involved reactions
of hydrophobic substrates and either (i) hydrophilic nucleophiles
as counterions of cationic surfactants, or (ii) very hydrophobic
nucleophiles, functional surfactants and cosurfactants. While the
kinetic effects observed[1,3] in the case of micelles with reactive
hydrophilic counterions are generally modest, remarkable rate
enhancements comparable, in some cases, to those of enzyme reactions have been reported[4-7] for functional micellar systems. The
use of effective functional micellar systems, however, often
implies troublesome synthetic problems.

Micelles of inert ionic surfactants with moderately hydrophobic functionalized counterions may be conceived as intermediate type of catalytic systems of potential interest due to the
fact that a variety of bi-or polyfunctional ionic organic reagents,
such as functionalized carboxylic acids, are available or readily
synthesized.

Recently, we approached the problem by studying the kinetic
effects upon the hydrolysis of p-nitrophenyl (PNPA, acetate, and
PNPH, hexanoate) esters in moderately alkaline solutions of
cetyltrimethylammonium (CTA$^+$, added as CTABr) and thiol, 1, or
imidazole, 2, functionalized carboxylates. Also, for comparison
purposes, we investigated the effects of derivatives 3, 4, and 5,
in the presence of micellar CTABr. Our aim in this work was to
test the catalytic effectiveness of micelles with functionalized
counterions and provide a description of the relative importance
of electrostatic and hydrophobic factors for the observed kinetic
effects.

$$C_{16}H_{33}-\overset{+}{N}(CH_3)_3 \qquad \begin{matrix} \diagup SH \\ (CH_2)_n \\ \diagdown CO_2^- \end{matrix} \qquad \begin{matrix} \diagup NH-CO-Im \\ (CH_2)_n \\ \diagdown CO_2^- \end{matrix}$$

$$CTA^+$$

$$\underline{1} \qquad\qquad \underline{2}$$

$$
\begin{matrix}
\underline{a}:\ n = 2 & \underline{a}:\ n = 2 \\
\underline{b}:\ n = 5 & \underline{b}:\ n = 5 \\
\underline{c}:\ n = 11 & \underline{c}:\ n = 10
\end{matrix}
$$

$$CH_3-(CH_2)_2-SH \qquad (CH_3)_3\overset{+}{N}-(CH_2)_2-SH\ I^- \qquad CH_3-(CH_2)_2-NH-CO-Im$$

$$\underline{3} \qquad\qquad \underline{4} \qquad\qquad \underline{5}$$

Im = 4-imidazolyl

EXPERIMENTAL

Thiol functionalized carboxylic acids 1b,c were obtained
from the corresponding bromo-substituted carboxylic acids via
Bunté salts, following standard procedures.[8] Imidazole functionalized acids 2 were obatined[4,7] by reaction of the corresponding
aminoacids with 4-imidazolylcarbamoyl chloride in pyridine and
compound 5 was likewise synthesized from propylamine. Thiocholine
iodide 4 was obtained as described[9] and other products were from
commercial sources purified following standard procedures. The
apparent pK_a of the thiol or imidazole functionalized reagents
(except that of thiocholine 4 [5,10], pK_a ca.7.8) were found by
spectrophotometric titration or estimated [4,7] in the range
10.5-10.8 in the presence of a 6-fold molar excess of CTABr, at
$\mu=0.2-0.07$ (borate, KCl). The procedure for the kinetic measurements
has been described elsewhere.[7,10] Special care was taken to avoid
oxidative degradation of the thiol functionalized reactants:
solvents were deoxygenated, purged with nitrogen and all manipula-
tions were done under anaerobic conditions. The SH titre was
determined by the Ellman's assay.[11]

RESULTS AND DISCUSSION

Micellar effects upon reactions of nucleophiles with esters
have been quantitatively explained by the pseudophase model by
considering the distribution of both reagents between aqueous and
micellar pseudophases and the second-order rate constants in each
pseudophase.[12,13] Dealing with reactive counterions one must take
also in account the competition between the reactive species and
other ions for micelles: the quantitative treatment must include
ion-exchange parameters, as recently discussed by Bunton et al.[3,14]
and Quina and Chaimovich.[15] The reactive counterion approach,[16]
i.e., the use of cationic micelles where only reactive ions are
present in both the micellar and the aqueous part of the medium,
simplifies the kinetic treatment but is not easily compatible with
the use of buffer solutions.

We carried out a set of kinetic experiments in 0.02M borate
buffer, 0.05M KCl, pH 8.7, using a constant molar ratio
(CTABr):(reactant, R)=(ca. 6) and constant ester concentration
(ca. 6×10^{-6}M), to define a series of rate-surfactant profiles.
At least for solutions ranging from c.m.c. ($3-5 \times 10^{-5}$) to
(CTABr)= ca.1×10^{-3}M, notwithstanding some changes in the molar
ratios of the various ionic species competing for micelles and
possible changes of the apparent pK_a's of the functionalized counterions, rate data were found to reasonably obey Equation (1):

$$k_\psi = \frac{(k_m/\overline{V})K_R K_S D_m + k_w}{(1 + K_S D_m)(1 + K_R D_m)} \qquad (1)$$

where k_ψ is the apparent second-order rate constant, k_m and k_w the
second-order rate constants in the micellar and aqueous pseudophases, K_S the association constant of the substrate to the micellized
surfactant (D_m), K_R is the apparent association constant of the
reactant to D_m and $\bar{V}$ is the molar volume of the micelles. While
k_ψ and k_w were measured directly and evaluated as described by
Chaimovich et al. [17], k_m and K_R were treated as adjustable parameters. The c.m.c. was generally estimated from the rate-surfactant
profiles and K_S values (M^{-1})[7] of 30 (PNPA) and 2000 (PNPH) were
used and eventually optimized to obtain the "best fit". Some rate-
-surfactant profiles are shown in Figure 1, solid lines are
calculated.

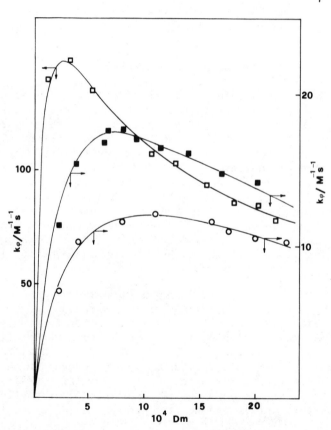

Figure 1.Rate-(CTABr + 2) profiles. Reactions of PNPH with:
O 2a, ■ 2b, and □2c. (CTABr) : (2)= 6.1.

Table I reports the kinetic parameters obtained from this set of kinetic runs together with the turnover rate constants estimated [9] from "burst" kinetics carried out using excess PNPH over the counterionic reactant. The k_m and k_w constants are not corrected for the fraction of dissociated functions (which is likely to be rather similar in each case) although, presumably, the effective nucleophilic species involved in the esterolysis are the thiolate or imidazolide anions.[7,10]

Table I. Kinetic Parameters for the Esterolysis of PNPA and PNPH at pH 8.7[a], 25°C

Reactant[b]	K_R^b/M^{-1}	$k_m^c \times 10/\overline{M}^{-1}s^{-1}$		$k_w \times 10/\overline{M}^{-1}s^{-1}$		$k^d \times 10^4/s^{-1}$ turnover
		PNPA	PNPH	PNPA	PNPH	
1a	100	7	1.5	4.0	1.5	<1
1b	750	3	2	3.1	2.2	<1
1c	5×10^4	7	3			
2a	500	0.4	0.2	0.11	0.40	>100
2b	900	0.4	0.2	0.21	0.72	>100
2c	7×10^3	0.8	0.5			≥120
3	450	4	0.8	10	7	
4	– [e]	– [e]		27	12	
5	600[f]	0.3	0.1[f]			

[a] In 0.02M borate buffer, 0.05M KCl. [b] (CTABr):(R)= 6.1-6.3. [c] Calculated using $\overline{V}$=0.37 1 mol^{-1}; see, however, references 13 and 9. [d] PNPH [e] Too small to measure: rate-(CTABr) profiles were insensitive to the presence of 4. [f] Very approximate value

The main indications from the data of the Table, although approximate, are the following. (a) The reactivity of counterions 1 and 2 increases with increasing chain length (n) due, essentially, to an increase in the apparent K_R and not to changes in k_m. Within both series, the k_m values are similar and close to k_w, thus indicating that the reaction occurs in a waterlike region of the micelles whatever the hydrophobic character of the counterion. (b) The increase of K_R with n is more pronounced in the series of thiol 1 than in that of imidazole 2 functionalized counterions. (c) Electrostatic effects are the obvious explanation for the absence of catalytic effect observed for solutions of CTABr and thiocholine 4. On the other hand, when one compares

the reactivity of 1a or 2a with that of their non-ionic counterparts 3 or 5, the effect of the electrostatic interactions is not
evident. In the case of the thiol functionalized pair, the non-
-ionic reactant 3 is even more effective than 1a, due to a larger
K_R value (other parameters being essentially similar). The above
comparisons may indicate that electrostatic effects are masked by
a strong competition of the functionalized counterions with other
ions, particularly Br^-, or/and the electrostatic attraction of the
counterions to CTA^+ micelles is balanced by a larger solubility in
water of the carboxylate ions relative to their non-ionic counterparts (K_R is a function[12] of the partition coefficient of the
reactant between micellar and aqueous pseudophases).

Experiments aimed at defining the relative importance of these
factors are now in progress. Preliminary results of kinetic runs
carried out using different (CTABr):(2a) ratios, or in the absence
of KCl, or in lower borate concentration buffers indicate that Br^-
strongly competes with 2a and binds to micelles to an extent
which is approximately half that of the functionalized counterion,
while Cl^- and borate ions are much less effective.

CONCLUSIONS

The main features of functionalized counterion micelles,
including effective catalytic turnovers, are comparable to those
of functionalized comicelles. The systems are, however, very
sensitive to the nature and concentration of other ionic species
present in the solution. A better undertanding of these factors
will provide information on the experimental conditions which
should be used for a full display of the catalytic effectiveness
of these micellar systems.

REFERENCES

1. J. H. Fendler and E. J. Fendler, "Catalysis in Micellar and
   Macromolecular Systems", Academic Press, New York, 1975.
2. U. Tonellato, in "Solution Chemistry of Surfactants", K. L.
   Mittal, Editor, Vol. 2, p. 541, Plenum Press, New York, 1979.
3. H. Al-Lohedan and C. A. Bunton, J. Org. Chem., 47, 1160 (1982)
   and references therein.
4. T. Kunitake, Y. Okahata, and T. Sakamoto, J. Amer. Chem. Soc.
   98, 7799 (1976).
5. R. A. Moss, G. O. Bizzigotti, and C.-W. Huang, J. Amer. Chem.
   Soc., 102, 754 (1980).
6. R. A. Moss, R. C. Nahas, and S. Ramaswani, J. Amer. Chem.
   Soc., 99, 627 (1977).

7. L. Anoardi, R. Fornasier, and U. Tonellato, J.C.S., Perkin II
   260 (1981).
8. J. L. Wardell, in "The Chemistry of the Thiol Group", S.
   Patai, Editor, Vol. 1, p. 163, Wiley, New York, 1974.
9. R. Fornasier and U. Tonellato, J.C.S. Faraday I, 1301 (1980).
10. R. Fornasier and U. Tonellato, Gazzetta, 112, 261 (1982)
11. A. F. S. A. Habeeb, Methods Enzymol., 25, 457 (1972).
12. K. Martinek, A. K. Yatsimirski, A. V. Levashov, and I. V.
    Berezin, in "Micellization, Solubilization, and Microemulsions",
    K. L. Mittal, Editor, Vol. 2, p. 489, Plenum Press, New York,
    1977.
13. C. A. Bunton, in "Solution Chemistry of Surfactants", K. L.
    Mittal, Editor, Vol. 2, p. 519, Plenum Press, New York, 1979.
14. C. A. Bunton, L. S. Romsted, and L. Sepulveda, J. Phys. Chem.
    84, 2611, (1980).
15. F. H. Quina and H. Chaimovich, J. Phys. Chem., 83, 1844,
    (1979).
16. C. A. Bunton and L. S. Romsted, in "Solution Behaviour of
    Surfactants: Theoretical and Applied Aspects", K. L. Mittal
    and E. J. Fendler, Editors, Vol. 2, pp. 975 - 991, Plenum
    Press, New York, 1982.
17. I. M. Cuccovia, E. H. Schröter, P. M. Monteiro, and H. Chaimo-
    vich, J. Org. Chem., 43, 2248 (1978).

SPECIFIC MICELLAR EFFECTS IN THE TEMPORAL BEHAVIOUR OF EXCITED

BENZOPHENONE: CONSEQUENCES UPON THE POLYMERIZATION KINETICS

P. Jacques, D.J. Lougnot and J.P. Fouassier

Laboratoire de Photochimie Générale, ERA n°386
Ecole Nationale Supérieure de Chimie
68093 Mulhouse Cedex, France

The temporal behaviour of triplet benzophenone (^{3}BP) in SDS micelles was investigated by nanosecond laser spectroscopy. Contrary to previous reports by us and others, our experimental results prove that the decay of the transient does not obey a first order kinetics when analyzed in the 500-600 nm wavelength range. The rather complex decay curves can only be accounted for by considering the scheme : $A \rightarrow B \underset{D}{\overset{C}{\rightleftharpoons}}$. The species are : A = the micellized triplet of BP, B = a micellized triplet radical pair, C = a free ketyl radical, D = a micellized singlet radical pair.

Such a scheme enables numerous discrepancies and some misinterpretations encountered in the literature to be explained. Among others, the rate constant for hydrogen abstraction is not so tremendously high as recently proposed. The average occupancy per micelle was varied for BP_0 from 0.2 to 20 and for ^{3}BP from 0.05 to 2. Surprisingly, neither selfquenching nor triplet-triplet annihilation were detectable. Information as to the location of the BP molecules is derived from this observation.

The mechanism proposed herein is basically important for explaining the results obtained in the polymerization of vinyl-monomers initiated by ^{3}BP in micellar assemblies.

INTRODUCTION

Much attention has been devoted to the photochemistry of ketones in homogeneous solution, the most wide-spread compound studied being undoubtly benzophenone (BP). Two review papers were
recently published[1,2] which underlined respectively :

- the various aspects of reaction (1) : mechanism, kinetics
and quantum yields,

$$(\underset{}{\geq}C = O) + XY \longrightarrow \geq\overset{\bullet}{C}OX + \overset{\bullet}{Y} \qquad (1)$$

(where X, Y are atoms or group of atoms)

- the energy wastage processes in ketones photochemistry, due
to reversion to the ground state of starting material from reaction intermediates.

One is thereby inclined to consider the photochemistry of
BP in a homogeneous phase as a well documented matter and as a
result this compound is taken as a probe for the study of environmental reactivity[3]. However, a novel emission was recently reported[4] and we stressed in a previous paper[5] that some points
still remain unclarified, let us mention : (i) the detailed mechanism of photoreduction of BP by benzhydrol[6] and (ii) the role
played by the photoproducts generated by the light excitation[5,7,8]
or by the analyzing light.

The photochemistry of BP in a microheterogeneous medium recently revealed to be a very interesting topic from a double point
of view : as a probe to investigate some processes in organized
medium and as a photoinitiator of general use in vinyl polymerization in micelles or emulsions.

Our purpose here is to present in the first part a clear description of the photophysical processes involved, which explains
the apparent discrepancies observed by several authors. In the
second part, we will discuss the implications of the proposed kinetic scheme when BP is used as a photoinitiator for vinyl polymerization.

EXPERIMENTAL PART

The kinetic data reported herein were obtained either from
mode locked (time resolution $\simeq$ 2 ns) or giant pulse ruby laser
spectroscopy devices (τ_r = 20 ns). The detection system was able
to record transient absorptions as weak as 2.10^{-3} in the submicro-
second time scale. Great care was taken to avoid the generation
of steady state photoproducts by inserting, in the analyzing light
beam, a glass filter with sharp cut off around 415 nm. The optical

density of the solution was colorimetrically controlled over the
duration of the experiments. Benzophenone of commercial grade
(Fluka) was recrystallized three times from ethanol/water mixtures, m.p = 48°C. SDS (Aldrich) was washed with ether and then recrystallized three times from 96% ethanol. The final purity was
verified by NMR spectroscopy. The solutions were degassed by argon bubbling. The experimental conditions of the photopolymerization under steady state illumination have been published elsewhere[14].

1 - Dynamics of ^{3}BP in SDS micelles

It is useful to make a brief survey of the significant information concerning the relaxation of ^{3}BP in SDS micelles. Laser excitation of BP results in two transients : one in the 100 ns time
scale, the other in the μs range. The assignment of the short decay is a delicate question and has led to apparent conflicting
views by several authors.

Hayashi et al.[9] were the first to mention the influence of a
magnetic field upon the relaxation kinetics, a fact which implies
the participation of a triplet radical pair. However, the overall
interpretation of the relaxation mechanism was somewhat confusing
due to erroneous deductions made at the unsuitable analyzing wavelength λ_a = 330 nm.

Scaiano and Abuin[10] assumed a nearly first order relaxation
kinetics but noted that the absorption spectrum of the transient
evolves gradually from a triplet one to a ketyl one (K$^\bullet$) as the
transient absorption decays. This observation suggested to these
authors the simultaneous participation of ^{3}BP and K$^\bullet$.

On the other hand, Braun et al[11] assumed a strictly first order relaxation and the transient was assigned to K$^\bullet$, the H abstraction being surprisingly so fast that the triplet was completely quenched at the end of the laser pulse. From quenching experiments of ^{3}BP phosphorescence in aqueous solutions with increasing concentrations of SDS a Stern-Volmer plot was reported indicating a quenching rate constant of 2.3×10^7 M^{-1}s^{-1}.

Preliminary results obtained by us[12] on the dynamics of BP
in SDS micelles with regards to photopolymerization (with or without monomer and H donor) suggested that the observed short lived transient had to be assigned to a triplet state. In a subsequent paper[5] we tried to reconcile these conflicting views by taking into account the possible participation of a triplet exciplex
between ^{3}BP and the surfactant chain. But the paper by Scaiano and
Abuin[10] provided the impetus for a detailed study of the relaxation
mechanism because it became clear that a single first order decay could
not explain definitively the whole experimental observations.

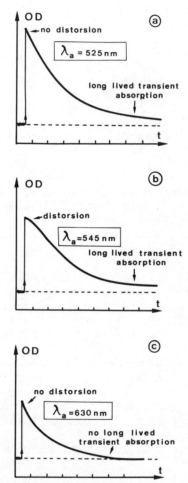

Figure 1. Dependence of the oscillograms on the analyzing wavelength λ_a (200 ns/division). [SDS] = 0.1 M.

In fact, inspection of Figure 1 shows clearly that the oscillograms present a "distortion" at the beginning of the relaxation (Figure 1b). This curvature, which is observed only within a relatively narrow range of λ_a, is quite small but nevertheless significant. It can be detected only with very sensitive devices, which explains that it was overlooked hitherto.

Let us now show that the following scheme :

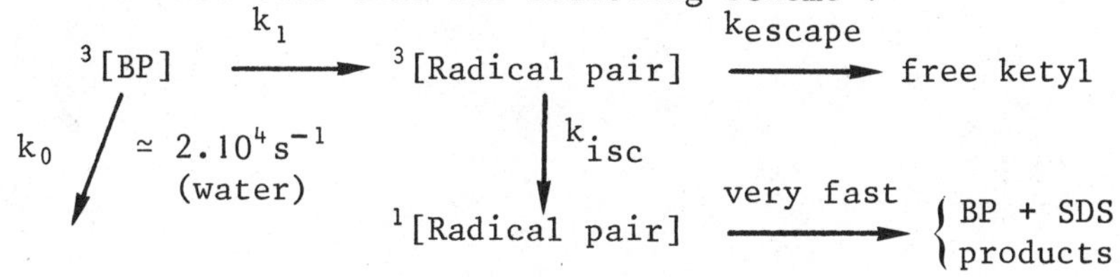

$$^3[BP] \xrightarrow{\ k_1\ } {}^3[\text{Radical pair}] \xrightarrow{\ k_{escape}\ } \text{free ketyl}$$

$k_0 \Bigg/ \simeq 2.10^4 s^{-1}$ (water) $\Bigg\downarrow k_{isc}$

$$^1[\text{Radical pair}] \xrightarrow{\text{very fast}} \begin{cases} BP + SDS \\ \text{products} \end{cases}$$

Scheme 1 : relaxation of ^3BP in SDS micelles

is able to account for the abovementioned distortion. The major difficulty is to deconvolute with sufficient precision the contribution of each species - ^3BP and radical pair - to the total optical density.

Operationally, the long lived transient absorption ($\tau > 1\ \mu s$) is substracted when it subsists at the end of the decay curve. Its origin can be safely assigned to free ketyl radical having escaped the micelles (note that at $\lambda_a = 630$ nm the absorption coefficient of the ketyl radical is rigorously equal to zero). Thus, we have to deal with the well known consecutive reaction

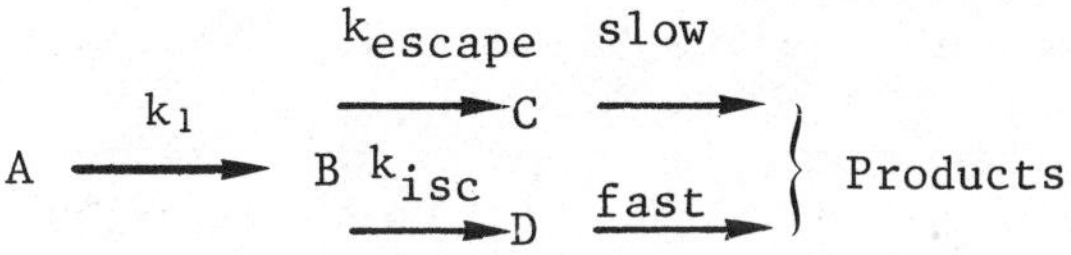

$$A \xrightarrow{\ k_1\ } B \begin{array}{c} \xrightarrow{k_{isc}} C \xrightarrow{k_{escape}\ \text{slow}} \\ \xrightarrow{\quad} D \xrightarrow{\text{fast}} \end{array} \Big\} \text{Products}$$

(k_0 can be neglected (vide infra)).

This system can be analytically solved and moreover - k_2 staying for the overall decay rate of the triplet radical pair ($k_2 = k_{isc} + k_{escape}$)- it is convenient to introduce the parameters ($\varepsilon_T = \varepsilon_{3BP}$) :

$$(1) \quad \rho = \frac{\varepsilon_K}{\varepsilon_T} \quad \text{and} \quad (2) \quad \alpha = \frac{k_1}{k_1 - k_2}$$

Thus starting with an initial concentration $^3BP_0$ corresponding to OD_0, the transient optical density is :

$$OD(t) = OD_0 [e^{-k_1 t} + \frac{\varepsilon_K}{\varepsilon_T} \frac{k_1}{k_1 - k_2} (e^{-k_2 t} - e^{-k_1 t})] \qquad (3)$$

This equation can be rewritten as

$$\frac{OD(t)}{OD_0} = (1 - \rho\alpha)e^{-k_1 t} + \rho\alpha\, e^{-k_2 t} \qquad (4)$$

It should be noted that

$$\lim_{t \to 0} \frac{d\left(\frac{OD(t)}{OD_0}\right)}{dt} = k_1(\rho - 1) \quad ; \text{ independent of } k_2 . \qquad (5)$$

moreover, since $\varepsilon_K \ll \varepsilon_T$ beyond $\lambda_a = 600$ nm, Equation (4) reduces to

$$\left(\frac{OD(t)}{OD_0}\right)_{\lambda > 600 \text{ nm}} = e^{-k_1 t} \qquad (6)$$

Thus from Equation (6) k_1 can be obtained ; then, Equation (5) provides value of ρ at a given wavelength λ_a. Both of these values are used in Equation (4) from which a value of k_2 is fitted by a trial and error method. This procedure is examined in much more details elsewhere[13], but Figure 2 provides evidence for the validity of the proposed method.

With solution of [BP] = 3.10^{-3} M and [SDS] = 0.1 M the numerical procedure gives the following values : the rate constant k_1 was found to be 3×10^6 s^{-1} ($\tau_{triplet} \simeq 330 \pm 30$ ns). This value is in accordance with the result of Scaiano[7] $k \simeq 3.10^6$ s^{-1}, somewhat different from the one published by Sakaguchi et al[9a] : 7.7×10^6 s^{-1} and in contradiction with the result of Braun et al.[11] : $\tau_T < 20$ ns.

The decay rate of the ketyl radical pair was fitted to $4.0 \pm 0.3 \times 10^6$ s^{-1} at $\lambda_a = 545$ nm for which ρ is maximum (τ radical pair $\simeq 250 \pm 20$ ns). Indeed, self consistency is observed when λ_a is varied.

The assumption made previously as to k_0 is thus validated. Moreover, the estimation of k_{esc} and k_{isc} were made from OD_0, OD_∞ and ρ according to the relationship :

$$k_{esc} = \frac{OD_\infty}{OD_0 \times \rho} k_2 = 3.4 \times 10^5 s^{-1}$$

and   (7)

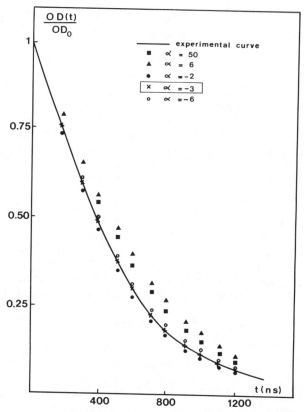

Figure 2. Numerical fitting of k_2 according to Equation (4).

$$k_{isc} = \frac{\rho OD_0 - OD_\infty}{\rho OD_0} k_2 = 3.7 \times 10^6 \text{ s}^{-1}$$

The value of k_{esc} is difficult to interpret rigorously since micellar assemblies are dynamic rather than static structures. The de-correlation of the radical pair may result either in the exit of the ketyl or in the exit of the alkyl chain radical, species which – in the present case – is part of the micellar assembly. Scaiano and Abuin[7] reported for the sum of these two exit rates, a value near 2.10^5 s^{-1} in complete agreement with the result obtained in the present work. However, the exact contribution of each radical remains unknown, though the main contribution probably comes from the ketyl. This point deserves further investigation since it is of crucial importance in the understanding of the polymerization ki-netics and efficiency (vide infra). At this point, some important deductions can be drawn from scheme I.

    (i) The magnetic effect is accounted for by the presence of a radical pair.

(ii) The abstraction of hydrogen from the alkyl chain by
^{3}BP occurs with a "normal" rate constant and not with
an unexplainable high rate as claimed by Braun et al[11].

(iii) We have stressed in a previous publication[5], the fact
that no fluorescence of the excited ketyl radical or
radical pair was detected, a conclusion apparently in
contradiction with the involvment of a ketyl in the
relaxation. This absence of K$^\bullet$ fluorescence – confirmed
by recent experiments – can be easily explained from
the value of k_1, k_2 : the concentration of the ketyl
radical pair is not important enough during the laser
pulse to induce any detectable signal. It is important
to note that upon addition of tetramethylethylene (TME),
its characteristic fluorescence is easily observed at
570 nm owing to the fact that the triplet is then entirely quenched within the pulse duration.

Finally, it should be noted that neither self-quenching nor
triplet-triplet annihilation were observed for solutions of <n>
up to 20 and when <n**> was varied from 0.02 to 0.25 (<n> = 1).

2 - Benzophenone as a photoinitiator of vinyl polymerization

a) General considerations on photopolymerization

One of the most attractive characteristics of emulsion radical polymerization is the production of polymers having high molecular weight with high reaction rate. However, very few studies
have been performed on photochemically initiated polymerization[16, 17], though this procedure appears as a very promising way to obtain
polymers at room temperature.

In a general way, radical photoinitiation proceeds as follows
(schemeII) : under U.V. light excitation, the photoinitiator (I)
is promoted to its first excited singlet state (^{1}I**) which via fast
intersystem crossing gets converted into its triplet state (^{3}I**).
This transient state can undergo either direct cleavage of the molecule (according to a Norrish type I photoscission process) or
hydrogen abstraction with amine compounds or hydrogen donors (DH)
such as tetrahydrofuran (THF) or alcohols : these reactions yield
reactive radical species which attack monomer molecules (M) and
initiate the polymerization. Moreover, the triplet transient state
can be quenched by the monomer by an energy transfer process which
does not undergo any chain initiation ; accordingly, the latter
process must be regarded as an unproductive pathway : the longer
the lifetime of the triplet state, the more efficient the quenching process. This deactivation path was observed to be very efficient not
only in solution and in bulk, but also in emulsion photopolymerization when the initiator is dissolved in the oil phase (Scheme III).

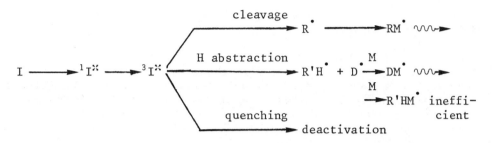

Scheme II : General scheme for a photopolymerization

As a consequence, the efficiency of the photoinitiator can be described by two quantum yields : the quantum yield of initiation (ϕ_i) which represents the number of starting polymer chains per photon absorbed (the term ϕ_i takes into account the overall process which yields the first monomer radicals RM$^\bullet$) and the quantum yield of polymerization (ϕ_m) which is the number of monomer units polymerized per photon absorbed. In the absence of chain transfer reactions, the degree of polymerization – DP_n –, ϕ_i and ϕ_m are related by the relationship : $\phi_m = \phi_i . DP_n$. Moreover, the rate R_p and the degree of polymerization are generally a function of four parameters : monomer concentration (M), optical density of the solution (OD), incident light intensity (I_0) and quantum yield of initiation (ϕ_i) according to Equation (8)

$$R_p = \frac{k_p}{k_t^{1/2}} (M) \sqrt{2.3\ I_0\ OD\ \phi_i} \quad ; \quad DP_n = \frac{k_p}{k_t^{1/2}} \frac{(M)}{\sqrt{2.3\ I_0\ OD\ \phi_i}} \quad (8)$$

(where k_p and k_t are the rate constants of propagation and termination of the polymerization reaction).

b) Use of BP as initiator in SDS micelles

Under UV light excitation, BP is promoted into its first excited singlet state, which then via fast intersystem crossing gets converted into its triplet state. On the basis of the results reported in part I three reaction pathways have to be considered (Scheme III). But only pathway (3) is of interest as regards photopolymerization.

In this context, two important observations are worth noting :

(i) Upon addition of MMA, the lifetime of triplet BP is drastically shortened : in fact, from an average occupancy of one molecule of monomer per micelle (upon the basis of an aggregation number of 65), ^{3}BP is quenched so much so that it becomes too short-lived to be investigated by nanosecond laser spectroscopy.

(ii) Addition of THF also shortens the triplet lifetime, and a concomitant increase of K° concentration is clearly observed.

At the present stage of our work, it is not possible to submit a definite and accurate analysis of the transient signal when MMA and THF are present together in the reactive medium. It should be noted that despite the severe quenching of BP by MMA, polymerization still occurs as shown by the following results obtained in steady illumination experiments.

c) Rate of polymerization

A typical mixture for micelle polymerization consists of 1 initiator, 1 H donor and 100 monomer molecules per micelle. Light irradiation yields monomer particles with higher rate of reaction and higher degree of polymerization than in bulk experiments : typical results are reported in Table 1. The ϕ_m values (number of monomer units polymerized per photon absorbed) are lower than those obtained in the presence of 2,2-dimethoxy-2-phenylacetophenone (DMPA)[15]. Moreover, the quantum yields of polymerization are lower in the presence of BP than those obtained with DMPA[13]. In the case of BP, a hydrogen donor (THF or amine derivative) is necessary to induce the reaction. Lastly, no magnetic field effect on the rate of polymerization is observed, in contrast with results obtained during the photopolymerization of MMA in the presence of DMPA[15]. This typical effect was first reported by Turro et al[16] while using dibenzoylketone as a photoinitiator. It seems probably due to the result of Zeeman splitting of the sublevels of the triplet radical pair formed, the external magnetic field being expected to reduce intersystem crossing and consequently to increase the yield of the escaped radicals which are the initiating species.

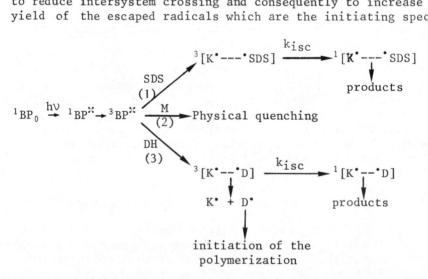

Scheme III. Synoptic scheme of the photopolymerization in SDS micelles with BP as photoinitiator.

Table 1 : Photopolymerization in micelle ([Monomer] = 0.5 M,
[SDS] = 0.5 M) and in bulk. Optical density : 0.15
at λ = 366 nm ; 1 : ethyl-4-aminobenzoate ; 2 : THF.

Monomer	Initiator	micelle			bulk	
		$R_p \times 10^5$ M.s^{-1}	ϕ_m	$10^3 \phi_i$	$R_p \times 10^5$ M.s^{-1}	$10^3 \phi_i$
MMA	BP + 1 (1% w/w)	15.1	16	2	3.8	5
MMA	BP + 2 (1% w/w)	12.9	13	6	4.5	13

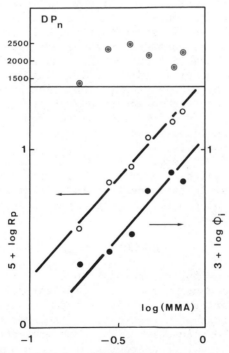

Figure 3. Photopolymerization in SDS micelles. Dependence of R_p,
DP_n and ϕ_i on the monomer (MMA) concentration. Initiator : BP,
absorbed energy : $0.95.10^{-5}$ Einstein.s^{-1}.ℓ^{-1}. The polydispersity of
the sample is about 2.

Due to the location of BP and THF, the "stabilizing" effect of
the micelle upon the triplet radical pair would be weak. Thus,
this radical pair is presumably shortlived (as in homogeneous solution) which would explain the absence of magnetic field effect
upon the rate of polymerization.

Figure 3 shows that R_p is a function of the monomer concentration whereas the degree of polymerization appears to be constant
over the range of values investigated (0.1-0.7 M).

From previous results, it can be seen that the quantum yields
of initiation (ϕ_i) of the polymerization are lower in micelle than
in bulk. The quantum yield ϕ_i can be calculated as a function of
the initial monomer concentration (using the relation $\phi_i = \phi_m/DP_n$).
Figure 3 shows that $\phi_i \propto [M]^{1.1}$. It is deduced that :

$$R_p \propto \sqrt{I_{abs}} \sqrt{\phi_i} \propto [M]^{1.1} \sqrt{I_{abs}} \qquad (9)$$

Since the number of monomer units per micelle is typically 100 and
the DP_n values obtained are of the order of 2000, one must imagine
that monomer micelles act as monomer generators towards a micelle
containing a growing macroradical : the particle size is expected
to increase during the course of the polymerization. From the
results reported above, it seems that the efficiency of the initiation process is not enhanced in micelle : this is probably due
to the specific interaction between ^3BP and the alkyl chain which
is detrimental to the yield of the initiating radical generation
(Schemes II,III). As a consequence, the higher rates of the polymerization in micelle could be accounted for by the specific role of
the surfactant assembly which reduces the probability to have simultaneously two radicals in the same "micelle" and thereby prevents the untimely termination of the growing polymer chains by
radical coupling (the terminology "micelle" cannot apply to this
kind of aggregate which is commonly referred to as a "particle"
by the polymerists. In fact, neither the diameters of these particles nor the optical properties of their solutions are compatible
with the concept of micelle).

CONCLUSION

A kinetic scheme for the relaxation of ^3BP in SDS micelles is
presented which accounts for all experimental results reported in
the literature. Moreover, it explains the misinterpretations and
the conflicting views with regard to the relaxation mechanism.
Among others it shows that a radical pair with the micelle component itself is involved. Consequently, the quantum yield of initiation (ϕ_i) for the photopolymerization in SDS micelles is severely
decreased compared with the bulk phase. Fortunately, the H abstraction rate from the alkyl chain, the escaping rate of $K^\bullet$ out of
the micelle are not high enough to exclude the formation of $D^\bullet$ ra-

dicals which are the reactive species. Despite these negative aspects the overall mechanism of polymerization is favoured by the confinement of the reactive radical into the micelle. Improvement could be expected in the use of perfluorated detergents but much more information is needed before being able to describe satisfatorily the processes involved in photopolymerization reactions : for example, nothing is known about the exchange of $K^\bullet$, $S^\bullet$ and $D^\bullet$ between the micelles.

ACKNOWLEDGEMENTS

We extend our thanks to Dr. J.C. Scaiano for sending us valuable information and a preprint[7].

REFERENCES

1. J.C. Scaiano, J. Photochem., 2, 81 (1973)
2. D.I. Schuster, Pure Appl. Chem., 41, 601 (1975)
3. N.J.Turro,"Modern Molecular Photochemistry", The Benjamin Cummings Publishing CO (1978)
4. N.J. Turro, M. Aikawa and I.R. Gould, J. Am. Chem. Soc., 104, 856 (1982)
5. D.J. Lougnot, P. Jacques and J.P. Fouassier, J. Photochem., 19, 1, 59(1982)
6. D.I. Schuster and P.B. Karp, J. Photochem., 12, 333 (1980)
7. J.C. Scaiano, E.B. Abuin and L.C. Stewart, J. Am. Chem. Soc. (in press)
8. J. Chilton, L. Giering and C. Steel, J. Am. Chem. Soc., 98, 1965 (1976)
9. a) Y. Sakaguchi, S. Nagakura and H. Hayashi, Chem. Phys. Letters, 72, 3, 420 (1980)
    b) Y. Sakaguchi, S. Nagakura and H. Hayashi, Chem. Phys. Letters, 82, 2, 213 (1981)
    c) Y. Sakaguchi and H. Hayashi, Chem. Phys. Letters, 87, 6, 539 (1982)
10. J.C. Scaiano and E.B. Abuin, Chem. Phys. Letters, 81, 2, 209 (1981)
11. A.M. Braun, M. Krieg, N.J. Turro, M. Aikawa, I.R. Gould, G.A. Graff and P.C. Lee, J. Amer. Chem. Soc., 103, 7312 (1981)
12. A. Merlin, D.J. Lougnot, P. Jacques and J.P. Fouassier, Makrom. Chem., 1, 687 (1980)
13. P. Jacques, D.J. Lougnot and J.P. Fouassier, J. Amer. Chem. Soc. Submitted for publication
14. A. Merlin, D.J. Lougnot and J.P. Fouassier, Polymer Bull., 2, 847 (1980)
15. D.J. Lougnot and J.P. Fouassier, Polymer Photochem. (in press)
16. N.J. Turro, M.F. Chow, C.J. Chung and C.H. Tung, J. Amer. Chem. Soc., 102, 7391 (1980)
17. A. Merlin and J.P. Fouassier, Polymer, 21, 1363 (1980).

QUANTITATIVE TREATMENT FOR SALT EFFECTS AND EQUILIBRIA SHIFTS

IN MICELLAR SOLUTIONS

R. Gaboriaud, G. Charbit* and F. Dorion*

Laboratoire de Chimie organique industrielle*
Laboratoire de Physicochimie des solutions
E.N.S.C.P. 11 rue Pierre et Marie Curie, Paris V, France

        The different models published in order to allow
a quantitative treatment of equilibria shifts in micellar media show some discrepancies and do not permit a
complete description of all the observed phenomena. We
have developed a new model in order to fit any experimental result with a good accuracy. This model is an a-
daptation of the pseudo-phase model, but it makes use
of an approach involving also electrochemical equilibria. Thus it combines advantages of both the ion exchange model and theories based on the surface potentials. The relationship proposed for the potential difference between phases is an extension of the formula
usually presented for the selective electrodes. This
has been done in order to make the model operational in
any case. The experimental work has been carried out
with various sodium alkylsulfates (chain length from C_8
to C_{14}), sometimes with a cosurfactant (alkanols from
C_4 to C_9) and weak base indicators in the presence of
various salts. The numerical values of the parameters
defined in the model appear to be independant of the indicator used, and may be applied to the treatment of
other phenomena (kinetics, cmc variations ...).

INTRODUCTION

Numerous measurements have been made in our laboratory concerning ionization equilibria of indicators and the kinetics of
nucleophilic transformations in reaction mixtures containing anionic or cationic surfactants and various electrolyte mixtures (a-
cids, bases, salts). Some preliminary results obtained in acidic
media have previously been published[1]. A large number of models
have been proposed in order to fit qualitatively and quantitatively , the experimental results obtained for reactions studied in
micellar systems[2-14]. Application of these models to our experimental results did not prove totally satisfactory; although they
did, in many cases, provide acceptable overall evaluations. By
employing these proposed models, we were able to note various in-
consistencies and insufficiencies, and we were thus led to adapt
these models in order to obtain improved compatibility with the
experimental results while preserving a sufficient thermodynamic
base. The model which we finally developed, permits a satisfactory
fit with the experimental results concerning both the kinetics and
equilibria studied in micellar media; in addition this model is
very useful in other fields. It permits, for example, correct predictions of variations of the critical micellar concentration of
a given surfactant in the presence of a mixture of electrolytes
when the composition of this mixture varies. In the same way, it
permits a good evaluations of the variations of the apparent pK
value (pK ap) under the same conditions. Conversely, measurements
of pK ap values are sufficient for determining most of the numerical parameters involved in the model.

The purpose of the present paper is to review the basis of
the model as well of its most immediate applications, particularly
the pK ap variations in the presence of electrolytes.

REVIEW

I. POTENTIAL DIFFERENCE BETWEEN PHASES.

The model assumes that each (i) solute, with charge z_i, is
in electrochemical equilibrium between the micellar phase (m) and
the aqueous phase (aq). The charge distribution between these two
phases generates a potential difference $\Delta\Phi$ defined by:

$$\Delta\Phi = \Phi_m - \Phi_{aq}$$

Because of the equilibrium and the resulting equality of the
the electrochemical potentials of (i) in each phase, one can write a relation between $\Delta\Phi$, the activities of (i) in the two phases
and the transfer activity coefficient $\Gamma_t(i)$ defined by:

$$RT \ Ln \ \Gamma_t(i) = \Delta G^\circ_t \ (i)_{aq \to m} = {}^m\mu^\circ(i) - {}^{aq}\mu^\circ(i)$$

In which $\Delta G^\circ_t \ (i)_{aq \to m}$ represents the standard free enthalpy of transfer of the (i) component from the aqueous phase (${}^{aq}\mu^\circ(i)$) to the micellar phase (${}^m\mu^\circ(i)$).

In the particular case of an anionic surfactant LM (amphiphilic anion L^-, counter-ion M^+) the above relationship becomes:

$$Ln \ \frac{(L^-)_{aq}}{(L^-)_m} = Ln \ \Gamma_t(L^-) - \frac{F}{RT} \ \Delta\Phi \tag{1}$$

or:

$$\Delta\Phi = \frac{RT}{F} \left[Ln \ \Gamma_t(L^-) - Ln \ \frac{(L^-)_{aq}}{(L^-)_m} \right] \tag{2}$$

where $(L^-)_{aq}$ and $(L^-)_m$ represent respectively the activities of the monomeric form of the amphiphile in the aqueous phase, and of the associated form in the micellar phase, each one of them being evaluated from a standard state defined in the corresponding phase. For the further development of the model, we will assume that the free enthalpy of the charged organic phase is not modified by the transformations taking place inside the bulk, so that the amphiphile activity in the micellar phase $(L^-)_m$ remains constant.

$$(L^-)_m = Const$$

Consequently, equation (2) can be written as:

$$\Delta\Phi = Const - \frac{RT}{F} \ Ln \ (L^-)_{aq} \tag{3}$$

One can recognize in relationship (3) a form very similar to the Nernst's equation. This analogy leads us to compare the micelle to an (L^-) specific microelectrode.

Furthermore, it has been clearly proved[15-18], that for c.m.c and higher concentrations, the activities of $(L^-)_{aq}$ and the counter-ion $(M^+)_{aq}$ obey a typical law:

$$(L^-)_{aq} \ (M^+)^b_{aq} = Const \tag{4}$$

This formula is better known in its linear form i.e the c.m.c variation versus the added amount of common ion electrolyte C_s

$$Ln \ (c.m.c) + b \ Ln \ (c.m.c + C_s) = Const \tag{5}$$

Taking into account equation (4), equation (3) may be written as:

$$\Delta\Phi = \text{Const} + \frac{RT}{F} \text{Ln } (M^+)^b_{aq} = \Delta\Phi° + b \frac{RT}{F} \text{Ln } (M^+)_{aq} \qquad (6)$$

This relationship is therefore valid when the solution contains only the amphiphile, but it still remains true when a common ion salt M^+X^- is added.

For practical uses, and especially because of the use of buffers, several salts are generally present in the solution, and this leads us to extend equation (6) to a mixture of electrolytes. We shall assume that the $\Delta\Phi$ variations can be expressed by the following relationship.

$$\Delta\Phi = \Delta\Phi°_{ref} + \frac{RT}{F} \text{Ln } \left[(M^+)_{aq} + s_1 (M_1^+)^{b_1}_{aq} + \ldots + s_i (M_i^+)^{b_i}_{aq} \right] \qquad (7)$$

This equasion will be used in the case of an aqueous solution of an anionic amphiphile in the presence of different cations M^+, $M_1^+, \ldots, M_i^+$. Each counter-ion is thus characterized by two numerical parameters s_i and b_i.

Except the b_i terms, this expression is similar to Eisenman's equasion which describes the sensitive electrodes response. The s_i terms (selectivities) are relative values and it is convenient to use an additional convention to fix them. That is the reason why in the former equation, the M^+ cation of the surfactant has been choosen as reference and its selectivity arbitrarily taken equal to unity ($s_{M^+} = 1$). The potential difference $\Delta\Phi°_{ref}$ is reffered to the same cation and its value is obtained when the sum inside brackets is equal to unity.

In the particular case of an excess of the M_iX salt, equation (7) may be written as:

$$\Delta\Phi° = \Delta\Phi°_{ref} + \frac{RT}{F} \text{Ln } s_i + \frac{RT}{F} \text{Ln } (M_i^+)^{b_i}_{aq} \qquad (8)$$

or:

$$\Delta\Phi° = \Delta\Phi°_{Mi} + b_i \frac{RT}{F} \text{Ln } (M_i^+)_{aq} \qquad (9)$$

where

$$\Delta\Phi°_{Mi} = \Delta\Phi°_{ref} + \frac{RT}{F} \text{Ln } s_i \qquad (10)$$

In conclusion, with a sufficient excess of the M_iX salt, the potential difference value $\Delta\Phi$ is related to the (M_i^+) activity in the aqueous phase. Moreover, $\Delta\Phi$ is constant if one keeps constant

$(M_i^+)_{aq}$.

II. PROTONATION EQUILIBRIA CONSTANTS

A simple application of the model consists in treating the protonation equilibria of indicators. Let us consider the following equilibrium:

$$I + H^+ \rightleftharpoons IH^+$$

As the I, H^+ and IH^+ are distributed among the two phases, one can define an apparent pK using the total concentrations of these species:

$$pK\ ap = \log \frac{IH^+}{I} + pH \qquad (11)$$

In the particular case of a strong extraction of the I and IH^+ by the micellar phase, the electrochemical equilibrium between phases implies a simple relation for the pK ap value:

$$pK\ ap = pK\ w + \log \frac{\Gamma_t(I)}{\Gamma_t(IH^+)} - \frac{F}{2,3\ RT} \Delta\Phi \qquad (12)$$

In the above mentioned equasion, pK w reflects the protonation equilibrium taking place in water with no added surfactant, with the condition that this equilibrium does really exist. Replacing $\Delta\Phi$ by its value defined in equation (7), it becomes:

$$pK\ ap = pK\ w + \log \frac{\Gamma_t(I)}{\Gamma_t(IH^+)} - \frac{F}{2,3\ RT} \Delta\Phi^\circ_{ref} - \log\Sigma s_i c_i^{bi} \qquad (13)$$

This equasion shows that the knowledge of the s_i and b_i parameters of each cation, should allow us to calculate the variations in the acidity shift due to micelle. As a matter of fact, for a given amphiphile-indicator system, one should note that:

$$pK\ w + \log \frac{\Gamma_t(I)}{\Gamma_t(IH^+)} - \frac{F}{2,3\ RT} \Delta\Phi^\circ_{ref} = Const \qquad (14)$$

Taking into account equation (14), equation (13) may be written as:

$$pK\ ap = Const - \log \Sigma\ s_i c_i^{bi} \qquad (15)$$

Consequently, it is now possible to give an accurate prediction of the micellar medium effects when adding various amounts of differents salts.

APPLICATIONS

Now we want to review some common systems and examine what kind of results may be predicted using the model.

I. PROTONATION EQUILIBRIA OF DYES

Two cases will be examined according to the presence or absence of inert salt.

When the medium does not contain electrolytes, the studies concerning the protonation equilibria usually lead to typical plots (see Figure 1a). At first sight these results appear to be convenient, in so far as one recognizes the classical representation of the same reaction carried out in aqueous medium. The logarithmic transformation of this curve should be a straight line with unit slope and an intercept equal to pK ap. In fact it is more of a curve (see Figure 1b) and consequently the pK ap value is never well defined since it constantly varies.

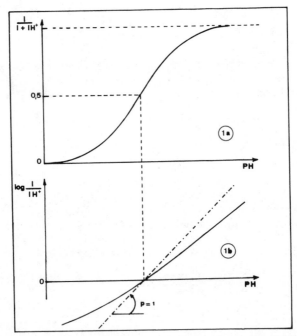

Figure 1. Typical plots for the ionization equilibrium of an indicator.

Thus when the I and IH^+ species are present in equal concentrations, it is incorrect to use the result as a pK ap value. One can only define this value as half-ionization pH.

This result could be anticipated by the model, since equation (13) shows that pK ap is a function of pH. In addition this equation shows that, if the potential difference $\Delta\Phi$ is fixed by an excess of electrolyte, then, the pH variations should have no influence on pK ap. Consequently, the pK ap value is determined by both the kind of the salt and its concentration.

Some experimental results with sodium dodecylsulfate as surfactant, trimethoxy 4, 4', 4'' triphenylmethanol as indicator and sodium chloride as inert salt are presented in Figure 2. One should observe that by adding sufficient amounts of salt, $\log(I/IH^+)$ becomes effectively a linear function of pH, and pK ap becomes a constant. Using other salts in various concentrations, similar straight lines are obtained, and the pK ap level depends on the kind of electrolyte.

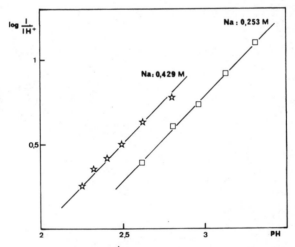

Figure 2. Plot of log (I/IH^+) of the trimethoxy 4, 4', 4'' triphenyl methanol (indicator), in the presence of dodecylsulfate (surfactant) and large excess of sodium chloride (electrolyte).

Determination of the b_i parameters. This can be done using the variations of pK ap with salt concentration. As a matter of fact, equation (13) predicts linear variation of pK ap versus $(-\log C_i)$ if the electrolyte concentration is in large excess with regard to both proton and couter-ion concentrations. The first of these two cations is always present in the solution as reagent.

Under this condition the b_i value wanted is the slope of the above mentioned straight line.

Experiments carried out with sodium dodecyl sulfate lead to the following values:

$$b_{Na^+} = 0,66 \qquad b_{H^+} = 0,69 \qquad b_{Li^+} = 0,9 \qquad b_{NH_4^+} = 0,7$$

$$b_{TEA^+} = 0,95 \qquad b_{Zn^{++}} = 0,5$$

(TEA$^+$ designates the tetraethylammonium cation $(C_2H_5)_4N^+$).

Using four different dyes, it has been found that these values do not depend on the indicator.

Determination of the s_i parameters. M^+ being the reference ion, implies that its selectivity is equal to unity ($s_{M^+}=1$). According to equation (9) and (12) the straight line intercept, q_i, in Figure (3) must be:

$$q_i = Const - \frac{F}{2,3\ RT}\ \Delta\Phi^{\circ}_{M_i}$$

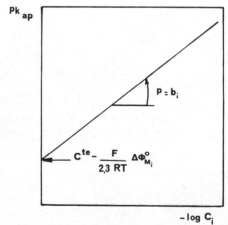

Figure 3. Typical plot of pK ap as a function of ($-\log C_i$) for the calculation of the s_i and b_i parameters.

The same treatment applied to each salt allows the calculation of the s_i parameters. Let q_{ref} be the intercept for the reference ion M^+, then it becomes:

$$q_{ref} - q_i = \frac{F}{2,3\ RT}\left[\Delta\Phi^\circ_{M_i} - \Delta\Phi^\circ_{ref}\right] = \log s_i$$

With various systems containing sodium dodecyl sulfate and
taking Na^+ as reference ion, we could determine:

" $s_{Na^+} = 1$ " $s_{H^+} = 2$ $s_{Li^+} = 0,8$ $s_{NH_4^+} = 1,5$

$s_{TEA^+} = 110$ $s_{Zn^{++}} = 2,5$

Just like for the b_i parameters, it appears that the experimental values found for the selectivities s_i do not depend on the
indicator.

Consequences. The knowledge of the coefficients s_i and b_i
permits the calculation of the variations of the potential difference between phases $\Delta\Phi$, in the presence of various analytical
concentrations of added salts. An exemple of such a calculation is
shown in Figure (4).
According to equation (13) it becomes possible, for each indicator, to correlate all the experimental results by a single
straight line. Such a correlation is presented in Figure (5) illistra-
ting experimental results for the system sodium tetradecylsulfate
trimethoxy 4,4',4'' triphenyl methanol.

II. OTHER SURFACTANTS

Within the family of sodium alkylsulfates, we have tested the
compounds of different chainlengths from octylsulfate C_8SO_4Na to
tetradecylsulfate $C_{14}SO_4Na$. The same mathematical treatment results
in the following conclusions :
- The s_i and b_i parameter values are slightly influenced by
the alkyl chainlength. For example, in the case of $C_{14}SO_4Na$ we have found :

$b_{Na^+} = 0.66$ $s_{Na^+} = 1$

$b_{H^+} = 0.69$ $s_{H^+} = 2$

$b_{Li^+} = 0.75$ $s_{Li^+} = 0.7$

$b_{NH_4^+} = 0.60$ $s_{NH_4^+} = 1.2$

$b_{TEA^+} = 0.98$ $s_{TEA^+} = 110$

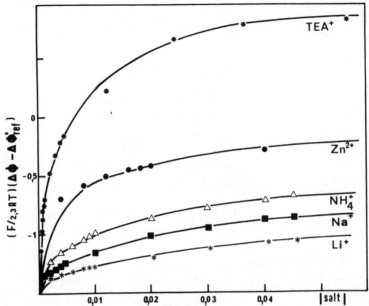

Figure 4. Effects of salts concentration on the variations of $\Delta\Phi$.
Sodium dodecyl sulfate as detergent, N,N dimethylparaphenylazoani-
line as indicator.

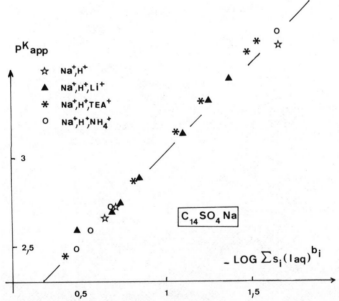

Figure 5. Correlation of experimental results for the variations
of pK ap value as a function of salts concentrations.

All the values given here arise from the mathematical treatment of the data available to date. We are presently trying to com
plete these data, and this should lead to refinement of these values which are open to improvement.

By combining the results relative to several indicators and
surfactants, we were led to the following conclusion : for a given indicator, the ratio of the transfer activity coefficients of
the two forms I an IH$^+$ does not vary by changing the alkyl chainlength of the amphiphile within the same chemical family of compounds :

$$\left[\frac{\Gamma_t(I)}{\Gamma_t(IH^+)}\right]_{D_i} \simeq \left[\frac{\Gamma_t(I)}{\Gamma_t(IH^+)}\right]_{D_j} \simeq \ldots$$

This result proves that for a given I$-$IH$^+$ system the transfer free enthalpies (from the bulk to the micellar phase) vary
very slightly from one amphiphile to another of the same family.
If one admits that C_8SO_4Na, in so far as solvent, is very similar
to $C_{10}SO_4Na$, $C_{12}SO_4Na$, $C_{14}SO_4Na$, then the above conclusion seems
to be logical.

Consequently for very similar media, that is to say for the
same value of the term ($\Sigma \ s_i C_i^{bi}$), the difference between the pK ap
values of an indicator measured in two amphiphiles, derives from
the difference existing between the $\Delta\Phi_M^\circ$ values.

The mathematical treatment presented here is simple provided
that the true concentrations of the salts in the bulk can be evaluated. When an electrolyte is in large excess with respect to the
others and the amphiphile, then its analytical concentration may
be related to its concentration in the bulk. In the opposite case
the problem remains and to solve it, two solutions are available.

$-$ a well improved electrode of the cation (H$^+$,Na$^+$,Li$^+$,...)
can be found. It is then sufficient to immerse the electrode in the
solution to measure the activity of each species.

$-$ this electrode does not exist. In this case, more complicated calculations must be undertaken based on the ions competition
at the surroundings of the micelle. This particular case goes beyond this article and will be the subject of further developments.

III. SURFACTANT-ALCOHOL MIXTURES

Actually, the system studied involves sodium dodecyl sulfate
as surfactant, dimethyl aminobenzene, trimethoxy 4,4',4'' triphe-

nyl methanol as indicators, and N alkanols from butanol to nonanol
as alcohols. This work results in the following conclusions:

i- for a given alcohol-surfactant ratio, the pK ap value decreases as the alcohol chain length increases.

ii- for a given alcohol, the pK ap value decreases with increasing alcohol concentration.

Others workers[20-22] have shown that the addition of alcohol in
amphiphile solutions results in releasing counter-ions in the bulk.
This effect increases with the concentration and the alkyl chain
length of the alcohol. These conclusions derived from structural
studies of the medium by means of physical techniques, cross-check
perfectly our results obtained by chemical probes.

As a matter of fact, the releasing of micelle counter-ions,
due to the addition of alcohol, implies an increasing concentration
of these ions in the bulk, and according to equation (6), an increa
se of the potential difference $\Delta\Phi$. The potential difference $\Delta\Phi$ is
negative so that the increase of the algebraic value is equivalent
to a decrease of the absolute value. This increase of $\Delta\Phi$ thus induces the observed depression of the pK ap value.

By applying the above described mathematical treatment to the
data concerning the four component system alcohol-surfactant-indi-
cator-electrolyte, we have obtained the same set of s_i,b_i parameters both with or without alcohol. Consequently an addition of electrolyte in a micellar solution seems to act in the same way on the
potential difference $\Delta\Phi$ either for a simple micelle or for a more
complicated system containing alcohol and amphiphile.

IV. CATIONIC SURFACTANTS

A similar study has been carried out on cationic surfactant
(dodecyl and cetyl trimethylammonium bromide). Even in this case,
the model appears to be valid. However , as expected, we found lar-
gers effects for halide than for alkaline ions, and these effects
are more differenciated with respect to the kind of the anion.

By employing the model, it becomes possible to clear up the
various cases of "pseudo-failure" of the exchange model[23-25]. For
this purpose it is unnecessary to invoke additional reaction pathway when the expected saturation effects are not observed. One
should note that the variability of the kinetic constants with the
amphiphile concentration is an experimental fact which cannot be
explained by the ion-exchange model. On the contrary these phenomena are in good agreement with the predictions of our model[26]. As a
matter of fact, an increase of the overall surfactant concentration

induces a corresponding increase in counter-ion concentration in the bulk which modifies the potential difference $\Delta\Phi$. The s_i and b_i parameters values deduced from experiment for the halide ion OH^- allow a good description of the phenomena. However we should emphasize that our model is not opposed to the exchange model since it involves the equilibrium:

$$\overset{+}{M_i}\, aq \;+\; \overset{+}{M_j}\, mic \;\rightleftharpoons\; \overset{+}{M_i}\, mic \;+\; \overset{+}{M_j}\, aq$$

which is implicitly imposed by the electrochemical equilibrium. The only difference between the two model is that ours does not set limits to the sum of the bound-ions activities $\Sigma\,(M_{i\,mic}^+)$ which only depends on the activities in the bulk because of the equilibrium (see equasion 1). In this sense, this sum is not directly connected to the number of available sites at the interface.

V. KINETIC OF PROTONATION EQUILIBRIUM.

A number of experiments have been undertaken with trimethoxy 4,4',4'' triphenylmethanol and sodium dodecylsulfate in strongly acidic media. Because of the weak concentrations used, all the kinetics are of pseudo-first order and we will represent by λ the corresponding apparent constants (s^{-1}). These results complete those obtained by Bunton and Huang in neutral and basic media[28-29].

The application of the model to these constants is quite similar to the above described treatment of the equilibrium constants. The typical plot of the λ logarithm as a function of pH, achieved in micellar medium (Figure 6), exhibits the two classical straight lines with -1 and $+1$ slopes for acidic and basic pH. The freezing of the pK_{R+} by a salt effect permits such a result. We have shown that for a given system, the $(pK\ ap - pK_w)$ difference may be determined if one knows the $(pK'\ ap - pk_w)$ difference for another system. In the same way, if it is impossible to have direct access to λ_{mic} from λ° measured in an homogeneous medium, one can obtain λ_{mic} only if λ'_{mic} is known; λ_{mic} and λ'_{mic} represent two apparent kinetic constants measured in two different micellar media.

This treatment shows that any modification of $\Delta\Phi$ is completely reflected in the kinetic constants; the addition of a salt results in a translation of the curve $\log\lambda = f(pH)$. Such is this translation that the reducing of the catalytic effect observed in the acidic medium is strictly equal to the increase of the inhibitory effect observed in basic medium.

The curve is principally composed of two linear parts with -1 and $+1$ slopes and the experimental results can be related by:

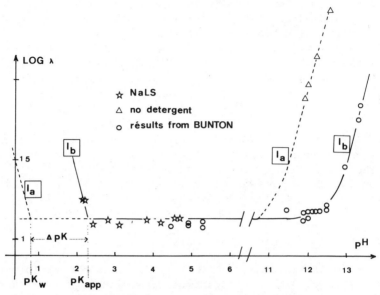

Figure 6. Variations of the apparent kinetic constant as a function
of the pH.(Results from Bunton, ref. 28 and 29).

$$\log \lambda = \log \left[k_{-1}(H^+) + k_o + k_1(OH^-) \right] + p \log \sum_i s_i C_i^{b_i}$$

If one writes the most simple reaction pathway, then, the a-
bove relationship can be easily understood.

$$I \xleftarrow{\quad k_{-1} \ (H^+) \quad} \xrightarrow{} IH^+$$
$$I \xleftarrow{\quad k_e \quad} IH^+$$
$$\xleftarrow{\quad k_1 \ (OH^-) \quad}$$

The central part of Figure 6, the zero slope straight line,
relates the thermal deionization of IH^+ which does not appear to
be influenced by changing the medium (transfer from water to micellar phase).

The linear part of +1 slope represents the deionization by
the OH^- ions; this reaction prevails when the pH level is high e-
nough. The IH^+ cation belongs to the micellar phase whose global
charge is negative; its attack is thus more difficult than in a-
queous medium. The addition of a salt decreases the $\Delta\Phi$ value and
consequently overcomes this difficulty. This implies in turn an
increase of the apparent kinetic constants.

The linear part of -1 slope is relative to the attack of the I species by the protons which, on the contrary, is made easier by the negative charge of the micelle. Salt additions reduce this favourable effect and induce a decrease of the apparent kinetic constants. This effect, which results only from the $\Delta\Phi$ variations, will have the same magnitude as in the previous case, if the salts concentrations are equal.

CONCLUSION

Based on the thermodynamical rules which govern the equilibria between solvents, the model presented here is a synthesis of theories worked out by other authors. The mathematical treatment which is derived from its development makes use of the variations of the potential difference $\Delta\Phi$, but it also takes into account the exchange terms. It thus improves the other models without excluding them.

The model appears to be efficient, since it allows a quantitative analysis of the micellar effect on both the protonation equilibrium and kinetic constants. This can be done, thanks to a set of single easily measurable parameters. The examples presented in this paper relate to system strongly extracted by the micellar phase since they undergo the most important effects. When the extraction is weak, the equations maintain their general form but, in addition, they include a reducing term due to weaker observed shifts.

Furthermore we have shown how the same parameters could provide a good description of the c.m.c variations as a function of added salt concentrations.

The advantage of the model is to utilize the concentrations in the bulk which are generally known or accessible, and a few parameters (two for each counter-ion). The model is not developed from a microscopical well defined structure and it only assumes that two separated phases can be statistically defined. A similar treatment could be developed from a more sophisticated model, but this would result in an increase in the number of adjustable parameters; this seems to us useless since the parameters introduced so far are adequate to describe correctly the experimental results.

REFERENCES

1. F. Dorion and R. Gaboriaud, J. Chim.Phys., 78, 555 (1981).
2. P. Mukerjee, K.L. Mysels and P. Kapauan, J. Phys. Chem., 71, 3262 (1967).
3. C.A. Bunton, M.J. Minch and J. Hidalgo J. Am. Chem. Soc., 95, 3262 (1973).
4. K. Shirama, Bull. Chem. Soc. Japan., 48, 2673 and, 49, 2731 (1976).

5. D. Pyzkiewicz, J. Am. Chem. Soc., 99, 1550 (1977).
6. P Mukerjee, J.R. Cardinal and R.N. Desai, in "Micellization, Solubilization and Microemulsions" K.L. Mittal, Editor, Vol 1, p.241, Plenum Press New York, 1977.
7. C.A. Bunton, K. Ohmenzetter and L. Sepulveda, J. Phys. Chem., 81, 2000 (1977).
8. L.S. Romsted, in "Micellization Solubilization and Microemulsions" K.L. Mittal, Editor, Vol 2, p.509, Plenum Press, New York, 1977.
9. C.A. Bunton, L.S. Romsted and H.J. Smith, J. Org. Chem., 43, 4299 (1978).
10. K. Martinek, A.K. Yatsimirsky, A.V.Levashov and I.V. Berezin, in "Micellization, Solubilization and Microemulsions" K.L. Mittal, Editor, Vol 2, p.489, Plenum Press New York, 1977.
11. F. H. Quina and H. Chaimovitch, J. Phys. Chem., 83, 1844 (1979).
12. M. Almgren, F. Greiser and J.K. Thomas, J. Am. Chem. Soc., 101, 279 (1979).
13. N. Funasaki, J. Phys. Chem., 83, 237 (1979) and 83, 1998 (1979)
14. C.A. Bunton, L.S. Romsted and L. Sepulveda, J. Phys. Chem., 84, 2611 (1980).
15. M. Koshinuma and T.Sasaki, Bull.Chem. Soc. Japan, 48, 275 (1975).
16. S.G. Cutler, P. Meares and D.G. Hall, J. Chem. Soc. Faraday Trans. I, 74, 1758 (1978).
17. T. Sasaki, M. Hattori, J. Sasaki and K. Nukina, Bull.Chem. Soc. Japan., 48, 1397 (1975).
18. K.M. Kale, E.L. Cussler and D.F. Evans, J.Phys. Chem., 84, 593 (1980).
19. G. Charbit, F. Dorion and R. Gaboriaud, Unpublished data.
20. N. Manabe, M. Koda and K. Shirama, J.Colloid. Interface.Sci., 77, 189 (1980).
21. A..C. Lawrence and J.T. Pearson, Trans. Faraday. Soc., 63, 495 (1967).
22. A.K. Jain, and R.P.B. Singh, J. Colloid. Interface. Sci., 81, 536 (1981).
23. F. Nome, A.F. Rubira,C. Franco and L.G. Ionescu, J.Phys. Chem., 86, 1881 (1982).
24. C.A. Bunton, J. Frankson and L.S.Romsted, J. Phys. Chem., 84, 2607 (1980).
25. C.A. Bunton, L.S. Romsted, and G. Savelli J. Am. Chem. Soc., 101, 1253 (1979).
26. R. Gaboriaud, F. Dorion. J. Lelievre and J. Lelievre, To be published.
27. R. Gaboriaud and J. Lelievre, These de Doctorat d'Etat, Paris 1982.
28. C.A. Bunton and S.K. Huang, J. Am. Chem. Soc., 94, 3536 (1972).
29. C.A. Bunton and S.K. Huang, J. Org. Chem., 37, 1790 (1972).

THE NICKEL(II)-PADA REACTION AS A SOLUBILIZATION PROBE IN ANIONIC

MICELLAR SOLUTIONS

J. R. Hicks and V. C. Reinsborough

Department of Chemistry
Mount Allison University
Sackville, New Brunswick, Canada, E0A 3C0

The kinetics of the Ni^{2+}/PADA reaction have been
used as a probe to examine solubilization structuring
and sites for benzene and dimethylphthalate in micellar sodium octylsulfate and sodium octanesulfonate
solutions. Benzene in small amounts relative to the
surfactant increased the rate enhancements of the
probe reaction in the cmc-to-twice-the-cmc range but
had no effect beyond this region. Both of the reactants experienced an increased attraction for the micellar surface region with benzene present. With
dimethylphthalate, neither Ni^{2+} nor PADA was more
strongly attracted into the micelles. The overall
effect on k_{obs} with dimethylphthalate as additive was
a decrease over the entire micellar range. Benzene
probably served to shorten the concentration range
over which micelles build up to a regular size and it
was mainly solubilized within the micellar core. Dimethylphthalate, on the other hand, was adsorbed into
the micellar surface region and hindered the surface
concentrative effect of the micelles upon the probe
reaction.

INTRODUCTION

Incorporation of organic matter into micelles, vesicles or micro-emulsions has been of much interest lately because of applications in such diverse areas as oil recovery, membrane adsorption and micellar structure determinations. Despite this widespread interest, solubilization, as this phenomenon is termed, is not adequately understood. The sites of solubilization in micelles for even the simplest substances are also mainly conjecture. Several spectroscopic techniques have been brought to bear upon this problem and results are not always in agreement. For example, benzene when added to micellar sodium dodecylsulfate solutions was found to be principally in the micellar surface region,[1] chiefly in the micelle core,[2] and distributed between micellar surface and core with most in the core.[3] Recently, much use has been made of the ESR,[4] fluorescence,[5-8] and C-13 NMR[6,9] techniques to pinpoint solubilization sites within micelles. No technique is without its problems and uncertainties. The first two usually require large molecular probes that are often larger than the surfactant monomer thereby calling into question the impartiality of the probe in the micelle formation process.[10] The last method, C-13 NMR, requires small chemical shifts or small changes in spin-lattice relaxation times to be resolved for which no adequate theory is yet available. For greater confidence in the results, the best approach with these methods is probably a combination of them.[6]

This work reports on a different type of experiment that might find use in locating solubilizates in micelles and in throwing light upon the solubilization process. In this micellar probe, the kinetics of a complexation reaction that takes place preferentially on the surface of anionic micelles is observed. The reactants are Ni^{2+} (aq) and the bidentate ligand, pyridine-2-azo-p-dimethylaniline (PADA) and each is partitioned between the bulk aqueous phase and the micellar surface according to independently measured distribution constants that allow the kinetics to be completely characterized.[11-15] The rate enhancements that are observed in micellar solution for this reaction are often thousand-fold, and are very sensitive to surfactant chain length, head group, and ionic strength in a predictable fashion. We wished to see how sensitive the reaction was to micellar changes brought about by solubilization. One advantage of this probe is that the guest reactants are low in concentration compared to the micellar surfactant and veil usually only a few percent of the micellar surface. The addition of the reactants generally does not affect the critical micelle concentration (cmc) which is some assurance that both the probing reactants are truly observing and not participating in the micellization process.

THE NICKEL(II)-PADA REACTION

EXPERIMENTAL

The small-volume stopped-flow apparatus with spectrometric detection and the experimental procedure have been described elsewhere.[11,15,16] One modification was the use of an 8-bit digital data logging system to process 64 or 128 points from each exponential transient to obtain a value of the observed kinetic constant (k_{obs}). Then, seven such values for each set of conditions were averaged.

The surfactants were obtained from Eastman Kodak, PADA from Sigma, and all other reagents from Fisher Scientific. They were used without further purification. The nickel and PADA concentrations were usually 10^{-3} and 10^{-5} mol dm^{-3} respectively to ensure pseudo-first order conditions. All kinetic and solubility data were obtained at 25°C.

RESULTS

Sodium octylsulphate (SOS) solutions from 0.05 mol dm^{-3} to 0.35 mol dm^{-3} (cmc 0.114 mol dm^{-3}) were doped with small amounts of benzene (2.5 x 10^{-3} to 2.0 x 10^{-2} mol dm^{-3}) and k_{obs} values for the Ni^{2+}/PADA complex formation reaction at 25° were then obtained in each of these solutions. These are displayed in Figure 1 where R is the rate enhancement or the ratio k_1^{app}/k_1^W where k_1^W is the forward rate constant in water (1.20 x 10^3 dm^3 mol^{-1} s^{-1}) for the metal-ligand complexation and k_1^{app} is given by

$$k_{obs} = k_1^{app} (Ni^{2+})_T + k_{-1}^M \qquad (1)$$

where $(Ni^{2+})_T$ is the total nickel concentration and k_{-1}^M is the backward rate constant for the reaction on the micelle surface[13].

Also in Figure 1 are displayed the R values for kinetic runs done with 0.020 and 0.030 mol dm^{-3} additions of dimethyl phthalate (DMP) in micellar SOS solutions. In Figure 2, the corresponding data for similar additions of benzene and DMP to the sodium octane-sulfonate micellar system (SOSO) are given.

The kinetic equasion relating k_{obs} to the partitioning coefficients when the reaction occurs predominantly in the micelle surface region is [13,14]

$$k_{obs} = k_1^M (Ni^{2+})_T / (CV_s \{1+(CK_{Ni})^{-1}\}\{1+(CK_{PADA})^{-1}\}) + k_{-1}^M \qquad (2)$$

where C is the concentration of micellar surfactant, i.e. the

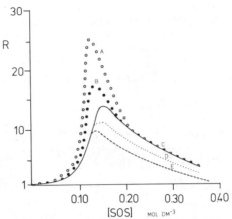

Figure 1.   Effect of benzene and DMP on R for the probe reaction in aqueous SOS solutions at 25.0°C. A, 0.020 mol dm^{-3} benzene; B, 0.014 mol dm^{-3} benzene; C, no additives; D, 0.020 mol dm^{-3} DMP; E, 0.030 mol dm^{-3} DMP.

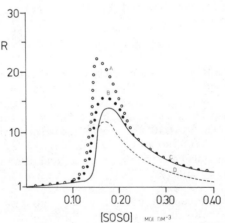

Figure 2.   Effect of benzene and DMP on R for the probe reaction in aqueous SOSO solutions at 25.0°C. A, 0.020 mol dm^{-3} benzene; B, 0.010 mol dm^{-3} benzene; C, no additives; D, 0.020 mol dm^{-3} DMP.

total surfactant concentration less than cmc; V_s, the reaction
volume per mole of micellized surfactant; and K_{Ni} and K_{PADA}, the
partitioning coefficients for each of the reagents between the bulk
water and micellar pseudo-phases. The distribution or partitioning
constants are defined as

$$K_{species} = (Species)_{bound}/C(species)_{free} \qquad (3)$$

K_{Ni} and K_{PADA} can usually be measured independently by the murexide technique[17,18] and solubility measurements[13,15] respectively.
With short-chained surfactants such as SOS and SOSO, the murexide
technique is only approximate[13,14] and K_{Ni} along with V_s values
are found by computer fitting the k_{obs} data to Equation (2). In
this analysis, it was assumed that $K_1^M \approx k_1^w$ and that the contribution of K_{-1}^M to Equasion (2) was negligible.

If the amount of additive compared to the amount of micellized
surfactant in solution were large, then an extra term incorporating
the excess volume per mole of additive and its concentration in the
micelle would have to be included in Equation (2). This was avoided
because, with the additive concentrations chosen for this study,
only in the vicinity of the cmc would this condition hold. Data
points in the vicinity of the cmc were not used for curve fitting.

The fitted values of K_{Ni} and V_s together with K_{PADA} obtained
through solubility measurements with the additives present are
given in Table I for the SOS system. Neither additive caused the
cmc (0.115 mol dm^{-3}) to be shifted so far as could be determined
through conductivity measurements. The relative error in K_{PADA}
is about 5% and for K_{Ni} and V_s, about 10%.

The choice was made to determine the kinetic parameters in
each case at constant additive concentration. A more appropriate
choice might have been to have kept constant the additive-to-mi-
cellar-surfactant ratio. After all, the micellar surfactant ratio increased about threefold over the range where the kinetic
parameters were obtained. However, it is difficult experimentally
to obtain K_{PADA} with this constraint of constant ratio of additive to micellar surfactant. Since the parameters changed only
slightly and in the same direction when additive concentrations
were increased severalfold, it was deemed sufficient to analyze
the results at constant additive concentrations. K_{PADA} was thus
determined from the solubilities of PADA in SOS solutions containing the same amount of additive.

Table I. Effect of Benzene and DMP on the Binding Constants and
Reaction Volume in Micellar SOS solutions.

SOS Solutions	K_{PADA} $(dm^3\ mol^{-1})$	K_{Ni} $(dm^3\ mol^{-1})$	V_s $(dm^3\ mol^{-1})$
no additive	370	11	0.55
2.5×10^{-3} M C_6H_6	385	12	0.50
7.5×10^{-3} M C_6H_6	415	13	0.55
1.4×10^{-2} M C_6H_6	450	17	0.60
2.0×10^{-2} M C_6H_6	480	23	0.60
2.0×10^{-2} M DMP	370	10	0.75
3.0×10^{-2} M DMP	370	10	0.90

With increasing benzene additions, both K_{PADA} and K_{Ni} increased in gradual fashion. With DMP additions, the solubility
of PADA in the micellar solutions showed no variation within
experimental error. When the k_{obs} data points were fitted to
equation (2), K_{Ni} values showed a similar constancy. However,
the reaction volume V_s underwent a marked increase with DMP additions.

The same trends were noted in the SOSO micellar system, e.g.
when 0.020 mol dm^{-3} DMP was added to SOSO solutions, V_s was increased to 0.85 $dm^3\ mol^{-1}$ from 0.70 $dm^3\ mol^{-1}$ while K_{PADA} and K_{Ni}
remained the same within experimental error.

The effect of the additives, benzene, DMP and cyclohexane,
on the rate enhancements in 0.10 mol dm^{-3} sodium dodecylsuflate
(SDDS) solutions were also tested (Figure 3).

                              DISCUSSION

Benzene-doped SOS and SOSO solutions showed greater rate enhancements for the Ni^{2+}/PADA reaction than for the pure surfactants
in the concentration range, cmc to twice the cmc. At higher surfactant concentrations, no difference in k_{obs} and consequently R
could be detected (Figures 1 and 2. Each curve represents 8–10
data points which were omitted for the sake of clarity. For the
same reason, several intermediate benzene-SOS and benzene-SOSO
curves are not shown.) Benzene additions to 0.10 mol dm^{-3} SDDS
also did not affect k_{obs}.

These same trends were noted in micellar sodium decylsulfate
(SDS) and sodium hexylsulfate (SHS) micellar solutions. Each of
these systems had its problems and was not examined extensively.

Benzene additions significantly changed the cmc of the SDS micellar
system and, in the SHS system, very little benzene could be solubilized. In the latter system, 0.015 mol dm^{-3} addition of benzene
to a 0.50 mol dm^{-3} SHS solution (cmc 0.4 mol dm^{-3}) caused k_{obs} to
be increased ·from 3.6 s^{-1} to 4.6 s^{-1}. Not enough data could be
collected for either the SHS or SDS micellar range to effect a complete kinetic anaylsis but it was clear that a similar pattern was
emerging.

From Table I, it is apparent that the increase in R in the
benzene-doped systems resulted from an increased attraction of each
of the reactants, Ni^{2+}$_{(aq)}$ and PADA, for the mixed micelles. This
increase in R was most dramatic when there was an excess of benzene
over micelles just beyond the cmc. As the ratio of benzene to micelles decreased, R gradually approached the pure surfactant values.

Accompanying this increase in R with increasing benzene additions was a shift in the maximum value of R to lower micellar concentration without any significant lowering of the cmc. This suggests that benzene, acting somewhat as a co-surfactant, served to
increase the surface potential of the micelle perhaps by promoting
the formation of larger SOS and SOSO micelles over the micelle
aufbau region than would normally form in the absence of benzene.
With short-chained surfactants, this range extends from roughly
the cmc to twice the cmc over which a problematic and persistent

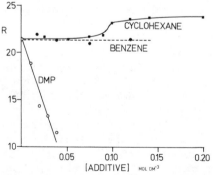

Figure 3. Effect of cyclohexane, benzene and DMP on R for the
probe reaction in 0.10 mol dm^{-3} SDDS.

decrease in the ultrasonic relaxation frequency has been noted.[19,21] If the micelle size were unchanging, then an increase in surfactant concentration in the micellar range should give rise to an increase in the relaxation frequency. However, an increase in micellar size and stability leads to a decrease in the relaxation frequency and, in the cmc-to-twice-the-cmc range, this could be the dominating influence on the relaxation frequency.

The unvarying k_{obs} with benzene additions beyond twice the CMC in the SOS and SOSO micellar systems points to a micellar core position for the solubilized benzene. Even if, at these concentrations, one or two benzene molecules per micelle were adsorbed into the micellar surface region as suggested by the ultrasonic data for the same systems,[3] the probe reaction would not "see" these as only a small fraction of the micellar surface was covered by the reactants. Additions of benzene to 0.15 mol dm^{-3} in 0.10 mol dm^{-3} SDDS solutions similarly had no effect on R (Figure 3). Cyclohexane is micellar core-solubilized[22,24] and, up to 0.10 mol dm^{-3}, additions of this compound made no difference to R (Cyclohexane was insufficiently solubilized in SOS and SOSO solutions to be added to these systems.) (The slight increase in R that occurred beyond 0.10 mol dm^{-3} cyclohexane probably reflected the formation of a larger or differently shaped SDDS micelle. The fact that this increase occurred at a 1:1 ratio of surfactant to additive may be significant but this was not pursued further.) Hence, the kinetic behaviour suggests that benzene like cyclohexane was most likely principally solubilized within the micellar core.

Reinforcement for this view comes from comparing the effect of adding DMP instead of benzene to these same three surfactant systems. DMP is solubilized in the micellar surface region.[25] At all micellar concentrations, R was reduced as would be expected with the reactants of the probe reaction competing for the same micelle surface inclusion. The kinetic analysis reveals that this reduction occurred not by either of the reactants being within experimental error significantly less attracted to the micellar surface but by the reaction volume being enlarged. The surfactant micelle itself was probably no larger nor any different with DMP in the neighbourhood. Otherwise surface potentials and hydrophobic forces would have been altered and this would have been reflected in changes in K_{Ni} and K_{PADA}. The presence of the adsorbed DMP simply swelled the reaction envelope thus hindering the reactant molecules in coming into contact. In other words, the DMP served to dilute the effective micellar surface concentrations of the reagents of the probe reaction. This is in contrast to the benzene and cyclohexane situations where the micellar core solubilization site of these additives allowed the probe reaction to proceed unfettered. The fact that the cmc did not change is some assurance that nothing more drastic, e.g. break-up into smaller micelles, occurred in these micelles.

ACKNOWLEDGEMENTS

Financial support came from the Natural Sciences and Engineering Research Council of Canada with technical assistance provided
by Dr. Alex Whitla, David Jobe and Dan Kelley.

REFERENCES

1. S. J. Rehfeld, J. Phys. Chem., 74, 117 (1970).
2. P. Mukerjee and J. R. Cardinal, J. Phys. Chem., 82, 1620 (1978).
3. D. J. Jobe, V. C. Reinsborough, and P. J. White, Can. J. Chem.
    60, 279 (1982).
4. T. Nakagawa and H. Jizomoto, Kolloid Z. Z. Polym., 250 (1972).
5. K. Y. Law, Photochem. Photobiol., 33, 799 (1981).
6. Y. Tricot, J. Kiwi, W. Niederberger, and M. Grätzel, J. Phys.
    Chem., 85, 862 (1981).
7. P. Lianos and R. Zana, J. Colloid Interface Sci., 84, 100 (1981).
8. M. Almgren and J. E. Löfroth, J. Colloid Interface Sci., 81,
    486 (1981).
9. D. L. Reger and M. M. Habib, J. Phys. Chem., 84, 77 (1980).
10. B. A. Lindig and M. A. J. Rodgers, Photochem. Photobiol., 31,
    617 (1980).
11. A. D. James and B. H. Robinson, J. Chem. Soc., Faraday Trans. 1,
    74, 10 (1978).
12. J. Holzwarth, W. Knoche, and B. H. Robinson, Ber. Bunsenges.
    Phys. Chem., 82, 1001 (1978).
13. P. D. I. Fletcher and V. C. Reinsborough, Can. J. Chem., 59,
    1361 (1981).
14. J. R. Hicks and V. C. Reinsborough, Aust. J. Chem., 35, 15
    (1982).
15. V. C. Reinsborough and B. H. Robinson, J. Chem. Soc. Faraday
    Trans. 1, 75, 2395 (1979).
16. B. H. Robinson, N. C. White, and C. Mateo, Adv. Mol. Relaxation
    Interact. Processes, 7, 321 (1975).
17. M. Fischer, W. Knoche, B. H. Robinson, and J. Wedderburn, J.
    Chem. Soc. Faraday Trans. 1, 75, 119 (1979).
18. M. Fischer, W. Knoche, P. D. I. Fletcher, B. H. Robinson, and
    N. C. White, Colloid Polym. Sci., 258, 733 (1980).
19. J. Gettins, P. L. Jobling, M. F. Walsh, and E. Wyn-Jones, J.
    Chem. Soc. Faraday Trans. 2, 76, 794 (1980).
20. D. A. W. Adair, V. C. Reinsborough, H. M. Trenholm, and J. P.
    Valleau, Can. J. Chem., 54, 1162 (1976).
21. D. A. W. Adair, V. C. Reinsborough, and S. J. Zamora, Adv. Mol.
    Relaxation Interact. Processes, 11, 63 (1977).
22. J. C. Ericksson and G. Gillberg, Acta Chem. Scand., 20, 2019 (1966).
23. G. Lindblom, B. Lindman, and L. Mandell, J. Colloid Interface
    Sci., 42, 400 (1973).
24. N. Kamenka, H. Fabre, M. Chorro, and B. Lindman, J. Chim. Phys.,
    74, 510 (1977).
25. T. Takenaka, K. Harada, and T. Nakanaga, Bull. Inst. Chem. Res.,
    Kyoto Univ., 53, 173 (1975).

THE APPLICATION OF SURFACTANTS IN THE SPECTROPHOTOMETRIC DETERMINATION OF METAL IONS: THE INTERACTION BETWEEN CATIONIC SURFACTANTS AND SOME ORGANIC DYES

L. Čermáková

Department of Analytical Chemistry
Charles University
Prague, Czechoslovakia 128 40

Tensides (surfactants) have recently been successfully used in spectrophotometric determinations of various ions. Cationic (and sometimes non-ionic) tensides are most often employed in the microdetermination of a number of cations. The presence of tensides considerably improves the spectrophotometric characteristics of metal complexes formed by reaction with metallochromic dyes. This beneficial effect of tensides was observed with many metallochromic dyes and most often with triphenylmethane dyes; the anionic complexes of the latter with various metal ions are advantageous, in presence of tensides, for determinations of metals either in aqueous media, or after extraction into organic solvents.

Despite extensive use, the mechanism of these sensitizations has not yet been elucidated. In order to help explain the effect of tensides, their interaction with some triphenylmethane dyes was studied, concentrating on the tenside behavior in aqueous solutions (micelle formation, solubilization, etc.). It was found that cationic tensides affect the dissociation of the ionizable groups of these dyes. The magnitude of the effect depends upon surfactant micellization and the structure of the dye molecule. As a result, the experimental conditions suitable for practical determinations are also affected by the critical micelle concentration of the tensides.

INTRODUCTION

The use of tensides in spectrophotometric microanalysis has brought about a great development in the application of chromophore chelating organic dyes (metallochromic indicators) to the determination of metal ions and has led to new analytical procedures for such determinations in aqueous solution and/or after extraction.

Spectrophotometry in the visible range as a method of quantitative analysis has been often used in the past years since the formation of binary colored chelate complexes occurs as a consequence of reaction between metal ions and dyes. The differences in the absorption spectra are determined by different electron arrangement in the molecular orbitals of the dye alone and in the presence of a metal ion, primarily by transitions in π-electron chromophore groups. The organic dyes used contain various more or less polar groups with donor atoms capable of dissociating protons. Generally, only some of these groups participate in the bonding to metal ions and the remaining groups enable the colored chelate to dissolve in water. If the dye molecule contains only a few of these groups and all the groups are bound in the chelate, then the product is generally insoluble in water and the colored chelate must be extracted into an organic solvent for spectrophotometric determination of the metal ion.

Although ion-associates have been infrequently used in analytical practice, reactions among large ions have been utilized for spectrophotometric determination of analytes after extraction.

Addition of gelatine as a protective colloid during spectrophotometric determinations has shown that gelatine exerts a pronounced effect on the absorption spectra of colored products[1] and suppresses side reactions of metal ions (e.g. hydrolysis). This marked the beginning of the application of similar substances, tensides, in spectrophotometry. The addition of gelatine and some other tensides to aqueous solutions of colored chelates leads to significant and favorable changes in the spectral characteristics of the solutions.

At present, the number of published works involving the use of a great variety of tensides in spectrophotometric inorganic analysis is greater than 200, thus illustrating the usefulness of these reactions. The only review of this field has been given by Hinze[2].

The use of tensides in photometry is varied due to their many unique properties. For example, solubilization ability of tensides can be utilized when a metal-dye complex is insoluble in water and thus an extraction procedure can be avoided (e.g. non-ionic tenside Triton X-100 has been successfully used for the purpose)[2]. On the

other hand, some tensides can be dissolved in organic solvents
(e.g. cationic tenside Septonex in chloroform[3,4]) and used for
extractive determination of various ions (such as complex anions).
This is possible because the anion can form ion associates with the
tenside cations. With valence-saturated, but coordination—unsaturated chelates containing coordinated water molecules, which are
usually difficult to extract into inert solvents, the presence of
tensides with quaternary ammonium surfactants enhances the extraction
due to the formation of higher-coordinated chelates[5].

However, the bathochromic and hyperchromic shifts in the
spectra of water-soluble metal-dye chelates, caused by the presence
of tensides, are primarily utilized in analytical practice since
these spectral characteristics are substantially improved. The
solubilization effect of tensides often also plays a role and, last
but not least, the basic reaction of the metal with the dye is also
favorably affected. In the presence of a tenside, the colored
chelate is formed in the more acidic pH region which often leads to
an improvement in the selectivity of the determination; the
reaction is usually faster, it is unnecessary to heat the reaction
mixture to develop fully the coloration, and, sometimes, the
presence of a tenside is necessary for the formation of the
chelate. The stability of the product color is also usually
greater in the presence of a tenside. Therefore, the coloration of
solutions containing tensides and prepared under optimal conditions
is used for quantitative analysis, provided that the absorbance of
the chelate formed is a linear function of the metal concentration.

The bathochromic shifts in the spectra of aqueous chelate
solutions depend on the dye and tenside used. The $\Delta\lambda$ value, i.e.
the shift in the absorption maximum wavelength for the metal-dye-
tenside ternary system relative to the metal-dye binary system,
amounts to 100 to over 200 nm. As an example, the absorption
spectrum of the Be(II)-ECR-Septonex system[6] is given in Figure 1.
As can be seen, there is a bathochromatic shift of $\simeq$ 100 nm in λ_{max}
of complex and a $\simeq$ 4- fold increase in the molar absorptivity of the
complex in the presence of the tenside.

Although the bathochromic shift for most metal-dye-tenside
systems described in the literature are not so large as above, the
effect of the background reagent blank is still sometimes virtually
completely eliminated, due to a great contrast in the coloration of
the chelate and of the blank solution. It can be seen in Figure 1
that the blank value is very low at λ = 600 nm.

Another example is given in Figure 2 depicting the absorption
curves for the Pd(II)-PG-Septonex system[7] and the corresponding
difference curves whose maxima are used for practical determination
of the metal ion. It can be seen from Figure 2 that the maximum of
the difference curve for the Pd-PG-Septonex system against the

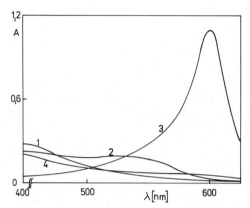

Figure 1. The absorption spectra of the Be(II)-ECR-Septonex sys-
tem. c_{Be} = 1.10^{-5}M, c_{ECR} = 3.10^{-5}M, c_T = 6.10^{-4}M, pH 6 (acetate
buffer), heated to 60°C for 10 min. Curve 1 represents the
absorption of reagent ECR alone at given pH; curve 2 describes the
formation of Be-ECR binary complex; curve 3, corresponding to the
Be-ECR-Septonex system demonstrates enormous bathochromic and
hyperchromic effects; and curve 4 corresponds to the blank solution
(i.e. ECR-Septonex). These curves were obtained against water as
the solvent on a Unicam SP-800 instrument (Pye Unicam, England),
with l = 1 cm.

blank lies at about 620 nm; however, if the calibration plot is
determined at λ = 650 nm, the blank value is zero and the sensitivity of the method remains essentially the same.

     The hyperchromic effect, i.e. the increase in the ε value
which is a characteristic constant for any colored product under
given conditions, leads to an increase in the overall sensitivity
of determination. The ε value for binary complexes increases several times in the presence of tensides and may attain values of the
order of 10^5 mol^{-1}l cm^{-1}. [For example, in the reaction of Ga(III)
with Chromal Blue G dye in the presence of cation active tenside
CPB, the λ_{662} value is 1.44 x 10^5 (ref. 8).] These procedures thus
rank among the most sensitive spectrophotometric methods ever developed for the determination of metals. The hyperchromic effect is
demonstrated in Figures 1 and 2 in the increased absorbance values

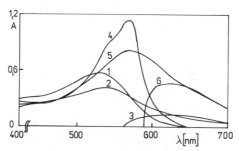

Figure 2. The absorption spectra of the Pd(II)-PG-Septonex system. $c_{Pd} = 2.10^{-5}$ M, $c_{PG} = 4.10^{-5}$ M, $c_T = 5.10^{-4}$ M, curve 1 - 3: pH 6.5, measured after 70 min. standing at laboratory temperature; 4 - 6: pH 5.5, measured after 20 min. standing at laboratory temperature. Curve 1: PG, 2: Pd-PG, 3: difference 2-1, 4: PG-Septonex, 5: Pd-PG-Septonex, 6: difference 5-4.

for the absorption bands of Be-ECR-Septonex and Pd-PG-Septonex systems.

The absorption bands of chelates in the presence of tensides may also differ in character. A sharp and narrow maximum shown in Figure 1 occurs less often; the bands are usually not as well resolved, are broadened to a varying extent and sometimes they are even split into several bands. However, even these less ideal bands have been successfully employed analytically.

The effects observed in the spectra of chelates also depend on the character of the tenside. The highest spectral effects are generally attained with cation active tensides of the type tetra-alkylammonium or pyridinium salts, with a halide counter-ion, most often Cl^- or Br^-. The condition for manifestation of a suffi-ciently high effect is the presence of at least one sufficiently long carbon chain in the tenside cation. For C < 8 in the hydro-phobic part of the molecule there are no effects on the spectra due to the fact that these tensides are not surface active. Salt effects have also been reported in some of these type systems[12].

Of the great number of cation active tensides, those of the tetraalkylammonium type, cetyltrimethylammonium chloride or bromide (CTMAC, CTMAB) are most often used (see Figure 3), as well as benzyltetradecyldimethylammonium chloride (Zehpiramine) and tri-caprylmethylammonium chloride (Capriquat). As representatives of

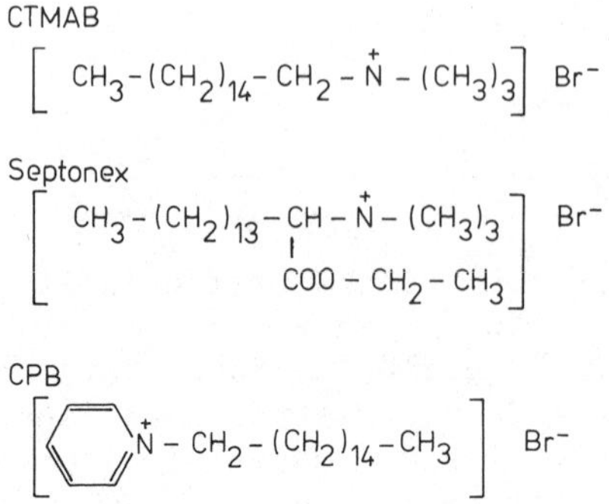

CTMAB

$$\left[CH_3-(CH_2)_{14}-CH_2-\overset{+}{N}-(CH_3)_3 \right] \; Br^-$$

Septonex

$$\left[\begin{array}{c} CH_3-(CH_2)_{13}-CH-\overset{+}{N}-(CH_3)_3 \\ | \\ COO-CH_2-CH_3 \end{array} \right] \; Br^-$$

CPB

$$\left[\text{⟨N⟩}-CH_2-(CH_2)_{14}-CH_3 \right] \; Br^-$$

Figure 3. Structural formulae of selected cationic active tensides.

the pyridinium type, cetylpyridinium chloride or bromide (CPC, CPB) can also be mentioned. In our laboratory, Septonex (CPTB, carbeth-oxypentadecyltrimethylammonium bromide) (Figure 3) yielded the best results in many cases. The results of a number of works indicate that the above effects of tensides are independent of the structure of the cationic active tenside used and the character of the ten-side hydrophilic group, but the magnitude of the effect depends on the length of the tenside chain[9]. For example, for the Be(II)-ECR chelate, the effect of Septonex ($\Delta\lambda = 88$ nm), CPB ($\Delta\lambda = 92$ nm) and CTMAB ($\Delta\lambda - 94$ nm) on λmax was followed[6]. However, the hyper-chromic effect caused by these tensides varied in a different order. Namely ε was equal to 9×10^4, 8.4×10^4 and 7.4×10^4 for Septonex, CTMAB and CPB, respectively. Such comparison of tenside effects may, however, lead to erroneous conclusions when a high and defined purity of the tensides is not established.

Non-ionic tensides, especially those of the polyoxyethylene type (with the $-(O-CH_2-CH_2)_n-$ group), have begun to be system-atically studied only recently[10,11]. For example, in the study of the reaction of chromazurol S (CAS) with some di- and trivalent

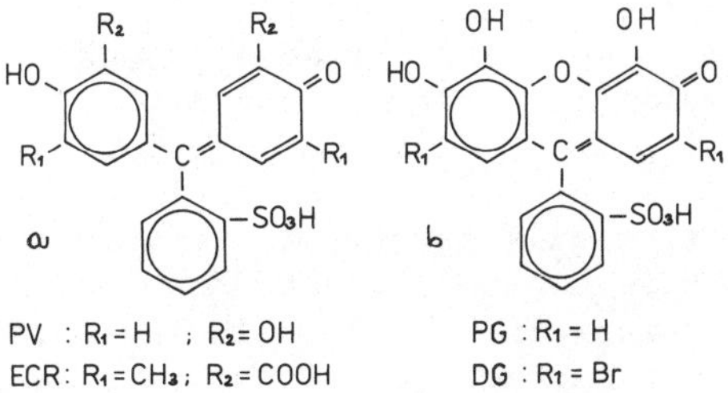

Figure 4. Structural formulae of selected triphenylmethane dyes: pyrocatechol violet, eriochromcyanine R, pyrogallol red and bromopyrogallol red.

metal ions in the presence of various nonionic tensides, ε value of up to 2.7×10^5 and bathochromic shifts of 200 to 250 nm were observed [10].

As follows from analytical work, the favorable effect of ten- sides has been observed with a great variety of organic dyes, e.g. azo dyes, phenoxazine dyes, etc. The greatest effects have so far been observed with triphenylmethane dyes. These substances also have the broadest analytical application.

The structural formulae of some of the most common representa- tives of these dyes are given in Figure 4. A common feature of all these dyes is the presence of four acid, polar substituents, i.e., $-SO_3H$ group and $-OH$ or $-COOH$ which are capable of dissociating a proton depending on the medium pH. A number of dyes from this group that are often used differ in their substituents and often do not contain the sulpho group (e.g. gallein and phenylfluorone). An advantageous grouping of these substituents with the oxygen donor atom leads to chelating properties of these substances with the formation of sterically favorable five-membered chelate rings. The quinoid system, involving π-electron chromophore groups, is the source of the color of these substances and the dyes are char- acterized by the corresponding absorption bands depending on the degree of dissociation of the individual ionizable groups (see Figure 1 for ECR and Figure 2 for PG). Coordination with a metal, forming a binary complex, is manifested by delocalization of the π-electrons in the dye molecule and by the appearance of absorption bands at longer wavelengths (see Figure 2, curve 2).

In applying the above spectral effects of tensides to analytical methods, most researchers have primarily tried to find and evaluate the optimum conditions for the determination of metals and to specify some undesirable effects; such as that of various strong electrolytes[12,21], buffer components[13], and the effect of the amount of metal present on the absorption band maxima[2].

However, substantially fewer authors have studied the composition and structure of the products formed. Even less information is available concerning the correlations of the observed spectral effects with the amphiphilic properties of the tensides. The micelle formation of tensides in the reactions of the metallochromic indicators was considered earlier[14], but these effects have been systematically discussed in relation to the critical micelle concentration of tensides, c_{cmc}, only recently[2]. The reaction mechanism has not yet been rigorously clarified; however, it can be assumed that tensides are involved in two possible types of interactions depending on the tenside concentration, c_T. They can react in the molecular form, when they behave in water as strong electrolytes with large cations, or as tenside molecules - micelle dynamic system.

If c_T is smaller than c_{cmc}, the tenside cation can form an ion-associate with a metal chelate by an electrostatic interaction through the dissociated polar groups of the dye not bound in the chelate. For example, water-insoluble products of Al(III) and Fe(III) with ECR were isolated in the presence of CTMAB, at a ratio of 1 : 3 : 3. The products are dissolved at c_T greater than c_{cmc} due to solubilization, not only in the presence of micelles of the same tensides, CTMAB, but also of a nonionic tenside[15].

Tensides also affect the spectra of dyes alone and cause changes in the maxima of the absorption bands of individual dissociation forms (as can be seen in Figure 1, curve 4 and Figure 2, curve 4), as observed earlier[2]. To clarify the interaction between dyes and tensides, we studied the effect of the Septonex tenside on the absorption spectra of simple triphenylmethane dyes[16], namely phenol red, PR (see Figure 4a, $R_1 - R_2 =$ H), bromophenol blue, BB (Figure 4a, $R_1 =$ Br, $R_2 =$ H) and bromocresol green, BG (Figure 4a, $R_1 =$ Br, $R_2 =$ CH$_3$). These dyes contain only a single -OH group and single -SO$_3$H group and have no chelating properties. Under analogous experimental conditions we then followed the behavior of chelating dyes ECR and DG[17].

The Septonex concentration was selected in the interval, $c_T \gtrless c_{cmc}$, 5×10^{-5} to 5×10^{-3}M. The critical micelle concentration of Septonex, c_{cmc}, was found to be[18] 7.7×10^{-4}M in the absence of strong electrolytes and 1×10^{-4}M in the presence of NaCl or NaNO$_3$.

The spectra of the dyes were recorded at a pH value at which one of the dissociation forms of the dye predominates (HR and R^{2-} for simple dyes, H$_3$R to R^{4-} for chelating dyes; the undissociated forms, H$_2$R and H$_4$R, were not studied).

It was found that at very low tenside concentrations, below 5×10^{-5}M, opalescence or even turbidity appeared due to the formation of insoluble ion-associates of the dye with Septonex[16]. The existence and stoichiometry of analogous ion-associates was later verified for analogous triphenylmethane dyes and similar tensides[9].

At c_T greater than 5×10^{-5}M the solutions remained transparent and, depending on the tenside concentration, various spectral effectss appeared, individual for each dye and its dissocation form[16,17]. However, it has been found that spectral shifts appear only at c_T smaller than or close to c_{cmc}, except with the completely dissociated dyes; if the tenside concentration is greater that the critical micellar concentration, the spectra do not keep on changing with increasing tenside concentration. Individual changes have been observed only with the R^{2-} and R^{4-} forms.

Hence it can be assumed that spectral changes with anions of chromophore dyes are given, at c_T smaller than c_{cmc}, by the formation of soluble ion-associates, by electrostatic bonding through various ionizable groups -OH or -COOH and -SO$_3$H. If tenside micelles are present simultaneously in the solution, the dye is blocked and its anions are part of the micelle. The contributions from various substituents of dyes in soluble or micelle solubilized ion-associates to these interactions would probably be found on the basis of further study of the changes in the acid base and optical properties of various dyes in the presence of sub- and super-critical tenside concentrations[19] and of an evaluation of the effect of hydrophobic interactions of tenside chains on individual groups of dye π-electron system[9].

On the basis of observed spectral effects we further followed the changes in the dissociation constants of dyes PR, BB, BG, ECR and DG as a function of the concentration of tenside Septonex. The spectrophotometric measurement of the $A = f(pH)$ dependence was used to determine individual mixed dissociation constants pK$_2$ for simple dyes[16] and pK$_2$ and pK$_3$ for chelating dyes (ref. 20). The experimental conditions were maintained constant, i.e. the pH was adjusted with the same buffers for simple and chelating dyes, a constant ionic strength was maintained using NaCl or NaNO$_3$ and final value of the dissociation constants were calculated based on 3 to 5 measurements.

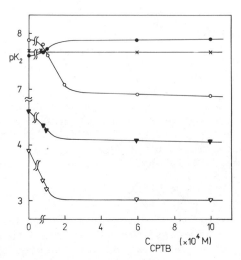

Figure 5. The plot of the pK_2 as a function of Septonex concentration for PR (●, x , ○), BB (▼) and BG (▽); I = 0.015M NaCl (○),
0.2M NaCl (x) and 0.5M NaCl (● , ▼, ▽).

It can be seen from Figure 5 for BB and BG and from Figure 6
for ECR and DG that the pK_2 value decreases with increasing
Septonex concentration up to the critical micelle concentration of
the tenside and then remains constant within experimental error.

The $pK_2 = f(c_T)$ dependence for PR given in Figure 5 also
demonstrates the salt effect: at a strong electrolyte concentration
of c_{NaCl} = 0.015M, the dissociation constant decreases with increasing tenside concentration; for c_{NaCl} = 0.2M it remains
virtually constant; and for c_{NaCl} = 0.5M the pK_2 value is higher
than that for the dye alone. The effect of strong electrolytes on
deprotonation of dyes affected by the presence of a tenside has
been observed in other systems and depends upon the nature of the
dye[16,20]. The salt effect, acting against the tenside effect, is
strongest for PR, as the weakest phenol in the group of simple
dyes.

On the other hand, the value of pK_3 for the dyes ECR and DG
(Figure 7) increase with increasing Septonex concentration and
again remain constant above the critical micelle concentration.

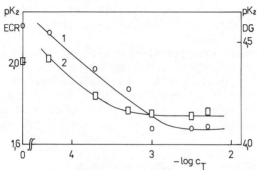

Figure 6.   The plot of the pK_2 as a function of Septonex concentration for ECR (curve 1) and DG (curve 2), I = 0.2M $NaNO_3$.

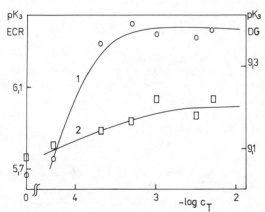

Figure 7.   The plot of the pK_3 as a function of Septonex concentration for ECR (curve 1) and DG (curve 2), I = 0.2M $NaNO_3$.

It can be concluded that all dissociation constants are affected by tensides, and the micelle formation has a principal importance for the degree of dissociation of the reagent donor heteroatom. The micelle formation thus exerts a strong effect on the experimental conditions for the formation of ternary colored systems of metal – dye – tenside and on their spectral effects; e.g., tensides influence substantially the regions of predominance of the individual dye forms capable of chelate formation[20].

Interactions of metal chelates with molecular or micellar tenside forms are therefore complex. To elucidate the mechanism of these reactions, various partial mechanisms will have to be studied, e.g. the salt effect, and other interactions that are less important from the point of view of the spectra, such as solubilization, interaction between the tenside and metal ion, etc.

ACKNOWLEDGMENTS

The author is grateful to Dr. J. Egermaierová and her husband for technical assistance in the preparation of this manuscript.

REFERENCES

1. M. Malát, Fresenius' Z. Anal. Chem., 187, 404 (1962).
2. W. L. Hinze, in "Solution Chemistry of Surfactants", K. L. Mittal, Editor, Vol. 1, p. 79, Plenum Press, New York; and references therein.
3. O. Božkov, L. Čermáková and M. Malát, Anal. Letters 12, 1259 (1979).
4. M. Malát, Fresenius' Z. Anal. Chem., 297, 417 (1979).
5. S. Noriki, Anal. Chim. Acta, 76, 215 (1975).
6. Z. Horníčkova, Thesis, Charles University, Prague 1976.
7. L. Čermáková, I. Fantová and V. Suk, Chem. Zvesti, 34, 357 (1980).
8. K. Uesugi and M. Miyawaki, Microchem. J., 26, 288 (1981).
9. S. B. Savvin, I. N. Marov, R. K. Chernova, S. I. Shtykov and A. B. Sokolov, Z. Anal. Chim., 36, 850 (1981).
10. S. B. Savvin, R. K. Chernova and L. M. Kudrjavceva, Z. Anal. Chim., 34, 66 (1979).
11. S. B. Savvin, I. N. Marov, R. K. Chernova, L. M. Kudrjavceva, S. N. Shtykov and A. B. Sokolova, Z. Anal. Chim., 36, 1461 (1981).
12. J. H. Callahan and K. D. Cook, Anal. Chem., 54, 59 (1982).
13. M. Škrdlík, J. Havel and L. Sommer, Chem. Listy, 63, 939 (1969).
14. V. Svoboda and V. Chromý, Talanta, 13, 237 (1966).
15. Y. Shijo and T. Takeuchi, Japan Analyst, 20, 980 (1971).
16. J. Rosendorfová and L. Čermáková, Talanta, 27, 705 (1980).

17. V. Škarydová and L. Čermáková, Collection Czech. Chem.
     Commun., 47, 776 (1982).
18. L. Čermáková, J. Rosendorfová and M. Malát, Collection Czech.
     Chem. Commun., 45, 210 (1980).
19. I. Burešová, V. Kubán and L. Sommer, Collection Czech. Chem.
     Commun., 46, 1090 (1981).
20. V. Škarydová and L. Čermáková, Collection Czech. Chem.
     Commun., 47, 1310 (1982).
21. H. Nishida, Bunseki Kagaku, 30, 509 (1981).

Part VI
Adsorption and Binding of Surfactants

STUDY OF THE HYDROPHOBIC AND THE ELECTROSTATIC INTERACTIONS IN

MICROPHASES CONCENTRATED IN SURFACTANT VIA ADSORPTION AT CHARGED

INTERFACES

D. Schuhmann, P. Vanel, E. Tronel-Peyroz and H. Raous

Laboratoire de Physico-Chimie des Systèmes Polyphasés
C.N.R.S.
Route de Mende – B.P. 5051 – 34033 Montpellier-Cédex
France

Similarities between micellization and adsorption are considered in the case of adsorption at electrodes. The absence of specific interactions between the adsorbate and the adsorbent is tested with some studies at the air-solution interface.

The additional leeway provided by the electrode potential proves to be a useful property for investigating the competition between the electrostatic and the hydrophobic interactions. Relevant data on micellar solutions concerning the effect of additives, the effect of the variation of the surfactant chain length and the role of counterions in micellization kinetics are discussed. The need of a new model predicting the adsorption behavior allowing a better understanding of the role of electrostatic interactions in micellization is stressed.

I – INTRODUCTION

Similarities between micellization and adsorption have been considered for a long time, especially in the case of adsorption at air-solution interface. Furthermore, parameters describing the local molecular interactions and dynamics were found closely the same for surfactant aggregates of different geometrical forms corresponding to different zones of the phase diagram [1]. The consistency between information obtained by transport techniques on structural properties of colloidal systems and that obtained on the structure of adsorbed layers was pointed out in another paper [2]. Other authors [3] stated that the classical electrochemistry of the electric double layer appropriate to materials that operate as electrodes provides a limiting description for non-electrode materials such as clays, inorganic oxides, insoluble salts, latex colloids and biosurfaces. When surfactant molecules dissolved in the same solvent are brought together at distances of the same order of magnitude in different systems, the same interactions are brought into play and it may be assumed that this is the reason why the equilibrium structures tend to be similar.
It is consequently tempting to think that a better understanding of the behavior of one particular system may provide new insights useful for describing the others.

The studies on aqueous solutions of ionic surfactants reported in the literature often deal with the micellization aspect which can be described as being due to the competition between the hydrophobic interactions and the repulsion between the polar or ionic groups of the surfactant [4,5]. The major remaining problem area concerns the electrostatic interactions [6], but at present, we are far from being able to provide a comprehensive description of the role they play in micellization processes, for example the counterion binding.

Following the approach defined above, it was a natural starting point [7] to analyze the role played by the electrostatic interactions in terms of the general electric double layer theory of charged interfaces, which has been adapted to the particular case of micelles by Stigter [8-12]. The superfical part of the micelle is comprised of the ionic heads and the interstitial counterions and has been termed by the above author the Stern layer. This term stresses the analogy with the inner or compact layer associated with the diffuse layer in the classical double layer model of the electrodes [13].

In connection with electrodes [14,15] or micelles [16] significant attention has been devoted to the improvement of the Gouy-Chapman theory of the diffuse layer with more elaborate treatments (in particular Monte Carlo computation) than those used in the classical theory based upon the Poisson-Boltzman equasion. On the other hand, direct investigations on micellar solutions using the various available techniques have provided a considerable body of

information concerning the size, aggregation number, degree of neutralization and relaxation times of micelles under various conditions. Some features of the latter properties quoted below have
lead the Göttingen school to consider the role of counterions.
However, at the moment, the information seems not to have been
sufficient to suggest a comprehensive model of the Stern layer in
micelles. The need for such a model will appear already from the
results reported in this paper.

At electrode surfaces but not at the micelle surfaces, the
potential can be varied at constant composition of the system under investigation. By comparison with air-solution interface or
with the micellar solutions, electrodes thus offer an additional
leeway and the hope that they could help in understanding the role
played by electrostatic interactions in packed structures of surfactant molecules seemed quite reasonable.

In this paper, we shall try to show that our results on the
adsorption at mercury of anionic surfactants compared with data
on micellar solutions are able to improve our understanding of both
systems or at least to lead to new assumptions allowing the elaboration of models of the micellar Stern layer which could perhaps
be checked by direct investigations on micellar solutions.
The relevant concepts and techniques concerning the studies
about adsorption on mercury are described in the following section.
Results obtained using sodium octylbenzene sulfonate and sodium
alkylsulphate solutions are then reported and discussed in the
light of relevant data concerning micellar solutions.

II - ADSORPTION ON MERCURY - CONCEPTS AND TECHNIQUES

An electrode is generally defined as a junction between an
electronic and an ionic conductor. Electron lack or excess may be
imposed at this interface using an external power supply and a potential decay appears between the electrode and the vacuum. It is
easier to choose as origin the potential adopted by a reference
electrode, reversible to an anion or to a cation present in the
solution. In the presence of surface active species in the latter,
interfacial tension γ is found to be dependent on the electrode
potential E. The plot $\gamma(E)$ measured for a particular concentration
of the surface active species is termed the electrocapillary curve;
γ can be determined easily and with accuracy in the case of mercury only, measuring for example the drop time. We have used the re-
lation [17] : $\log \gamma_1 / \gamma_2 = K \log t_1/t_2$ where t_1 and t_2 are the drop
times, respectively for two interfacial tensions γ_1 and γ_2 and K is
a constant depending on the radius of the capillary, the height
of the column of mercury and   the interfacial tension. The value

of γ_2 can thus be evaluated knowing the value γ_1 corresponding to a reference solution and the ratio t_1/t_2 . The experiments were[8] carried out using a specialized apparatus described elsewhere .

The Gibbs formula was adapted to electrodes [19]. In the case of anionic adsorbates, one has :

$$\Gamma^- = - (\partial\gamma/\partial\mu_{salt})_E + (1)$$

where μ_{salt} is the chemical potential of the salt in solution and E^+ the potential with respect to a reference electrode reversible to the cations. For example, in the study of alkylsulphate adsorption, the experiments were carried out by using a calomel 0.1M KCl electrode connected to the cell by a bridge containing a sodium alkylsulphate solution [20]. This reference electrode is not reversible to the cations. The experimental potential values were corrected by using the Grahame's technique [21] to obtain a value differing from E^+ only by a constant.

The electrode charge density σ^M is given by the Lipmann equation :

$$\sigma^M = -(\partial\gamma/\partial E)_{\mu \ salt} \quad (2)$$

The isotherms $\Gamma^-(c)$ may thus be defined at constant charge or constant potential. They are not reported here because we could not deduce new qualitative information from them. In contradiction, it will be shown in this paper that plots representing $\Gamma^-(\sigma^M)$ at constant concentration may be very informative.

The solutions were prepared using the procedures usually adopted in double layer investigations to remove impurities from water and from solution. It must be stressed that the effect of impurities on measurements is much smaller with a dropping mercury electrode than that observed at other interfaces. The mercury surface is periodically renewed, the drop times being of the order of seconds. In return, slow adsorption cannot be studied with this technique. The absence of any kinetic effect is easily verified by using several capillaries giving rise to different drop times.

The adsorption at the air-solution interface was studied by measuring the surface tension with a Prolabo tensiometer [22]. The interest in comparing adsorption data obtained at a nonpolarized mercury electrode and at the air-solution interface derives from the works of Frumkin and his school [23] and also other authors [24] concerned with adsorption of organics at both interfaces.

For example, aromatic molecules without any aliphatic chain adsorb more on mercury than at the free surface, showing interactions between π electrons and mercury. On the contrary, in the case of aliphatic molecules, identical isotherms were observed, and it was concluded that the configuration of the adsorbed molecules is the same at both interfaces. It has thus been reliably established for a long time that aliphatic polar molecules such as alcohols are adsorbed at mercury as at the free surface with their polar side turned toward the solution [25].

III - RESULTS ON ANIONIC SURFACTANTS AND DISCUSSION

1 - Comparison of the adsorption at two interfaces

Figure 1 presents variation of the interfacial tension for an uncharged mercury electrode and of the surface tension with the logarithm of the surfactant concentration, for the case of sodium octylbenzene sulfonate (SOBS) solutions without salt addition[26].

In passing, it should be noted that the c.m.c. can be deduced from electrocapillary experiments. Measurements corresponding to different constant values of charge or potential should provide

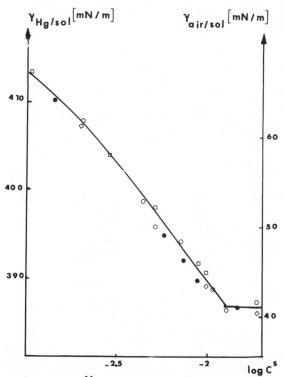

Figure 1. Interfacial (σ^M = o) and surface tensions for SOBS solutions as a function of the logarithm of the surfactant concentration.
  o = our results; ● □ literature data (see ref.26);
  ◊ interfacial tension.

the same value for the c.m.c. and this condition offers an addition-nal test for checking the reliability of the results.

One of the ordinate axis is shifted with respect to the other, the value of the shift being chosen so that the two curves appear to be identical within the limits of accuracy. The derivative of the function $\gamma(\ln c)$ provides the isotherm. Parallelism between the two corresponding curves and consequently the identity of isotherms was also observed with all the other anionic surfactants investi-gated [20]. Thus, as expected, no specific interaction with the mo-del electrode is observed and the chain length was sufficently long for the values of the adsorption excess investigated to prevent the molecules to be adsorbed flat.

The curves representing the variation of γ with the logarithm of concentration for different values of the electrode potential are reported elsewhere for SOBS [26] and for sodium Dodecyl-, (SDS) Decyl- and octylsulphate solutions [20]. The isotherms did not pro-vide any information qualitatively useful here, Γ was found to be approximately a linear function of log c and the plots were used only for obtaining the reliable value of Γ at the c.m.c.

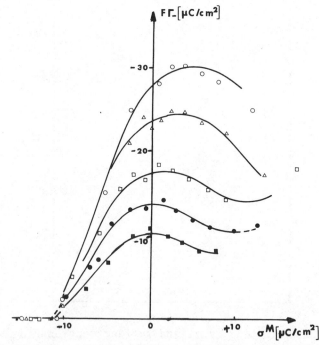

Figure 2. Variations of the adsorbed charge $F\Gamma^-$ as a function of the electrode charge σ^M, for different concentrations of SOBS.
■ : 5.10^{-4}M ; ● : 10^{-3}M ; □ : 2.10^{-3}M ; △ : 5.10^{-3}M ; o : $1.2.10^{-2}$M

2 - Comparison of the structure of adsorbed layers and micel-
    les

    In this review of our results, we report in Figure 2 the va-
riation of the adsorbed charge $(F\Gamma^-)$ with the electrode charge
σ^M for different concentrations,c,of SOBS only.
    The discussion of all the features of these curves is beyond
the objective of this paper but we will stress that these features
can be predicted neither by the available models of the diffuse
layer nor by those which satisfactorily describe the adsorption
of polar organics. However, the value of Γ(c.m.c., σ^M = o) will
be discussed in detail. As a first approximation, the Stern model
of the double layer may be adopted, the surfactant ions being mainly located in the compact part, the electrical neutrality being
ensured by the diffuse part of the double layer. Assuming that
the proportion of counterions β in the Stern layer is the same as
in micelles and equal to 0.59 [27], using the Gouy-Chapman theory
for calculating the anionic and cationic excesses in the diffuse
double layer, a very small value of the diffuse anionic excess is
found (2% of the experimental excess). Because a negligible correction was thus obtained, it was not necessary to use a more elaborate treatment of the diffuse layer. The effective area per surfactant molecule in the adsorbed layer can thus be calculated, and
is found to be $58 \pm 2\mathring{A}^2$. Assuming that micelles are spherical
with a radius of $20.8\pm 3\mathring{A}$ and an aggregation number of 85 ± 5 [28],
an area per surfactant head in micelles of $63 \pm 6\mathring{A}^2$ is obtained.

    This procedure for calculating the headgroup area is suppor-
ted by the block model of micelles proposed recently by Fromherz [29].
"This new packing model combines the features of the droplet model
which conveniently describes the thermodynamical data and the bilayer fragment model the structural data". The blocks are assumed
to be composed of three surfactant molecules, the block assembly
being equivalent to the one corresponding to a bilayer. The mean
sphere would have a radius of 1.2 chain length if it was not reduced by slight bending of the chains without affecting the one by
the binding of counterions [29]. It thus seems reasonable to assume
that the sphere radius is equal to the chain length as in the
naive "star" model.

    From the surface excess values obtained by studying SDS solu-
tions [20], the corresponding value of the effective area is found
to be $57 \pm 2\mathring{A}^2$. From steady state flourescence quenching measurements, Almgren and Swarup reported very recently [30] a value of
about 59 $\mathring{A}^2$ for micelles without salt addition to the solution.
The agreement between these values suggests that the layer adsorbed from solutions at the c.m.c. is as closely packed as are micelles, which is consistent with the block packing model. This result
also confirms that at the c.m.c. few adsorbed surfactant ions are
in the diffuse layer.

3 - The abrupt desorption and the micelle charge density

The most striking feature of the curves shown in Figure 2 is that for an electrode charge less than - 10 $\mu C/cm^2$ a total desorption occurs abruptly, the adsorption excess being nil at more negative charges. A characteristic desorption charge was also found for sodium alkylsulphates of different chain lengths [20]. In contrast, such behavior was never observed previously when investigating other kinds of adsorbates such as inorganic anions, tetraalkylammonium cations, normal alcohols etc ... In the case of these anions, Γ decreases smoothly when the electrode is made increasingly negative. The same behavior is observed at positive charges for hydrophobic cations and at both sides for polar species.

To our knowledge, results such as those plotted in Figure 2 but corresponding to long chain cationic surfactants have not been reported previously. Preliminary data[31], not yet published, obtained with alkyl betain solution at a p.H. value where this surfactant may be assumed to be cationic shows mutatis mutandis a positive characteristic desorption charge. When the p.H. value is 7, the alkyl betains behave as zwitterions and the variation of Γ with the electrode charge is similar to the one usual for polar organics. The appearance of a desorption charge thus seems a specific feature of the ionic surfactants.

The surface charge density of micelles σ_{mic} can be calculated from the literature data quoted above and is found to be - 10 $\pm$ 3 μC cm^{-2} [25] . Such an equality between the desorption charge and the micellar charge was also observed with three other anionic surfactants previously investigated[20]. This equality strongly supports the conclusion that during the micellization process of anionic surfactants, at least the final steps are controlled by the balance between hydrophobic and electrostatic interactions. The desorption charge for SOBS is equal to the one observed for decylsulphate solutions. This is in agreement with the hydrophobicity of the aromatic nucleus. However, it was observed that the longer the chain length, the weaker the desorption or micellar charge σ_{des} ($\Delta|\sigma_{des}| = -1$ $\mu C/cm^2$ per carbon atom).

4 - The surface potential and the effect of additives

Figure 3 shows the results obtained for the surfactants investigated, when the electrode potential is chosen as the electric variable. It is seen that a characteristic desorption potential E_{des} also exists and varies in the opposite direction ($\Delta|E_{des}| \sim 35$ mV per carbon atom), which is now understandable from the above consideration of competition between electrostatic and hydrophobic forces. The electrode potential then expresses

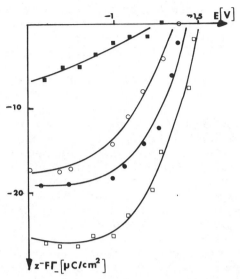

Figure 3. Variation of the adsorbed charge $F\Gamma^-$ as a function
of the electrode poentiel E versus a S.C.E. reference electrode

□ : $NaC_{12} SO_4$ $5.10^{-3}M$;   ● : $NaC_{10} SO_4$ $10^{-2}M$;

○ : $NaC_8 SO_4$ $3.10^{-2}M$;   ■ : $NaC_{12}SO_4$ $10^{-3}M$ + t-BuOH 10%

better the effect of these forces than does the charge. This conclusion is supported by further results concerning the addition
of ter-butanol to a SDS solution. It is well known[32] that many
thermodynamic properties of water-ter-butanol mixtures in the absence or presence of a soluble salt exhibit drastic changes at
a composition near 3% in alcohol where all the water molecules
solvate those of terbutanol. We have found that for concentrations
higher than the critical one (5 and 10%), the desorption charge
changes abruptly from - 8 to - 3 $\mu C/cm^2$ [33]. However, the corresponding curve reported in Figure 3, if it is compared to the curves
corresponding to some alkylsulphates without addition of cosurfactant, shows that the desorption potential changes only slightly,
approximately taking the value which would characterize a chain
with eleven carbon atoms. This small decrease is easily explained
by a decrease of the hydrophobic effect due to the change of solvent.

The surface potential may be caused by the presence either
of charges or of dipoles. The occurence of such dipolar potentials
is well established for explaining many aspects of the double

layer on metals, in the absence or presence of neutral organic adsorbate [34]. It is also known that the variation of the surface potential at the free surface of aqueous solutions when a normal
alcohol or other aliphatic polar species are adsorbed is proportional to the amount adsorbed [35]. This variation is easily explained by the contribution to the potential due to the adsorbed
dipoles.

If the small effect of ter-butanol on E_{des} can be ascribed to
a bulk process, its large influence on the desorption charge seems
due to a change in the interfacial structure correlated to the polarity of the alcohol. The micelle structure being controlled by
the surface potential, the micelle would respond to an addition
of alcohol by a change of its degree of neutralization until the
variation of the potential component due to the charge is exactly
opposite to the component due to the dipoles.

These results and their interpretation are fully in line with
those of Almgren and Swarup [30,36]. They found that the size of SDS
micelles changes on increasing the volume of the hydrophobic constituents, either by addition of hydrocarbons or by addition of Na
tetra-decyl-sulphate so that the number of ionic heads per surface
area is about the same as in the original micelle. Addition of
alcohols resulted in a decrease of the charge density in a way that
was a function of the mole fraction of alcohol in the micelle but
independent of which alcohol was used. Cationic surfactants with
one long alkyl chain induced a micelle growth that left the surface
area per SDS headgroup constant. Symmetrical tetraalkylammonium
counterions and nonionic surfactants of the ethylene oxide dodecyl
ether type acted qualitatively as spacers between the SDS headgroups.
These results were explained by competition between the electrostatic and the other interactions. We may conclude from the results
and the discussion reported above that the changes observed in the
micelle structures are mainly due to the restoration of the original surface potential.

5 - Comparison with the indirect determination of the surface
    potential

Combining these results with ours, it seems clear that it is
the latter property which indeed plays a prime role in micellization as far as electrostatic interactions are concerned. Furthermore, the above discussion shows that, in addition to the surface
charge, it would be very informative to determine the surface potential of micelles.

This was attempted by the Göttingen school [37], according to the suggestion by Hartley and Roe[38] to use pH indicators adsorbed onto charged micelles. The apparent shift of pK detectable in the micellar
solutions as compared to pure aqueous solutions was attributed to
a change in the local interfacial proton activity and this pH shift
was related to an interfacial potential ψ by the Boltzmann's law.
Fernandez and Fromherz [37] used coumarin dyes as flourescent pH
indicators, incorporated in micelles by means of long paraffinic
chain substitutes. The potential found in this way for SDS micelles
(- 134 mV) was considerably larger in magnitude than the value
detected by others [39] using chainless umbelliferone as indicator
(- 27 mV). This discrepancy was explained by the fact that the latter was not located at the micelle surface. From an evaluation of
the effective permittivity (ε = 32), the above authors have deduced that their own indicator was in the expected location.

This assertion can be tested by comparing their results to
ours, given the analogy between the behaviors of the two systems.
The absolute values cannot be compared because in the case of
electrodes, the measured potential has a component, the so-called
potential χ which cannot be evaluated. However, the variation, for
example, with the chain length of the surfactant can be compared
since the χ contribution may be assumed to be constant.

The micellar surface potential evaluated by the procedure
described above varies from about - 95 mV for 10 carbon atoms
in the chain length of the alkylsulphate to about - 155 mV for
14 carbon atoms when no salt is added to the solution. The variation is thus - 15 mV per carbon atom. Comparing it with the corresponding variation of about - 35 mV found for E_{des} , it may be concluded that the absolute value of the potential given by the
pH indicator is too low even if it is very much greater than that
obtained using the chainless umbelliferone.

This discrepancy can be explained either by assuming that the
indicator is not exactly located at the surface or that the Boltzmann's law is no longer valid in the presence of surfactant ions.
Using the Poisson-Boltzmann equation, the above authors calculated
the surface charge. Of course, the values so obtained vary in the
same direction as the surface potential when the chain length
increases. This is not consistent with the results reported above.
This discrepancy and the fact that the curves in Figure 2 cannot
be described by the classical theories of the double layer suggest
that the Gouy-Chapman theory cannot be applied to micelles without
great caution even if finite ion sizes and the dependence of the
permittivity on the electric field strength are taken into
account[16,41].

The consistency between our results and those of Almgren
and Swarup concerning the effect of alcohols suggests that it would
be interesting to study layers adsorbed at a mercury electrode
with the same systems as theirs. However, the interpretation in
terms of dipole potentials already seems reasonable. It is then
tempting to examine if the variations of the surface potential and
of the surface charge density in opposite directions could be explained using the same concept.

6 - A tentative discussion on the role of counterions

The pairing of a surfactant anion with a counterion might
be highly favored at a negatively charged interface as compared
to the solution bulk. The equilibrium constant could depend on
the chain length and thus influence a dipolar potential which
would be created by the interfacial pairs near the electrode or
the micelle surface. It seems to us that the best way to clarify
this question would be to elaborate a new model at the double
layer able to predict the behavior expressed by the curves in Figure 2.

At the moment, the actual effect of the counterions within
the micelle surface is unknown but is probably of prime importance
This statement is consistent with the Göttingen school's works
concerning micellization kinetics. The theory of Aniansson and
Wall [42] though now classical [43] cannot explain all the experimental features of the relaxation time of the slow process. However,
a modification of this theory [44] taking into account the counterions in the law of mass action could describe the data. Moreover,
a third very fast relaxation time has been observed using electric
field jump (~ 50 ns), p jump and ultrasonic [47] techniques and is
attributed to a change of the degree of neutralization of the micelles. We think from the above discussion in agreement with this
interpretation that a change of the thermodynamical properties
near the micellar surface due to pressure or to a high electrical
field gives rise to a change in the eqilibrium surface potential
and thus in the equilibrium surface charge.

A kinetic mechanism cannot be deduced from steady state
techniques. However, given the influence of the potential on the
structures, one is lead to think that other transient properties
might also depend on it. It seems reasonable to assume in the first
steps of micellization, neutral preaggregates grow, for example
by subsequent incorporations of one surfactant, one counterion
and so on. The electrostatic interactions would not play any role
during these steps which could occur again until a packed neutral
structure is formed. One may think that further growth would involve
the replacement of counterions by new surfactant monomers. The
surface charge density would then increase until the mean surface
potential attains the critical value E_{des}.

One would then expect that the rate constant of the second part of the aggregation steps decreases when the step order increases and cancels when the surface potential of the aggregate is equal to E_{des} . We are not able to evaluate the transient surface potential whenever a relevant model of the double layer is not available. Subsequently, however, it would be interesting to examine what would be the effect of a variation of the rate constant during the growth, assuming a reasonable law describing the rate constant decrease.

IV - CONCLUSION

The adsorption of anionic surfactants at a mercury electrode as a function of the surface charge density or of the potential cannot be described by the usual models of the double layer. However, comparing the data with those corresponding to micellar systems, a quantitative analysis could be performed for two particular values of the surface charge. At zero charge, the structure of the layer adsorbed from solutions at c.m.c. is identical to that of micelles. When the electrode charge density is equal to the corresponding value for micelles, a total desorption occurs. The addition of alcohol to the solution seems to have the same effect in both systems, explained by the appearance of an interfacial potential due to the alcohol dipoles. The electrical parameter controlling the equilibrium between the electrostatic and the hydrophobic interactions is shown to be the interfacial potential. The variation of the desorption charge with the chain length is not understood and stresses the need for a new model of the double layer able to predict the adsorption behavior of ionic surfactants. The discrepancy observed between the adsorption results and those corresponding to the determination of surface potential using pH indicators incorporated in micelles supports this conclusion.

REFERENCES

1.   B. Lindman and H. Wennerström, Top. Current Chem., 87, 1 (1980)
2.   D. Schuhmann, Proc. of the 4th Physico-Chemical Hydrodynamics Conference, New York, June 1982 - Annals N.Y. Academ. Sci. in press.
3   T. H. Healy and L.R. White, Adv. Colloid Interface Sci., 9, 303 (1978).
4   G. C. Kresheck in "Water, A Comprehensive Treatise", F. Franks, Editor, Vol.4., p.95, Plenum Press, New York, 1975.

5.   H. Shinoda, T. Nakagawa, B.I. Tamamushi and T. Isemura, in
      "Colloid Surfactants, Some Physicochemical Properties",
      Academic Press, New York, 1963.
6.   H. Wennerström and B. Lindman, Phys. Reports, 52, 1 (1979).
7.   B. Lindman, C. Lindblom, H. Wennerström and G. Gustavsson in
      "Micellization, Solubilization and Microemulsions", K.L. Mittal
      Editor, p. 195, Plenum Press, New York, 1977.
8.   D. Stigter, J. Phys. Chem. 68, 3603 (1964).
9. ' D. Stigter, J. Colloid. Interface. 47, 473 (1974).
10.   D. Stigter, J. Phys. Chem., 78, 2480 (1974).
11.   D. Stigter, Ibid, 79, 1008 (1975).
12.   D. Stigter, Ibid, 79, 1015 (1975).
13.   O. Stern, Z. Elektrochem., 30, 508 (1924).
      H. Gustavsson and B. Lindman, J. Am. Chem. Soc., 97, 3923 (1975).
14.   E. Henderson, Symposium on Double Layer, San José (Ca) 1981.
15.   E. Henderson and L. Blum, J.Electroanal. Chem., 132, 1 (1982).
16.   See among others, the contributions in these proceedings by J.
      Frahm and S. Diekmann and by B. Jönsson, P. Linse and T. Akes-
      son.
17.   R. G. Barradas and F. M. Kimmerlé, Canad. J. Chem., 45, 109
      (1967)
18.   E. Verdier and P. Vanel, J. Chim. Phys., 67, 1412 (1970).
19.   D. M. Mohilner in "Electroanalytical Chemistry", J. Bard. Ed.,
      Vol.1, p. 241, Marcel Dekker, New York, 1966.
20.   G. Naficy, P. Vanel, D. Schuhmann, R. Bennes and E. Tronel-
      Peyroz, J. Phys. Chem., 85, 1037 (1981).
21.   D.C. Grahame and B. A. Soderberg, J. Chem. Phys., 22, 449 (1954)
22.   N. Van Mau and G. Amblard, J. Colloid Interface Sci., 91, 138
      (1983).
23.   B.B. Damaskin, O. E. Petrii and V. V. Batrakov, "Adsorption
      of Organic Compounds on Electrodes", Plenum Press, New York,
      1971.
24.   S. Bordi and G. Papeschi, J. Electroanal. Chem., 20, 297 (1969).
25   A. N. Frumkin and B. B. Damaskin, Pure Appl. Chem., 15,
      263 (1967).
26.   P. Vanel and H. Raous, C.R. Acad. Sci. Paris, 295 II, 857 (1982).
27.   N. Kamenka, B. Lindman, K. Fontell, M. Chorro and B. Brun,
      C. R. Acad. Sci. Paris, 284C, 403 (1977).
28.   R. G. Paquette, E. C. Lingafelter and H. V. Tartar, J. Am. Chem.
      Soc., 65, 686 (1943).
29.   P. Fromherz, Ber. Bunsenges. Phys. Chem., 85, 891 (1981).
30.   M. Almgren and S. Swarup, J. Phys. Chem., 86, 4212 (1982).
      See also their contribution in these proceedings.
31.   G. Haouche, B. Brun, N. Kamenka, D. Schuhmann and P. Vanel,
      Comm. to the 33th Meet. of the Intern. Soc. of Electrochem.,
      Lyon, Sept. 1982.
32.   F. Franks and D. J. G. Ives, Quart. Rev., 20, 1 (1966).

33. G. H. Naficy, D. Schuhmann, P. Vanel and E. Verdier, C.R. Acad.
     Sci. Paris, 289C, 169 (1979).
34. S. Trasatti, in "Modern Aspects of Electrochemistry", B. E.
     Conway and J. O'M. Bockris, Editors, Vol. 13, p. 81, Plenum
     Press, New York, 1979.
35. A. Frumkin et al, Dokl, Akad. Nauk SSSR, 158, 706 (1964).
36. M. Almgren and S. Swarup, J. Phys. Chem., 87, 876 (1983).
37. M. S. Fernandez and P. Fromherz, J. Phys. Chem., 81, 1755 (1977).
38. G. S. Hartley and J. W. Roe, Trans. Faraday. Soc., 36, 101 (1940).
39. M. Montal and C. Gitler, Bioenergetics, 4, 363 (1973).
40. J. Frahm, S. Diekamnn and A. Haase, Ber. Bunsenges., Phys.
     Chem., 84, 566 (1980).
41. J. Frahm and S. Diekmann, J. Colloid Interface Sci., 70, 440
     (1979).
42. E. A. G. Aniansson and S. W. Wall., J. Phys. Chem., 78, 1024
     (1974).
43. E. A. G. Aniansson, S. M. Wall, M. Almgren, H. Hoffman, L. Kiel-
     mann, W. Ulbricht, R. Zana, J. Lang and C. Tondre, J. Phys.
     Chem., 80, 905 (1976).
44. E. Lessner, M. Teubner and M. Kahlweit, J. Phys. Chem., 85,
     1529 (1981).
45. H. H. Grünhagen, J. Colloïd Interface Sci., 53, 282 (1975).
46. S. K. Chan, V. Herrmann, W. Ostner and M. Kahlweit, Ber.
     Bunsenges. Phys. Chem., 81, 60 (1977).
47. S. Diekmann, Ibid, 83, 528 (1979).

THERMODYNAMICS OF BINDING CATIONIC AND ANIONIC SURFACTANTS TO BINARY AND TERNARY MIXTURES OF PROTEINS

B. K. Sadhukhan and D. K. Chattoraj

Department of Food Technology & Biochemical Engineering
Jadavpur University
Calcutta-700032, India

Recently Chattoraj and co-workers have derived simplified equations for the calculation of the change of standard free energy ($\Delta G°$) due to the transfer of a mole of a surfactant from the aqueous solution to the boundary region of a protein. This study, previously used to analyze the data pertaining to binding of long chain amines and sodium dodecyl sulphate to a single protein, has now been extended for the study of binding of these surfactants to binary and ternary mixtures of bovine serum albumin, gelatin and casein at several pH, temperatures and ionic strength. The values of the maximum binding ratio (Γ_a^m) for the protein mixtures are always found to be less than those calculated from the values of Γ_a^m for individual protein based on the additivity rule. The standard free energy changes ($\Delta G°$) per kilogram of protein mixtures due to the biopolymer-surfactant binding interactions have been calculated for the change of amphiphile concentration in the bulk solution from zero to unit mole fraction. This free energy change always refers to the real or hypothetical states of binding saturation so that different values of $\Delta G°$ become comparable to each other. The order of magnitudes of $-\Delta G°$ and Γ_a^m are also found to be similar. The deviation of $\Delta G°$ from their respective ideal values based on additivity rules at various weight fractions of a protein in the mixture has been graphically represented. This deviation has been explained in terms of the free energy of protein-protein interaction in the mixture in the presence of a surfactant in solution. From the values of $\Delta G°$ at different temperatures, the standard entropy change ($\Delta S°$)

and enthalpy change ($\Delta H°$) respectively for protein-
protein interaction have been calculated.

INTRODUCTION

Lipids present in the living cells interact frequently in vivo
with proteins, carbohydrates and water. Using different types of
physico-chemical techniques, interactions between cationic and
anionic surfactants with a protein have been extensively investigated in vitro[1-7]. These studies in various types of model systems
are important for understanding relatively complex interaction of
lipids with various proteins actually occurring in the living cells.
Recently, based on the extensive data on equilibrium dialysis, detailed thermodynamic aspects of binding cationic and anionic ligands
to bovine serum albumin and gelatin have been investigated by
Chattoraj and co-workers[8-10]. In most of these studies, however,
binding of a detergent in vitro to a single protein has been considered. Very little investigation has been carried out so far on
the ligand-protein and protein-protein interactions in multicomponent
biopolymer systems. Based on osmotic pressure data, Scatchard and
co-workers[11] made an attempt to present a thermodynamic interpretation of the interaction of serum albumin and γ-globulin in aqueous
media as a function of ionic strength, pH and temperature. Sen,
Mitra and Chattoraj[10] have recently reported their preliminary
data on the binding of detergents to gelatin, BSA and gelatin-BSA
mixtures.

In the present paper, results of the investigation on the
extent of binding of cationic and anionic detergents to casein,
gelatin, and BSA have been compared with the data of binding of
these surfactants to the binary and ternary mixtures of these proteins at several pH, ionic strengths, temperatures, and biopolymer
compositions. Attempts have also been made to interpret the results
on the basis of thermodynamics of protein-surfactant and protein-
protein interactions.

EXPERIMENTAL

Whole casein was isolated from bovine milk by high speed
centrifugation at 78,000 g for 30 minutes in the MLW JANETZKI
ultracentrifuge (type VAC-601) using the method described by Berlin
et al.[12]. The product was lyophilized and stored in a desiccator
The dry weight of casein so prepared was obtained by heating a
definite amount of casein in a vacuum oven at 105°C for 24 hours.
All calculations have been carried out on the basis of dry weight
of casein. Sufficient amount of gelatin (E. Merck grade) was
dissolved in hot distilled water. The solution was then cooled and
dialyzed exhaustively against distilled water to remove soluble

impurities. After dialysis, the solution was dried at 105°C. Dried
material was powdered and used for binding experiments. Crystalline
bovine serum albumin (BSA) used in the binding experiments was obtained from Sigma Chemicals, USA. Polyvinylpyrrolidone (PVP) used
was of BDH analytical reagent grade.

The cationic detergent cetyltrimethylammonium bromide (CTAB)
was of analytical grade supplied by Deosan Ltd., London. The anionic detergent, sodium dodecyl sulphate (SDS) was obtained as a
gift from Dr. B. N. Ghosh of the University of Calcutta. The standard purity of these two samples has been reported earlier[13]. The
critical micellar concentrations (CMC) of CTAB and SDS, determined
by conductometric method, were found to be 9.5×10^{-4}(M) and 8.23
$\times 10^{-3}$(M) respectively which agree satisfactorily with the literature values[14]. The purity of the cationic and anionic detergents
was confirmed from the observation that there was no minimum in
the variation of surface tension of water with increasing concentration of the detergents. The salts used in these experiments,
to maintain definite pH and definite ionic strength, were of
analytical grade and were used without further purification. A
large amount of stock buffer solution of given pH and ionic strength
was prepared by dissolving the required amount of the salt in an
appropriate volume of water. This was treated as solvent throughout the experiment. A stock detergent solution of the required
concentration (usually 10^{-4}M/l) was prepared by dissolving the
requisite amount of detergent in an appropriate volume of the
solvent. The detergent solutions of various concentrations were
prepared by appropriate dilution of the stock detergent solution.

The binding of a detergent to a protein or a mixture of
several proteins was studied using the equilibrium dialysis technique described in earlier publications[8,10,13]. For the binding
experiments with a binary mixture of BSA and casein, a required
amount of dry solid casein was poured to a dry dialyzing casing
(3/4 inch diameter and 8000 molecular weight cut off supplied by
Arthur Thomas and Co., Philadelphia, USA) and then 2 ml of BSA
solution was poured into it. The bag, knotted appropriately was
then put inside a bottle containing 25 ml solution of the detergent
of known concentration. The bottle was shaken in a thermostatic
shaker until dialysis equilibrium was reached within 48 hours. The
concentration (C) in moles per liter of the detergent in the dialysate was determined by the dye partition technique[15,16]. Dyes used
were sodium salt of disulphine blue VN-150 and methylene blue for the
estimation of CTAB and SDS respectively. Similar procedures were
followed in general for the studies of binding of surfactants to
ternary mixtures of casein, gelatin and serum albumin. Binding
experiments were carried out as a function of the concentration of
the detergent at different compositions of the mixture, pH, ionic
strength and temperature. The pH 6.0 and 8.0 were maintained by
phosphate buffer, and the pH 4.0 was maintained by adding HCl to

NaCl solution of appropriate ionic strength.

The moles of surfactant bound per kg of the individual or mixed proteins were calculated[13] using Equation (1).

$$\Gamma_a = \left(\frac{C_t - C}{C_p}\right) \left(\frac{V_i + V_o}{V_i}\right) \times 10^3 \tag{1}$$

C_t and C refer to the initial and equilibrium concentrations of the detergent in moles per liter, respectively. C_t was calculated with respect to the total volume $V_i + V_o$. V_i and V_o are the respective initial volumes (in liter) of the solutions inside and outside the dialyzing bag at the beginning of the experiment and C_p is the weight of total proteins in grams present in V_i liter of the solution.

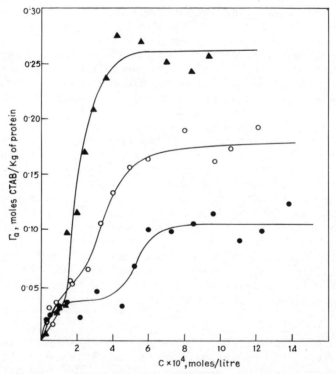

Figure 1. Γ_a vs. C plot for binding CTAB to (casein +BSA) mixture at pH 6.0, μ = 0.125 and temperature 30°C. ▲ , Casein + BSA (1:1); ○, Casein + BSA (1:2); ●, Casein + BSA (1:3).

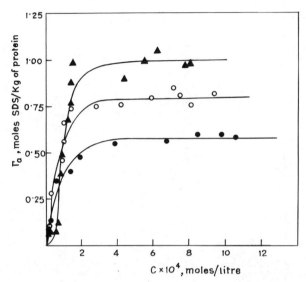

Figure 2. Γ_a vs. C plot for binding SDS to (Casein + BSA) mixture at pH 6.0, $\mu = 0.125$ and temperature 30° C. ●, Casein + BSA (1:1); ○, Casein + BSA (1:2); ▲, Casein + BSA (1:3).

RESULTS AND DISCUSSION

The isotherms for binding CTAB and SDS to casein–BSA mixtures of respective weight ratios 1:1, 1:2 and 1:3 are shown in Figures 1 and 2. The temperature, pH and ionic strength were fixed at 30°C, 6.0 and 0.125 respectively, in these binding experiments. Initially Γ_a increases gradually with increase of C until the binding ratio reaches a maximum value Γ_a^m which remains unchanged with further increase of C. The isotherms obtained for the protein mixtures at other pH, ionic strangths and biopolymer compositions all show similar features with respect to the variation of Γ_a with C. This type of behavior has been observed earlier for binding surfactants to pure casein[17] and pure gelatin[10]. In the case of BSA-amine and BSA SDS binding isotherms, multistep variation of Γ_a against C has been observed by Sen et al.[8,9]

Values of Γ_a^m for binding CTAB and SDS to casein–BSA mixtures for different weight ratios of the proteins are included in Tables I, II and III. Γ_a^m is found to alter significantly

Table I. Maximum Binding Ratio and Standard Free Energies of Binding CTAB-Mixed Protein Complex at Different pH, Ionic Strength and Temperatures.

System			Casein + BSA (1:1)			
Temperature (°C)	pH	Ionic strength	Γ_a^m moles CTAB/ kg Protein	$-\Delta G°$ KJ/kg	$(\Gamma_a^m)_i$ Moles CTAB/ kg Protein	$(-\Delta G°)_i$ KJ/kg
30	6	0.125	0.260 ± 0.013	9.08 ± 0.63	0.332	12.2
30	6	0.0625	0.058 ± 0.010	2.01 ± 0.09	0.202	7.74
30	6	0.0125	0.394 ± 0.016	14.1 ± 1.4	0.548	20.1
30	4	0.125	0.145 ± 0.0006	5.02 ± 0.46	0.191	7.45
30	8	0.125	0.084 ± 0.017	2.68 ± 0.36	0.152	5.53
45	6	0.125	0.227 ± 0.016	8.29 ± 0.75	0.372	13.5
65	6	0.125	0.430 ± 0.022	16.4 ± 0.84	6.45	218

Table II. Maximum Binding Ratio and Standard Free Energies of Binding SDS-Mixed Protein System at Various pH, Ionic Strength and Temperatures.

Temperature (°C)	System pH	Ionic strength	Γ_a^m moles SDS/ kg Protein	$-\Delta G°$ KJ/kg	$(\Gamma_a^m)_i$ moles SDS/ kg Protein	$(-\Delta G°)_i$ KJ/kg
				Casein + BSA (1:1)		
30	6	0.125	0.560 ± 0.040	24.1 ± 2.1	1.49	52.5
30	6	0.0625	0.383 ± 0.032	13.0 ± 1.1	1.49	50.2
30	6	0.0125	0.263 ± 0.022	9.83 ± 0.84	0.685	25.7
30	4	0.125	0.143 ± 0.018	5.48 ± 0.42	2.59	90.4
30	8	0.125	0.550 ± 0.055	20.5 ± 2.0	-	-
45	6	0.125	0.110 ± 0.020	3.89 ± 0.08	0.842	31.4
65	6	0.125	0.252 ± 0.017	8.96 ± 0.50	0.802	28.9

Table III. Maximum Binding Ratio and Standard Free Energies of Binding Ligand-Mixed Protein Complexes at pH 6.0, Ionic Strength 0.125 and Temperature 30°C.

System		Γ_a^m	$-\Delta G°$	$(\Gamma_a^m)_i$	$(-\Delta G°)_i$
Composition of the mixture	Ligand	Moles ligand/ kg protein	KJ/kg	moles ligand/ kg protein	KJ/kg
Casein + BSA (1:2)	CTAB	0.173 ± 0.017	5.98 ± 0.42	0.291	10.6
	SDS	0.780 ± 0.038	28.5 ± 0.79	1.64	57.8
Casein + BSA (1:3)	CTAB	0.105 ± 0.010	3.72 ± 0.24	0.272	9.88
	SDS	0.971 ± 0.055	33.9 ± 1.2	1.72	60.3

but not always regularly with the change in the composition of the protein mixture. One can also calculate the ideal value $(\Gamma_a^m)_i$ for a protein mixture using Equation (2) based on the additivity rule.

$$(\Gamma_a^m)_i = W_1 (\Gamma_a^m)_1 + W_2 (\Gamma_a^m)_2 \tag{2}$$

Here symbols 1 and 2 stand for casein and BSA respectively. W_1 and W_2 are the respective weight fractions of casein and BSA in the mixture. $(\Gamma_a^m)_1$ and $(\Gamma_a^m)_2$ are the maximum ratios in moles per kg of pure casein and pure BSA for binding CTAB or SDS respectively whose values at pH 6.0, ionic strength 0.125 and temperature 30°C are already known[8,9,17]. This equation is expected to remain valid if the two proteins present in the mixture do not interact with each other. Values of $(\Gamma_a^m)_i$ are presented in Tables I, II and III for various systems studied.

It is of considerable interest to note that the experimental values of Γ_a^m in Tables I, II and III are always significantly lower than the corresponding values of $(\Gamma_a^m)_i$. This lowering from the ideal value indicates that casein and BSA in the aqueous mixture interact with each other significantly in the presence of CTAB or SDS so that a fraction of the sites of the proteins becomes unavailable for binding the surfactants from the aqueous media. From Tables I, II and III, it is also obvious that the nature and magnitude of the protein-protein interaction in the presence of cationic detergent are entirely different from those observed when anionic surfactant is present.

The experiments on binding CTAB and SDS to casein-BSA mixture in 1:1 weight ratio have been carried out at different pH, ionic strengths and temperatures respectively. Values of Γ_a^m evaluated for different binding isotherms presented in Tables I - III are always found to be considerably less than $(\Gamma_a^m)_i$. The extent of protein-protein interaction signified by the difference $(\Gamma_a^m)_{ideal} - \Gamma_a^m$ depends on pH, ionic strength and temperature of the media. Further, we also note with interest that Γ_a^m for SDS (vide Table II) varies more or less regularly with change of ionic strength of the medium indicating a major role of the electrostatic effect and the salting out effect in controlling the binding interaction. Such regularity is,however, absent for CTAB binding (vide Table I) which means that besides electrostatic and salting out effects, other types of cooperative and non-cooperative interactions control the protein-ligand binding to a large extent. The pH effect in both the cases is not straight-forward thus indicating many competing factors involved in protein-surfactant interaction.

It has been previously shown that BSA at pH 6.0 and ionic
strength 0.125 binds 0.217 and 12.3 moles of CTAB per kg of protein at 30°C and 65°C, respectively, at the state of binding
saturation[9]. The large number of peptide groups serving as binding sites for the cationic surfactant remain hidden in the globular structure of BSA at 30°C and become thus exposed when the
protein is heat-denatured at 65°C so that Γ_a^m increases enormous-
ly[9]. At the same pH and ionic strength, casein[17] binds only
0.448 and 0.615 moles of CTAB per kg of the protein at 30°C and
65°C, respectively, in spite of the fact that the biopolymer exists
in the denatured states at both the temperatures. Unlike BSA,
casein contains a large number of pyrrolidine groups in place of
ordinary peptide groups since the chemical analysis indicates the
presence of large amount of proline and hydroly-proline residues
in their polypeptide chains[18]. It has been shown[17] that pyrrolidine
groups of polyvinyl pyrrolidone (PVP) do not bind CTAB at all. We
may also propose here that BSA, by possibly forming complex with
casein at 65°C, has lost its capacity to bind CTAB to a large
extent because of the presence of the pyrrolidine group in the
complex. In support of this, experimental studies indicate that

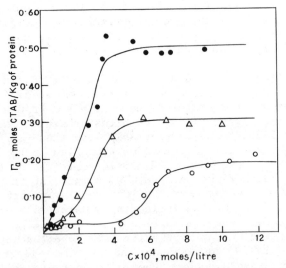

Figure 3. Γ_a vs. C plot for binding CTAB to (Casein + BSA + Gelatin) mixture at pH 6.0 and $\mu = 0.125$. ○ , Casein + BSA + Gelatin
(1:1:1), 30°C; ●, Casein + BSA + Gelatin (1:1:1), 65°C; △, Casein
+ BSA + Gelatin (1:2:1), 30°C.

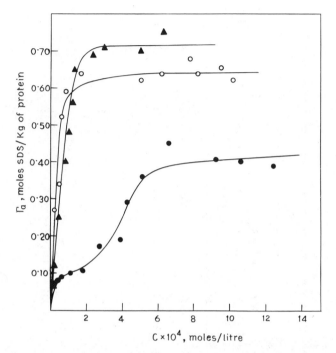

Figure 4. Γ_a vs. C plot for binding SDS to (Casein + BSA + Gelatin) mixture at pH 6.0 and μ = 0.125. $\circ$, Casein + BSA + Gelatin (1:1:1), 30°C; $\bullet$, Casein + BSA + Gelatin (1:1:1), 65°C; $\blacktriangle$, Casein + BSA + Gelatin 1:2:1), 30°C.

BSA-PVP mixture in 1:1 weight ratio does not bind CTAB at all even when C is relatively high[19].

The isotherms for binding CTAB and SDS to the ternary mixture of casein, BSA and gelatin in various weight ratios have been presented in Figures 3 and 4 respectively, at 30°C and 65°C. Ionic strength and pH in all these experiments were kept fixed at 0.125 and 6.0 respectively. Values of Γ_a^m obtained from these isotherms are presented in Table IV. Γ_a^m is observed to depend significantly on the weight ratio of the three proteins.

Table IV. Maximum Binding Ratio and Standard Free Energies of Binding Ligand-Mixed Protein Complexes at pH 6.0 and Ionic Strength 0.125.

System			Γ_a^m moles ligand/ kg protein	$-\Delta G°$ KJ/kg	$(\Gamma_a^m)_i$ moles ligand/ kg protein	$(-\Delta G°)_i$ KJ/kg
Composition of the mixture	Ligand	Temperature (°C)				
Casein + BSA + Gelatin (1:1:1)	CTAB	30	0.180 ± 0.020	6.11 ± 0.46	0.447	15.3
	CTAB	65	0.490 ± 0.022	18.8 ± 1.0	4.30	145
	SDS	30	0.632 ± 0.031	24.8 ± 1.2	2.06	69.0
	SDS	65	0.411 ± 0.030	16.0 ± 1.2	1.15	40.8
Casein + BSA + Gelatin (1:2:1)	CTAB	30	0.302 ± 0.080	10.9 ± 1.2	0.388	13.4
	SDS	30	0.704 ± 0.038	27.4 ° 1.8	2.03	68.6

Ideal values ($\Gamma_a{}^m)_i$ of binding a surfactant to the three proteins in the mixture can similarly be computed from the equation,

$$(\Gamma_a{}^m)_i = W_1(\Gamma_a{}^m)_1 + W_2(\Gamma_a{}^m)_2 + W_3(\Gamma_a{}^m)_3 \qquad (3)$$

Here number 3 stands for gelatin whose weight fraction and maximum binding in the mixture are W_3 and ($\Gamma_a{}^m)_3$. Values of ($\Gamma_a{}^m)_3$ at pH 6.0, ionic strength 0.125 and temperatures 30°C and 65°C have already been determined by Sen et al.[10] so that ($\Gamma_a{}^m)_i$ can be computed. Values of $\Gamma_a{}^m$ are always found to be less than the ideal value of ($\Gamma_a{}^m)_i$ computed from Equation (3) based on the additivity rule (vide Table IV). This large difference again indicates that the three proteins are in significant inter-action with each other at the state of binding saturation. From this table it is also apparent that the nature of variation of $\Gamma_a{}^m$ with W_2 in the presence of CTAB and SDS are different. This indicates that the nature and charge of the surfactant ion has a large contribution to the protein-protein interaction in the ternary mixture. From Figures 3 and 4 we also note with inter-est that $\Gamma_a{}^m$ for CTAB is increased when temperature is increased from 30°C to 65°C, but the reverse is true for SDS. These results again indicate the significant role played by the charge of the surfactant ions in controlling the intermacromolecular inter-action.

The change in the free energy (- ΔG) due to the binding of Γ_a moles of a surfactant per kg of protein as a result of the change in the bulk concentration of the detergent from zero to hypothetical unit mole fraction has been calculated for various systems using the equation[8-10,13,17],

$$\Delta G = - RT \int_0^{X_a} \frac{\Gamma_a}{X_a} dX_a + RT \ \Gamma_a \ln X_a \qquad (4)$$

The integral in this equation can be calculated graphically after replacing the mole fraction X_a of the surfactant by $C/55.51$, and using the series of experimental data of Γ_a and C. Equation (4) based on the unitary scale has been used by Chattoraj and co-workers for the calculation of the change of free energy due to the binding of cationic and anionic surfactants to DNA, BSA, gelatin and casein. The form of the equation in the molar or molal scale was deduced originally by Bull on thermodynamic grounds[20,21]. It can be shown that for a strictly cooperative binding process, the integral term of Equation (4) becomes zero and the equation under this circumstance will assume the form given by Tanford[1,13] for the calculation of the free energy change

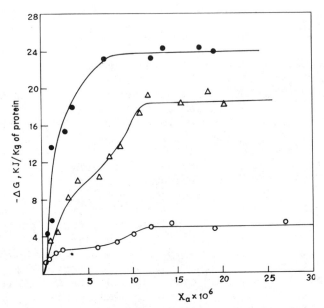

Figure 5. $- \Delta G$ vs. X_a plot for SDS binding to Casein + BSA (1:1) mixture at $\mu = 0.125$ and temperature 30°C. O, pH 4.0; •, pH 6.0; △ , pH 8.0.

due to the ligand protein binding interaction.

ΔG thus calculated from Equation (4) for the different systems studied are found to increase with increase of the mole fraction of the surfactant concentration in the bulk until it reaches a maximum value $- \Delta G^m$ when Γ_a reaches the value Γ_a^m and a state of saturation is established[8-10,13]. Typical plots of ΔG versus X_a are shown in Figure 5.

In the unsaturated state, the fraction of the bound sites of the mixed proteins is Γ_a / Γ_a^m. Hence the slope of the linear plot of ΔG against Γ_a / Γ_a^m represents the standard free energy change $(\Delta G°)$ per kg of the protein mixture for the hypothetical coverage of the unit fraction of sites[8-10,13] so that

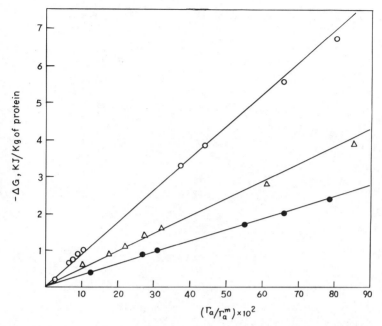

Figure 6. - ΔG vs. $\Gamma_a / \Gamma_a^{\ m}$ plot for CTAB binding to Casein + BSA (1:1) mixture at $\mu = 0.125$ and temperature 30°C. $\triangle$, pH 4.0; $\circ$, pH 6.0; $\bullet$, pH 8.0.

$$\Delta G = \Delta G° \quad \frac{\Gamma_a}{\Gamma_a^{\ m}} \qquad (5)$$

Typical linear plots of ΔG versus $\Gamma_a / \Gamma_a^{\ m}$ are shown in Figure 6. Values of $\Delta G°$ for binding cationic and anionic surfactants to the binary and ternary mixtures of proteins are given in Tables I - IV. It may be pointed out that at the state of saturation, Γ_a is equal to $\Gamma_a^{\ m}$ so that $\Delta G°$ is equal to ΔG^m. Values of ΔG^m evaluated from the plot of ΔG versus X_a (vide Figure 5) are thus found to be close to $\Delta G°$ obtained from the slope of the plot of ΔG against $\Gamma_a / \Gamma_a^{\ m}$. It may also be noted with interest that the order of the magnitudes of $-\Delta G°$ and $\Gamma_a^{\ m}$ are similar. Thus the changes in the standard free energy for different systems evaluated on the basis of a unified thermodynamic scale[8-10] may be used to compare the reactivities of the protein mixtures for binding surfactants under various experimental conditions. In Figures 7 and 8 $\Delta G°$ for the binary and ternary mixtures of proteins respectively are plotted as functions of weight fraction of BSA for binding CTAB and SDS.

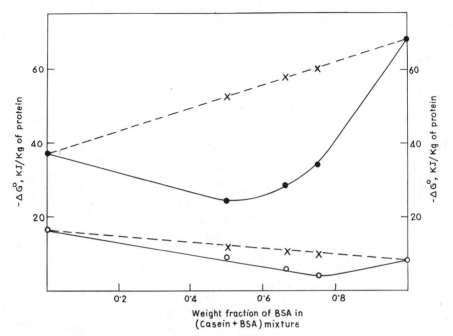

Figure 7. Plot of $-\Delta G°$ vs. weight fraction of BSA in (Casein + BSA) mixture at pH 6.0; μ = 0.125 and temperature 30°C. ---, ideal system (CTAB or SDS); o, CTAB (experimental); ● , SDS (experimental).

It may be pointed out here that $\Delta G°$ calculated for individual and mixed protein systems are expressed per unit quantity of protein undergoing binding interaction. Only under this circumstance, $\Delta G°$ for different systems are comparable to each other. Magnitude of $\Delta G°$ is independent of the bulk concentration of the surfactant in the solution. Equation (4) used for the calculation of the free energy change is based on the Gibbs adsorption equation valid for the two phase model of the biocolloid-water system[9,13]. $\Delta G°$ includes the effects arising from all kinds of interaction between the ligand and the proteins. Recently Jones and Manley[22] calculated the change in the free energy due to the binding of alkyl sulphates to lysozyme using the concept of binding potential initiated originally by Wyman[23]. They have also used the mass action model for the binding interaction in which protein-solvent is assumed to form a single phase. The free energy changes expressed by these worker per mole of the surfactant transferred from the bulk to the bound phase of the appropriate amount of protein significantly depends on the bulk concentrations of the ligand in the solution.

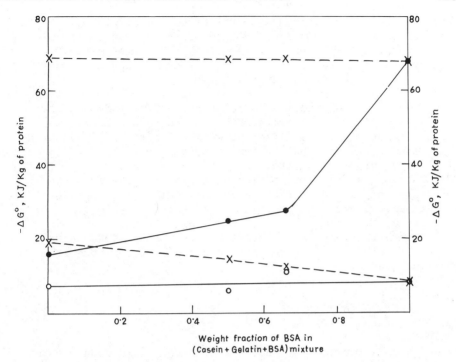

Figure 8.   Plot of $-\Delta G°$ vs. weight fraction of BSA in (Casein + BSA + Gelatin) mixture at pH 6.0, μ = 0.125 and temperature 30°C. ---, ideal system (CTAB or SDS); ○ , CTAB (experimental); ● , SDS (experimental).

If the proteins present in the mixture do not interact with each other, then the ideal free energy of mixing $(\Delta G°)_i$ can be calculated using Equation (6) based on the additivity rule.

$$(\Delta G°)_i = W_1(\Delta G°)_1 + W_2(\Delta G°)_2 + W_3(\Delta G°)_3 \qquad (6)$$

Hence $(\Delta G°)_1$ $(\Delta G°)_2$, $(\Delta G°)_3$ are the standard free energy changes for binding a surfactant individually to casein, BSA and gelatin respectively.   Values of these quantitites for different proteins under various experimental conditions have already been reported[8-10].   Values of $(\Delta G°)_i$ for casein-BSA-gelatin mixtures and for casein-BSA mixtures $(W_3 = 0)$ calculated using Equation (6) are also presented in Tables I-IV.

We again note with interest that the magnitude of $(\Delta G°)_i$ is always higher than $\Delta G°$. One can thus conclude that $\Delta G° - (\Delta G°)_i$ represents the standard free energy change for the protein-protein interaction in the mixture. The value of this difference is always positive since $-\Delta G° < - (\Delta G°)_i$. The positive sign indicates

that the binding of a ligand to the biopolymer mixture is partially
hindered by the protein-protein interaction. Even in the presence
of strong protein-protein intereaction, the ligand-protein binding
is spontaneous since ΔG under these circumstances is always
negative (vide (Figures 5 and 6).

Values of $\Delta G°$ and $(\Delta G°)_i$ for binding CTAB and SDS to binary
and ternary mixtures of proteins have also been compared in Figures
7 and 8 respectively at different weight fractions of BSA at a
fixed ionic strength, pH and temperature. The nature and magnitude
of the deviation of $\Delta G°$ from the ideal value in these figures for
CTAB and SDS differ grossly from each other. This indicates different roles played by CTAB and SDS in controlling the protein-
protein interactions in the binary and ternary systems.

From a thermodynamic consideration, the standard enthalpy
change $\Delta H°$ for ligand-protein binding can be calculated from
the equation[8-10,13]

$$\frac{d\ (\Delta G°/T)}{d\ (1/T)} = \Delta H° \tag{7}$$

Here T stands for the absolute temperature. This equation on
integration becomes

$$\frac{\Delta G°_2}{T_2} - \frac{\Delta G°_1}{T_1} = \Delta H°_{av}\ (\frac{1}{T_2} - \frac{1}{T_1}) \tag{8}$$

provided $\Delta H°_{av}$ (written in place of $\Delta H°$) does not vary for a pair
of temperature T_1 and T_2 close to each other for which correspond-
int standard free energies are $\Delta G°_1$ and $\Delta G°_2$ respectively. $\Delta H°_{av}$
is then regarded as average standard enthalpy change at the
average temperature $\frac{1}{2}(T_1 + T_2)$ corresponding to the average free
energy change $\frac{1}{2}(\Delta G°_1 + \Delta G°_2)$. When $\Delta G°_{av}$ and $\Delta H°_{av}$ are calculated
from the binding data at two different temperatures, the average
standard entropy change $\Delta S°_{av}$ can be calculated from the relation[8-10,13]

$$\Delta G°_{av} = \Delta H°_{av} - T_{av}\ \Delta S°_{av} \tag{9}$$

The values of $\Delta H°_{av}$, $\Delta G°_{av}$ and $\Delta S°_{av}$ at the average of two temperatures close to each other for binding CTAB and SDS to binary and
ternary protein mixtures have been calculated using Equations (8)
and (9). These are presented in Table V. The data in this table
indicate that the contributions of $\Delta H°_{av}$ and $T\Delta S°_{av}$ to $\Delta G°_{av}$ are
significant so that electrostatic, enthalpic, hydrophibic and many
other types of co-operative and non-cooperative interactions are

Table V. Standard Entropy and Enthalpy Changes for Detergent-Mixed Protein Binding at pH 6.0 and Ionic Strength 0.125.

Protein	Ligand	Average Temperature (°K)	$-\Delta G°_{av}$ KJ/kg of Protein	$\Delta H°_{av}$ KJ/kg of Protein	$T_{av}\Delta S_{av}$ KJ/kg of Protein	$(-\Delta G°_{av})_i$ KJ/kg	$(\Delta H°_{av})_i$ KJ/kg	$(T_{av}\Delta S°_{av})_i$ KJ/kg
Casein + BSA (1:1)	CTAB	310	8.66	25.1	33.8	12.8	-19.9	-7.06
-Do-	CTAB	328	12.3	-121	-109	115	3046	3150
-Do-	SDS	310	14.0	431	445	41.9	-73.0	-31.0
-Do-	SDS	328	6.40	-76.6	-70.2	30.3	0.550	39.9

involved in the process of binding surfactant to the heterogeneous binding sites of a protein mixture.

Since values of $(\Delta G^\circ)_i$ for the ideal mixtures of proteins are also calculated at different temperatures (vide Tables I-IV), the standard enthalpy and entropy changes $(\Delta H^\circ_{av})_i$ and $(\Delta S^\circ_{av})_i$ for binding surfactant to the ideal binary or ternary mixtures of proteins may also be calculated using Equations (8) and (9) following similar procedure. These are also included in Table V. We can also assign $\Delta H^\circ_{av} - (\Delta H^\circ_{av})_i$ and $\Delta S^\circ_{av} - (\Delta S^\circ_{av})_i$ as the standard changes in enthalpy and entropy due to the protein-protein interaction in the biopolymer mixture in the presence of CTAB and SDS respectively. The values of these quantities are however average and approximate since these are based on the validity of Equations (8) and (9). As evident from the data presented in Table V, these average values of enthalpies are either positive or negative depending upon the solution parameters. The magnitudes of both of these average parameters are high. From this, it can be qualitatively concluded that all types of co-operative and non-cooperative forces are responsible for the protein-protein interaction so that both enthalpy and entropy changes for the interaction become significant.

ACKNOWLEDGMENT

The financial help of CSIR, New Delhi to one of us (B.K.S.) is acknowledged with thanks.

REFERENCES

1. C. Tanford, "The Hydrophobic Effect: Formation of Micelles and Biological Membranes", 2nd Edition, John Wiley and Sons, Inc., New York, London, 1980.
2. J. Steinhardt and J. A. Reynolds. "Multiple Equilibria in Proteins", Academic Press, New York, 1969.
3. K. Hiramatsu, C. Ueda, K. Iwata, K. Arikawa and K. Aoki, Bull. Chem. Soc., Japan, 59 (2), 268 (1977).
4. J. A. Reynolds, H. Polet and J. Steinhardt, Biochemistry, 6, 937 (1967).
5. H. P. Lundgren, R. W. Elans and R. A. O'Connel, J. Biol. Chem., 149, 183 (1943).
6, B. Jirgensons, Biochem. Biophys. Acta, 317 (1), 131 (1973).
7. H. B. Bull, J. Amer. Chem. Soc., 67, 10 (1945).
8. M. Sen, S. P. Mitra and D. K. Chattoraj, Indian J. Biochem. Biophys., 17, 370 (1980).
9. M. Sen, S. P. Mitra and D. K. Chattoraj, Colloids and Surfaces, 2, 259 (1980).

10. M. Sen, S. P. Mitra and D. K. Chattoraj, Indian J. Biochem. Biophys., 17 405 (1980).
11. G. Scatchard, G. Allen and W. Jeanetta, J. Phys. Chem., 54, 783 (1954).
12. E. Berlin, P. G. Kliman and M. J. Pallansch, J. Colloid Interface Sci., 34, 488 (1970).
13. R. Chatterjee and D. K. Chattoraj, Biopolymers, 18, 147 (1979).
14. P. Mukerjee and K. J. Mysels, "Critical Micellar Concentrations", National Bureau of Standards, Washington, D.C. 1971.
15. P. Mukerjee, Anal Chem., 28, 870 (1956).
16. M. K. Biswas and B. M. Mondal, Anal. Chem., 44 (9), 1936 (1972).
17. B. K. Sadhukhan and D. K. Chattoraj, Indian J. Biochem. Biophys., in press.
18. B. H. Webb and A. H. Johnson, "Fundamentals of Dairy Chemistry", The AVJ Publishing Co., Inc., Westport, Conn., 1965.
19. B. K. Sadhukhan, "Thermodynamics of hydration and binding interaction of proteins and colloidal aggregates in the presence and absence of fat, electrolytes and detergents", Ph.D. thesis, Jadavpur University, Calcutta, India, (1982).
20. H. B. Bull, Biochem. Biophys. Acta, 19, 464 (1956).
21. H. B. Bull, Arch. Biochem. Biophys., 68, 102 (1957).
22. M. N. Jones and P. Manley, J. Chem. Soc. Faraday Trans. I, 75, 1736 (1979).
23. J. Wyman, J. Mol. Biol., 11, 631 (1965).

AN INTERNAL REFLECTION INFRARED SPECTROSCOPIC STUDY OF AOT

ADSORPTION ONTO SOLID SURFACES

Kevin McKeigue and Erdogan Gulari

Department of Chemical Engineering
The University of Michigan
Ann Arbor, Michigan 48109, USA

The adsorption of Aerosol-OT onto thallium-bromoiodide and oxidized germanium surfaces was studied using internal reflection infrared spectroscopy. This technique provides a means for accurate in-situ measurements of the surface concentration of the adsorbed species. For adsorption onto oxidized germanium surfaces from micellar solutions of AOT in water and heptane, we present adsorption isotherms at 25 C as well as the results of our kinetic studies. Multi-layered adsorption is shown to occur in these systems. We observed that after an initial rapid adsorption of approximately one-half to one monolayer, the further build-up of adsorbed AOT follows first order kinetics. Equilibrium amounts of adsorbed AOT range from 3 to 6 monolayers and are found to be weakly dependent on the bulk AOT concentration. Adsorption onto the thallium bromoiodide surface is found to occur only in the presence of added salt and to result in submonolayer surface coverage at equilibrium.

INTRODUCTION

The adsorption of surfactants from solution onto solid surfaces is of interest both because of its commercial applications in such areas as ore flotation and tertiary oil recovery, and because of its fundamental importance in understanding the solution and interfacial behavior of surfactants. Infrared spectroscopy has for a number of years been a principal tool in the study of surface adsorption. Its value as

1271

an experimental probe of the adsorption process lies in its
ability to be used to not only measure the total amount of
adsorbed material, but also to provide information about the
chemical environment of the adsorbed molecules.

The literature on infrared spectroscopy of adsorbed molecules
is extensive, including several texts[1,2]. Much of this body of
literature consists of studies of the transmission spectra of high
surface area solids onto which adsorption has occured. With a
large enough surface area, it is possible to obtain a detectable
signal when the sample is placed in a vacuum or a relatively low
pressure gas environment. However, in the presence of liquid
solution, the signal from the adsorbed molecules is essentially
masked by the strong absorption by the liquid itself, making such
studies extremely difficult experimentally [3,4,5]. The strong
infrared absorbance of water makes studies of adsorption from
aqueous solutions virtually impossible by these methods.

The technique of internal reflection spectroscopy (IRS) is
free of such difficulties. This is mainly because the infrared
light does not pass through the sample as in transmission
spectroscopy, but instead only interacts with that portion of the
sample located within several hundred to several thousand
Ångstroms of the solid surface. A detailed mathematical analysis
of the theory of internal reflection spectroscopy is given by
Harrick [6]. Using this technique, infrared spectra of the material
in contact with a solid surface can be obtained. Infrared light
enters a prism of high refractive index which is in contact with
the sample. This is shown schematically in Figure 1. The optics
are arranged so that the beam approaches the inside of the prism
at greater than the critical angle and is therefore totally
reflected back into the prism. However, the reflection phenomenon
results in the establishment of an evanescent wave at the
solid/liquid interface. The amplitude of the evanescent wave
decreases exponentially into the liquid with a decay constant d_p.
Absorption of energy from the evanescent wave results in an
infrared spectrum of the material in the region in which the wave
exists. The magnitude of d_p, and thus the effective "penetration
depth", is governed by the relative refractive indices of the
solid and the liquid, and the angle of incidence of the beam. The
value of d_p is generally in the range of one to several thousand
Angstroms for such an experiment. Through multiple internal
reflections, enough of the interfacial region is sampled so that a
reasonable signal is obtained.

The feasibility of using this technique has been demonstrated
by a number of investigators [7,8]. In the past few years numerous
studies of protein adsorption onto solids have been performed
using this technique [9,10]. However, internal reflection
spectroscopy has been virtually overlooked as a means of studying

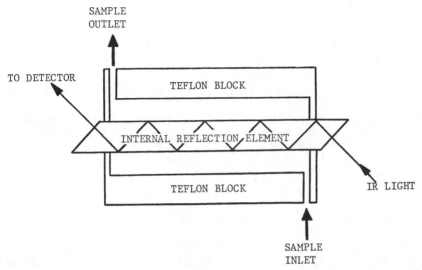

Figure 1.   Schematic diagram of the internal   reflection liquid
sample cell used in this study.

the adsorption of surfactants from solution. In this paper, we
intend to demonstrate the usefulness of the technique in the study
of adsorption of surfactants at the solid/liquid interface.

The   systems   chosen were   solutions of   the   surfactant
Aerosol-OT (Sodium   di-2-ethylhexylsulfosuccinate)   in   three
solvents: heptane, water, and 1% NaCl brine. We monitored the
adsorption of AOT onto two materials: thallium bromoiodide and
surface oxidized germanium.

EXPERIMENTAL

Apparatus

A Digilab FTS-20C Fourier transform infrared   spectrometer
equipped with a wide band Mercury-Cadmium-Telluride detector was
used to obtain all infrared spectra. The instrument was operated
in a single beam mode at a resolution of $4cm^{-1}$. From 300 to 2000
scans were taken for each spectrum, each scan taking approximately
0.6 seconds.   The beam condensing optics necessary for IRS
were obtained from Harrick Scientific Co., as were the germanium
and thalliumbromoiodide internal reflection elements (IRE's).

The liquid IRS cell was designed and built in our laboratory and is shown schematically in Figure 1. It consists of two Teflon blocks with surfaces machined parallel and flat. In each block, a channel approximately 40mm x 8mm was milled to contain the liquid sample. The cell was sealed by sandwiching the 50mm x 10mm x 2mm IRE between the blocks and applying pressure through a clamping device. Connections to the cell were through polyethylene tubing and Teflon Swagelok fittings. Therefore, aside from the IRE itself, a solution placed in the cell would come in contact only with relatively inert materials. This greatly facilitated the use of strong acids in cleaning and oxidizing the IRE surfaces. Experiments were carried out under both static and flow conditions.

Materials

The AOT used in this study was Fluka purum grade with a specified purity of 98%. It was further purified through a series of solvent extractions [11] and dried for several months in a dessicator over phosphorous pentoxide. Water was purified using a Millipore RT water purification system, resulting in a resistivity of 2.0 M ohm/cm or better. The heptane was Phillips Petroleum pure grade (99%). All other chemicals were reagent grade and were used without further purification.

Procedure

Clean surfaces of thallium bromoiodide were prepared by mechanical grinding and polishing of the internal reflection elements. The germanium internal reflection elements were prepared by oxidizing their surfaces with chromic acid cleaning solution followed by rinsing with nitric acid to remove residual chromium ions. This was followed by multiple rinsing with distilled deionized water. The entire procedure was performed after the cell had been assembled with the IRE in place. This procedure results in a thin oxide layer on the germanium IRE. A typical spectrum of a cleaned, oxidized germanium surface is shown in Figure 2. This spectrum indicates the presence of several surface oxides of germanium: GeO, GeO_2 and amorphous GeO_2 [6]. Careful monitoring of the cleaning/oxidizing process made it possible to reproduce the amount of surface oxide present. However, we encountered some difficulty in reproducing the relative amounts of the different oxides of germanium present. This problem and its implications will be discussed later in this paper.

Quantitative determinations of surface coverage require a knowledge of the surface area of the internal reflection elements.

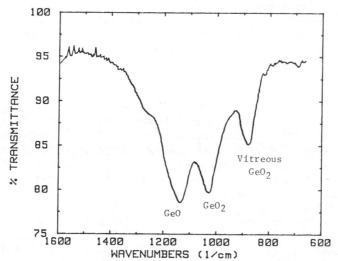

Figure 2. Transmittance spectrum of an oxidized germanium internal reflection element showing the different oxide forms present on the surface.

This can be accomplished either by measuring the absorbance of a substance which has been deposited as a monolayer, such as calcium stearate[7] or by measuring the absorbance of submonolayer coverages of substances whose adsorption isotherms are known for the given solution/surface system. Following the latter procedure, we monitored the adsorption of stearic acid from solutions in CCl_4 until equilibrium was reached. Previously reported adsorption isotherms[8] were used in the calculation of the surface area of the IRE.

Before the start of an adsorption experiment, the cell was rinsed with distilled water and dried in a vacuum oven at 120 C and 10^{-3} Torr. A spectrum was obtained of each surface prior to its use in an experiment. All the surfaces used were free of any detectable water or organic material.

The procedure for a typical adsorption experiment consisted of i) obtaining a spectrum of the surface in air to be used as a reference; ii) injecting the AOT solution into the cell and iii) obtaining spectra at 5 minute intervals for the first hour and thereafter at 20 minute intervals. Desorption experiments were carried out by rinsing the cell with approximately 100 ml of pure solvent and then slowly pumping the pure solvent through the cell for the remainder of the experiment. All experiments were performed at room temperature which ranged from 22 C to 25 C.

Data Analysis

The amount of AOT adsorbed on the surface was followed by monitoring the area of the absorbance peak due to the sulfonate group at 1045 cm^{-1} and the peak due to the C = O stretching of the ester group at 1740 cm^{-1}. All areas were obtained by numerical integration of the absorbance spectra. Absorption coefficients for these peaks were measured using transmission spectra. It was assumed that surface adsorption would not significantly alter the values of the absorption coefficients. This assumption is entirely reasonable for the 1740 cm^{-1} peak because the C = O bond is not expected to be strongly involved in adsorption interactions with the surface. This presumption is supported by the lack of a detectable difference in the location of the peak in solution and adsorption spectra. In the case of the heptane solutions of AOT, excellent agreement in the amount of surface coverage (± 5%) was obtained by using the two different peaks. This suggests that while the sulfonate group is probably involved in surface interactions, the absorption coefficient of the sulfur-oxygen bonds appears to be relatively unaffected. Therefore, in aqueous solutions where interference from the neighboring O-H band, (which was unstable because of drifts in the humidity), prevented us from utilizing the 1740 cm^{-1} band, the results obtained using the 1045 cm^{-1} band should prove equally reliable.

The amount of AOT on the surface can be obtained from the area of the absorbance peaks by using:

$$A = \alpha C_s - \alpha C_b \, d_{eff} \qquad (1)$$

where A is the integrated absorbance of the peak, α is the integral absorption coefficient, C_s is the surface concentration in moles per unit area, C_b is the molar concentration in the bulk solution, and d_{eff} is the effective length of travel of the evanescent wave through the sample. The angle of incidence of the beam and the refractive indices of the sample and the IRE determine the value of d_{eff}, which for our conditions was approximately 1200 Å. The second term on the right hand side of Equation 1 is an approximate correction for the absorbance from the AOT in solution. We are neglecting the fact that as material is adsorbed, less of the bulk solution is sampled by the evanescent wave. However, for our experimental conditions, the error arising from this approximation is less than 2 %.

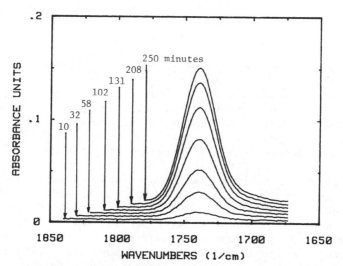

Figure 3.   Increase in absorbance of the C=O peak during the
adsorption process for a solution of 2.0 wt% AOT in heptane and an
oxidized germanium surface.

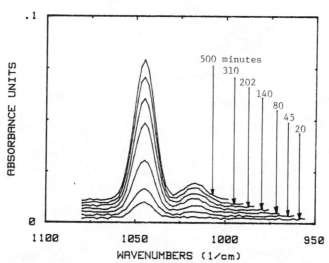

Figure 4.   Increase in the sulfonate absorbance peak during the
adsorption process.   The system is   0.8 wt% AOT in water with an
oxidized germanium substrate.

RESULTS AND DISCUSSION

Kinetics

Adsorption experiments using oxidized germanium surfaces were performed for solutions of AOT in heptane ranging from 0.1 to 5.0 wt % AOT and for aqueous AOT solutions in the range of 0.05 to 1.2 wt % AOT. For each concentration, infrared spectra were obtained at suitable intervals to follow the build-up of the adsorbed surfactant layer over time. Figure 3 is typical of the results obtained, showing the increase in absorbance of the C = 0 bond of AOT for a solution of 2.0 % AOT in heptane. In this figure, the ordinate of each spectrum has been displaced for the sake of clarity. Each spectrum has had the solvent background subtracted and has been numerically smoothed using a standard three point smoothing procedure. A similar plot showing the increase in absorbance of the sulfonate peak over time is shown in Figure 4. In this case, the sample was a 0.8 wt % solution of AOT in water. Results qualitatively similar to Figure 4 were obtained

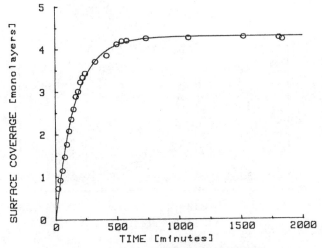

Figure 5. The amount of adsorbed AOT plotted as a function of time for the system 0.8 wt% AOT in water on oxidized germanium. The solid line is the least squares fit to first order kinetics.

for aqueous solutions throughout the AOT concentration range studied.

The amount of adsorbed AOT was determined from the area under the absorbance peaks by applying Equation 1. By assuming a value for the surface area that the sulfonate head group occupies, the surface concentration, C_s, can be represented in terms of surface coverage expressed in equivalent numbers of monolayers. The surface area occupied by the head group of an AOT molecule in a water in oil microemulsion has been reported as 55 Å^2 in the limit of large microemulsions[12]. Similarly, the surface area occupied by a sulfonate group adsorbed at a water/air interface was found to be 52 Å^2 [13]. We chose to use this value because it probably represents a good approximation of the closest physically realizable packing of sulfonate head groups at a surface. Thus, the number of adsorbed monolayers we report will represent a lower limit of the number of adsorbed layers. By expressing surface coverage in this manner, we are not suggesting that adsorption occurs by the successive buildup of one complete monolayer at a time. We are merely using this unit as a convenient measure of the average thickness of the adsorbed layer.

A plot of the surface coverage vs. time is shown in Figure 5 for a solution of 0.8 wt % AOT in water. Again, the behavior observed in this figure is representative of all our adsorption experiments. For each experiment, the surface coverage vs. time data were fit to an integrated first order rate equation of the form:

$$C_s = C_b K_{eq} [1 - \exp(k_1 (t - t_o) / K_{eq})] \qquad (2)$$

where: C_s = surface concentration of AOT
 C_b = bulk concentration of AOT
 K_{eq} = adsorption equilibrium constant equal to C_s/C_b
 $\phantom{K_{eq} =}$ at equilibrium
 k_1 = rate constant
 t = time
 t_o = time at the start of the reaction

The adsorption rate constant, k_1 , when defined in this manner and fit to the data was found to be 1.46 monolayer-liter/min-gmol $\pm$ 20%, regardless of bulk AOT concentration for the aqueous solutions and 1.84 monolayer-liter/min-gmol $\pm$ 10% regardless of bulk AOT concentration for the solutions of AOT in heptane. The rather slow rate of adsorption we observed is not surprising. There have been no other reported studies of the kinetics of adsorption for similar systems; however, in obtaining adsorption isotherms for alkylbenzene sulfonates adsorbed onto mineral oxide surfaces, Scamehorn and co-workers noted that from one to two days

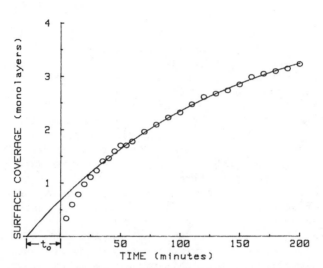

Figure 6. Adsorption of AOT by an oxidized germanium surface from
a 0.8 wt% solution of AOT in water. Note the deviation from first
order kinetics at the start of the process.

were required for equilibration of their samples. For our systems
approximately 12 hours were required for the adsorption to be 99%
completed.

The appearance of the term t_o in Equation 2 stems from the
observation of a deviation from first order kinetics at the start
of each adsorption experiment. Several experiments were performed
in order to obtain more detailed information regarding the initial
phase of adsorption. Figure 6 shows the lack of adherence to
first order kinetics at the start of the adsorption process.
Initially, the rate of adsorption observed is at least ten times
higher than the first order kinetics rate. The duration of this
initial phase of adsorption is approximately 20 to 30 minutes,
after which time we observe excellent agreement with first order
kinetics. Correlation coefficients for the fit, neglecting the
first few data points, were typically 0.97 to 0.99. We included
the adjustable parameter t_o in our integrated rate expression
(Equation 2) as an approximate correction for the initial rapid
adsorption.

The amount of AOT deposited on the surface during the initial adsorption phase was approximately one half to one monolayer. Possibly, what was observed was an initial rapid adsorption onto the most active surface sites. This initial rate is likely to be strongly influenced by the rate of diffusion making it pointless to attempt to fit a rate expression to the data in this regime. However, in the first order rate regime, even for our most dilute solutions, the calculated rate of diffusion to the surface was nearly 100 times greater than the fastest rate of adsorption. This indicates that after the initial regime, the adsorption we observed was definitely surface reaction rate limited.

Desorption experiments showed that surface depletion also follows first order kinetics with essentially the same rate constants. However, we observed that the desorption was not complete. As long as a week after the start of the desorption process, there remained an amount of AOT on the surface equivalent to 0.2 to 0.35 monolayers. This supports our suggestion of a rapid strong adsorption of AOT at the most active surface sites. This strongly adsorbed AOT remained on the surface even after thorough rinses with methanol and heating to 60 C.

Adsorption Isotherms

The equilibrium surface coverage is shown in Figure 7 plotted against AOT concentration for both the AOT/heptane and the AOT/water systems. For the AOT/heptane system, we observed very little concentration dependence of the surface coverage, with the amount being fairly constant at about 5.7 monolayers. The aqueous system shows slightly more of a dependence on AOT concentration. The surface coverage appears to increase slightly in the micellar AOT region from about 3.7 to 4.0 monolayers. However, we estimate the experimental uncertainty in these points to be about 20%. Therefore, the slight increase in surface coverage with concentration observed in this region might not be meaningful.

In a thorough study[13,14,15] of the adsorption of sodium alkylbenzene sulfonates on alumina and kaolinite, Scamehorn and co-workers observed adsorption to be independent of concentration above the critical micelle concentration (CMC) for isomerically pure surfactants. All of our experiments were carried out at AOT concentrations significantly above the CMC. Thus the shape of the adsorption isotherms shown in Figure 7 is in agreement with Scamehorn's results. While most reports of surfactant adsorption studies indicate submonolayer coverage, a number of investigators have reported bilayer adsorption in the case of anionic surfactants in polar media[16,17,18].

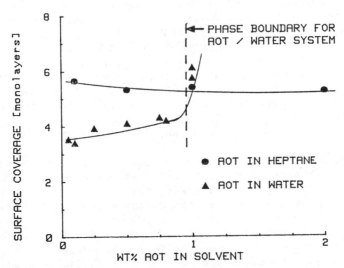

Figure 7.   Adsorption isotherms at 25 C for the systems AOT in heptane and AOT in water on an oxidized germanium surface.

        The physical basis for the multilayered adsorption we observe is not clear. However, given the nature of the surfactant involved, multiple adsorbed layers are not unreasonable. The hydrophobic part of the AOT molecule consists of two branched hydrocarbon tails. Micelle formation by AOT in water is therefore geometrically unfavorable. Only in dilute aqueous solutions (< 1% AOT at 20 C) does AOT exist solely as normal micelles. Above this concentration, lamellar liquid crystals are formed[19,20] . The solid surface at which adsorption occurs might then allow for the formation of a structurally favorable lamellar phase at lower AOT concentrations. Such structures could form on a surface more easily than in solution simply because the disruptive effect of thermal motion would be lessened as a result of the rigidity of the surface. Thus, in aqueous solutions, the existence of multiple adsorbed layers is plausible. Because the lamellar structure should become less rigid with each additional layer, we would not expect unbounded adsorption. Thermodynamically, the hydrophobic and hydrophilic interactions of AOT with the solvent would favor an even number of layers in water and an odd number of layers in non-polar solvents.

The argument we've advanced for adsorption from aqueous solutions is not as satisfying in the case of non-polar solvents. Reversed micelles are highly favored geometrically in non-polar media. The tendency of AOT to form liquid crystals in non-polar media is low, with the L_2 (reversed micellar) region usually extending to 80% AOT. However, we can again point to the stabilizing effect of the solid surface as one factor in promoting the formation of lamellar structures. In addition, the possibility of strong electrostatic interactions between the AOT head groups and the solid surface can not be ruled out. The low dielectric constant of heptane causes the solvent to be essentially invisible as far as electrostatic interactions are concerned.

At approximately 1% AOT, we observed a sharp increase in the surface coverage accompanied by a significant increase in the scatter in our experimental data. This behavior is easily explained by the fact that the phase boundary between the L_1 (micellar AOT) and the L + LC (micellar AOT and lamellar liquid crystals) occurs at about 1 wt % AOT[19,20]. We observed that this phase boundary is very temperature sensitive, thus explaining the scatter in our data and our inability to obtain reproducible results near the phase boundary.

We would like to point out that we did have some difficulty in reproducing the data points shown in Figure 7. Several attempts at reproducing a particular point resulted in a value different by as much as 50% from the previously obtained value. However, these discrepancies could always be traced to a difference in the relative amounts of GeO and GeO_2 as determined from the IR spectrum of the surface (Figure 2). It appears that GeO is a more favorable surface for AOT adsorption than GeO_2. Each of the data points shown on Figure 7 results from the agreement of at least two experiments within experimental uncertainty, for surfaces that had, as well as could be determined, the same relative amounts of GeO and GeO_2. These results suggest that AOT adsorption is likely to be strongly influenced by the chemical nature of the surface and that the adsorption isotherms presented should be viewed in this light.

Adsorbed AOT Spectra

One of the advantages which the technique of infrared spectroscopy has over other probes of adsorption at the liquid/solid interface is that, in addition to kinetic data and adsorption isotherms, it can provide information about the chemical environment of the adsorbed molecules. The interactions of a particular atom of a molecule, such as in the case of a hydrogen bond to one of the oxygens in a sulfonate group, will

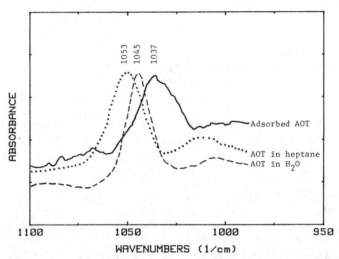

Figure 8. Infrared absorbance spectra of the sulfonate groups of AOT in three different environments: i) micellar AOT in water, ii) reversed micellar AOT in heptane, and iii) AOT adsorbed onto an oxidized germanium surface. The first two spectra were obtained from infrared transmission experiments. The third spectra was obtained by using spectral deconvolution techniques on the spectrum obtained at the start of an adsorption experiment from an aqueous solution of AOT (i.e., the contribution of the 1045 cm peak has been subtracted out). The third spectrum is expanded by about a factor of 60 relative to the first two spectra.

lower the energy of the bond and result in a shift in the vibrational spectrum to a lower wavenumber. Figure 8 shows how the sulfonate absorbance peak is affected by the environment of the AOT molecule.

The peak furthest to the left at 1053 cm^{-1} for reversed micellar AOT in heptane, corresponds to the strongest sulfur-oxygen bonds of the three. Eicke[21] has argued that a small amount of water is probably incorporated in a reversed micelle in an apolar medium. He suggested that a minimum molar ratio of one water molecule to every three AOT molecules was necessary for micelle formation. This ratio was confirmed experimentally by Bedwell[22]. In Eicke's model, 2/9 of the sulfonate oxygens would be hydrogen bonded and the remaining 7/9 would interact strongly only with the cations. The absorbance peak at 1053 cm^{-1} is attributed to these oxygens which are not involved in hydrogen bonding. This relatively broad peak probably overshadows the absorbance peak at 1045 cm^{-1} resulting from the 2/9 of the sulfonate oxygens which are hydrogen bonded.

In a normal micelle, the sulfonate oxygens should all be hydrogen bonded to water molecules. These hydrogen bonds lower the vibrational frequency of the sulfur-oxygen bonds and result in a shift in the absorbance peak to 1045 cm^{-1} (Figure 8). The further shift to 1037 cm^{-1} for the adsorbed AOT indicates that the sulfonate oxygens are involved in bonding to the surface which is stronger than the hydrogen bonds to water which occur in a micelle.

The fact that we observe three distinct peaks for AOT in these different environments allows us to obtain more information about the adsorption process by examining spectra obtained over the course of an adsorption experiment such as those shown in Figure 4. In Figure 9 we have plotted the sulfonate absorbance peak we observed at the start (t = 3 minutes) of the adsorption process for a micellar solution of AOT in water. We observe a broad peak with a maxima located at 1045 cm^{-1} and a distinct shoulder located to the right of the main peak. The deconvolution of this spectrum shows that it is made up of two component peaks, one at 1045 cm^{-1}, corresponding to sulfonate groups in an aqueous environment, and a peak at 1037 cm^{-1} corresponding to the surface-bound sulfonate groups. Even at the very start of adsorption the area of the 1045 cm^{-1} peak is twice that of the 1037 cm^{-1} peak. This suggests that only one of the three oxygens in each sulfonate group is strongly bound to the solid surface. During the course of an adsorption experiment we found that the area of the 1037 cm^{-1} peak gradually increased until it accounted for 30 to 40 % of a monolayer coverage after which time it remained constant. The peak at 1045 cm^{-1} on the other hand continued to increase until equilibrium was achieved.

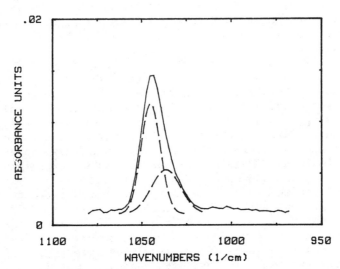

Figure 9.   Spectrum obtained at the start of adsorption from a 0.8
wt% solution of AOT in water. The dashed lines indicate the two
component peaks of the spectrum obtained by deconvolution of the
spectrum (solid line).

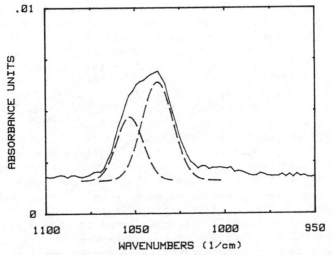

Figure 10.   Spectrum obtained at the start of adsorption from a
solution of 2% AOT in heptane.   The dashed lines indicate the two
component peaks obtained by spectral deconvolution.

In the case of adsorption from an AOT/heptane solution we observe that initially the 1037 cm^{-1} peak is larger than the 1053 cm^{-1} peak (see Figure 10). After about 60% of a monolayer coverage is achieved, the 1037 cm^{-1} peak stops increasing while the 1053 cm^{-1} peak continues to grow until equilibrium is reached. In contrast to the aqueous soltions, in heptane it appears that roughly two of the three sulfonate oxygens in the primary AOT monolayer are strongly interacting with the solid surface.

In both solvents we observed that the majority of infrared absorption by the sulfonate groups of the adsorbed AOT occured at the same wavenumber as in the bulk solution. This suggests that in both solvents, the majority of the adsorbed AOT is in a similar or possibly the same structural form as the micellar AOT in solution.

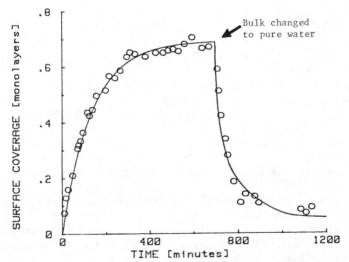

Figure 11.   Amount of AOT adsorbed on a thalliumbromoiodide surface vs. time. The solution was 0.1 wt% AOT in 1.0% NaCl brine at the start of the process. After 700 minutes, pure water was pumped through the cell resulting in desorption.

Adsorption onto Thalliumbromoiodide

The ionic thalliumbromoiodide crystal, which has very different surface properties than oxidized germanium, was found to have a much lower affinity for adsorbing AOT. We observed no detectable adsorpton of AOT onto thalliumbromoiodide from solutions of AOT in heptane (0.1 to 5.0 wt %) or from aqueous solutions of AOT (0.05 to 1.2 wt%). We estimate the detectability limit of our equipment to be about 0.05 monolayer. However, the addition of NaCl to the system resulted in adsorption as shown in Figure 11. As in the case of the germanium surfaces, the adsorption followed a first order rate expression. The rate constant was found to be approximately 20% larger than for the germanium systems. Desorption was also found to follow first order kinetics with some AOT remaining strongly bound to the surface. One major difference between the two surfaces is that adsorption onto thalliumbromoiodide was found to occur in submonolayer amounts.

ACKNOWLEDGEMENTS

We would like to thank Brian Bedwell for purifying the AOT used in this study and for engaging in many helpful discussions. Financial support of this work by NSF grants CPE-79-16518, CPE-80-10800, CPE-81-07724, and DMR-81-00130, is gratefully acknowledged.

REFERENCES

1. L.H. Little, "Infrared Spectra of Adsorbed Species", Academic Press, New York (1966).
2. M.L. Hair, "Infrared Spectroscopy in Surface Chemistry", Marcel Dekker, New York (1967).
3. M.J.D. Low and M. Hasegawa, J. Colloid Interface Sci. 26, 95 (1968).
4. M. Hasegawa and M.J.D. Low, J. Colloid Interface Sci. 29, 593 (1969).
5. M. Hasegawa and M.J.D. Low, J. Colloid Interface Sci. 30, 378 (1969).
6. N.J. Harrick, "Internal Reflection Spectroscopy", Harrick Scientific Corp., Ossining, New York (1967).
7. G.L. Haller and R.W. Rice, J. Phys. Chem. 74, 4386 (1970).
8. R.T. Yang, M.J.D. Low, G.L. Haller and J. Fenn, J. Colloid Interface Sci. 44, 249 (1973).
9. R.M. Gendreau and R.J. Jakobsen, Applied Spectroscopy 32, 326 (1978).
10. R.J. Jakobsen and R.M. Gendreau, Artificial Organs 2, 183 (1978).

11. L.J. Magid, private communication (1981).
12. H.F. Eicke and J. Rehak, Helv. Chimica Acta 59, 2883 (1976).
13. J.F. Scamehorn, R.S. Schechter and W.H. Wade, J. Colloid Interface Sci. 85, 463 (1982).
14. J.F. Scamehorn, R.S. Schechter and W.H. Wade, J. Colloid Interface Sci. 85, 479 (1982).
15. J.F. Scamehorn, R.S. Schechter and W.H. Wade, J. Colloid Interface Sci. 85, 494 (1982).
16. B. Tamamushi, in "Colloidal Surfactants", K. Shinoda, B. Tamamushi, T. Nakagawa and T. Isemura, Editors, Academic Press, New York (1963).
17. T.F. Tadros, J. Colloid Interface Sci. 46, 528 (1974).
18. G. Goujon, J.M. Cases and B. Mutaftsciev, J. Colloid Interface Sci. 56, 587 (1976).
19. B. Tamamushi and N. Watanabe, Colloid and Polymer Sci. 258, 174 (1980).
20. P. Ekwall, L. Modell and K. Fontell, J. Colloid Interface Sci. 33, 215 (1970).
21. H.F. Eicke and H. Christen, Helv. Chimica Acta 61, 2258 (1978).
22. B. Bedwell and E. Gulari in "Solution Behavior of Surfactants", K.L. Mittal and E.J. Fendler, Editors, Vol. 2, pp. 833-846, Plenum Press, New York (1982).

RELATION BETWEEN ADSORPTION ON A METAL SURFACE AND MONOLAYER

FORMATION AT THE AIR/WATER INTERFACE FROM AMPHIPHILIC SOLUTIONS

T. Arnebrant, T. Nylander, P.A. Cuypers*,
P.-O. Hegg and K. Larsson
Department of Food Technology, University of Lund
Lund, Sweden and *Department of Biophysics, Faculty
of Medicine, Rijksuniversiteit Limburg, Maastricht
The Netherlands

The adsorption curves of sodium dodecyl sulphate,
monoolein and sodium caseinate versus concentration on
chromium have been determined by ellipsometry using a
new method of evaluation, which gives quantitative information. The chromium was treated to give a hydrophobic surface, and the critical surface tension of
wetting was about 45 mN/m according to contact angle
determinations. The adsorption data were related to
monolayer formation at the air/water interface according to surface tension measurements with the drop/
volume technique. A close relation was found between
the adsorption on the hydrophobic chromium surface and
the surface tension reduction as a function of concentration in the case of the lipid type of amphiphiles.
The protein, however, showed an adsorption maximum at
much lower concentration than the maximum in the surface tension reduction. The amounts adsorbed for the
micelle forming amphiphiles studied reach a plateau
after the maximum, which corresponds to monomolecular
layers.

INTRODUCTION

In order to understand mechanisms behind the adsorption of milk components ('fouling') or processing equipment we have started systematic studies using ellipsometry on the absorption on solids of different well-defined preparations of milk proteins, and mixtures corresponding to whey, skim milk and full-milk. It is natural to assume that the structure of the first adsorbed monolayer is of critical significance, resulting either in continuing deposition of the actual amphiphiles or in a surface passive with regard to fouling. Much is known on the formation of amphiphilic monolayers at the air/water interface, such as the conditions for formation of gaseous, liquid-condensed or solid films and their relations with bulk composition of the amphiphile-water phases. Due to experimental difficulties, however, less is known about the film formation at solid/water interfaces. This work was started in order to determine the relation between the successive steps of amphiphile monolayer formation at the air/water and metal/water interfaces. Such relations might then be used for prediction of adsorption on solids from the extensive data on air/water interfacial behaviour. As far as we know, no such comparison has been made earlier.

In order to correspond to the actual biological system we have used amphiphiles of lipid and protein type. We used one lipid which forms a micellar solution and another which forms a liquid-crystalline phase with water. Casein is a protein which is known to behave similar to simple surfactants in many respects, e.g. exhibiting self-association[2], and therefore sodium caseinate was used.

EXPERIMENTAL SECTION

Materials

Sodium lauryl sulphate, SDS, "specially pure" (declared purity 99%, with regard to real purity see Ref.[3,9])BDH Chemicals Ltd., Poole, England, and sodium-caseinate, food grade, DMV Zuid Nederlandse Melkindustrie B.V., Veghel, Holland, were used. The monoolein sample (Sigma) was checked by thin layer chromatography. Both the SDS and the sodium-caseinate were dissolved in double distilled water, and the caseinate solutions used were all freshly prepared. Glass slides with vacuum deposited chromium were kindly provided by Bengt Ivarsson, Department of Applied Physics, IFM, University of Linköping, Sweden. The metal film thickness was a few thousands Å.

Methods

The surface tension was measured according to the drop-volume technique using the following procedure[4,5]: A drop of a certain volume was formed, and the time for detachment of the droplet was recorded. The surface tension was calculated according to Tornberg[5] and the surface tension was plotted as a function of time. The equilibrium values of the surface pressure, i.e. the surface pressure attained after 40 minutes, were used. An aerometer was used for determination of the densities of the solutions.

Adsorption was measured using an automated Rudolph ellipsometer[6]. The chromium-plated slide was made hydrophobic by repeated washing with a detergent solution, double distilled water, and finally chloroform. The slide was placed in the sample holder, and the cuvette was filled with 4.5 ml double distilled water which was continuously stirred by a magnetic stirrer. The temperature of the system was allowed to be stabilized for 10 minutes, and then the optical properties of the chromium surface were measured. After the addition of 0.5 ml surfactant solution, the ellipsometer was read every 7 seconds, until the adsorption was complete. This procedure was repeated for each concentration on two different slides. The adsorbed amount was calculated according to a new method . The values of the partial specific volumes for SDS and caseinate used were those for a phospholipid and albumen, respectively . The partial specific volume of an SDS micellar solution has been reported to be 0.85 by Husson[8].

The ultra-pure liquids used for determination of the critical surface tension of wetting were kindly supplied by Dr. R.B. Baier (Calspan Corp., Buffalo, N.Y.)

All experiments were carried out at 24.5°C.

RESULTS AND DISCUSSION

The surface tension reduction versus concentration of SDS solutions is shown in Figure 1. Samples of SDS are extremely sensitive in aqueous solution, and the corresponding alcohol is easily formed. It is known that the presence of lauryl alcohol[9] and homologues, such as sodium myristyl sulphate, result in the occurrence of a minimum in the surface tension curve at the CMC[10,11], and our commercial sample exhibits this expected behaviour. From Gibbs equation of adsorption, it is known that there is a linear decrease in surface tension as a function of log (conc.) up to CMC, and above the CMC the surface tension is nearly constant. Discontinuities below the CMC, however, have recently been reported (cf. Ref.[12]). Our data indicate a similar feature. The decrease in surface tension with SDS concentration is considered to correspond to

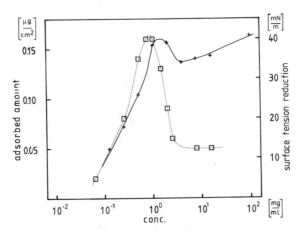

Figure 1. Amount adsorbed of SDS on a hydrophobic chromium surface versus concentration (squares) and reduction in surface tension (crosses) attained after 40 minutes.

successive increase in close-packing of the amphiphilic molecules
at the air/water interface.

    One critical factor in studies of adsorption on solids is the
knowledge of the nature of the solid surface. Unfortunately, this
is seldom recognized, and adsorption data are reported for surfaces which are not characterized well enough to allow interpretation from a surface structural point of view. We have used the
empirical method by Zisman (cf. Ref.13) to determine the critical
surface tension of wetting, and the results are given in Figure 2.
The critical surface tension is about 45 mN/m, and different surfaces show negligible variations in their wettability. The surface
which was chloroform treated, as described above, was also plasma-
cleaned, using a radio frequency glow discharge apparatus, which is
known to remove organic surface contaminations efficiently[14]. As expected a hydrophilic surface was obtained, as obvious from water
wettability. It is thus probable that the chloroform cleaning
technique we have applied results in a surface covered by a chloro-

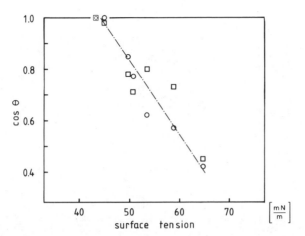

Figure 2. Determination of the critical surface tension of wetting according to Zisman plot (cf. Ref. [13]). The liquids used are
from the right: glycerol, formamide, thiodiglycol, methylene
iodide, tetrabromoethane, bromonaphtalene, and o-di-bromobenzene.
Squares and circles correspond to measurements on two different
chromium surfaces.

form film firmly bound (the surface tension of chloroform is 27
mN/m). With such hydrophobic surfaces it would be expected that the
amphiphilic molecules would orient on the surface in the same way
as at the air/water interface, i.e. with the polar head groups directed towards the water phase.

The adsorbed amount of SDS versus concentration is shown in
Figure 1, together with the surface tension reduction. There are
similarities between the two curves. The variations between adsorption measurements on two different metal surfaces prepared
identically are illustrated in Figure 3. The location of the maximum is the same, although the variations in amount adsorbed are
relatively large (the largest variation is about 20%). The measured

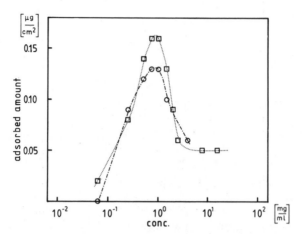

Figure 3. A comparison of the amount adsorbed of SDS on two
chromium plates, prepared in the same way (the difference is
assumed to be due to variations in the metal surface topography).

amount of adsorbed SDS corresponds to a liquid-condensed monolayer
at the maximum, and after that, the film expands with a factor
about three to give a minimum in the adsorption curve.

    During the course of this work an extensive study of the ad-
sorption of surfactant molecules on mineral oxides was reported by
Scamehorn et al.[15,16]. Although they assume that the surfactant
adsorbs with the polar head anchored towards the surface, our data
can be explained by their theory[15] (see also[19]). Thus different
behaviour in different regions of the adsorption curve can be observed below CMC, which according to their interpretation[15] corresponds to a low concentration region, where hemimicelles are formed on the surface. The occurrence of a maximum at the CMC,
followed by a minimum in the case of mixtures of surfactant homologues was also demonstrated experimentally and convincingly explained from an adsorption model based on adsorption of monomers
only[16,19].

There is a significant shift in the maximum for the amount of adsorbed SDS, compared to the maximum in surface tension reduction. Also this can be explained according to the work by Scamehorn et al.[16], who observed that an increase in the solution/solid ratio shifted the maximum to lower concentrations, and this was also explained from variations in monomer concentrations of the components[16]. The solution/surface area ratio in our adsorption measurements by ellipsometry was about 20 times larger than that for the surface tension measurements, which thus can explain the observed shift.

We have also examined the adsorption of monoolein versus concentration, but these measurements were quite limited for experimental reasons (monoolein forms large liquid-crystalline aggregates above its water solubility of about 10^{-6} M). The general features of the behaviour are similar to that of SDS. Thus, when the solubility level is reached a condensed monolayer is formed at the air/water interface (with a surface tension reduction of about 40 mN/m), and also the amount adsorbed on the metal surface is consistent with the formation of a monolayer.

The relations between surface tension reduction and adsorption on a metal surface from caseinate solutions are shown in Figure 4. Although the behaviour of a protein must be expected to be quite complex compared to simple lipid amphiphiles, the general features are the same.

The so-called casein micelles are complex molecular aggregates, with a wide size distribution (cf. Refs.[17,18]). A caseinate solution consists of relatively small micelles (about 12 monomers), which exist in dynamic equilibrium with the monomers as an ordinary micellar solution[18]. The strong increase in surface tension reduction up to a plateau value shown in Figure 4 takes place at about the same concentration as that which has been interpretated as a kind of CMC[18]. The adsorption curve has a similar shape, but the steep increase starts already at about ten times lower concentration. The amount adsorbed reaches a plateau as the simple amphiphiles discussed earlier. The amount adsorbed at this plateau corresponds to a monomolecular layer, and obviously this monolayer makes the surface passive with regard to further adsorption. The formation of films at the air/water surface from a caseinate solution is mainly a reversible process, contrary to the adsorption on a metal surface, and the extent of irreversibly adsorbed molecules should be expected to result in a corresponding shift towards lower concentration.

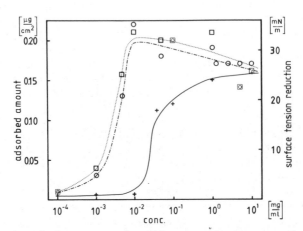

Figure 4. Adsorbed amount of sodium caseinate on a hydrophobic chromium surface versus concentration (squares and circles correspond to two plates) and surface tension reduction (crosses) attained after 40 minutes.

ACKNOWLEDGEMENTS

This work has been financed by a grant from the Swedish Board for Technical Development. Professor P.-O. Glantz supported us in the contact angle determinations.

REFERENCES

1. G.L. Gaines, Jr.,"Insoluble Monolayer at the Liquid-Gas Interface", Wiley/Interscience, New York (1966).
2. D.F. Waugh, in "Milk Proteins", H.A. McKenzie, ed., vol. 2. p. 3, Academic Press, New York (1971).
3. K.S. Birdi, Anal. Biochem. 74, 620 (1976).
4. E. Tornberg, J. Colloid Interface Sci. 60, 50 (1977).
5. E. Tornberg, J. Colloid Interface Sci. 64, 391 (1978).

6. P.A. Cuypers, "Dynamic Ellipsometry Biochemical and Biomedical Application", Thesis, Rijksuniversiteit Limburg, Maastricht, Holland (1976).

7. P.A. Cuypers, J.W. Corsel, M.P. Janssen, J. Kop, W.Th. Hermens and H.C. Hemker, these proceedings.

8. F. Husson, "An X-ray Low-Angle Study of Micelles", Thesis CNRS, Gif-sur-Yvette (1964).

9. B.R. Vijayendran, J. Colloid Interface Sci. 60, 418 (1977).

10. G.D. Miles and L. Shedlovsky, J. Phys. Chem. 48, 57 (1944).

11. P.H. Elworthy and K. Mysels, J. Colloid Interface Sci. 21, 331 (1966).

12. J.-C. Eriksson, Finska Kemistsamfundet Meddelar, in press.

13. R.E. Baier, E.G. Shafrin and W.A. Zisman, Science 162, 1360 (1968).

14. D.F. O'Kane and K.L. Mittal, J. Vac. Sci. Technol. 11, 567 (1974).

15. J.F. Scamehorn, R.S. Schechter and W.H. Wade, J. Colloid Interface Sci. 85, 463 (1982).

16. J.F. Scamehorn, R.S. Schechter and W.H. Wade, J. Colloid Interface Sci. 85, 479 (1982).

17. D.G. Schmidt and T.A. Payens, Surface Colloid Sci. 9, 165 (1976).

18. E. Tornberg, J. Sci. Fd.Agric. 29, 762 (1978).

19. F.J. Trogus, R.S. Schechter and W.H. Wade, J. Colloid Interface Sci. 70, 293 (1979).

QUANTITATIVE ELLIPSOMETRY OF PROTEIN ADSORPTION AT SOLID-LIQUID INTERFACES

P.A. Cuypers, J.W. Corsel, M.P. Janssen, J.M.M. Kop, H.C. Hemker and W.Th. Hermens

Department of Biophysics
University of Limburg
Maastricht, The Netherlands

Ellipsometry makes it possible to measure directly the refractive index and thickness of thin layers of material on a reflecting surface.
The calculation of the adsorbed mass from the measured refractive index and thickness is done using the recently developed mass equations[1]. These equations are based on the Lorentz-Lorenz relation between the molar refractivities, refractive index and the density of a binary mixture of substances.
As an illustration of the application of this technique in biochemical research, the results obtained for adsorption of several proteins on different surfaces are discussed. Unrealistically high values of refractive indices are sometimes observed but can be explained in terms of the changes in the optical parameters of the adsorbing system, indicating possible swelling shrinking or penetration, which is shown by computer simulation.

INTRODUCTION

In many fields of research there is an increasing need to study the reactions taking place at interfaces. In our field of research, i.e. blood coagulation, many proteins react at phospholipid-water interfaces[2].

Ellipsometric measurements can be carried out at liquid-solid, solid-air, or air-liquid interfaces. The first ellipsometric studies to measure thickness and refractive index of organic material were done already by Rothen et al. in 1942. Other authors like Trurnit (1954), Vroman (1969), Mathot (1969)[3,4,5,6] and others published ellipsometric data on organic material. However, in most of these studies the assumption was made that the refractive index of the adsorbed layer was a constant. We showed in several papers that in most of the proteins we studied the assumption of a constant refractive index is not true[1,7,8]. This paper describes the adsorption of several proteins on chromium and on chromium covered with stacked monolayers of phospholipids. In order to illustrate different types of adsorption we choose fibrinogen as a big rod-shaped molecule, which binds very well to chromium, prothrombin which interacts with phospholipids via calcium and bloodcoagulation factor V_a which is supposed to bind via hydrophobic parts of the molecule with the phospholipids. Automatization enables us to measure at intervals of about 4 seconds, this makes it possible to follow (slow) dynamic processes like adsorption, desorption, phase transitions, swelling, shrinking etc.

MATERIALS AND METHODS

The phospholipid used was 1,2 di-oleoyl-sn-glycero-3-phosphoserine (18:1/18:1-PS), prepared by enzymatic synthesis from the corresponding glycero-phosphocholine[9].

Bovine prothrombin ($M = 73000$) was prepared according to the method of Owen et al.[10].

Fibrinogen, human, 99,9% clottable grade L was obtained from Kabi Sweden ($M = 330000$).

Blood coagulation factor V_a ($M = 174000$) was purified according to a method of Esmon[11] modified by Lindhout[12].

Chromium coated glass slides were manufactured by Stabilix, The Hague, Holland.

Sparkleen was purchased from Fisher Scientific Company U.S.A. All other chemicals used were Merck P.A.

Stacking of monolayers or multilayers

Stacking was done with a preparative Langmuir-trough (Lauda, Type FW-1) according to the method of Langmuir and Blodgett. Unless mentioned otherwise the trough was filled with double distilled water and 5 µM $CaCl_2$. On this aqueous subphase, a mono-molecular film of phospholipids was spread by adding 100 µl of a solution containing 2 mg of phospholipid per ml chloroform. The surface pressure was held constant at 40 dyne/cm. A chromium coated glass slide cleaned with Sparkleen and chloroform was mechanically dipped into this trough and subsequently redrawn, both at a speed of 2 mm per minute, so that a double layer of phospholipid is deposited on the slide at each repeated dip.

Protein adsorption

The adsorption of fibrinogen was measured on two differently prepared chromium slides. First adsorptions were performed on slides that were treated for 10 minutes at 25° with chromic acid, after cleaning with Sparkleen and chloroform (Chromium type I). Secondly adsorptions were performed on slides cleaned with Sparkleen and chloroform (Chromium type II). These chromium slides were placed in the cuvette filled with buffer: 0.01 M Tris-HCl pH=7.0. After equilibration at $25^o \pm 0.1$ degree for 3 minutes in the well stirred, temperature controlled cuvette, several µl of a concentrated protein solution were added. Final fibrinogen concentration was 10 µg/ml. Prothrombin, final concentration (10 µg/ml) was adsorbed on chromium coated glass slides, stacked with 4 layers 18:1/18:1-PS. Factor V_a, final concentration (2 µg/ml) was adsorbed on chromium coated glass slides, stacked with one layer 18:1/18:1-PS. Buffer used for prothrombin was 0.05 M Tris-HCl pH=7.5 10 mM $CaCl_2$ 0.1 M NaCl. The same buffer without $CaCl_2$ was used for V_a.

Ellipsometry

The ellipsometer, a modified Rudolph and Sons type 43303-200E, is an optical instrument that measures the changes in polarization of light due to reflection (figure 1). The changes are measured by two adjustable polarizers indicated in Figure 1 as polarizer (P) and analyzer (A). The measurement consists of finding the positions of p and a corresponding to minimal light intensity reaching the photodiode. The change in polarization due to reflection is dependent on the optical properties of the reflecting surface. If the surface is covered with phospholipidlayers or a protein layer, then the optical properties of the reflecting surface are changed. This change in optical properties results in changes in the positions of P and A.

A complete description of the instrument has been given earlier[7,8] as well as the method of computation[13,14]. The computation can be summarized as follows.

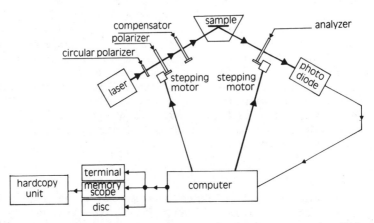

Figure 1   Schematic representation of the automated ellipsometer.

The ratio R_p/R_s, where R_p is the reflection coefficient for
light polarized parallel to the plane of incidence and R_s is
the reflection coefficient for light polarized perpendicular
('senkrecht') to the plane of incidence, is given by

$$R_p/R_s = \tan \psi \cdot \exp(i\Delta) \tag{1}$$

where ψ and Δ can be directly determined from the readings of
A and P respectively and $i = \sqrt{-1}$.

n and d of the phospholipid
layers stacked on chromium
slides were analysed according
to the system presented in
Figure 2. The reflection coefficients R_p and R_s are dependent upon the angle of incidence ϕ_1 (68°), the wavelength
of the light λ (632,8 nm), the
refractive indices n_1, n_2 and
n_3 and the thickness d_2. Equation (1) can be written as

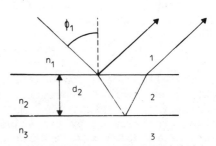

Figure 2 Chromium surface(3)
covered with a layer of
phospholipid(2) in buffer(1)

$$C_1(\exp D)^2 + C_2(\exp D) + C_3 = 0 \tag{2}$$

where C_1, C_2 and C_3 are complex functions of the refractive
indices, ψ and Δ and

$$D = - 4 \pi i n_2 \sqrt{1-(n_1 \cos\phi_1/n_2)^2} \; d_2/\lambda \qquad (3)$$

The value of n_1 is determined by refractometry and the (complex) value of n_3 is determined ellipsometrically for the chromium slide in buffer, before it is coated with phospholipid. Substituting these values, and an arbitrary (real) value for n_2, in equation (2) will generally yield a complex value for d_2. The correct value for d_2 must however be real, so equation (3) is solved by an iterative procedure in which n_2 is adjusted such that the complex part of d_2 minimizes. Proteins adsorbed on phospholipid were analysed according to the system presented in Figure 3. Equations (2) and (3) remain valid but the complex functions C_1, C_2 and C_3 now also depend upon n_3 and d_3. Values of n_3 and d_3 are determined by ellipsometric measurement before the protein is added to the cuvette.

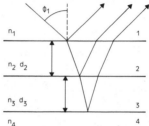

Figure 3   Chromium surface (4) covered with a layer of phospholipid (3) and a layer of protein (2) in buffer (1).

Calculation of the adsorbed mass from the refractive index, n, and thickness, d, of an adsorbed layer.

The equations for calculation of adsorbed mass from thickness and refractive index have been derived elsewhere[1]. The mass of protein in an adsorbed mixed layer of thickness d (Å) is given by

$$m = \frac{0.03 \; d \; f(n)}{\dfrac{A_p}{M_p} - V \dfrac{n_b^2 - 1}{n_b^2 + 2}} (n - n_b)$$

where

$$f(n) = \frac{n + n_b}{(n^2 + 2)(n_b^2 + 2)}$$

In this expression
- m = mass ($\mu g/cm^2$)
- d = thickness (Å)
- n = refractive index of the adsorbed layer
- n_b = refractive index of the buffer
- A_p = molar refractivity of the protein
- M_p = molecular weight of the protein
- V^p = partial specific volume of the protein

These expressions have to be used if the value of the refractive index n is between n_b and n_{max} where n_{max} is the refractive index of the pure protein ($n_{max} = \pm 1.60$ depending on the protein).

If the refractive index is higher than n_{max}, the mass can be calculated by

$$m_p = \frac{0.1\, d.M_p}{A_p} \frac{n^2 - 1}{n^2 + 2}$$

The values of the molar refractivity can be calculated from the molecular structure of the adsorbed substance[1], being A = 16070 cm^{-3} for fibrinogen, A = 17260 cm^{-3} for prothrombin. As an approximation we used the A value of prothrombin for the V_a molecule.

RESULTS

Fibrinogen Adsorption on Chromium type I. The cuvette was filled with buffer and the optical properties of the chromium surface, positions of P and A, were measured. After about 200 seconds 0.07 mg of fibrinogen was added (70 μl of 1 mg/ml solution).

After the moment of addition, the polarizer and analyzer change fast and an end level is reached after about 600 seconds for both analyzer and polarizer. Further addition of fibrinogen does not cause any further changes. The results interpreted in terms of thickness, refractive index and mass Figure 4 show that the refractive index does not change much during the adsorption. The thickness follows the saturation behaviour of a monolayer adsorption. After about 4000 seconds the thickness stabilizes around 130 Å. The refractive index is then n = 1.38. The calculated mass shows the same behaviour as the thickness does. The end mass value is 0.47 μg/cm^2.

Fibrinogen Adsorption on Chromium type II. The adsorption of fibrinogen on chromium type II is done under the same experimental conditions as chromium type I. This adsorption Figure (5) shows that during the first 150 seconds the thickness grows at constant refractive index n = 1.8. Then the refractive index drops while the thickness still increases. The thickness grows until about 68 Å while the refractive index drops to n = 1.48. From this point on the layer gets thinner and optically denser. The layer stabilizes at d = 35 Å and n = 1.72. The calculated mass shows a regular monolayer type adsorption which reaches its maximum before the layer gets thinner m = 0.56 μg/cm^2.

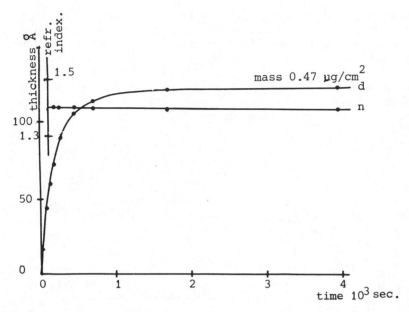

Figure 4   Adsorption of fibrinogen on chromium type I surface
(10 µg/ml).

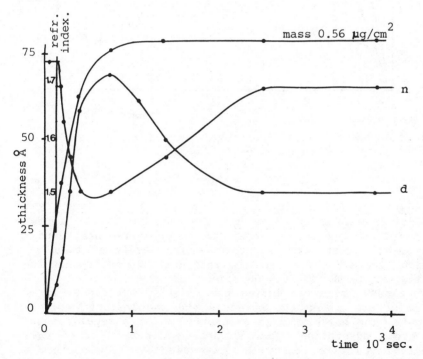

Figure 5   Adsorption of fibrinogen on chromium type II surface
(10 µg/ml).

Adsorption of Prothrombin on 4 18:1/18:1-PS Layers Stacked on Chromium. Four monolayers of di $C_{18:1}$ PS were stacked on
a chromium coated glass slide. This slide was placed in the
cuvette filled with buffer and the prothrombin was added (10
µg/ml). The adsorption data show that the refractive index
varies between n = 1.35 and n = 1.37 Figure 6. The average
thickness is about 110 Å. The mass shows a monolayer type
adsorption with a final value of m = 0.24 µg/cm^2.

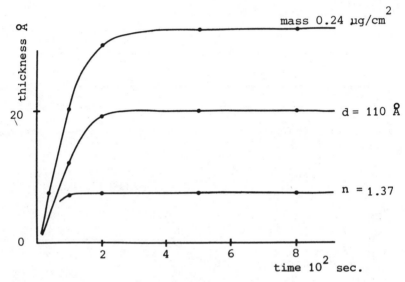

Figure 6   Adsorption of Prothrombin on 4 18:1/18:1-PS layers
            stacked on chromium (10 µg/ml).

Adsorption of Bloodcoagulation Factor V_a on one Layer
18:1/18:1-PS Stacked on Chromium. The adsorption of factor
V_a (2 µg/ml) was performed under the same experimental conditions as the prothrombin adsorption. The result of this adsorption Figure 7 shows that the refractive index is changing
slightly during the first 5000 seconds.
The thickness increases during the first 2000 seconds. After
this point the thickness decreases and the refractive index
increases. The calculated mass adsorbed shows a regular monolayer type adsorption. The refractive index of the layer
increases even after the total mass adsorbed has reached an
equilibrium value m = 0,16 µg/cm^2. The value of the refractive
index of the adsorbed layer is, n = 2.3 after 10^4 seconds.

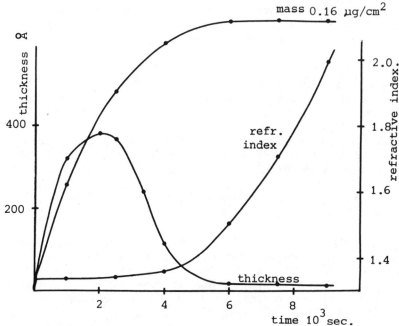

Figure 7   Adsorption of factor V_a on one layer of di $C_{18:1}$ PS stacked on chromium (2 μg/ml).

COMPUTER SIMULATIONS AND DISCUSSION

The adsorption of factor V_a and the adsorption of fibrinogen on a chromium type II surface show unrealistically high values of the refractive index. The refractive index of a hundred percent solution should be about n = 1.6 according to our mass equations.
To interpret these high values we did the following computer simulations.
We started with a four layer system such as:

buffer: n_1 = 1. 3335

protein: n_2 = 1.45 d_2 = 90   Å m_2 = 0.60 μg/cm^2

phospholipid: n_3 = 1.50 d_3 = 100 Å m_3 = 0.94 μg/cm^2

chromium: n_4 = 3.00 k_4 = -2.50

This system can be described by three fixed ψ and Δ values:
1. the chromium buffer system which gives the starting values
   $ψ_s$ and $Δ_s$

2. the chromium surface covered with the phospholipid layer in
   buffer (ψ_p and Δ_p)
3. the total system (ψ_t and Δ_t)

In our simulations ψ_s and Δ_s and the ψ_t and Δ_t values were kept
constant. In this system we choose arbitrary values for the
refractive index and thickness of the phospholipid layer. Any
arbitrary refractive index and thickness of the phospholipid
layer results in a new set of ψ_p and Δ_p for the chromium
surface covered with phospholipid in buffer. Using this new ψ_p
and Δ_p we calculated the refractive index and thickness of the
protein layer that resulted in the same end ψ_t and Δ_t. For
reasons of simplicity we used the same M/A and V_{20} for both
protein and phospholipid. Each chosen set of n and d values of
the phospholipid layer resulted in a new set of n and d values
of the protein layer.
The X axis in Figure 8 shows the phospholipid mass, while the
sum of phospholipid- and protein mass was kept constant at m =
1.54 $\mu g/cm^2$. The Y axis shows the refractive index of the
protein layer. The different lines in the figure indicate
lines of equal refractive indices of the phospholipid layer.
To interpret this graph we plotted the values of the refractive index of the protein layer as a function of the refractive index of the phospholipid layer at constant phospholipid
mass this means a constant X value. This results in Figure 9.

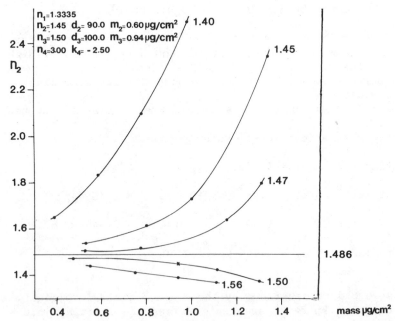

Figure 8   Relation between the optical parameters of layer two
(protein) and layer three (phospholipid).

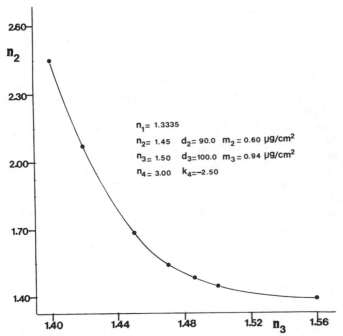

Figure 9   Relation between the refractive index of layer two (protein) and layer three (phospholipid) at constant mass ratio of these layers.

This graph clearly shows that if a very high refractive index of the protein is found, this refractive index can be decreased drastically by assuming that the refractive index of the phospholipid layer was increased during the protein adsorption. This explanation is also supported by the experimental observations that adsorption on more swollen phospholipid layers results in higher refractive index values of the protein layer.
The high refractive index of n = 1.78 of the adsorbed fibrinogen on chromium type II can also be explained by a very small change in the optical parameters of the chromium type II surface that might not be quite stable in buffer especially if their are high local electrolyte concentrations caused by the adsorption.

REFERENCES

1. P.A. Cuypers, J.W. Corsel, M.P. Janssen, J.M.M. Kop, W.Th. Hermens and H.C. Hemker. J. Biol. Chem. 258, 4 (1983).
2. R.F.A. Zwaal. Biochim. Biophys. Acta 515, 163 (1978).
3. A. Rothen, K. Landsteiner. J. Exp. Med. 76, 437 (1942).
4. H.J. Trurnit, Arch. Biochem. Biophys. 51, 176 (1954).

5. L. Vroman, A.L. Adams. Surface Science 16, 438 (1969).
6. C. Mathot, A. Rothen. Surface Science 16, 428 (196).
7. P.A. Cuypers, "Dynamic Ellipsometry", Thesis University Limburg, Maastricht, The Netherlands (1976).
8. P.A. Cuypers, W.Th. Hermens and H.C. Hemker, Anal. Biochem., 84, 56 (1978).
9. P. Comfurius and R.F.A. Zwaal, Biochim. Biophys. Acta, 488, 36 (1977).
10. W.G. Owen, G.T. Esmon and C.M. Jackson, J. Biol. Chem, 249, 594 (1974).
11. C.T. Esmon, J. Biol. Chem., 254, 964 (1979).
12. M.J. Lindhout, J.W.P. Govers-Riemslag, P. van de Waart, H.C. Hemker and J. Rosing, Biochemistry, 21, 5494 (1982).
13. F.L. McCrackin, E. Passaglia, R.R. Stromberg and H.L. Steinberg, J. Res. Natl. Bur. Std. (US), 67A, 363 (1963).
14. A. Vasicek, "Optics of thin films", North-Holland, Amsterdam (1960).

ASSOCIATION OF SURFACTANTS IN DILUTE AQUEOUS SOLUTIONS:

EFFECT ON THEIR SURFACE PROPERTIES

D. Exerowa and A. Nikolov

Institute of Physical Chemistry
Bulgarian Academy of Sciences
Sofia 1040, Bulgaria

An interpretation is made for association (pre-micellization) of surfactants in dilute solutions on the basis of measured dependences of surface tension decrease, $\Delta\sigma$, on bulk concentration C_s for aqueous solutions of ionic (n-decanoic acid, Na-alkyl sulphates) and nonionic (saturated alcohols) surfactants. The isotherms obtained have, after initial linear section (Henry's region), a horizontal part where the surface tension is constant. This section is discussed as the range in which the premicellization occurs. The measurements of the surface tension decrease are carried out by the spheretensiometric method. At low $\Delta\sigma \sim 1$ dyn/cm – the accuracy of the method is $\pm 2.10^{-3}$ dyn/cm. Another peculiarity in the course of the isotherm, found after the premicellization range, is interpreted as a phase transition in the adsorption layer: destruction of the adsorbed premicelles at the solution/air interface. The change in the structure of the adsorption layer is also shown by the dependence of surface potential on surfactant concentration in the case of alkyl sulphates. The relation between the bulk and surface properties of surfactant aqueous solutions and their ability to form black foam films (especially the bilayer Newton formations) is also discussed.

INTRODUCTION

As is well known, the amphiphilic character of surfactant molecules determines the specific physico-chemical properties of their aqueous solutions. The fact that majority of them form micelles has been well investigated. The surface and bulk properties of these solutions change suddenly in the region of CMC.

The phenomenon preceding CMC - the premicellization (association)[1-8] has not, however, been studied satisfactorily until now. Association is also possible in surfactant solutions having no CMC. It can be expected that as a result of this process both the bulk and surface properties of these solutions will also change[5-8]. The investigation of the surfactant ability to associate is an important problem, the solution of which is both of theoretical and practical importance[5,9,10].

The most sensitive and reliable method of CMC determination is by measurement of surface tension of surfactant aqueous solution. When the conventional methods of surface tension measurement are used[11] to investigate the premicellization, the isotherms obtained very often differ significantly[12] by reason either of insufficient sensitivity or of non-equilibrium values of $\Delta\sigma$ vs C_s (C_s - surfactant bulk concentration, $\Delta\sigma$ - decrease in surface tension).

The spheretensiometric method of Scheludko and Nikolov[13] for surface tension determination with sensitivity of $\pm 2.10^{-3}$ dyn/cm allows one to measure $\Delta\sigma$ vs C_s with sufficient accuracy that makes it applicable to very dilute surfactant solutions (two orders lower than CMC).

The most important factor determining the theoretical accuracy of $\Delta\sigma$ determination ($\delta\Delta\sigma$) by this method is the accuracy with which the sphere radius can be measured (R = 0,0563cm$\pm 10^{-4}$cm). $\delta\Delta\sigma$ is estimated as in Ref. 14: $\delta\Delta\sigma \approx \delta$ R/R $\Delta\sigma$.

The premicellization in dilute aqueous sodium dodecylsulphate solutions (n-NaDoS) has long been studied[4,6,12,15-17] but nevertheless it is still a matter of discussion[18].

The removal of impurities and the hydrolysis of NaDoS are of paramount importance for obtaining a reliable surface tension isotherm. Details about the measuring techniques and the purity of the sodium alkyl sulphates used are given in Ref. 7.

EXPERIMENTAL RESULTS

The isotherms $\Delta\sigma$ vs C_s for NaDoS, measured by the spheretensiometric method at 22°C, are plotted in Figure 1.

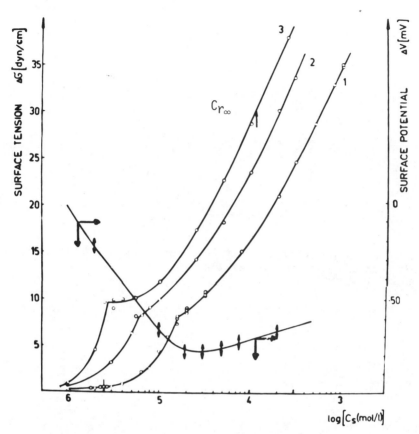

Figure 1.   Dependence of surface tension on the NaDoS concentration.   Curve 1 – C_{NaCl} = 10^{-1} mole/liter; curve 2 – C_{NaCl} = $3,5.10^{-1}$ mole/liter; curve 3 – C_{NaCl} = 5.10^{-1} mole/liter. Dependence of surface potential (ΔV) on the NaDoS concentration at C_{NaCl} = 10^{-1} mole/liter are given in double arrows. All measurements were carried out at 22°C.

The selected electrolyte concentrations maintain a constant solution ionic strength required for the investigation of the properties of black foam films[9] which are described below.

All three curves have a disconuity in slope in the region of 8-10 dyn/cm and have been drawn through the points using a square spline regression analysis.

A noticeable horizontal region is found in curve 1 at the lower concentrations of $2,5.10^{-6}$ mole/l to 3.10^{-6} mole/l where $\Delta\sigma \approx 0,5$ dyn/cm and $d\Delta\sigma/dC_s$ = 0. This region of $\Delta\sigma$ vs C_s is shown in Figure 2. Double arrows in Figure 1 show the surface

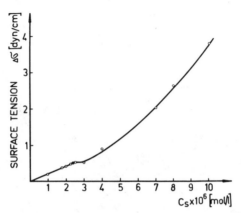

Figure 2.   Dependence of surface tension on the NaDoS concentration when $C_{NaCl} = 10^{-1}$ mole/liter.

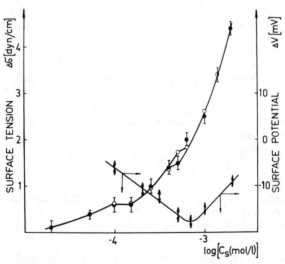

Figure 3.   Dependence of surface tension (at 22°C) and surface potential as a function of concentration of NaOS when $C_{NaCl} = 10^{-1}$ mole/liter (at 20°C). The values of $\Delta\sigma$ by Wilhelmy method are given in black dots and those by the spheretensiometric method in circles; the values of ΔV obtained by Volhardt and Wüsteek (Kolloid.Zh. 36, 1121 (1974)) are given in double arrows.

potential changes: ΔV vs C_S at 10^{-1} mole/1 NaCl. Analogous
investigations of $\Delta\sigma$ vs C_S and ΔV vs C_S at $C_{NaCl} = 10^{-1}$ mole/1
were also made with sodium octyl sulphate (NaOS). The results
obtained are shown in Figure 3.

It is worth noting that the measurements of $\Delta\sigma$ vs C_S for
NaOS were made in two ways: by the Wilhelmy method using a
scratched glass plate (black dots), and the spheretensiometric
method (circles). It is evident that both methods give very similar
values (an experimental scatter of ± 0,15 dyn/cm is found for the
values obtained by Wilhelmy method). The surface potential
changes are plotted by double arrows in Figure 3.

Measurements of $\Delta\sigma$ vs C_S were also made, using the spheretensiometric method, for the acidified aqueous solutions (C_{HCl} =
7.10^{-2} mole/1) of chromatographically pure decanoic acid (Merck
product) at 7°C and 25°C[19]. The results of these measurements
are presented in Figures 4 & 5.

It can be seen that the behavior of the $\Delta\sigma$ vs C_S relation is
similar to that of NaOS and NaDoS isotherms.

The isotherms $\Delta\sigma$ vs $\log C_S$ of the saturated fatty alcohols
n-pentanol, n-hexanol and n-heptanol (chromatographically pure
"Serva" products) obtained by the spheretensiometric method at 20°C
are shown in Figure 6.

The solutions were prepared using doubly distilled water with
surface tension of 72,85 dyn/cm measured by the spheretensiometric
method at 20°C.

DISCUSSION

The isotherms of investigated ionic surfactants (NaDoS,
NaOS and decanoic acid) over a wide concentration range and at
different electrolyte concentrations and temperature show very
close trends. These results are discussed in detail below.

In the lower concentration range, the relation $\Delta\sigma$ vs C_S has
linear region where $\Gamma = kC_S$ (Γ - number of molecules adsorbed
per unit surface). When the linear region is extrapolated to zero
concentration it should cross the origin of the coordinate system.
Any crossing of the positive ordinate would mean considerable
amount of surface active impurities of greater surface activity
than that of the basic component. For example, from the linear
region in the $\Delta\sigma$ vs C_S curve for decanoic acid (Figure 5) at 25°C
when plotted by the least squares method, the following values are
obtained for the linear regression coefficients: slope: $7,6.10^{4\pm}$

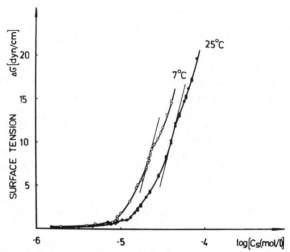

Figure 4. Dependence of surface tension on the decanoic acid concentration when C_{HCl} = 7.10^{-2} mole/liter at 7°C (circles) and 25°C
(black dots).

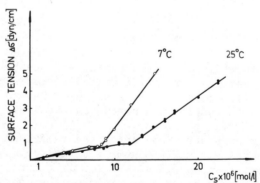

Figure 5. The data from Figure 4 only at low concentration (up to
$\Delta\sigma \approx 5$ dyn/cm); the concentrations are given in linear scale.

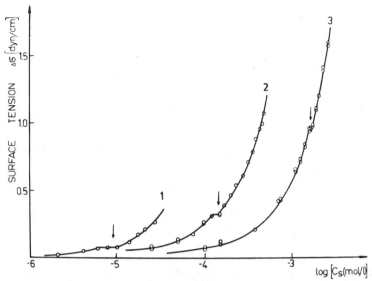

Figure 6.   Dependence of surface tension on the fatty alcohol concentration.   Curve 1 - n-heptanol; curve 2 - n-hexanol; curve 3 -
n-pentanol at 20°C

$3,5.10^3$; intercept : $1,53.10^{-2} \pm 9,04.10^{-8}$, and correlation coefficient $0,982 \pm 5.10^{-3}$.   Furthermore, the accuracy with which
the curve slope is determined makes the determination of the
experimental accuracy of the methods possible ($\pm 3.10^{-2}$ dyn/cm).

The linear portion of the $\Delta\sigma$ vs C_s plots is followed by
a short horizontal region where $d\Delta\sigma/dC_s \approx 0$ (Figures 2,5).   This
sector was first found by Exerowa and Scheludko[22]. The horizontal
part is followed by a steep increase in the curve slope after
which a drastic decrease in $d\Delta\sigma/d \log C_s$ is observed (Figures 1,3,
4).   A similar change of the slope has also been noticed by Zimmels
and others[5,10] for solutions of Na-oleate and K-oleate and by
Dervichian[21] for decanoic and undecanoic acid solutions.   Figure
7 illustrates the $\Delta\sigma$ vs C_s - curves for the latter two cases.

The slope changes in $\Delta\sigma$ vs $\log C_s$ plots and the minimun in
the ΔV vs $\log C_s$ plots for NaDoS and NaOS (Figures 1, 3) can be
interpreted as phase transition in the adsorption layer. In support of this assumption one should notice the clearly expressed
scattering in the data for $\Delta\sigma$ vs $\log C_s$ in curves 1, 2, 3 in
Figure 1 in the zone of $\Delta\sigma \approx$ 8-10 dyn/cm.   The surface phase
transition can be explained if we assume that not the individual
molecules but molecular aggregates, e.g., flat premicelles, are
adsorbed prior to the phase transition. They are most likely
complexes formed by two negatively charged ions for the case of
NaDoS-$(DoS^-)_2$ and NaOS-$(OS^-)_2$ and by two undissociated decanoic

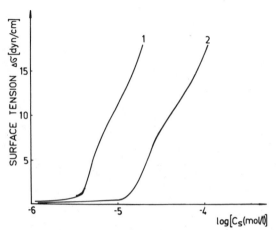

Figure 7. Dependence of surface tension on the undecanoic acid –
1 and decanoic acid – 2 concentration measured by Dervichian[21] at
20°C and plotted as Δσ vs log C_s.

acid molecules. The flat premicelles adsorbed at the interface
cannot attain complete surface saturation. Evidently closer packing
can be reached by adsorption of individual straight molecules. For
this reason a disintegration of adsorbed premicelles into individual
molecules commences from the concentration corresponding to the
adsorption monolayer saturation by the flat premicelles. The disintegration of the premicelles at the interface can also result in a
decrease in surface adsorption. The latter is observed experimentally (Figures 1, 3, 4); the slope dΔσ/d log C_s~Γ of the curves
before the phase transition is greater than the slope after the
transition. The decrease in the number of the adsorbed molecules
is due to the fact that the adsorbed premicelles are electrically
neutral before the bend in the curve Δσ vs log C_s (e.g. in NaDoS;
Δσ ≈ 8–10 dyn/cm, Figure 1). Therefore the electrostatic forces of
repulsion between the individual premicelles at the surface are
either very small or completely absent (the electrolyte concentrations in the solution are sufficiently high to suppress the Debye
atmospheres). The electrical neutrality in the adsorption plane
(coinciding with the water-air interface) is disturbed when the
premicelles are disintegrated. Overall neutrality is preserved
but only at the expense of the Na-ions penetration in the adsorption layer[23], i.e., the formation of a double electric layer at
the interface. Repulsion forces become important between the

individual ions and the number of adsorbed molecules per unit surface reduces. Evidently with further increase in the surfactant
bulk concentration the number of adsorbed ions will increase. The
charge of the double electrical layer and the surface potential,
ΔV, thus determined, will also increase in the same manner and
this is observed in Figures 1 and 3.

After establishing the surface adsorbed premicelles disintegration as the cause of the phase transition, it is necessary to
explore when and where they are formed.

The time taken for reaching the equilibrium surface tension
value depends on the bulk concentration. As an example we shall
examine the time required for reaching of steady surface tension
for NaDoS aqueous solutions at $C_{NaCl} = 5.10^{-1}$ mole/l. At low
concentration (2.10^{-6} mole/l) this time is of the order of 6 hours
(Figure 8, curve 1) and it continuously increases to several dozens
of hours in the zone of the surface phase transition (Figure 8,
curve 2). At higher concentrations ($C_s = 2,5.10^{-4}$ mole/l Figure 8,
curve 3) the time necessary for reaching a constant value of σ
again drops to several hours. These measurements have been carried
out by the sphere-tensiometric method with sensitivity of $\pm 2.10^{-3}$
dyn/cm.

Similar dependence of σ on time for saturated fatty acids in
aqueous medium has been found by Dervichian[21].

Baret and Roux[24] have shown that the time for reaching a
steady surface tension value is determined by surfactant diffusion
to the interface, and for this reason the $1/\Delta\sigma$ vs $1/\sqrt{t}$ plot is
linear. An analogous picture is obtained in our case when $C_s <$
$2,5.10^{-6}$ mole/l (Figure 9, curve 1). For $C_s > 2,5.10^{-6}$ mole/l, the
results show a broken straight line (Figure 9, curve 2). This is
an indication that initially particles of the same size diffuse
to the interface, but subsequently, particles of different size
become important in the diffusion process.

In fact we know from the diffusion theory that the diffusion
equation solutions depend not on $1/\sqrt{t}$ but on the combination
$(1/D)t^{1/2}$ where D is the diffusion coefficient. For this reason
the breaking of the straight line is an indication of different
diffusion coefficients of adsorbed particles during the first and
the second stage of the process. It is quite natural to assume
that initially monomers (single ions or molecules) are adsorbed
and subsequently the premicelles adsorption occurs. As $D \sim 1/R$,
where R is the equivalent radius of the particle, the data in
Figure 9 show that the particle radii of the new phase are approximately twice as large as that of the monomers.

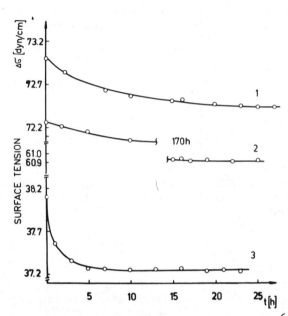

Figure 8. σ vs time plot. Curve 1 – C_{NaDoS} = 2.10^{-6} mole/1; curve 2 – C_{NaDoS} = 3.10^{-5} mole/1; curve 3 – C_{NaDoS} = 2,5.10^{-4} mole/1 at C_{NaCl} = 5.10^{-1} mole/1 and T = 22°C.

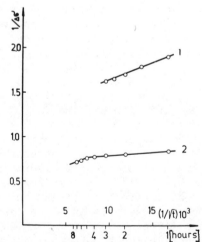

Figure 9. Dependence of the decrease of surface tension on time in plots of $1/\Delta\sigma$ vs $1/t^{1/2}$ – on the top of abscissa, t is given in minutes, below in hours. Curve 1 – C_{NaDoS} = 1,5.10^{-6} mole/liter; curve 2 – C_{NaDoS} = 4.10^{-6} mole/liter at C_{NaCl} = 10^{-1} mole/liter at 22°C.

This is an indication that $(DoS^-)_2$ premicelles are formed in the case of NaDoS. Thus the kinetic σ vs time curves show that associates are formed in the bulk at very low concentrations ($d\Delta\sigma/dC_s \approx 0$) and their adsorption starts after the horizontal region in the $\Delta\sigma$ vs C_s isotherm. Conclusion that the association takes place in the bulk was also made by others[6,8], while measuring NaDoS solution electrical conductivity.

Association at low concentrations as shown by means of the surface tension isotherms has also been observed in aqueous solutions of saturated fatty alcohols (Figure 6).

It may be assumed in this case that more complex aggregates are formed and the association can be presented by a general equation of the type:

$$mH_2O + (ROH)_{n-1} + ROH \rightleftharpoons (ROH)_n(H_2O)_m$$

The existence of associates in aqueous ROH solutions has been proven by Franks and Ives[27]. The ROH aqueous solution isotherms were not obtained at high concentrations due to evaporation at the solution interface. Regardless of the preventive measures taken, it was found that after about two hours σ began to increase slowly.

It is very difficult to prove by direct method (e.g. optical ones) existence of associates in the bulk of the solution at low concentrations ($10^{-6} - 10^{-5}$ mole/liter) because of the lower association degrees. One possibility for assessing the degree of association in the case of NaDoS is to calculate the diffusion coefficient from σ vs time relationship. Thus curve 2 in Figure 9 was used to calculate the hydrated $(DoS^-)_2$ ion diffusion coefficients. The following values were obtained: $1,6.10^{-6}$ cm^2.sec^{-1} for monomers, and $1,04.10^{-6}$ cm^2.sec^{-1} for the aggregates, and hydrate radii were found to be 13Å and 20 Å.

Black Foam Films and the Role of the Adsorbed Layers

The surface tension isotherms were used to understand the role of the adsorbed layers in the formation of the black foam films (BF). The latter are formed in the thicker thermodynamically unstable films in the form of black spots, see, e.g., Refs. 22, 28 and 32.   These thinner areas in the film grow and merge until the whole film becomes black. The ability of black spot formation is associated with some new facts unaccounted for in the Deryaguin-Landau-Verwey-Overbeek (DLVO) theory for thicker (polymolecular) films. The quantitative measurements show the existence of two types of black foam films: common (CBF) and Newton (NBF) as two

different equilibrium states of the black film.[9,28,30] They
reach very low thicknesses (especially the Newtonian ones) of
nearly molecular dimensions, where they deomonstrate their specific
properties. This fact alone suggests the relation between their
features and adsorption at the solution/air interface. We shall
try to explain the formation of the two types of black foam films
in terms of the condition of the adsorbed layer at the bulk interface of the same solutions from which the films have been obtained.

The use of microscopic foam films (of a radius about 10^{-2}
cm[9,22,28]) is quite convenient in the investigation of the origin
of the black spots. This model enables us to work at relatively
low surfactant concentrations where their specific effects begin
to show. In our previous papers we have introduced and investigated the different aspects of the concentration where the black
spots appear, the so-called C_{bl} concentration[9,28,29,31]. This
concentration proved to be an important quantitative surfactant
characteristic. As far as the formation of black spots is concerned, it is a probability process, and the determination of C_{bl}
is made by investigating the $\Delta N/N$ relation where N is the number
of films investigated and ΔN is the number of films where black
spots are formed (regardless of their life time[9]). Figure 10
illustrates such a plot for a common NaDoS black film *(each circle
on the curve being the result of about 500 measurements). It can
be seen that at the onset of the curve the dependence grows steeply
after which (at $C_s = 10^{-5}$ mole/1 NaDoS and $\Delta N/N = 0,9$) it slightly changes to $C_s = 3,5.10^{-4}$ at $\Delta N/N = 1$. We may conclude from
the comparison of the $\Delta N/N$ (Figure 10) curve with the adsorption
isotherm for 10^{-1} mole/1 NaCl (Figure 1, curve 1) that the formation of the common black spots commences after the deviation from
the ideal state (Henry's region in isotherm $\Delta\sigma$ vs C_s). The probability of CB-spot formation steeply increases with the surfactant
concentration, or the saturation of the adsorbed layer and at $C_s =
10^{-5}$ mole/1 (saturation $\Gamma/\Gamma_\infty = 0,4$), $\Delta N/N$ becomes 0,9.

Probability of unity for black spot formation is $3,5.10^{-4}$
mole/1, which is very close to the concentration where the
adsorbed layer becomes saturated ($C_{\Gamma_\infty} = 5.10^{-4}$ mole/1).

The pictures in Figure 10 illustrate only partially the
black spot and black film formation. The probability for the
appearance of black spots, as well as the probability of their
extension (a) and coalescence (b) until the whole film becomes

---

*When the black foam films are investigated in the measuring
cell, there was always a sufficient reservoir of the surfactant
molecules in the doubly concave drop. Furthermore, black film
measurements are made after surface saturation, depending on the
time taken to reach equilibrium σ values.

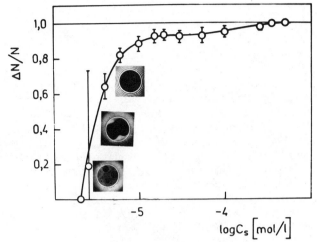

Figure 10. Probability $\Delta N/N$ for CB-spot formation as a function of the concentration of NaDoS (electrolyte concentration 10^{-1} mole/liter, film radius 10^{-2} cm, T = 22°C).

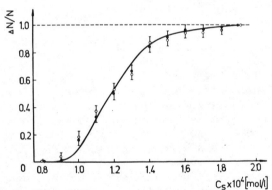

Figure 11. Probability $\Delta N/N$ for NB-film formation as a function of the concentration of NaDoS (electrolyte concentration 5.10^{-1} mole/l) at 22°C: ●,○ experimental data for foam film radii of $2.5.10^{-3}$ cm and 5.10^{-2} cm respectively; - theoretical curve according to Refs.34, 35.

black (c) increases when increasing the saturation in the adsorbed layer. In case of higher electrolyte concentrations (e.g., $3,5.10^{-1}$ mole/l, curve 2, Figure 1), the formation of the common black spots occur at a higher surfactant concentration corresponding to more closely packed adsorbed layer. This is probably due to the fact that when increasing the electrolyte concentration, the electrostatic component of disjoining pressure becomes lower[9,22].

For the formation of NB spots or films, even with the least probability, the adsorbed layer should already be saturated $(d\Delta\sigma/d \log C_s - const)$[9,30,34,35]. For the formation of the Newton film not only the disjoining pressure barrier should be overcome, but it is also necessary that the surfactant concentration in the solution should be close to $C\Gamma_\infty$. Figure 11 illustrates $\Delta N/N$ vs C_s (each experimental point, being a result from over 100 independent tests). It is seen that the experiment yields a steep relation. It is shown that $\Delta N/N$ vs C_s does not depend on the grey foam film radius[35]. The curve is plotted on the basis of the NBF stability theory[33-35]. These studies indicated that the adsorbed layer is very important for the CBF and NBF formation. Regardless of the new steps taken, the problem cannot be, however, regarded as finally solved. Of interest is, for example, the investigation of the nature of the black points (tiny diffusely - seen spots with radii of about 5.10^{-4} to 10^{-3} cm)[35].

In many cases it is observed that the Newton black spot forms from these black points, they may prove to be the "nucleus" for the NB-film. The black points can be observed at concentrations at which surface phase transition (Figures 1, 3 and 4) appears in the adsorbed layer. The formation of black points and black spots coincide for the CBF.

CONCLUSION

The precise investigation of surface tension isotherms of several aqueous surfactant solutions suggests that there is a surfactant molecule association in the bulk of the solution at very low concentrations. The association effect in aqueous surfactant solution has long been known[1-8] as mentioned in the Introduction. This work has shown an association effect at very low concentrations (immediately after the solution ideal state, in bulk as well as on the surface - Henry's region) which affects the isotherm and results in a peculiar behavior unexpected for low surfactant concentrations. The dependence of the association effect on the chain length, the electrolyte concentration, the time taken for reaching of σ and the behavior of the ΔV vs C_s relation suggest that the interpretation of the $\Delta\sigma$ vs C_s isotherms is correct. This approach would enable the finding of a relation between the bulk and surface surfactant properties as well as a relation with the black foam film formation. We consider this as an interesting opportunity worth developing in the future.

ACKNOWLEDGEMENT

The authors acknowledge the help of Dr. M. Schwuger, Head of the Physical Chemistry Division of Henkel, Düsseldorf for supplying pure NaDoS and NaOS.

REFERENCES

1. J. W. McBain, Kolloid Z., 12, 256 (1913).
2. P. Ekwall, Kolloid Z., 80, 77 (1937).
3. J. Stauff, Kolloid Z., 96, 246 (1941).
4. M. E. L. McBain, W. B. Dye and S. A. Jonson, J. Am. Chem.
    Soc., 61, 321 (1939). M. E. L. McBain, J. Colloid Sci.,
    10, 223 (1955).
5. Y. Zimmels and I. J. Lin, Colloid Polym. Sci., 254, 594 (1974).
    Y. Zimmels, I. J. Lin and J. P. Friend, Colloid Polym. Sci.,
    253, 404 (1975).
6. P. Mukerjee, K.J. Mysels and C. I. Dulin, J. Phys. Chem.,
    62, 1390 (1958). P. Mukerjee, Adv. Colloid Interface Sci.,
    1, 241 (1967). P. Mukerjee, J. Phys. Chem., 69, 2821 (1965).
7. A. Nikolov, G. Martynov and D. Exerowa, J. Colloid Interface
    Sci., 81, 116 (1981).
8. L. P. Panicheva and Z. N. Markina, Kolloid. Zh., 43, 671
    (1981).
9. D. Exerowa, A. Nikolov and M. Zacharieva, J. Colloid Interface
    Sci., 81, 419 (1981).
10. K. P. Ananthapadmanabhan, Ph.D. Dissertation, Henry Krumb
    School of Mines, Columbia University, 1979.
11. J. F. Padday, in "Surface and Colloid Science", E. Matijevic,
    Editor, Vol. 1, p. 101, Wiley - Interscience, New York, 1969.
12. S. Vijayan, D. R. Woods and H. Vaya, Canad. J. Chem. Eng.,
    55, 718 (1977). S. Vijayan, D. R. Woods and H. Vaya, Canad.
    J. Chem. Eng., 56, 103 (1978). S. Vijayan, D. R. Woods and
    D. Lowe, Canad. J. Chem. Eng., 57, 496 (1979).
13. A. Scheludko and A. Nikolov, Colloid Polym. Sci., 253, 404
    (1975).
14. A. Nikolov, God. Sof. Univ., Chim. Fac., 72, 151 (1977/1978).
15. D. Eagland and F. Franks, Trans.Faraday Soc., 61, 2468 (1965).
16. F. Franks and H. Smith, J. Phys. Chem., 68, 3581 (1964).
17. Van Voorst Vader, Trans. Faraday Soc., 57, 110 (1961).
18. S. Vijayan, C. Ramachandran and D. R. Woods, Canad. J. Chem.
    Eng., 58, 485 (1980).
19. A. Nikolov and D. Exerowa, submitted for publication.
20. A. Frumkin, Z. Phys. Chem., 116, 466 (1925).
21. D. C. Dervichian, Kolloid Z., 146, 96 (1956).
22. D.Exerowa and A. Scheludko, in "Proc. 4th Intern. Congr. Sur-
    face Activity, Brussels", Vol. II, 1097 (1964).
23. B. Pethica and A. Few, Discussions Faraday Soc., 18, 258
    (1954).
24. J. E. Baret and R. A. Roux, Kolloid Z., 225, 139 (1968).
25. D. Exerowa and A. Nikolov, paper presented at the VIIth
    European Chemistry Conferences at Interfaces, Finland, 1980.
26. A. Posner, J. Anderson and A. Alexander, J. Colloid Interface
    Sci., 7, 623 (1952).
27. F. Franks and D. Ives, Quart. Rev. London, 20, 1 (1966).
28. A. Scheludko, Adv. Colloid Interface Sci., 1, 391 (1967).

29. D. Exerowa, Isv. Khim. Inst., BAN, 11, 739 (1978).
30. K. J. Mysels, K. Shinoda and S. Frenkel, "Soap Films",
     Pergamon Press, New York, 1959.
31. J. S. Clunie, J. F. Goodman and B. T. Ingram, in "Surface and
     Colloid Science", E. Matijevic, Editor, Vol. 3, p. 167, Wiley,
     New York, 1971.
32. D. Exerowa, Chr. Christov and I. Penev, in "Foams", R. J.
     Akers, Editor, p. 109, Academic Press, London, 1976.
33. D. Kashchiev and D. Exerowa, J. Colloid Interface Sci., 74,
     501 (1980).
34. D. Exerowa, D. Kashchiev and B. Balinov, in "Microscopic
     Aspects of Adhesion and Lubrication", J. M. Georges, Editor,
     p. 107, Elsevier Scientific Publishing Company, Amsterdam,
     1982.
35. D. Exerowa, B. Balinov, A. Nikolova and D. Kashchiev, J.
     Colloid Interface Sci., (in press).

POLYDISPERSE NON-IONIC SURFACTANTS: THEIR SOLUTION CHEMISTRY AND EFFECT ON THE WETTABILITY OF SOLID SURFACES

G.G. Warr, P. Scales, F. Grieser, J.R. Aston,
D.R. Furlong and T.W. Healy

Department of Physical Chemistry
University of Melbourne
Parkville, Victoria, 3052, Australia

The behaviour of a series of polydisperse nonyl
phenol ethoxylates in aqueous solution and their
influence on the wettability of hydrophobic and
hydrophilic solid surfaces has been studied. Surface
tension and ultrafiltration techniques have been used
to study the characteristics of the non-ionic
surfactants in solution over a wide concentration
range. The effect of adsorption of the nonyl phenol
ethoxylates on the wettability of hydrophilic and
methylated quartz plates was monitored by contact
angle measurements. The results suggest that monolayer coverage of the solid surface occurs at about
the cmc of the surfactant solution, and that the
orientation of the adsorbed molecules is such that
the ethoxy part of the amphiphile is directed towards
the aqueous phase, independent of the natural
wettability of the original surface. A phase
separation model has been used to help define the
contributing effects of the individual components of
polydisperse mixtures for the systems studied. The
model predicts that at concentrations above the cmc,
the micelles are enriched with the shorter ethoxy
chain components of the mixture, while the aqueous
phase is enriched with the longer ethoxy chain
components of the mixture. The model also helps in
both a quantitative and qualitative understanding
of some of the results obtained in the surface
tension and ultrafiltration experiments.

INTRODUCTION

Non-ionic surfactants have found wide industrial application as frothers, emulsifiers and foaming agents[1]. An example of their importance is the ability to modify the surface properties of solids in fine coal recovery processes such as froth flotation and oil agglomeration[2-4].

A large proportion of commercial non-ionic surfactants consist of an alkyl chain bonded to a polyethoxy chain of variable length[5]. The method of preparation generally leads to a statistical distribution of the number of CH_2CH_2O units in the hydrophilic chain, so that the surfactant solution is always a mixture of similar species with varying solution properties, hydrophobicities and surface activities.

In the present study the solution chemistry, surface activity and tendency to modify the hydrophobicity of solid surfaces has been examined with a view to understanding how the individual components of the mixture at any given total concentration control the solution and interfacial properties of (mixed) nonyl phenol ethoxylated surfactants.

EXPERIMENTAL SECTION

(i) Polydispersity of the Non-ionic Surfactants

The non-ionic surfactants used in this study were poly-oxyethylene nonyl phenols, kindly supplied by I.C.I. (Australia), as research samples. Gas chromatographic analyses of the samples indicated that they consisted of a nonyl phenol base, with a distribution of polyethoxylate chain lengths which closely approximated a Poisson distribution about a mean ethoxylate number[2]. In this paper the mean value of the distribution is used to name the mixture i.e. $\overline{N}5$ refers to a nonyl phenol ethoxylate sample which has a mean of 5 ethoxy units per nonyl phenol moiety.

(ii) Surface Tension Measurements

Surface tension measurements were made at 20 ± 1°C by the Du Noüy ring method fully corrected for wall effects[6].

(iii) Ultrafiltration Measurements

Ultrafiltration experiments were performed using an Amicon cross-flow filtration apparatus through a precleaned 3nm pore diameter Amicon XM-50 membrane with a pressure head of 0.1MPa. The XM-50 filter is a neutral, hydrophilic membrane made of an unspecified copolymer with a nominal molecular weight cut off at 50,000.

Twenty millilitres of a 500ml solution were collected from each experiment to minimise changes in the solution composition due to any preferential filtration effects. Filtrate and filtrand concentrations were analysed by UV absorption at ~275nm (ϵ_{max} = 1540 ± 50M^{-1}cm^{-1}). No noticeable concentration change due to adsorption on the membrane or flow retardation of the surfactant was observed in submicellar solutions of $\overline{N}$30.

(iv) Contact Angle Measurements

Static three-phase contact angles were measured using the captive bubble technique. The contact angle of an air bubble on a prepared quartz plate in the presence of various aqueous concentrations of non-ionic surfactant was measured by a method described in detail previously[7].

Both the advancing and receding contact angles were determined, and following the method of Wolfram and Faust[8] an equilibrium contact angle was calculated to express the results. The preparation of plates and the contact angle apparatus have been described elsewhere[7].

RESULTS AND DISCUSSION

The surface tension (γ) of the air/solution interface as a function of concentration (C_T) for a series of ethoxylated nonyl phenols is shown in Figure 1. It can be seen that as the mean ethoxy number ($\overline{N}$) increases, the cmc increases and the minimum surface tension rises in accordance with the reported behaviour of nonyl phenol ethoxylates and similar non-ionic surfactants[9]. The cmc values (in moles dm^{-3}) were found to vary with the mean number of ethoxy units according to Equation (1),

$$\ln \text{cmc}_{\text{mix}} = -10.566 + 0.092\overline{N} \tag{1}$$

This is in reasonable agreement with previous cmc measurements on nonyl phenol ethoxylates[10].

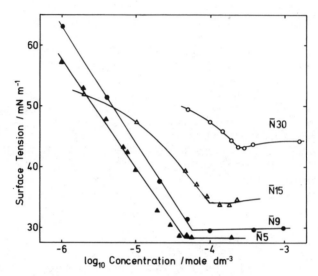

Figure 1. The liquid/vapour surface tension as a function of
concentration for a series of nonyl phenol ethoxylates.

        Surface tension measurements also provide information about
the molecular density at the air/water interface. Using the
Gibbs convention, the maximum adsorption density at the air/
solution interface, Γ_{max}, can be determined from slope of the
γ–log C_T curves at the cmc[1]. The values of area/molecule
obtained from Γ_{max} are presented in Figure 2.

        Most of the surface tension-concentration profiles exhibit
behaviour resembling that of a monodisperse surfactant. It is
only at large mean ethoxy chain lengths (such as $\bar{N}30$) that
surface tension minima, for example, are observed.

        A minimum in the surface tension profile of a monodisperse
surfactant is often explained by invoking the presence of a
single impurity species. This is not such a convincing
explanation for polydisperse surfactant mixtures since the

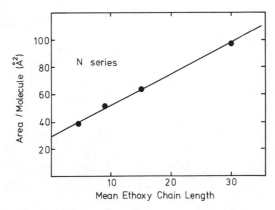

Figure 2. The average molecular cross–section area for some
nonyl phenol ethoxylates as a function of their mean ethoxy
chain length.

multicomponent species can themselves be considered "impurities".
Further, the minimum is not evident for the lower N mixtures,
although these are also polydisperse. To understand the N30
behaviour requires a consideration of the solution chemistry of
the surfactant mixtures. Some of this information can be
obtained from ultrafiltration studies, and theoretical
calculations.

The ultrafiltration experiments on N30 (Figure 3) indicate
that there is a distinct break in the filtrate vs. filtrand
concentration plot at the cmc of N30 as determined from surface
tension results. There is also a small, but measurable, increase
in the filtrate concentration at filtrand concentrations above
the cmc. The increase may be due either to an increase in the
monomer concentration in equilibrium with micelles above the cmc,
or to the leakage of aggregates through the membrane,
or a combination of both. In a previous ultrafiltration study,
a similar result was reported[11]. Dye solubilization studies[11]
confirmed that the increase in the ultrafiltrate concentration
was primarily due to increasing monomer concentration of the
filtrand.

The observed increase in the monomer concentration can
actually be predicted using a phase separation – ideal mixing model
for micellization in these polydispersed surfactant systems. By

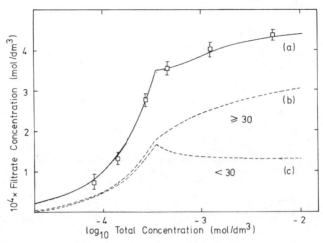

Figure 3. The concentration of ultrafiltrate as a function of the filtrand for $\bar{N}30$, with theoretical prediction (all lines) of monomer concentration, (a) Total monomer activity of all components, (b) Monomer activity of all components with chain length $\geqslant 30$, (c) Monomer activity of all components with chain length < 30.

extending Clint's[12] treatment for a binary system to a multicomponent one, it is possible to derive an expression for the monomer concentration of each species above the cmc[13]. If we make the usual assumption that no micelles are present below the cmc, then the monomer concentration of the ith component (C_{mon}) in this regime is simply,

$$c_{mon}^i = \alpha_i \, C_T \tag{2}$$

where α_i is the mole fraction of the ith species in the mixed solute, and C_T the total concentration, and above the cmc,

$$c_{mon}^i = \frac{[(C_T - C_{\mu i} + C_{Mi})^2 + 4\alpha_i \, C_T \, (C_{\mu i} - C_{Mi})]^{\frac{1}{2}} - C_T - C_{Mi} + C_{\mu i}}{2 \left[\dfrac{C_{\mu i}}{C_{Mi}} - 1 \right]} \tag{3}$$

The quantity $C_{\mu i}$ is defined as the cmc of a mixture of all species in the system except i. It is related to the cmc of the ith component C_{Mi} by,

$$\frac{1 - \alpha_i}{C_{\mu i}} = \left(\frac{1}{cmc_{mix}} - \frac{\alpha_i}{C_{Mi}}\right) \tag{4}$$

$C_{\mu i}$ is only introduced here to arrive at a useful representation of C_{mon}^i.

It can be shown[13,14] that for multicomponent systems the cmc of the mixture (cmc_{mix}) can be expressed by Equation (5).

$$\sum_{all\ i} \frac{\alpha_i}{C_{Mi}} = \frac{1}{cmc_{mix}} \tag{5}$$

From these equations, and from the observation[9] that the cmc of Poisson distributed mixtures of components are virtually the same as the cmc of the pure component with the same number of ethylene oxides as the mean of the mixture it is possible to calculate the total monomer concentration in solution both above and below the cmc of the mixed system. The results of these calculations are shown in Figure 3. Also shown are the concentration profiles summed over all components with less than 30 ethoxy units, and also those with 30 or more ethoxy units.

This phase separation – ideal mixing model predicts the observed increase in monomer activity above the cmc, and that the mixed micelles are richer in the more hydrophobic (shortest ethoxy chain) members of the distribution[‡]. The equilibrium solution sampled by ultrafiltration is, for $C_T >$ cmc, enriched in the more hydrophilic members of the distribution.

These latter theoretical results can be of use in under-standing the minimum in surface tension observed for $\overline{N}30$ (Figure 1). Enrichment of the bulk solution by components with more than 30 ethoxy units will cause the air/solution interface to be similarly enriched. Since the reduction of the surface tension by long ethylene oxide chain surfactants is not as great as for shorter chain species, the change in the component distribution at the air/solution interface will result in a rise in the surface tension above the cmc to produce a minimum in the curve at the cmc.

[‡]Analogue of Raoult's Law in micellar systems.

Calculations for surfactants with $\bar{N} < 30$ also predict a slight increase in the monomer concentration above the cmc, although not as pronounced as for the higher mean chain lengths. The enrichment of the bulk solution with components greater than the mean is also less pronounced for the lower mean chain lengths, which is consistent with the lower $\bar{N}$ series surfactants not displaying a pronounced surface tension minimum (Figure 1).

Another ramification of the change in concentration ratio of the bulk phase components above the cmc is that for adsorption on a solid surface, the distribution of components on the surface will also be affected. Thus changes in the wettability of the surface may occur above and below the cmc. To investigate this possibility in more detail, contact angle measurements were made on hydrophilic and hydrophobic (methylated) quartz plates[‡]. The behaviour of the equilibrium contact angle as a function of the concentration of $\bar{N}9$ and $\bar{N}30$ is shown in Figures 4 and 5. For both surfaces the equilibrium contact angle is substantially changed below the cmc, but is virtually constant above the cmc at about $28 \pm 5°C$.

Adsorption studies of $\bar{N}9$ and $\bar{N}30$ onto methylated and untreated hydrophilic silica powders have shown[15] that monolayer coverage of the surface has occurred at $C_T > $ cmc. For $\bar{N}30$ it was also found that the adsorption density decreased above the cmc, which is consistent with the larger, longer chain components (see Figure 2) of the mixture enriching the surface layer, for reasons discussed above.

The constant contact angles above the cmc indicate that changes in the surface layer do not influence the wettability of the solid surface. In some respects this may be anticipated, because the contact angle at about the cmc of $\bar{N}9$ is, within experimental error, the same as that for the much longer ethylene oxide chain length mixture, $\bar{N}30$, at monolayer coverage.

These contact angle results also show that at monolayer coverage of the hydrophobic and hydrophilic surfaces, the surfaces are indistinguishable from a wettability point of view. One can conclude from this observation that the molecular orientation of the adsorbed surfactant is the same on the two substrates. Considering the relatively low contact angle at surface saturation[16], it seems reasonable to suggest that at monolayer coverage the adsorbed molecules are oriented with their hydrophobic moiety directed to the surface as much as possible and their ethylene oxide portions protruding into the aqueous phase.

[‡]The process of methylation of quartz plates with trimethyl-chlorosilane is discussed elsewhere[7].

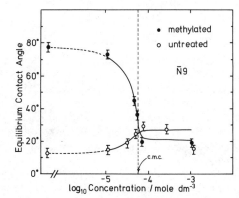

Figure 4. Equilibrium contact angles as a function of the $\overline{N}9$
concentration on hydrophilic and methylated quartz plates.

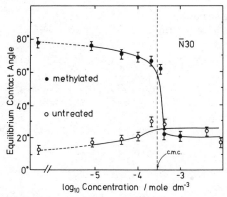

Figure 5. Equilibrium contact angles as a function of the $\overline{N}30$
concentration on hydrophilic and methylated quartz plates.

ACKNOWLEDGEMENTS

F.G. acknowledges the award of a Queen Elizabeth II Fellowship. This work was sponsored by grants from the National Energy Research Development and Demonstration Programme and the Australian Research Grants Scheme.

REFERENCES

1.   M. J. Rosen, "Surfactants and Interfacial Phenomena," Wiley-Interscience, New York, 1978.
2.   J. R. Aston, M. J. Deacon, D. N. Furlong, T. W. Healy and A. C-M. Lau, in "Proceedings of the 1st Australian Coal Preparation Conference," A. R. Swanson, Editor, pp. 358-378, Newcastle, 1981.
3.   P. Becher, in "Non-ionic Surfactants," M. J. Schick, Editor, pp. 604-625, Marcel Dekker Inc., New York, 1967.
4.   A. M. Schwartz, J. W. Perry and J. Berch, "Surface Active Agents and Detergents," Volume II, Interscience, New York, 1958.
5.   N. Shachat and H. L. Greenwald, reference 3, pp. 8-43.
6.   I. Metcalfe and L. R. White, (1982), J. Colloid and Interface Sci., submitted for publication.
7.   R. N. Lamb and D. N. Furlong, J. Chem. Soc. Faraday Trans. I, 78, 61 (1982).
8.   F. Wolfram and R. Faust, in "Wetting, Spreading and Adhesion", J. F. Padday, Editor, Academic Press, 1978.
9.   E. H. Crook, D. B. Fordyce and G. F. Trebbi, J. Phys. Chem., 67, 1987 (1963).
10. P. Becher, reference 3, p. 487.
11. H. Schott, J. Phys. Chem., 68, 3612 (1964).
12. J. H. Clint, J. Chem. Soc. Faraday Trans. I, 71, 1327 (1975).
13. G. G. Warr, F. Grieser and T. W. Healy, (1983), J. Phys. Chem., in press.
14. F. Harusawa and M. Tanaka, J. Phys. Chem., 85, 882 (1981).
15. J. R. Aston and D. N. Furlong, Colloids and Surfaces, 4, 121 (1982).
16. T. Wakamatsu and D. W. Fuerstenau, Trans. AIME, 254, 123 (1973).

CONDITIONS OF PHASE SEPARATION, BOTH AT THE INTERFACE AND IN SOLUTION: THE ADSORPTION ISOTHERM AND THE CONSEQUENCE OF CRITICAL PHENOMENA ON THE BEHAVIOUR OF THE SYSTEM

M. Privat and R. Bennes

Laboratoire de Physico-chimie des Systèmes Polyphasés
C.N.R.S., Route de Mende, B.P. 5051
34033 Montpellier-Cédex, France

The effects of modifications in the bulk and surface phases of a binary mixture in equilibrium with its vapor on the form of the surface adsorption isotherm have been studied; particular attention has been paid to the influence of deviations from ideality and of the temperature.

One observes, depending on the temperature as compared to the critical temperatures of the surface and the solution, that a simultaneous demixion both in the surface and in the solution is possible. It may lead to a system evolving in a different manner from the one simply separating into two phases.

The monolayer model was used in a part but not in the whole work to describe the surface phase.

INTRODUCTION

An attempt [1] has been made to study, as phenomenologically as possible, the reciprocal consequences of thermodynamical accidents in solution and at the interface. Any irregularity in the properties of the solution has repercussion on the adsorbed phases; conversely, any anomaly in the properties of the interface çan possibly have a bearing on those of the bulk solution phase [2].

BASIC PRINCIPLES

1. Equilibrium.

What characterizes the surface phase at equilibrium when it exists between the bulk and surface phases are the facts that its composition is generally different from that of the bulk phase while the chemical potentials of a given constituent (i) are the same in both phases, i.e.,

$$\mu_i^\sigma = \mu_i^\alpha \quad \text{and therefore}$$

$$x_i^\sigma = f(x_i^\alpha) \quad \text{"adsorption isotherm of i".}$$

2. Stability conditions with respect to the diffusion [3] :

for the bulk phase (α) $\qquad \partial\mu_i^\alpha / \partial x_i^\alpha > 0$

for the surface phase (σ) $\qquad \partial\mu_i^\sigma / \partial x_i^\sigma > 0$

3. Utility of the isotherm for studying the stability of the two phases :

$$\frac{dx_i^\sigma}{dx_i^\alpha} = \frac{dx_i^\sigma}{d\mu_i^\sigma} \frac{d\mu_i^\alpha}{dx_i^\alpha}$$

In this expression it is seen that the slope of the isotherm is the ratio of two terms, the sign of which constitutes a criterion of stability of the two phases in presence of one another.

1 - ISOTHERM WITH AN INFLEXION POINT   (Figure 1)

In this case :

$$\frac{\partial^2 x_2^\sigma}{\partial(x_2^\alpha)^2} = 0 \quad \text{(with} \quad \frac{\partial^3 x_2^\sigma}{\partial(x_2^\alpha)^3} = 0 \text{)}$$

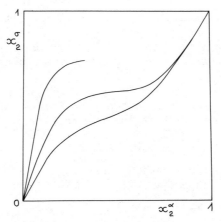

Figure 1.Type 1 isotherm.

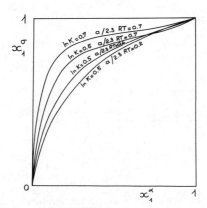

Figure 2. Verification of type 1 isotherms.

At the critical point of the bulk phase where :

$$\frac{\partial \mu_2^\alpha}{\partial x_2^\alpha} = \frac{\partial^2 \mu_2^\alpha}{\partial (x_2^\alpha)^2} = 0$$

the slope is zero. Intuitively, beyond the critical temperature and for reason of continuity, we have an inflexion point with a positive slope. An inflexion point with a negative slope must appear when one reaches the region of separate phases.

Vérification. Using Everett's theoretical results [4] of x_2^σ for the regular solutions one obtains the results summarized in Figure 2.

2 - ISOTHERM WITH AN EXTREMUM

The condition is $\dfrac{\partial x_2^\sigma}{\partial x_2^\alpha} = 0$

that is $\dfrac{\partial \mu_2^\alpha}{\partial x_2^\alpha} = 0$

Then, as $\dfrac{\partial^2 x_2^\sigma}{\partial (x_2^\alpha)^2} > 0$, the extremum is a maximum and corresponds

to a change of sign of $\partial \mu_2^\alpha / \partial x_2^\alpha$ and a phase separation in the bulk phase (cf. Figure 3).

3 - ISOTHERM WITH A SLOPE TENDING TO INFINITY

If $\dfrac{\partial x_2^\sigma}{\partial x_2^\alpha}$ tends to infinity (with $\dfrac{\partial^2 x_2^\sigma}{\partial (x_2^\alpha)^2} \neq 0$), probably

$\dfrac{\partial \mu_2^\sigma}{\partial x_2^\sigma}$ tends to zero and we have a phase separation in the surface phase (cf. Figure 4).

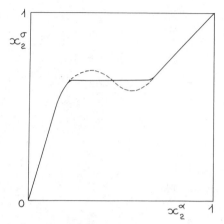

Figure 3. Type 2 isotherm. Case of a demixion in the bulk phase.

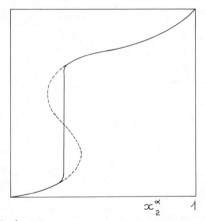

Figure 4. Type 3 isotherm. Case of a demixion in the surface phase.

Verification of type 2 and 3 isotherms

Using the explicit expressions for the chemical potential [5] :

$$\mu_2^\sigma = \mu_2^0 + A_2^{\sigma 0}\, \gamma_2^\sigma - A_2^\sigma \gamma + RT\ln a_2^\sigma$$

$$\mu_1^\sigma = \mu_1^0 + A_1^{\sigma 0}\, \gamma_1^0 - A_2^\sigma \gamma + RT\ln a_1^\sigma$$

$$\mu_2^\alpha = \mu_2^0 + RT\ln a_2^\alpha$$

$$\mu_1^\alpha = \mu_1^0 + RT\ln a_1^\alpha$$

one obtains :

$$\frac{\partial x_2^\sigma}{\partial x_2^\alpha} = \frac{x_2^\sigma}{x_2^\alpha}\,(1 - A_2^{\sigma 0}\,\Gamma_{2,1})\,(1 - x_2^\alpha\,\frac{\partial \ln f_2^\alpha}{\partial x_1^\alpha})/(1 - x_2^\sigma\,\frac{\partial \ln f_2^\sigma}{\partial x_1^\sigma})$$

$$\frac{\partial x_1^\sigma}{\partial x_1^\alpha} = \frac{x_1^\sigma}{x_1^\alpha}\,(1 - A_1^{\sigma 0}\,\Gamma_{1,2})\,(1 - x_1^\alpha\,\frac{\partial \ln f_1^\alpha}{\partial x_2^\alpha})/(1 - x_1^\sigma\,\frac{\partial \ln f_1^\sigma}{\partial x_2^\sigma})$$

and one finds the condition for an extremum as

$$1 - x_2^\alpha\,\frac{\partial \ln f_2^\alpha}{\partial x_2^\alpha} = 0$$

i.e. <u>the condition of a bulk phase separation</u> [3].
and, as condition for an infinite slope

$$1 - x_2^\sigma\,\frac{\partial \ln f_2^\sigma}{\partial x_2^\sigma} = 0$$

i.e. the <u>condition of a surface phase separation</u>[3].

Another condition for an extremum is

$$A_2^\sigma\,(\Gamma_{2,1})_{\lim} = 1$$

4 - COMBINATION OF THE TWO INSTABILITY EFFECTS WITH RESPECT TO THE DIFFUSION

In the bulk phase it is well known that the separation into two phases is the consequence of the evolution of the system towards a new equilibrium state as a result of internal constraints whose forces are the difference in the chemical potentials within the system, tending to destroy the former equilibrium. It is not unreasonable therefore to imagine that in another type of system, the same internal forces could lead to a non equilibrium stationary state.

This perhaps simply means that the values of μ and the deviation from ideality, which make possible the situation envisaged in this diagram, must lead to a new state of the system in which the surface phase would act a part more extended (cf. Figure 5).

Verification. The example of binary regular solutions of molecules with the same size :

Given the formula for f_2^σ and f_2^α in this case [6],

$$RT \ln f_2^\sigma = \alpha l (x_1^\sigma)^2 + \alpha m (x_1^\alpha)^2$$

$$RT \ln f_2^\alpha = \alpha (x_1^\alpha)^2$$

it is possible to calculate the regions of the molar fraction within which the system is unstable with respect to the diffusion.

The lengths of these regions are given by:

$$\left[0,5 \left(1 - \sqrt{1 - 2 \frac{RT}{\alpha}} \right) \right] \leqslant x_2^\alpha \leqslant \left[0,5 \left(1 + \sqrt{1 - 2 \frac{RT}{\alpha}} \right) \right]$$

$$\left[0,5 \left(1 - \sqrt{1 - 2 \frac{RT}{\alpha l}} \right) \right] \leqslant x_2^\sigma \leqslant \left[0,5 \left(1 + \sqrt{1 - 2 \frac{RT}{\alpha l}} \right) \right]$$

1 - Bulk phase. The first formula defines the length of the constant plateau of demixion in the phase diagram (or the $x_2^\alpha - x_2^\alpha$ diagram) for the bulk phase : it is symmetrical with respect to $x_2^\sigma = 0,5$ and all the greater when α has a larger value (great deviation from ideality) and/or when T diminishes.

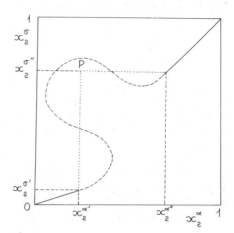

Figure 5. Type[4] isotherm. Case of a simultaneous demixion both in the bulk and in the surface phases.

2 - *Surface phase.* The second formula defines the length of the plateau of demixion for the surface phase. In this phase the critical temperature is $T_c^\sigma = \ell^o T_c^\alpha$, i.e. $T_c^\sigma < T_c^\alpha$. Therefore an interfacial instability cannot occur without a corresponding phenomenon in the bulk solution.

As a summary : $T < T_c^\sigma < T_c^\alpha$ a phase separation can occur simultaneously both in the bulk solution and at the surface : fourth type (cf. Figure 5).

$$T_c^\sigma < T < T_c^\alpha \quad : \text{ second type (cf. Figure 3)}$$

$$T > T_c^\alpha \quad : \text{ first type (cf. Figure 1).}$$

3 - *Simultaneous observation of phase separation in the surface and bulk phases.* To observe a possible simultaneous phase separation, it is necessary that these two instability regions (Δx_2^σ and Δx_2^α) should have a common zone where the isotherm is likely to exist, and this will be all the more likely when the two regions of concentrations are large.

A complete mathematical approach being quite out of the question, so some numerical examples have simply been taken:

$\alpha = 2,8 \times 2$ RT $l = 2/3$ $\Delta x_2^\alpha = 0,8$ $\Delta x_2^\sigma = 0,68$

$\alpha = 1,56 \times 2$ RT $l = 2/3$ $\Delta x_2^\alpha = 0,6$ $\Delta x_2^\sigma = 0,2$

$\alpha = 1,5 \times 2$ RT $l = 2/3$ $\Delta x_2^\alpha = 0,58$ $\Delta x_2^\sigma = 0$

These show that the two instability regions have every chance of overlapping in a zone where the isotherm exists, i.e., when deviations from ideality are (values of α) sufficiently large.

Within this region there must be a smaller region corresponding to the mechanical instability of the system, since the latter is situated within the binodal curve. Hence in a certain number of limited cases, the isotherm could pass through a region of mechanical instability, which could involve a modification of the curvature at the interface.

CONCLUSION

All this leads to the following remarks:

1 - A simultaneous phase separation both in the surface and bulk phases, if it does exist and if it is not masked by another phenomenon, such as a crystallization for instance, can only take place at temperatures below T_c in the case of an upper consolute solution temperature (UCST) or the inverse for a lower consolute solution temperature (LCST), at least in the case of regular solutions.

2 - The nature of this phenomenon cannot be predicted by a simple thermodynamical analysis and one may well wonder whether it would lead to a sort of "critical wetting"[2] or, why not, to "micellization".

REFERENCES

1. M. Privat and R. Bennes, J. Colloïd Interface Sci., 90, 454 (1982).
2. J.W. Cahn, J. Chem. Phys., 66, 3667 (1977).
3. R. Defay and I. Prigogine, "Chemical Thermodynamics", 5 ed. Longmans, 1969.
4. D.H. Everett, Trans. Faraday Soc., 60, 1803 (1964); 61, 2478 (1965).
5. J.C. Eriksson, Ark. Kemi, 25, 331 (1965); 25, 342 (1965); 26, 49 (1966).
6. R. Defay, I. Prigogine, A. Bellemans and D.H. Everett, " Surface Tension and Adsorption", p. 171, Longmans, 1966.

CAROTENOID FILMS AT THE AIR/WATER INTERFACE

E. Chifu and M. Tomoaia-Cotişel

Department of Physical Chemistry
"Babeş-Bolyai" University, Arany J. 11
3400 Cluj-Napoca, Romania

The surface properties of seven all-trans ca-rotenoid compounds in single-component films at the air/water interface, obtained by compression iso-therms (surface pressure versus molecular area), are discussed in terms of both the rotating rigid plate model of carotenoid molecules previously pro-posed by us and the magnitude of the intermolecular forces.

The behaviour of binary mixtures of some ca-rotenoids and egg lecithin in spread films at the air/water interface was also studied. The collapse surface pressure as well as the free molar energy of mixing in the two-dimensional solution are ex-plored and the dependence of these parameters on monolayer composition is interpreted. The results are discussed both in terms of the packing of the two components and according to the different theo-ries on molecular interactions in mixed monolayers.

INTRODUCTION

The influence of carotenoids, universally distributed in the
living world, on the structure and stability of some biological
membranes has prompted us to study the carotenoid films at fluid
interfaces, both in pure state [1-7] and in mixtures with other components of biological interest.[8-11] Among our studies performed
on the model of monomolecular films, this paper deals with the
results of the researches on the monolayer properties of seven
all-trans carotenoids: canthaxanthin, isozeaxanthin, β - cryptoxanthin, lutein, β - apo - 8'- carotenal, ethyl ester of
β - apo - 8'- carotenoic acid,and retinyliden - oxazolone in
single-component films at the air/water interface. Also, two -
component films of carotenoid (canthaxanthin, β - cryptoxanthin
or lutein) and egg lecithin are presented. Such studies of pure and
mixed films are a first step in understanding the behaviour of these
compounds of biological importance at interfaces.

SINGLE-COMPONENT FILMS

The properties of seven carotenoids (see Figure 1), in
monomolecular spread films at the air/water interface were investigated by means of the compression isotherms : surface pressure
(π, mN/m) versus molecular area (A, nm^2) at ambient temperature (22 ± 2°C). The surface pressure was measured by the
Wilhelmy method,[4,7] within an error of ± 0.5 mN/m; the compression
curves being reproducible within 0.02 - 0.04 nm^2/molecule. Details
on the experimental measurements were given by us previously.[1-5,7]

In Figure 2, the compression isotherm is plotted as an example for canthaxanthin film at the air/water interface. The extrapolation to $\pi = 0$ of the linear portion corresponding to high
pressures gives the limiting molecular area A_O. By using the
slope of this portion the surface compressional modulus, C_s^{-1} mN/m,
was calculated employing the formula: $C_s^{-1} = - A_O (\partial \pi / \partial A)_T$.
Also, from the compression isotherms the collapse pressures π_c
were estimated as the highest pressures to which the spread films
can be compressed without detectable expulsion of the molecules
to form a new collapsed bulk phase - marked on the isotherms
by slope change at high surface pressures (Figure 2), as well as
the corresponding collapse areas A_C.

The experimental data obtained in the study of these substances in the spread film (see Table I) can be correlated to
the molecular dimensions using the rotating rigid plate model for-
merly proposed by us.[6,7]

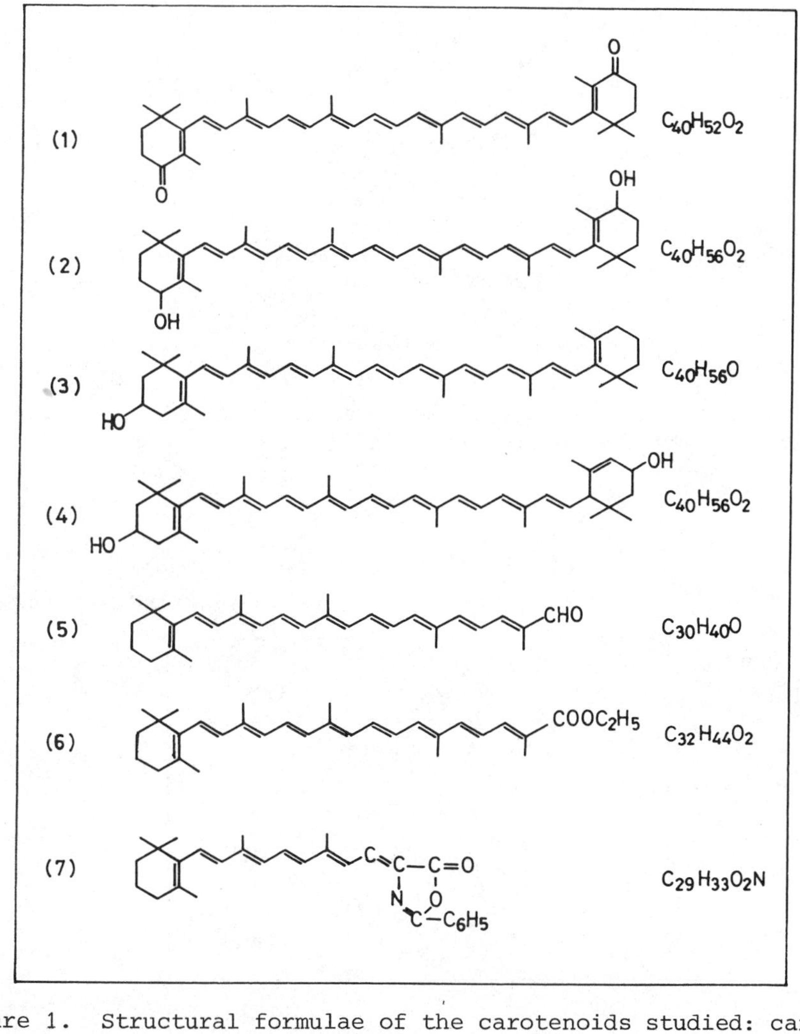

Figure 1. Structural formulae of the carotenoids studied: canthaxanthin, (1); isozeaxanthin, (2); β - cryptoxanthin, (3);
lutein, (4); β - apo - 8'- carotenal, (5); ethyl ester of the
 β - apo - 8'- carotenoic acid, (6) and retinyliden oxazolone,
(7).

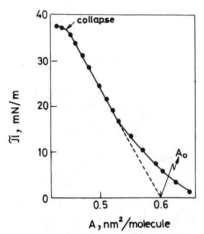

Figure 2. Surface pressure versus area isotherm of canthaxanthin pure film at the air/water interface.

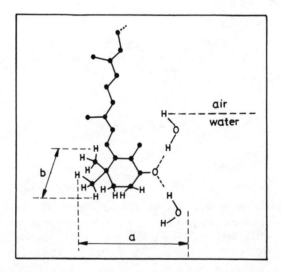

Figure 3. The shape of one half of the canthaxanthin molecule and the dimensions of its hydrated polar head group.[6]

Without giving again a detailed description of the model, we should point out that it assumes the delocalized π-bond system of the polyene chain (see Figure 1), which makes the molecule rigid and allows no rotational transitions about the chemical bonds of this chain. In the case of simple σ-bonds, a free rotation about such bonds has been assumed and the most stable conformation has been considered, corresponding to the minimum molecular cross-section on a direction perpendicular to the longitudinal axis,[12] by taking into account the possibilities of steric hindrance as well as the intramolecular repulsion. For the bond lengths and bond angles in the carotene skeleton molecules, the values found by X-ray diffraction in the case of canthaxanthin have been used.[13] As for the other atoms, the internuclear distances and bond angles have been approximated on the basis of bond lengths given in chemical tables, as well as of the hybridization type. For the terminal atoms, Pauling's covalent radii have been considered.

The carotenoid molecule is assumed to be anchored in the water phase by its hydrated polar head group. As an example, we present the shape and the dimensions of the canthaxanthin molecule oriented in the monomolecular film at the air/water interface (Figure 3). For the sake of simplicity, only half of the molecule was considered. The length a and width b of the hydrated polar group was calculated from the internuclear distance and bond angles, as described earlier[6] (see Table I).

On the basis of the above given assumptions, the molecules can be assimilated to a rigid plate, having its longitudinal axis (that of the polyene chain) perpendicular to the interface. Assuming that the rigid plate molecule performs a free rotation about its vertical axis, the rotating rigid plate can be assimilated to a cylinder. As for the significance of this free rotation, it might be a real one, but a dynamic random orientation of the head group axis can also be imagined. Assuming a tetragonal close packing of the rotating rigid plates, the area necessary for a molecule is $A_4 \cong a^2$. The compact packing, corresponding to the parallel orientation of axes a, of the same rigid plates (non-rotating this time) leads to values of the molecular area given by: $A_p = a.b$.

Table I shows the experimental values from the compression isotherms A_0, A_c, π_c, and C_s^{-1}, for the seven carotenoids studied at the air/water interface, as well as the calculated a, b, A_4 and A_p, [1-7] as defined above.

From this table it should be noted that A_4 is in a very good agreement with the A_0 experimental value for compounds 1-3 5 and 6; thus, a tetragonal close packing of the rotating molecules at the A_0 value can be assumed. In the case of compounds 4 and 7 the experimental values A_0 are higher than A_4.

Table I.   Dimensions of Carotenoid Molecules and Characteristics of the Monolayer.

Compound	a (nm)	b (nm)	A_0 (nm²)	A_4 (nm²)	A_c (nm²)	A_p (nm²)	π_c (mN/m)	C_s^{-1} (mN/m)
1, canthaxanthin	0.78	0.49	0.60	0.61	0.44	0.38	36	145
2, isozeaxanthin	0.77	0.45	0.60	0.59	0.39	0.35	40	110
3, β- cryptoxanthin	0.63	0.38	0.42	0.40	0.23	0.24	37	82
4, lutein	0.64	0.60	0.63	0.41	0.38	0.38	38	96
5, β- apo - 8'- ca-rotenal	0.72	0.42	0.54	0.52	0.47	0.30	13	100
6, ethyl ester of the β- apo - 8'- ca-rotenoic acid	0.68	0.42	0.46	0.46	0.35	0.29	18	85
7, retinyliden - oxazolone	0.75	0.40	0.63	0.56	0.48	0.30	7	30

Compound 4 has a bipolar nature (Figure 1) and a dimerization process due to intermolecular hydrogen bonds seems to be likely in its monolayer.[3] It is worth mentioning that the A_4 value for the rotating dimer molecule is higher than twice the A_4 value of the monomer, which could explain the high A_o value. This dimerization does not affect A_p. As for compound 7, it has no polar terminal group in the air phase, but its free chain is much shorter than those of the other compounds studied. Consequently, the intermolecular forces (consisting mainly of London - type dispersion forces) are not strong enough to ensure the parallel orientation of the chain axes, i.e. their perpendicular orientation to the air/water interface. This fact, leading to a random inclination of the chain axes, could explain the high A_o value.

Generally, the experimental collapse areas A_c are higher than A_p - corresponding to the oriented closest packing - due to the random orientation of the head group axes. Thus, in the case of the carotenoids studied (compounds 1, 2, 5-7) one can presume the free rotation of the molecule to be hindered before the collapse, but the parallel orientation of the head group axes cannot generally occur. However, a few exceptions are noted, namely in the case of compounds 3 and 4, where A_c equals A_p.

As for collapse pressure π_c, its values are high in the case of molecules with long chain (compounds 1-4), as the dispersion intermolecular forces are much stronger than in the case of molecules with short chains (compounds 5-7). Therefore, the stability of the monolayer is determined mainly by these forces. The lowest π_c values occur with compound 7, which has the shortest molecular chain in this series. It is likely that the polarity of the head groups is also important in the complex phenomenon of the film stabilization.

In Table I, it can be noted that the surface compressional modulus C_s^{-1} values for carotenoid monolayers at the air/water interface range within the limits allowed for the state of liquid film, that is either condensed liquid (compounds 1-6) or expanded liquid (compound 7).[14] The surface compressional modulus C_s^{-1} seems to be determined first of all by the structure of the polar head group and by the orientation of the free chain. It is high if the polar group is rigid enough (compounds 1-6) and much smaller in the case of bulky and deformable head groups (compound 7); for the latter, the low C_s^{-1} value may also be caused by the above-mentioned random inclination of the free chains just as in the case of the molecular area A_o.

Thus, comparing the calculated magnitudes with the experimental data, our rotating rigid plate model for carotenoid molecules is able to explain the main features of the carotenoid monolayers.

TWO-COMPONENT FILMS

Among our researches on mixed monolayers,[3,8-11] the two-component films containing one pure carotenoid (canthaxanthin, β - cryptoxanthin or lutein) and one molecular species of lecithin (egg lecithin, EL) in different mole fractions will be presented here. The monolayers of the pure components and of the mixtures were investigated by means of the compression isotherms at the air/water interface, as described before.[10] Care was taken so that all the monolayers were spread at the same initial area (about 1.20 nm^2/ molecule) avoiding thus the artifacts of spreading process.[15] The compression isotherms were recorded with the compression speeds ranging between 0.02 - 0.09 nm^2. molecule^{-1}. min^{-1}. These compression rates ensured that the internal equilibrium was achieved in the film, as indicated by the constancy of the surface pressure for each value of the area.

The experimental data are plotted as surface pressure versus mean molecular area for the following systems: canthaxanthin - EL (Figure 4), β - cryptoxanthin:EL (Figure 5) and lutein - EL (Figure 6).

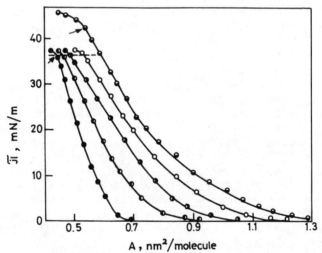

Figure 4. Surface pressure versus area of monolayers of canthaxanthin : egg lecithin (EL) at various mole fractions of EL. x_{EL}, ● : 0.0 ; ◐ : 0.25 ; ◑ : 0.5 ; ○ : 0.75 ; ◓ : 1.0.

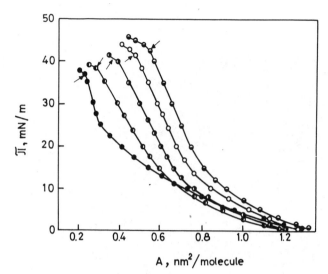

Figure 5. Surface pressure – mean area curves of β-cryptoxanthin : egg lecithin monolayers at different compositions. (Symbols as in Figure 4.)

It should be mentioned that the parameters characteristic to the compression isotherms for pure films of carotenoids are listed in Table I. The compression isotherm for the pure EL film is similar to those previously reported by us,[12] namely $A_o = 0.92$ nm^2/molecule, $A_C = 0.54$ nm^2/ molecule, $\Pi_c = 42.5$ mN/m and $c_s^{-1} = 102$ mN/m. The values of the collapse surface pressures marked in Figures 4 - 6 by arrows show that the carotenoids are more easily collapsable in the single-component film as compared to the film of pure lecithin.

In order to understand the molecular interactions that may occur in such surface solutions, it is necessary to determine at first the degree of miscibility of the components in the two-dimensional state. The first criterion of defining the miscibility is based on the application of the additivity rule of molecular areas.

It should be noted that in the canthaxanthin:egg lecithin and β - cryptoxanthin : EL systems, the mean area values follow the additivity rule, the negative deviations being only found in the

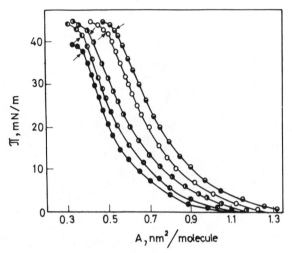

Figure 6. Surface pressure - area curves of mixed monolayers of lutein and egg lecithin. (Symbols as in Figure 4.)

case of the latter at low surface pressure (i. e. 10 mN/m).[3] In contrast to β - cryptoxanthin : EL and canthaxanthin : EL monolayers, the lutein : EL system showed negative deviations from the additivity rule (see full line in Figure 7) at all the surface pressures, especially at small molar fractions of egg lecithin.

The condensing effect observed in the lutein : EL films can be interpreted, at least in part, by the geometrical fit between the components in mixed monolayers at the air/water interface. This process might be accompanied by altered molecular interactions resulting in enhanced intermolecular hydrophobic interactions,[16] or even specific interactions between polar groups.[15,17]

Taking into account the fact that on the basis of the additivity rule one cannot decide whether the mixture is ideal or the components are immiscible in the film, a complementary criterion to determine miscibility consists in applying the two-dimensional phase rule. This rule[18] in the case of plane interfaces at constant temperature and external pressure and in absence

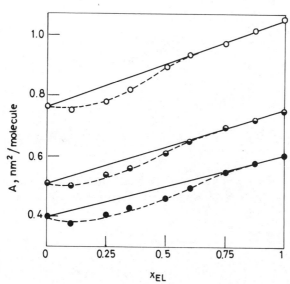

Figure 7.   Mean molecular area of monolayers of lutein and egg
lecithin as a function of the mole fraction of EL, at three values
of the surface pressure. Π : 5 mN/m (O); 20 mN/m (◐);
35 mN/m (●).

of chemical reactions is :

$$w = (c - \varphi) - (\Psi - s) \qquad\qquad (1)$$

w - being the system variance, c - the number of chemical components distributed in φ - bulk phases and Ψ - contiguous surface
phases; the surface phases coexist on s - types of surface.

     We now proceed to a discussion of the equilibrium between
the surface film and the collapsed bulk phase, applying in this
respect the two - dimensional phase rule. If both components are
miscible in the film, then any of the types of surface (air/water,
collapsed phase/water, collapsed phase/air) bears only one surface phase. This results in s = 3 and Ψ = 3. Taking into account
that c = 4 (water, carotenoid, EL , air - assumed to be one

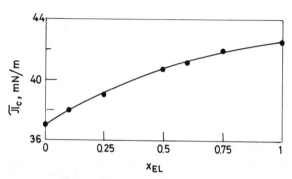

Figure 8. Collapse pressure of mixed β- cryptoxanthin : egg lecithin monolayers as a function of the mole fraction of EL.

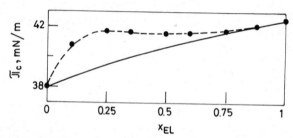

Figure 9. Collapse pressure of two - component films of lutein :
egg lecithin monolayers as a function of the composition.

component for brevity[19,20] and $\psi = 3$ (water, air, collapsed phase) according to Equation (1), the variance of the system is w =
1, and as a consequence the surface collapse pressure varies as
a function of the monolayer composition : $\overline{\Pi}_c = \overline{\Pi}_c(x_i)$.

     If the two components are immiscible in the monolayer, then
one of the types of surface, namely - air/water, carries simul-

taneously two surface phases, thus, $\psi = 4$, and according to Equation (1) w = 0. Therefore, the system is invariant and there is only one value of the surface collapse pressure (namely, π_c = const.) compatible with the coexistence of the two surface phases in the mixed film.

Comparing Figures 4, 5 and 6, it is found that the collapse pressure of the canthaxanthin : EL system (see dashed line in Figure 4) is independent of the film composition, and equals the one of the pure film of canthaxanthin; whereas in the case of β - cryptoxanthin or lutein with EL the collapse pressure (see arrows in Figures 5 and 6, respectively) depends on the composition of the two-dimensional mixture.

Taking into account the above theoretical discussion, it is natural to assume that the components are immiscible in the mixed canthaxanthin : EL film at the air/water interface, yet they are miscible in the systems of xanthophyll : EL (i.e. β - cryptoxanthin : EL and lutein : EL). It should be mentioned that, on the basis of the phase rule, immiscibility of canthaxanthin and egg lecithin has been determined at high surface pressures near the collapse, but at low pressures the two components may be miscible in the monolayer and the subsequent compression lead to phase separation prior to collapse.[15,21]

The distinct behaviour as concerns the miscibility of the three carotenoids studied (canthaxanthin, β - cryptoxanthin and lutein) with egg lecithin at the air/water interface is probably related to their different molecular structures (Figure 1). As canthaxanthin is immiscible with EL while the two xanthophylls (β - cryptoxanthin and lutein) are each miscible with EL, the hydroxyl group on the carotene skeleton seems likely to be of importance in the complex phenomenon tending to stabilize the mixed films.

In order to decide on the behaviour of the xanthophyll : EL films, the collapse surface pressure, π_c , of these systems versus composition is plotted in Figures 8 and 9.

The dependence on concentration of the collapse pressure in the ideal case[22] is:

$$x_1 \exp \frac{(\pi_c - \pi_{c1})A_{c1}}{kT} + x_2 \exp \frac{(\pi_c - \pi_{c2})A_{c2}}{kT} = 1 \quad (2)$$

where, π_{ci} is the collapse pressure of component i in the pure film, while A_{ci} represents the corresponding collapse molecular area, k and T are Boltzmann's constant and absolute temperatu-

re, respectively. When the collapse pressures of the two components are not very different, the exponentials in Equation (2)
may be expanded in series and the following approximate relation is
obtained:

$$\overline{\pi}_c = (x_1 A_{c1} \overline{\pi}_{c1} + x_2 A_{c2} \overline{\pi}_{c2})/(x_1 A_{c1} + x_2 A_{c2}) \quad (3)$$

It should be noted that for β-cryptoxanthin:EL monolayers,
the collapse pressures follow Equation (3) for ideality (see
full line in Figure 8). In contrast, for the lutein : EL system,
the collapse pressures show positive deviations (see dashed line
in Figure 9) from the ideal behaviour at compositions also rich
in xanthophyll. These results for lutein : EL films are consistent
with the observations presented in relation to the negative deviations of the mean molecular area as a function of composition at
the different surface pressures considered (Figure 7).

For the xanthophyll : EL systems, we further evaluated the
excess free molar energy of mixing [23-25] in the monolayer by the
following Equation:

$$\Delta G_M^E = N \int_0^{\overline{\pi}} (A - x_1 A_1 - x_2 A_2) \, d\overline{\pi} \quad (4)$$

the A values having the known significance and N is Avogadro's
constant. The upper limit for the integration may be arbitrarily
selected and Equation (4) is applied to any surface pressure $\overline{\pi}$,
but smaller than the one corresponding to the collapse of each
pure component. Thus, the ΔG_M^E values were directly calculated
graphically using the compression isotherms of the pure and mixed
films for xanthophyll : EL systems (Figures 5 and 6).

For the sake of comparison, we have plotted in Figure 10
the excess free energy of mixing as a function of composition
for β - cryptoxanthin : EL (O) and lutein : EL (●) films
at the same high surface pressure ($\overline{\pi}$ = 35 mN/m).

The small negative ΔG_M^E values found in the case of
β - cryptoxanthin : EL film (Figure 10; O) are in accord with
the results obtained by the collapse pressure method (Figure 8)
and these observations suggest a quasi - ideal behaviour of this
mixed film at the air/water interface.

In the lutein : EL system, the negative values of the excess
free energies of mixing are higher than twice the corresponding
values for β - cryptoxanthin : EL. For the former the experimental
curve (Figure 10 ; ●) shows a minimum at a composition again rich

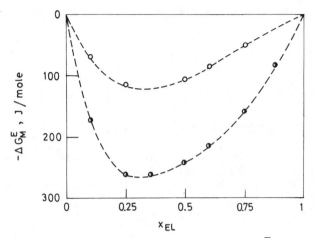

Figure 10. Excess free energies of mixing, ΔG_M^E versus composition for the following two-component films of β - cryptoxanthin : EL (O), and lutein : EL (●) at the air/water interface.

in lutein. These results for lutein and egg lecithin system are in accord with the previous ones obtained both by the collapse pressure method (Figure 9) and by the additivity rule of molecular areas (Figure 7). The enhanced surface collapse pressures in mixed monolayers of lutein : EL, as well as the negative values of the excess free energy of mixing indicate a high stability of the two-dimensional solutions, and probably stronger interactions than in an ideal mixture.

The differences in the mixing behaviour of lutein from that of β - cryptoxanthin might be ascribed to the different molecular structures of the two xanthophylls, i.e. the bipolar nature of lutein as well as the presence of an α - ionone ring on its xanthophyll structure.

REFERENCES

1. E. Chifu, M. Tomoaia-Cotişel, Z. Andrei and E. Bonciu, Gazz. Chim. Ital., 109, 365 (1979).
2. E. Chifu, M. Tomoaia, E. Nicoară and A. Olteanu, Rev. Roumaine Chim., 23, 1163 (1978).
3. E. Chifu and M. Tomoaia-Cotişel, (1981), unpublished data.

4.   E. Chifu, M. Tomoaia and A. Ioanette, Gazz. Chim. Ital., 105,
     1225 (1975).
5.   M. Tomoaia-Cotişel, E. Chifu, V. Tămaş and V. Mărculeţiu,
     Rev. Roumaine Chim., 25, 175 (1980).
6.   J. Zsakó, E. Chifu and M. Tomoaia-Cotişel, Gazz. Chim. Ital.,
     109, 663 (1979).
7.   M. Tomoaia-Cotişel, Ph. D. Thesis, Babeş-Bolyai University of
     Cluj-Napoca, 1979.
8.   M. Tomoaia-Cotişel and E. Chifu, Gazz. Chim. Ital., 109,
     371 (1979).
9.   E. Chifu, M. Tomoaia-Cotişel and Z. Andrei, Stud. Univ.
     Babeş-Bolyai Chem., 24 (2), 63 (1979).
10. E. Chifu and M. Tomoaia-Cotişel, Rev. Roumaine Chim., 24,
     979 (1979).
11. E. Chifu and M. Tomoaia-Cotişel, Rev. Roumaine Chim., 27,
     27 (1982).
12. M. Tomoaia-Cotişel, J. Zsakó and E. Chifu, Ann. Chim. (Rome),
     71, 189 (1981).
13. O. Isler, "Carotenoids", Birkhäuser Verlag, Basel - Stuttgart,
     1971.
14. J. T. Davies and E. K. Rideal, "Interfacial Phenomena", 2nd ed.
     p. 265, Academic Press, New York, 1963.
15. G. L. Gaines, Jr., "Insoluble Monolayers at Liquid - Gas In-
     terfaces", Interscience, New York, 1966.
16. D. A. Cadenhead and F. Müller-Landau, J. Colloid Interface
     Sci., 78, 269 (1980).
17. O. A. Roels and D. O. Shah, J. Colloid Interface Sci., 29,
     279 (1969).
18. R. Defay, I. Prigogine, A. Bellemans and D. H. Everett, "Sur-
     face Tension and Adsorption", p. 77, Longmans, Green, London,
     1966.
19. M. C. Phillips and P. Joos, Kolloid - Z. Z. Polym., 238,
     499 (1970).
20. N. Nakagaki and N. Funasaki, Bull. Chem. Soc. Japan., 47,
     2094 (1974).
21. K. Tajima and N. L. Gershfeld, in "Monolayers", E.D. Goddard,
     Editor, p. 165, Adv. Chem. Ser. No. 144, American Chemical
     Society, Washington, D.C., 1975.
22. P. Joos, and R.A. Demel, Biochim. Biophys. Acta, 183,
     447 (1969).
23. F. C. Goodrich, in "Proceedings of the Second International
     Congress of Surface Activity", J. H. Schulman, Editor, Vol. 1,
     p. 85, Butterworths, London, 1957.
24. I. S. Costin and G. T. Barnes, J. Colloid Interface Sci., 51,
     106 (1975).
25. K. Fukuda, T. Kato, S. Machida and Y. Shimizu, J. Colloid
     Interface Sci., 68, 82 (1979).

COMPARISON OF INTERFACIAL ACTIVE PROPERTIES

OF GLYCOLIPIDS FROM MICROORGANISMS

S. Lang, A. Gilbon, C. Syldatk and F. Wagner

Institute of Biochemistry and Biotechnology
Technical University of Braunschweig
D-3300 Braunschweig, West-Germany

Six microbial glycolipids were shown to have good
surfactant properties. They caused significant lowering
of interfacial tension between different aqueous salt
solutions and n-hexadecane at all values of pH between
3 and 9. Among the nonionic fructose- and sucrose-cory-
nomycolates from Arthrobacter paraffineus ATCC 15591,
the sucrose lipid reached a minimum interfacial tension of 2 mN/m at a concentration of 10mg/l in synthetic
deposit water. The critical micelle concentrations of
the lactonic and acidic sophorose lipids from Torulopsis bombicola ATCC 22214 didn't exceed 10mg/l. The anionic rhamnose lipids from Pseudomonas fluorescens
ATCC 15453 and Pseudomonas sp. MUB reduced the interfacial tension to about 1 mN/m at critical micelle concentrations of 5 to 30mg/l, depending on salt concentrations and pH values. Different influences of higher
temperatures up to 90°C on surfactant properties could
be detected.

INTRODUCTION

The growth of microorganisms on hydrocarbons is often associated with the production of surfactants. These metabolites are involved in the mechanism for the initial interaction of hydrocarbons with the microbial cell. Among them glycolipids with both nonionic and anionic characteristics have been found. Besides other biosurfactants, these are of interest for an enhanced oil recovery or an improved oil spill clean up.

Cultivating Rhodococcus erythropolis on n-alkanes nonionic, cell wall bound trehalose-6,6'-dicorynomycolates[1] and trehalose-6-mono-corynomycolates[2] were produced. The measurements of interfacial active properties, hydrolysis rate in deposit water, and flooding experiments on oil containing sand stones showed that these glycolipids could be useful products for the tertiary oil recovery.[2-4] Examples of anionic glycolipids are the extracellular rhamnose lipids produced by diverse Pseudomonas species[5,6] when grown on hydrocarbons. Microbial glycolipids are also formed by Arthrobacter paraffineus in media without any lipophilic carbon substrates but with carbohydrates under nitrogen limited growth conditions.[7,8] In addition Torulopsis bombicola, growing on glucose and co-oxidizing octadecane, excretes an lactonic sophorose lipid.[9]

However, only a few studies have been reported on the surface and interfacial active properties of these compounds. In order to determine the properties of such substances, some nonionic and anionic glycolipids were prepared by our as well as other reported methods and their interfacial tensions were measured.

EXPERIMENTAL

Interfacial tensions against n-hexadecane (Merck-Schuchardt, West-Germany) were measured automatically by the ring method with a Lauda autotensiomat (Fa. Lauda-Wobser KG, Königshofen, W.-Germ.) until a constant power level was reached. For each experiment, the substances were added to 20ml of various aqueous solutions obtaining concentrations from 10^{-1}mg/l to about 100mg/l. Keeping in mind that surfactants often show high sensitivity against salt solutions[10], comparative measurements were made in distilled water, 5% NaCl solution (w/v), synthetic deposit water (NaCl, 100g/l; $CaCl_2$, 28g/l; $MgCl_2$, 10g/l; w/v, pH 6,0) and Teorell-Stenhagen buffers[11] pH 3 and pH 9 with about 10 to 11% salt concentration. The salt composition of the synthetic deposit water corresponds to an average value of north-german deposit water.

The critical micelle concentration (cmc) was determined by plotting the interfacial tension against the concentration of the substances.

RESULTS

Nonionic Glycolipids from Arthrobacter paraffineus ATCC 15591

Using the method of Azuma and Yamamura[12] with growing cultu-
res of Arthrobacter paraffineus ATCC 15591 in a sucrose medium,
two cell wall associated glycolipids could be detected. After extraction of the whole culture broth by organic solvents and purification of the crude product by column chromatography using silica gel, nonionic fructose- and sucrose-monocorynomycolates were
isolated.
Figures 1 and 2 show the dependence of the interfacial tension in
water, 5% NaCl solution and synthetic deposit water against n-hexadecane on the concentration of these compounds. In the case of
fructose lipid, the interfacial tension and the critical micelle
concentration were relatively insensitive to the electrolyte concentration. But the measurements with the more polar sucrose lipid indicate that the critical micelle concentration decreases in
the presence of salts. Beyond it the interfacial tension was lowered to about 2 mN/m, whereas the minimum values were 20 to 25
mN/m for fructose lipid.

Glycolipids from Torulopsis bombicola ATCC 22214

A lactonic sophorose lipid was prepared by a modification of
the method of Tulloch et al[9]. Cultivating Torulopsis bombicola
ATCC 22214 in a glucose medium supplemented with octadecane, the
yeast utilizes the carbohydrate as carbon source and co-oxidizes
the hydrocarbon to 17-L-hydroxyoctadecanoic acid. The main product
of the resulting glycolipids is an extracellular lactonic component
with two acetate groups, as shown in Figure 3. The interfacial active properties of this substance and of the acidic, deacylated
sophorose lipid, which was obtained after alkaline saponification
of the original lacton, are shown in Figures 3 and 4. For light
acidic (pH 6 of deposit water) as well as light alkaline (pH 9)
conditions, both surfactants reduced the interfacial tension to
about 7 mN/m and 4 mN/m, respectively, against n-hexadecane. The
values for distilled water solutions show a difference of 12 mN/m.
The critical micelle concentrations didn't exceed 10mg/l in all
cases.

Glycolipids from two Pseudomonas species

From the literature it is known that Pseudomonas fluorescens
ATCC 15453 produces two extracellular rhamnose lipids[6] during
growth on n-alkanes with long chains of 12 to 15 carbon atoms.
After application of the described methods for this bacterial cul-
tivation and the subsequent product isolation, the interfacial

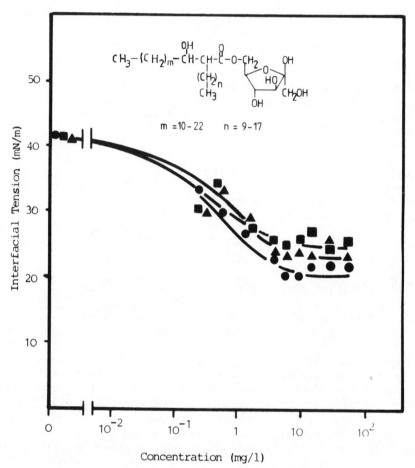

Figure 1. *Interfacial tension of fructose lipid from* <u>Arthrobacter</u> <u>paraffineus</u> *ATCC 15 591 against n-hexadecane in distilled water* (●); 5% *NaCl solution* (▲); *and synthetic deposit water** (■) *at* 40°C.

*composition: NaCl, 100g/l; CaCl$_2$, 28g/l; MgCl$_2$, 10g/l

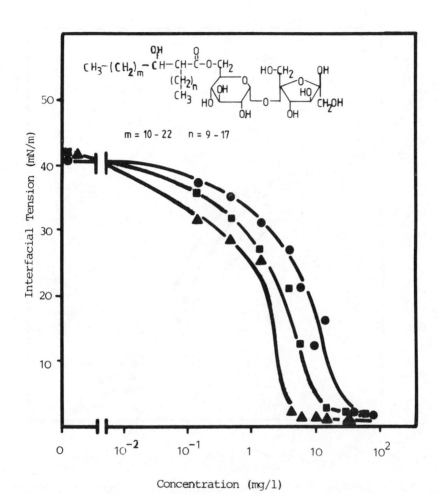

Concentration (mg/l)

Figure 2. Interfacial tension of sucrose lipid from Arthrobacter
paraffineus *ATCC 15 591 against n-hexadecane in distilled water
(●); 5% NaCl solution (▲); and synthetic deposit water* (■) at
40°C.*

*composition: NaCl, 100g/l; CaCl$_2$, 28g/l; MgCl$_2$, 10g/l.

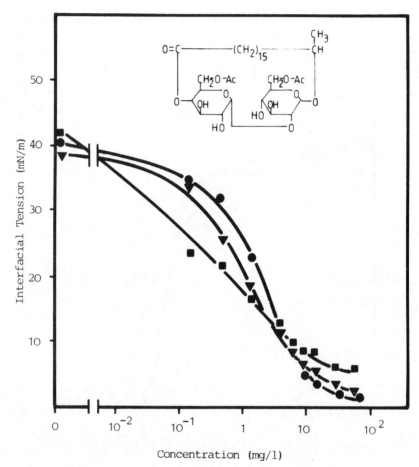

Figure 3. Interfacial tension of lactonic sophorose lipid from Torulopsis bombicola ATCC 22 214 against n-hexadecane in distilled water (●); synthetic deposit water* (■); and Teorell-Stenhagen buffer pH 9** (▼) at 40°C.

* composition: NaCl, 100g/l; CaCl$_2$, 28g/l; MgCl$_2$, 10g/l
**supplemented with NaCl, 10% (w/v)

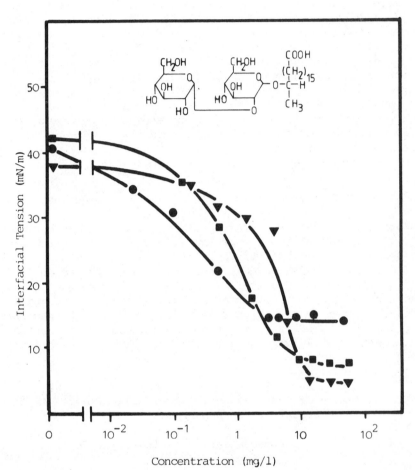

Figure 4. Interfacial tension of acidic sophorose lipid from Toru-lopsis bombicola ATCC 22 214 against n-hexadecane in distilled water (●); synthetic deposit water* (■); and Teorell-Stenhagen buffer pH 9** (▼) at 40°C.

* composition: NaCl, 100g/l; CaCl$_2$, 28g/l; MgCl$_2$, 10g/l
**supplemented with NaCl, 10% (w/v)

tension of 2-0-α-L-rhamnopyranosyl-β-hydroxy-decanoyl-β-hydroxy
decanoic acid was determined in distilled water, synthetic deposit
water and Teorell-Stenhagen-buffer pH 9 in order to demonstrate
possible dependence on salt concentrations or on pH conditions.
Figure 5 indicates that the pH alters only slightly the interfacial behaviour. However, higher salt concentrations lower the cmc
value from 30 to about 10mg/l compared to distilled water solution.
These effects are intensified in the case of a second more hydrophilic, anionic rhamnose lipid produced by Pseudomonas sp. MUB
grown on n-alkanes under nitrogen limitation[13]. Increasing of the
salt concentration reduces the cmc value from 20mg/l to 5mg/l
(Figure 6), and in alkaline conditions, the cmc increases from
5mg/l (pH 3) to about 30mg/l (Figure 7). The structure of this gly-
colipid corresponds to 2-0-α-L-rhamnopyranosyl-α-L-rhamnopyranosyl-
β-hydroxy-decanoyl-β-hydroxy decanoic acid, as also described in
the case of other Pseudomonas species.[5,6] But only poor data about
physical properties of this lipid are reported in the literature.

Influence of Temperature on Interfacial Active Properties

For the application of surfactants in the enhanced oil recovery, it is important to know if the interfacial active properties
are dependent on temperature. Thus, the interfacial tensions of
all glycolipids were measured in the range from 20°C to 90°C. The
results, summarized in Table I, indicate that for four of these
surfactants, the interfacial tension values in the system synthetic deposit water/n-hexadecane either are reduced or stay constant
with increasing temperature. Only the sucrose lipid and lactonic
sophorose lipid show increase in interfacial tension with temperature.

Table I. Influence of Temperature on Interfacial Tension in the
System: Synthetic Deposit Water[1]/n-Hexadecane.

Glycolipid	Concentration (mg/l)	Interfacial Tension (mN/m) at				
		20°	40°	60°	75°	90°
fructose lipid[2]	30	25	25	24	23	21
sucrose lipid[2]	30	3	2	3	7	10
lactonic sophorose[3] lipid	15	8	8	12	15	14
acidic sophorose[3] lipid	15	9	8	3	3	3
rhamnose lipid[4]	15	3	3	3	3	3
rhamnose lipid[5]	40	1	1	1	1	1

[1] composition: NaCl, 100 g/l; CaCl$_2$, 28 g/l; MgCl$_2$, 10 g/l
[2] from Arthrobacter paraffineus ATCC 15591
[3] from Torulopsis bombicola ATCC 22214
[4] from Pseudomonas fluorescens ATCC 15453
[5] from Pseudomonas sp. MUB

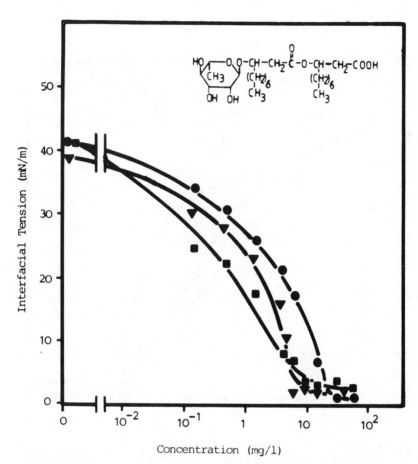

Figure 5. *Interfacial tension of a rhamnose lipid from* <u>Pseudomonas</u> <u>fluorescens</u> *ATCC 15 453 against n-hexadecane in distilled water* (●); *synthetic deposit water** (■); *and Teorell-Stenhagen buffer pH 9*** (▼) *at 40°C.*

* *composition: NaCl, 100g/l; CaCl₂, 28g/l; MgCl₂, 10g/l*
***supplemented with NaCl, 10% (w/v)*

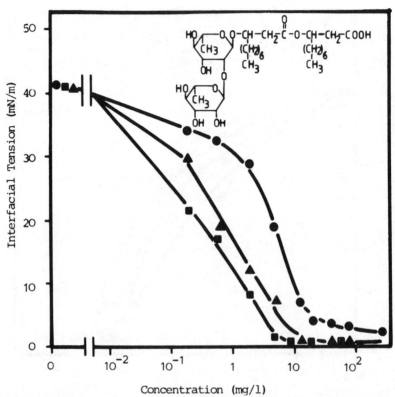

Figure 6. Interfacial tension of a rhamnose lipid from *Pseudomonas sp. MUB* against n-hexadecane in distilled water (●); 5% NaCl solution (▲); and synthetic deposit water* (■) at 40°C.

*composition: NaCl, 100g/l; $CaCl_2$, 28g/l; $MgCl_2$, 10g/l

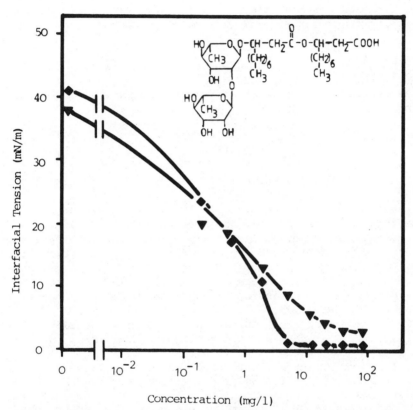

Figure 7. Interfacial tension of a rhamnose lipid from Pseudomonas
sp. MUB *against n-hexadecane in Teorell-Stenhagen buffers pH 3**
(◆) and pH 9* (▼) at 40°C.

*supplemented with NaCl, 10% (w/v)

DISCUSSION

Because of interfacial tension values still orders of magnitude above what is likely to be required in the case of synthetic surfactants, these glycolipids appear unsuitable for use in enhanced oil recovery. But surprisingly microbial trehalose lipids, reducing the interfacial tension in the same range as the presented glycolipids, increased the mobilization of oil in oil containing sand stones from north-german reservoirs.[2-4] Therefore flooding experiments being in progress, will give more information.

ACKNOWLEDGEMENT

We express our appreciation to Mrs. Andrea Walzog for her excellent technical assistance.

REFERENCES

1. *P. Rapp, H. Bock, V. Wray and F. Wagner, J. Gen. Microbiol.* 115, 491 (1979)
2. *F. Wagner, H. Bock and A. Kretschmer, in "Fermentation", R. M. Lafferty, Editor, p. 181, Springer Verlag, Wien, 1981*
3. *F. Wagner, P. Rapp, W. Lindörfer, W. Schulz and W. Gebetsberger, German patents DE-26 46 505, DE-26 46 506, DE-26 46 507 (1976)*
4. *P. Rapp, H. Bock, E. Urban, F. Wagner, W. Gebetsberger and W. Schulz, Dechema-Monographie,* 81, *177 (1977)*
5. *K. Hisatsuka, T. Nakahara, N. Sano and K. Yamada, Agr. Biol. Chem.,* 35, *686 (1971)*
6. *T. Suzuki and S. Itoh, German patent DE-OS 21 50 375 (1971)*
7. *T. Suzuki, H. Tanaka and S. Itoh, Agr. Biol. Chem.,* 38, *557 (1974)*
8. *S. Itoh and T. Suzuki, Agr. Biol. Chem.,* 38, *1443 (1974)*
9. *A. P. Tulloch, A. Hill and J. F. T. Spencer, Can. J. Chem.,* 46, *3337 (1968)*
10. *K. Oppenlaender, M. H. Akstinat and H. Murtada, Tenside Detergents,* 17, *57 (1980)*
11. *T. Teorell and E. Stenhagen, Biochem. Z.,* 299, *417 (1939)*
12. *J. Azuma and Y. Yamamura, France patent F-OS 2 399 845 (1977)*
13. *C. Syldatk, diploma thesis, Technical University Braunschweig (1981)*

STUDY OF THE INTERACTION BETWEEN SURFACTANTS AND POLYACRYLAMIDE OF VARIOUS HYDROLYSIS DEGREE

J. Sabbadin, J. Le Moigne and J. François

Centre de Recherches sur les Macromolécules, CNRS

6, rue Boussingault, 67083 Strasbourg-Cedex, France

In this work we have reported some principal results concerning the interaction between unhydrolyzed or partially hydrolyzed polyacrylamide and sodium dodecyl sulfate. Conductivity, viscosity and solubilization measurements do not reveal changes of the micellar properties in presence of unhydrolyzed polyacrylamide solutions. For hydrolyzed polyacrylamide we cannot expect association phenomena with the surfactant. We observe by viscosimetry the same decrease in the expansion of polymer chain with added NaCl or SDS until the CMC of SDS. But the salt effect of hydrolysed polyacrylamide upon the CMC appears only after 8.10^{-3} mole/l of charged groups COO^-. The aggregation number of micelles decreases from 65 in solution without polymer to 50 with hydrolyzed polymer (35%) for concentrations higher than 10^{-3} g/cm^3.

INTRODUCTION

The study of the interaction between hydrosoluble polymers and surfactants is of both fundamental and industrial interest. The fundamental questions are : what is the influence of the polymer on the micellar properties of the detergent (critical micellar concentration, aggregation number, solubilization properties) ; how does the surfactant modify the polymer conformation (viscosity, radius of gyration) ; and finally is there an association and complex formation between the two types of molecules ? The most studied case is the system where the polymer is non ionic

1377

(polyethyleneoxide, polyvinylpyrrolidone...) and the detergent anionic (sodium dodecyl sulfate, SDS) or cationic (cetyltrimethyl-ammonium bromide, CTAB). The numerous works dealing with this type of system have revealed the existence of a surfactant-polymer complex exhibiting a polyelectrolytic character [1] . Saito[2,3] has also investigated the properties of the complex formed by the association of the polyacrylic acid and a non ionic detergent such as ethoxylated nonyl phenol. To our knowledge, few data are available about mixtures of polyelectrolyte and surfactants, and particularly charged soaps[4] . Nevertheless, such a system is used in the process of enhanced oil recovery and some workers have recently discussed the difficulties arising from the interactions between the polymer and the anionic surfactant of the microemulsion[5].

The purpose of this paper is to report some principal results concerning the interaction between partially hydrolysed polyacrylamide, which is one of the most used polymer in oil recovery process, and different surfactants. In the first part, we will discuss the case of unhydrolyzed (non ionic) polyacrylamide (PAM) which slightly interacts with any kind of surfactant. The most part of our results are obtained with sodium dodecyl sulfate (SDS). In the second part, we will consider the role played by the charges distributed along the polymer chain on the interactions with an anionic surfactant and the hydrolyzed polyacrylamide (PAMh).

MATERIALS AND METHODS

One unhydrolysed polyacrylamide sample $(M=5.10^6)_4$ was obtained form Polysciences Inc. and the other one $(M=2.10^4)$ was prepared in our laboratory.

The copolymer acrylamide acrylic acid, AD17, 27, 37 and 60 were obtained from Rhône Poulenc $(M=6.10^6)$. The hydrolysed samples (HC and LC) were Polysciences Inc. products $(M=2.10^5)$. The hydrolysis degree or the acrylic acid content determined by potentiometry and elemental analysis were found to be equal to 7,17,27,32,15 and 35% for AD17, 27, 37 and 60, LC and HC samples. These polymers were used in their basic form and the pH of the solutions was between 8-9.

The polyethylene oxide sample was obtained from Toyosoda $(M=6.10^6)$.

We used a SDS sample from SERVA and a cetyltrimethylammonium bromide (CTAB) sample from Aldrich.

The changes in the polymer conformation were investigated by viscometry and ultracentrifugation. We used an automatic capillary viscosimeter and a SPINCO Model E Analytical Ultracentrifuge.

The CMC in the presence of polymer was obtained from :

  • conductivity measurements (Wayne Kerr B221 Universal bridge)

  • measurements of solubility limits of anthracene by using the method described by Schwuger [6] . Solutions of various soap concentrations were prepared with a large excess of anthracene stirred during 48 hours, and the amount of dissolved anthracene was there measured with a UV Beckmann spectrophotometer at 254nm. This method was only used for low polymer concentrations of not too high viscosity.

  • measurements with a Shimadzu spectrophotometer of the wavelength of the maximum absorption of benzene (10^{-2}M) dissolved in the polymer-SDS solutions. This wavelength depends on the polarity of the benzene microenvironment [7] and changes when the benzene molecules pass from water into the micelles.

  • the aggregation number of the micelles were determined from the analysis of the fluorescence decay curve of pyrene according to the method described by Atik et al. [8] , under conditions where two pyrene molecules can form an excimer in some micelles (ratio of the micelle and probe concentrations around 1). The curve analysis was carried out with a Commodore microcomputer and provided the aggregation number, N.

RESULTS AND DISCUSSION

Unhydrolysed polyacrylamide

In Figure 1 are reported the results of conductivity measurements of polyacrylamide solutions at different polymer concentrations in the presence of SDS and CTAB. The change in the slope of the curves is obtained for the same critical concentration than in pure water for two samples of very different molecular weight. One experiment carried out with solutions containing 0,01N NaCl does not reveal a change in the CMC due to the presence of polymer. But, in all the cases, we observe a decrease in the specific conductivity of the surfactant with increasing polymer concentration ; this effect being more important with higher molecular weight sample (Fig. 1). This fact could be explained by a decrease either of the ionization or of the mobility or of the number of the ionic species. A lowering of the ionization is not probable because it is in disagreement with the constancy of the CMC. Because of the influence of the molecular weight of the polymer, one could think that increa-

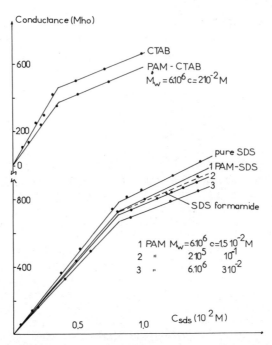

Figure 1. CMC determination by conductivity measurements.

sing the solution viscosity leads to a decrease of the ions mobility. Nevertheless, we do not observe any difference in the
specific conductivity of NaCl in pure water and in polymer
solutions, and one generally considers that the local viscosity
of the solvent is not modified by the polymer. We must remark that
in the surfactants, with long paraffinic tails and micelles
of great size, a viscosity effect could be expected to be more
pronounced than for salts of small ions. A fixation of some
surfactant molecules onto the polymer chain could explain
the conductivity results but an increase of the CMC should be
observed and it is not the case. On the other hand, we have investigated the influence of formamide on SDS conductivity and
our results reported in Figure 1 show the same behavior as
the polyacrylamide and we have also remarked that the formamide
does not change NaCl conductivity. One could then assume that
the polymer effect is due to the amide functions and to its influence on the water structure around the paraffinic tails of
the surfactant molecules. The complete elucidation of these
observations needs further investigation.

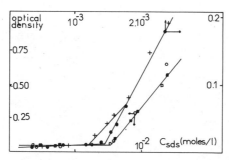

Figure 2. Solubility limit of anthracene versus SDS concentration: o in pure water, ■ water with polyacrylamide $M_w=5x10^6$, ▫ water with polyacrylamide $M_w=2x10^4$; ● water 0,1N NaCl, + water 0,1N NaCl with polyacrylamide $M_w=5x10^6$.

The study of the amount of solubilized anthracene in solutions of polyacrylamide in pure water as a function of the SDS concentration confirms the independence of the CMC on the polymer concentration (Figure 2). Above the CMC, the increase of the solubilization is identical to that observed in pure water and the size of the micelle is probably similar. Some measurements of the aggregation number of the surfactant in the presence of polyacrylamide (Fluorescence decay of pyrene) confirm this assumption. In presence of salt (Fig. 2), the anthracene solubilization curves present the features of those obtained by Schwug er for polyethyleneoxide with two critical concentrations. We suggested in a previous work [9] that the surfactant could form a complex with polyacrylamide in the presence of salt. But our viscosimetric investigations, reported below, are not in agreement with this hypothesis and such an observation only signifies a slight influence of the polymer on the micellization process but cannot be intepreted in terms of a complex formation. The existence of two critical points does not systematically indicate the formation of a complex. Some additional investigations are in progress in order to verify the existence of micelles just above the first critical concentration.

Table I. Aggregation Number, N, of SDS by Measurements of the
Fluorescence Decay of Pyrene as a Function of the Molar
Concentration of SDS, C_{SDS}

	$C_{SDS}=8.10^{-3}M$	$C_{SDS}=2.10^{-2}M$	$C_{SDS}=10^{-1}M$
SDS in pure water		N = 65±5	N = 70±5
SDS + PAM ($M_w=6.10^6$, Cp=3.10^{-2}M) in pure water		N = 60±5	N = 65±5
SDS in water 0,1N NaCl	85±5	–	90±5
SDS + PAM ($M_w=6.10^6$, $C_p=3.10^{-2}$M) in water 0.1N NaCl	70±5		

In Figure 3, we have presented the variation in the reduced viscosity of polyacrylamide and polyethylene oxide solutions as a function of SDS concentration. Whatever the salinity, the viscosity of polyacrylamide remains constant. In the same manner, the sedimentation coefficient determined by ultracentrifugation experiments does not vary with the SDS concentration (Table II). On the contrary, Figure 3 shows considerable increase in the polymer viscosity in the case of polyethyleneoxide. The binding of surfactant molecules onto this polymer chain leads to the formation of a complex, the electrochemical behavior of which resembles that of polyelectrolyte. A characteristic feature of a polyelectrolyte is the strong dependence of its size on the ionic strength of the solution. In pure water, (Figure 3) we observe a considerable increase of the reduced viscosity of POE due to the binding of the surfactant micelles onto the polymer, the limits of this range are in good agreement with the results of other workers [6,10]. This effect becomes progressively weaker when the salt concentration increases, by screening the electrostatic interactions between charged micelles fixed along the chain, but it remains important even for 0,1N NaCl. Ultracentrifugation experiments show also a change in the sedimentation coefficient of the polyoxyethylene in the same range of SDS concentration for 0,1 N NaCl (Table II).

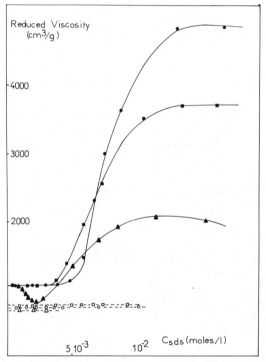

Figure 3. Reduced viscosity of SDS polyethylene oxide ●,■,▲, and SDS polyacrylamide solutions o,□,Δ, for pure water 0,01 N NaCl, and 0,1 N NaCl respectively as a function of SDS concentration.

Table   II.   Sedimentation   Coefficients,   S,   of   Polyacrylamide
              and Polyethylene oxide at Different SDS   and NaCl Con-
              centrations from Ultracentrifugation Experiments

	$C_{SDS}=0$	$C_{SDS}=6\ 10^{-3}M$
PAM($M_w=6.10^6$, $C_p=7.10^{-4}$g cm^{-3})		
0,01N NaCl	$9.1\ 10^{-13}$SV	$8.9\ 10^{-13}$SV
0,1N NaCl	$9.4\ 10^{-13}$SV	8.10^{-13}SV
POE($M_w=6.10^6$, $C_p=8.10^{-4}$g cm^3)		
0,1N NaCl	$4.5\ 10^{-13}$	$9\ 10^{-13}$SV

The comparison between the experiments carried out with polyacrylamide and polyethylene oxide samples of the same molecular
weight   and   under   the   same   conditions   definitely   shows   that
polyacrylamide   interacts   very   slightly   with   surfactants   and
does not form complexes irrespective of salinity. The addition of
other   types   of   surfactants   (cationic   or   nonionic)   does   not
change the polymer viscosity.

Hydrolysed Polyacrylamide

For the hydrolyse polyacrylamide, being an ionic polyelectrolyte, we do not expect association phenomenon with an anionic detergent. But, some decrease in the chain expansion may occur if
the surfactant molecules act as added salts by screening the repulsions   between   the   charged   groups   along   the   polymer.   On
the   other   hand,   it   is   well   known   that   the   CMC   of   an   ionic
detergent   depends   on   the   ionic   strength   of   the   medium.   Our
purpose, here, is to study these two effects. We should add here
that in this work we used the polymer in its basic form (sodium
salt) and did not investigate the pH effect.

We measured the CMC of SDS in the presence of polymer by conductivity   measurements.   Some   experiments   carried   out   with
high molecular weight polymer samples do not reveal any variation
of the CMC until a polymer concentration equal to 4.10^{-4} g cm^{-3}
We know from other investigations [11] that the ionization coefficient of   these   polymer   is   near   unity,   when   the   counterion
is Na$^+$.   The   ionic   strength   can   then   be   calculated   directly
from the molar concentration of the carboxylic groups. For an ionic
strength of the most hydrolysed polymer solution at 4.10^{-3}g cm^3,
we can expect a CMC equal to $6.5\ 10^{-3}$ mole|1, if we take into

account the literature data for the NaCl effect. This difference could be explained in terms of the nonuniform repartition of the counterions and DS⁻ anions in a polyelectrolyte solution. Indeed, the local ionic strength in the domains where the DS⁻ anions are localized is much lower than the global ionic strength. In order to verify this hypothesis we have studied low molecular weight samples of hydrolysed polyacrylamide of two different hydrolysis degrees. With this molecular weight (2.10^5) it is possible to investigate the range of rather high polymer concentration without problems arising from the high viscosity of the solutions. Indeed this polyelectrolyte effect should disappear when the polymer concentration is high enough to obtain again a homogeneous repartition of the counterions in the solution. In Figures 4 and 5, we are presentive the results of these measurements. Figure 4 shows that the presence of the polymer does not change the general features of the curves and the break is probably well related to micellization. Figure 5 gives the variation of the CMC determined from the preceding curves as a function of the concentration of charged groups and can be compared with the effect of NaCl and $NaCH_3COO$. For low polymer concentration, the CMC remains constant as for the high molecular weight polymers. But for a concentration of charged groups higher than 8.10^{-3} mole/l, an abrupt decrease of the CMC appears and above 2.10^{-2} mole/l the curves join again those corresponding to simple salts. On the other hand, the effect is normally more pronounced for the polymer with higher hydrolysis degree, without significant influence of the molecular weight. Thus, the differences with respect to the behavior of a simple salt must be due to the polyelectrolytic character of the polymer.

In Figure 6, we present the results of the study of the spectroscopic properties of benzene solubilized in solutions of SDS in pure water, in water containing 0.1N NaCl and in water containing 3.10^{-3} g/cm³ polyacrylamide of high hydrolysis degree. As discussed by different authors[7,12] the UV spectrum of solubilized benzene depends on the polarity of its microenvironment. Particularly, the wavelength of maximum absorption changes when the micellization occurs. These measurements provide values of the CMC of SDS, which are in good agreement with other experiments. Let us note that λ_{max} increases more slowly with the SDS concentration in the presence of polymer than in pure water. Since the λ_{max} is known to increase when the solvent polarity decreases, this could signify that the environment of the solubilized species is more polar in presence of PAMh. Mukerjee et al.[12] have shown that the ratio of absorbance at a wavelength of 3.6nm higher the peak and absorbance at the peak decreases when the dielectric constant of the microenvironment increases. In the case of polymer solutions, the variation of this ratio with the SDS concentration

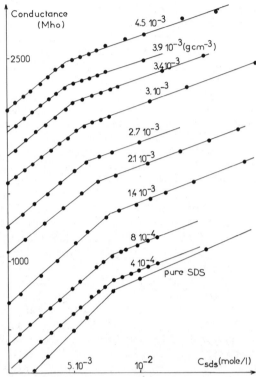

Figure 4. Variation of the conductance of the polymer–SDS solutions as a function of the SDS concentration at different polymer concentrations.

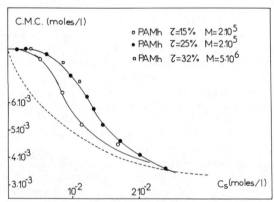

Figure 5. Variation of the CMC of SDS as a function of the concentration of charged groups COO^-Na^+. The dotted line corresponds to the effect of NaCl and $NaCH_3COO$.

is not significantly different from that obtained in pure water and it seems difficult to draw any conclusions from the study of this parameter.

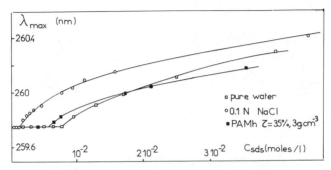

Figure 6. Variation of the maximum absorption wavelength of solubilized benzene as a function of SDS concentration.

In Table III, we have reported the values of the aggregation number of micelles measured by the analysis of fluorescence decay of solubilized pyrene [8] . In spite of the experimental difficulties arising from the high viscosity of the solutions, these values are significantly lower than 65 and 90, values determined under the same conditions in pure water and water containing 0.1M NaCl respectively. However, one could expect an increase of this aggregation number with increasing polymer concentration. In order to take into account the observed phenomena, variation of the CMC and decrease of the aggregation number, we could propose a model based upon the competition between polyelectrolyte and micelles for the attraction of the Na^+ ions in their diffused layer. Some additional investigations are in progress. According to some previous measurements the amount of solubilized anthracene [9] in hydrolysed polyacrylamide solutions is less important than in water. This result could also be explained in terms of smaller size of the micelles.

Table III. Aggregation Number as a Function of the Polymer – Concentration. PAMh $\tau=35\%$, $M_w=2.10^5$ $C_{SDS}=2.10^{-2}M$

$C_p g/cm^3$	0	$6,8.10^{-3}$	$1,3.10^{-3}$	$2,7.10^{-3}$	$4,5.10^{-3}$	$5,4.10^{-3}$	$6,8.10^{-3}$
$N\pm5$	65	52	37	42	50	50	48

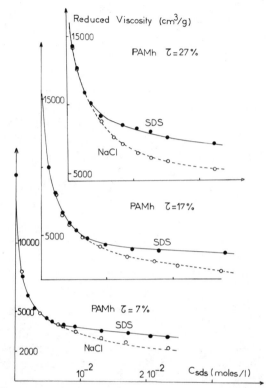

Figure 7. Reduced viscosity of solution of PAMh (C$_p$=10^{-4}mole/l) of different hydrolysis degrees ; τ as a function of SDS or NaCl concentration.

Some examples of the variation of reduced viscosity of hydro-lyzed polyacrylamide are given in Figure 7, as a function of NaCl and SDS concentrations. It is well known that the confor-mation of a polyelectrolyte changes from a more or less extended rod to a statistical coil when the salinity increases. Experimen-tal studies show a dependence of the intrinsic viscosity on the salt concentration according to $[\eta] \propto 1/\sqrt{C_s}$. Until the SDS concentration approximately equals its CMC, the effect of the surfactant is very similar to that of Na Cl with the expected variations, as shown in Figure 8. On the contrary, we can note a departure from this behavior above the CMC. This phenomenon can also be easily understood by taking into account the abrupt

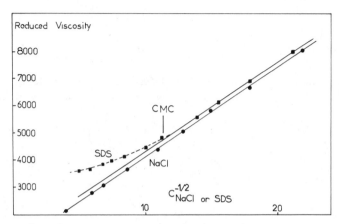

Figure 8. Reduced viscosity of PAMh ($\tau = 17\%$) plotted as a function of $C_{NaCl}^{-\frac{1}{2}}$ or $C_{SDS}^{-\frac{1}{2}}$.

change in the variation of the ionization of the SDS molecules with concentration when the micellization occurs.

REFERENCES

1. T. Isemura and A. Imanishi, J.Polymer Sci., 33, 337 (1958)
2. S. Saito and T. Taniguchi, J.Colloid Interface Sci., 44, 114 (1973).
3. S. Saito, Colloid and Polym.Sci., 257, 266 (1979).
4. S. Trushenski, AIch E Improved Oil Recovery Symposium, Kansas City, April 1976.
5. V.K. Bansal and D.O. Shah in "Micellization, Solubilization and Microemulsions", Volume 1, K.L. Mittal, Editor, Vol.1, pp. 87-113, Plenum Press, New York, 1977.
6. M.J. Schwuger, J.Colloid Interface Sci., 43, 491 (1973).
7. S.J. Rehfeld, J.Phys.Chem., 74, 117 (1970).
8. S.S. Atik, M. Nam and L.A. Singer, Chim.Phys.Letters, 67, 75 (1979).
9. J. Sabbadin and J. François, Colloid Polym.Sci., 258,1250 (1980).
10. B. Cabane, J.Phys.Chem., 81, 1639 (1977).
11. T. Schwartz and J. François, Die Makromol.Chem., 182,2757,

2775 (1981).
12. P. Mukerjee, J.R. Cardinal and N.R. Desai, in "Micelliza-
    tion, Solubilization and Microemulsions", K.L. Mittal, Editor,
    Vol. 1, pp 241-261, N.Y. Plenum Press, 1977.

ACKNOWLEDGEMENTS

This work has benefited from grants of the French DGRST. The authors wish to thank J. Dayantis and G. Pouyet for the ultracentrifugation experiments.

STUDIES ON THE INTERACTION OF SODIUM SACCHARIN WITH ALKYLTRIMETHYL-
AMMONIUM BROMIDES

S.S. Davis, P.E. Bruce, P. Daniels and L. Feely

Department of Pharmacy
University of Nottingham
University Park, Nottingham NG7 2RD, UK

The interactions occurring between large organic
ions of opposite charge are of importance in analytical
separations and pharmaceutical formulation. The main
equilibria that can occur include ion pair and
complex coacervate formation. The mechanism
responsible can be explained in terms of ion-
association reinforced by a strong hydrophobic effect.
Saccharin sodium (SS) is used widely in pharmaceutical
products and can enhnace the uptake of cationic drugs
by ion-pair formation, however at higher concentrations
insoluble salts can be formed. The alkyltrimethylammonium bromides (ATAB) are known to undergo gel
formation with organic ions, eg salicylate. The
present study was undertaken to investigate the
stoichiometry of interaction of SS with ATAB's of
different chain length and the nature of the viscoelastic gels formed at high concentrations
($> x 10^{-3}$ mol dm^{-3}). An automated conductimetric
titrimeter was used to follow the interactions. The
end point for complex coacervate formation was obtained
from a change in gradient in the titration curve.
All chain lengths of ATAB ($C_{12, 14, 16, 18}$) gave 1:1
complexes. The solubility product decreased by a
constant factor of 4.3 for each additional methylene
group. This compares well with the published values
for the free energy change per CH_2 group from other
complexation studies.

The viscosity of the 1:1 SS-ATAB systems was
measured by U-tube and rotational viscometry. At high

1391

concentrations of the two interacting components
(> 10^{-2} mol dm^{-3}) viscoelastic gels were formed. The
logarithm of the concentration of SS-ATAB to give an
arbitrary viscosity (1 cP at $25^{\circ}C$) was found to be
linearly related to alkyl chain length. The nature
of the structures responsible for viscoelastic gel
formation is discussed with reference to the work
of Gravsholt who proposed the formation of rod-shaped
micellar aggregates.

INTRODUCTION

Large organic ions of opposite charge can interact in
aqueous solution to form complex species such as ion pairs and
coacervates. These interactions are important in pharmaceutical
formulation and analytical science. The mechanisms of ion pair
formation and complex coacervation have been studied in detail by
Davis and colleagues[1-7] for both 1:1 and 2:1 interactions.
Association has been described in terms of ion association
reinforced by a strong hydrophobic effect. Various environmental
and constitutional factors can affect the equilibria, these
include molecular size, structure and charge as well as ionic
strength and temperature and additives like alcohols and ureas.
Recently Tomlinson and Davis[7] have demonstrated that complex
coacervation occurring with two large interacting ions can be
described quite well by Sinanoglu's[8] solvophobic theory. This
shows that the interactions with respect to temperature, added
electrolyte and co-solvents, are dependent upon the relative
magnitudes of surface and solvation energy terms and the sizes of
the interacting ions.

The formation of association species can affect the passage
of ionised solutes across synthetic and biological membranes.
Wilson et al[9] showed that ion-pair formation between the
dianionic drug sodium cromoglycate and the cationic preservative
dodecylbenzyldimethyl ammonium chloride altered the rate and
extent of corneal penetration of both large ions upon their co-
administration. Additionally, using a complex coacervate of the
two large ions in the applied dose form (in which the complex is
in equilibrium with the ion pairs) a change in the overall
disposition of both large ions to the various components of the
eye was observed. Similarly, the enhanced absorption of quinine
and chlorpheniramine with anionic agents such as saccharin sodium
and sodium dodecyl sulphate from the gastrointestinal tract has
been attributed to ion-pair formation and the surface activity of
ion-pair complexes[10].

The formation of a complex coacervate is usually heralded
by a sudden change in the turbidity of the solution containing

the two ions[1] and the eventual separation of an oily layer[4] unless one of the interacting ions is surface active and can solubilise the complex added in excess. In contrast, it has been reported that hexadecyltrimethylammonium bromide (HTAB) interacts with some organic ions (such as the salicylates, benzoates and tosylates) to form viscoelastic gels. Binuclear aromatic compounds such as the naphthols also give viscoelastic solutions with HTAB at concentrations below the critical micelle concentration of the surfactant in water[14].

The present investigation was carried out to study the interaction between an anionic material of pharmaceutical importance (sodium saccharin) and alkyltrimethylammonium bromides (ATAB) of different chain lengths. The effect of alkyl chain length on the interaction process and the rheological properties of the resultant gel formed have been given particular attention.

EXPERIMENTAL

Sodium saccharin was of British Pharmacopoeial quality obtained from British Drug Houses. Hexadecyltrimethylammonium bromide (HTAB) from ABM Chemicals; dodecyltrimethylammonium bromide (DTAB) from Sigma Chemicals Ltd, decyltrimethylammonium bromide (DETAB) from Pfaltz and Bauer. Tetradecyltrimethyl-ammonium bromide (TTAB) and octadecyltrimethylammonium bromide (OTAB) were synthesised from the corresponding alkyl bromides and trimethylamine and were recrystallised several times from a benzene:ethanol mixture[15]. The high purity of the commercial materials was checked by GLC analysis[16] and the determination of critical micelle concentrations (cmc) and comparison with literature values. Water, doubly distilled from an all glass still was used in all experiments. The specific conductance was about 1×10^{-6} ohm^{-1}cm^{-1}. Conductimetric titrations were made using a conductivity bridge (Wayne-Kerr B642) which was continuously and automatically balanced and was connected to a dip type conductivity cell (Mullard) contained in a beaker thermostated to $\pm$ 0.1°C. Delivery of titrant was by means of a motorized automatic syringe (Braun). Full details of the technique and its use in complexation studies have been given by Mukhayer et al[17]. The complexation end point was given by an abrupt change in the relation between specific conductance and volume of titrant. The pH of the solutions was such that the sodium saccharin (pKa = 2.0) was in the totally ionized form. Concentrations of saccharin were chosen so that the concentrations of the quaternary ammonium compound required for complexation were well below their critical micelle concentrations in water. The complexation was defined in terms of the solubility product K_s for 1:1 stoichiometry.

$$K_s = [SS][C_nTAB] \qquad\qquad (1)$$

All experiments were carried out at low concentrations of the interacting species so that the concentrations of the interacting species could be taken as equivalent to their thermodynamic activities[6].

The viscosities of the complex systems, in the form of 1:1 mixtures of the two ionic species, were evaluated using a U-tube viscometer (type C) at 25°C. The viscometer was calibrated using double distilled water. The rheological properties of saccharin-ATAB gels were measured using a rotational viscometer (Epprecht, Rheomat 15) with cone and plate geometry (diameter 53 mm, cone angle 2°59').

RESULTS

Solubility product values for the interaction of SS with C_nATAB analogues of different chain lengths are given in Table I.

Table I. Interaction of Sodium Saccharin with Alkyltrimethyl-ammonium Bromides at 25°C.

ATAB Compound	K_s	Gradient*	C.M.C.[18] (mol dm^{-3})
C_{10}	(1.76×10^{-4})		6.5×10^{-2}
C_{12}	9.67×10^{-6}	1.00	1.5×10^{-2}
C_{14}	5.59×10^{-7}	0.99	3.5×10^{-3}
C_{16}	3.18×10^{-8}	0.92	8.5×10^{-4}
C_{18}	1.69×10^{-9}	1.01	

* gradient - plot of log (SS) versus log (ATAB) for 4 concentrations

In all cases a plot of log (SS) versus log (ATAB) gave linear relations with gradients close to unity indicating a 1:1 stoichiometry for the interaction between the two ions. Figure 1 shows the change of solubility product with increasing alkyl chain length for the cationic species. This linear relationship can be represented by an equation of the form

$$\log K_s = a + bn \tag{2}$$

where a is a constant and b is −0.63 (equivalent to a factor of 4.3 for linear co-ordinates). The standard free energy of complex formation is given by the expression

$$\Delta G_f^{\,o} = RT \ln K_s \tag{3}$$

Thus each methylene group contributes −1.44 kT or −3.5 KJmole^{-1} to the complexation process at 25°.

The viscosities of equimolar mixtures of the two ions for different concentrations are given in Figure 2. At low concentrations the viscosity is similar to that of pure water but at a critical concentration the viscosity rises dramatically. These critical concentrations are very much higher than the concentration required to form complexes ($\log C_s = (\log K_s)^{\frac{1}{2}}$). The data in Figure 2 can be analysed in two ways to investigate the effect of chain length on the viscosity of the complex. An arbitrary viscosity (1 cP) can be selected and the concentration of the two ions required to give that viscosity can be plotted against alkyl chain length (Figure 3). Alternatively, the linear portion of the rapidly rising part of the viscosity curve can be extrapolated back to intersect with the horizontal portion of the log viscosity versus log concentration plot to give a concentration value. Both parameters are linearly related to alkyl chain length (Figure 3).

At higher concentrations of the two interacting species (10^{-1} mol dm^{-3}) thick viscoelastic gels were formed that gave complex rheograms when tested under continuous shear with cone and plate geometry (Figure 4).

DISCUSSION

The interaction between SS and the various ATAB's follows a 1:1 stoichiometry. An increase in the alkyl chain length of the cation (that is an increase in hydrophobicity) results in an enhanced interaction and a reduction in the solubility of the complex as measured by the solubility product. Each additional methylene group contributes a constant increment of −3.5 KJmol^{-1} to the free energy of the complexation process. Similar relations between the free energy of complexation and alkyl chain length have been reported by Davis and others for 1:1 and 2:1 interactions between large ions[6]. For the alkylsulphate: phosphonium system each additional methylene group in the anion provided a constant increment at 25°C of −3.20 KJmol^{-1} to the free energy of the complexation process. Thus the interaction between SS and ATAB's can be described in terms of ion association reinforced by a strong hydrophobic effect.

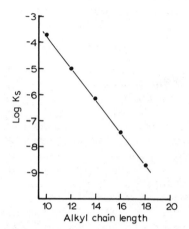

Figure 1.   The relation between the 1:1 solubility product (K_s) and alkyl chain length (n)

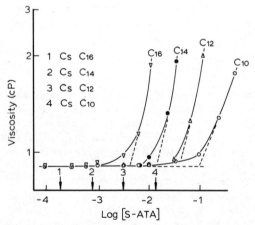

Figure 2.   The relation between viscosity and concentration for 1:1 complexes of saccharin sodium and alkyltrimethyl-ammonium bromides

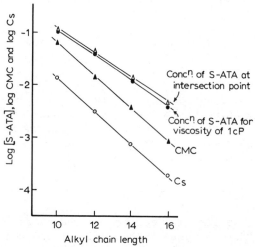

Figure 3.   The relation between the critical concentrations for
            viscosity change and alkyl chain length for 1:1
            complexes between sodium saccharin and alkyltrimethyl-
            ammonium bromides

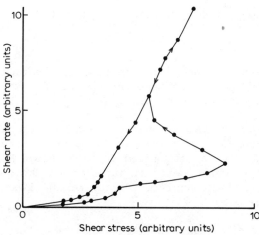

Figure 4.   Rheogram for 1:1 complex between saccharin sodium and
            hexadecyltrimethylammonium bromide at 10^{-1} mol dm^{-3}

Tomlinson and Davis[7] have considered the interaction between ions of opposite charge in terms of the solvophobic theory developed by Sinanoglu and Obduhur[8]. This can be used to examine the present system.

Consider the reversible association between the two ions

$$SA^- + ATA^+ \rightleftharpoons S.ATA$$

where each species is assumed to be in the Henry's Law region. According to the solvophobic theory the standard free energy change for this association (ΔG_f^O) can be viewed as consisting of two parts; a hypothetical interaction in a gas phase (no solvent effect) and a solvent effect

$$\Delta G_f^O = \Delta G_f^O \text{ (gas)} \quad \Sigma \Delta G^O_{solv,j} \tag{4}$$

where $\Sigma \Delta G^O_{solv,j}$ is the net free energy change for transfer of species j from the gas phase to the solvent (ie the sum of the free energies of all overall species - solvent interactions). Sinanoglu[8] has described the solvent effect on molecular associations in terms of a species solvent interaction term ($\Delta G_{int,j}$); a cavity term ($\Delta G_{cav,j}$) which accounts for the free energy change required for the creation of a cavity to accommodate species j in the solvent and an entropy effect term ($RT\ln(RT/PV)$) where P is the atmospheric pressure and V the molar volume of the solvent. The <u>net</u> solvent effect can be written[7] as

$$\Delta G_{solv}^O = \Delta G_{vdw} + \Delta G_{es} + \Delta G_{cav} + \Delta(RT\ln(RT/PV)) \tag{5}$$

where $\Delta G_{vdw} + \Delta G_{es} = \Delta G_{int}$ \hfill (6),

such that the terms on the left-hand side of equation (6) refer to the energies of solute-solvent interaction due to van der Waals and electrostatic forces respectively.

The individual terms in equations (4) and (5) can be expressed in measurable quantities. For example, the electrostatic interaction can be estimated from the Debye-Huckel theory for ions[7]. $\Delta G_{cav,j}$ can be written in terms of the molecular surface area (A) and the surface tension of the liquid (γ) in which association occurs

$$\Delta G_{cav,j} = NA_j\gamma + 4.84N^{\frac{1}{3}} (K^e-1)V^{\frac{2}{3}}\gamma \tag{7}$$

where N is Avogadro's number and K^e is an energy term associated with the formation of a cavity.

Thus

$$\Delta G_f^{\ O} = \Delta G_{f(gas)}^{\ O} + \Delta G_{vdw} + \Delta G_{es} + NA\gamma + 4.84\ N^{\frac{1}{3}}\ (K^e - 1)\ V^{\frac{2}{3}}\ \gamma$$

$$+ \Delta(RTln(RT/PV)) \hspace{4cm} (8)$$

where A refers to the contact surface area of the associated species.

The ΔG_{cav} term is seen to increase linearly with the reduction in total molecular surface area upon association (equivalent to the contact area between species). Thus for the present system where the electrostatic (ion–ion) terms are independent of the chain length of the ATAB a linear relation should exist between log K_s and the non-polar surface area of the cation, provided that the alkyl chain is not coiled in the cavity. Figure 1 shows this to be the case. Thus the energetics of interaction for large ions of opposite charge are largely dominated by the gain in energy due to the decrease in cavity size upon association[7].

The SS–ATAB complexes do not separate out as turbid phase above the complexation point, but form clear solutions that eventually can form viscoelastic gels at higher concentration of the two interacting ions. Others have reported gel formation by alkylammonium salts and organic anionics[11-13], although Barry and Russell[19] reported the separation of a coacervate phase when the dye Amaranth interacted with ATAB's of similar chain length to those in the present study. However, they also mentioned that the viscosities of solutions (sic) of long chain surfactant with dyes were higher than those with short-chain surfactants. A similar observation can be made in the present study (Figure 2); indeed there are striking linear relations between the logarithm of the concentration of S-ATA complex required to produce a given viscosity and alkyl chain length.

The phase structure, nuclear magnetic resonance and rheological properties of viscoelastic sodium dodecyl sulphate and alkyltrimethylammonium bromide mixtures was described by Barker et al[20] who suggested that the rheological behaviour arose from the occurrence of cylindrical micelles. Under static conditions the cylinders would become randomly oriented by Brownian motion but the cylinders would become aligned under shear. The cylinders would resume a random configuration when shear was ceased and the energy released would give the elastic contribution.

Wan[21] has reported that salicylic acid interacted with HTAB to give an increase in viscosity and Gravsholt[11] obtained a similar effect using various substituted benzoic acids. He proposed that some rather specific secondary structure existed in the systems which differed fundamentally from the ordinary spherical or rod-like micelles formed by HTAB, since such aggregates did not show any viscoelastic effect in pure HTAB even at high concentrations (0.5 mol dm^{-3}). More recently Ulmius et al[12] have used proton magnetic resonance and linear dichroism induced in a shear gradient to examine such systems further. They found that the viscoelastic behaviour of dilute solutions correlated with the formation of large rod-shaped micellar aggregates.

Strong ion binding to ionic micelles is one of the important factors in determining the energetics of micelle formation. For a rod-shaped micelle the surface charge density is larger than the corresponding spherical one and the counterion binding is consequently larger. If in addition to the electrostatic effects (which can be considered to be the same for all counterions) there is an attractive chemical contribution to ion binding as in the case for the mixture of large ions of opposite charge then large rod-like aggregates are to be expected[12].

It may be concluded that the viscoelastic gels formed by SS-ATAB systems are of a similar nature to other ATAB gels formed using large organic amphiphiles. The rapid rise in viscosity occurs at concentrations well above that necessary for the formation of a 1:1 complex (coacervate). The critical concentrations for increase in viscosity shown in Figure 2 are thought to indicate the formation of rod-like micelles of HTAB that can build to form long range periodic structures. In the formation of these structures two competitive equilibria will exist. At low concentrations the saccharin anion will interact with the ATA cation as governed by the solubility product, consequently the thermodynamic activity of each species in solution will fall. However, the saccharin anion will also bring about a reduction in the critical micelle concentration of the ATAB, counteracting the reduction in ATAB activity due to complexation. This counterion effect will increase in proportion to the size of the cation[22]. Once formed, the micelles of ATAB will solubilize the complex S-ATA species (probably in the palisade layer of the micelle)[16]. The large rod-like micelles so created provide the viscoelastic gel structure[23]. The relations between the log CMC for the ATAB's and log C_s (= (log K_s)$^{\frac{1}{2}}$) for the S-ATA complexes, and alkyl chain length, are shown in Figure 3. The contributions of the methylene group to micelle formation and complexation are identical indicating the similarities between the two processes of phase separation and the role of hydrophobic interaction. Such relations between complexation and micellar equilibria have been discussed by Tomlinson and Davis[7] who pointed out that they were useful in predicting complexation in pharmaceutical formulations.

REFERENCES

1. G.I. Mukhayer and S.S. Davis, J. Colloid Interface Sci., 53 224 (1975).
2. G.I. Mukhayer and S.S. Davis, J. Colloid Interface Sci., 56 350 (1976).
3. G.I. Mukhayer and S.S. Davis, J. Colloid Interface Sci., 59 350 (1977).
4. G.I. Mukhayer and S.S. Davis, J. Colloid Interface Sci., 66 117 (1978).
5. E. Tomlinson and S.S. Davis, J. Colloid Interface Sci., 66 335 (1978).
6. E. Tomlinson, S.S. Davis and G.I. Mukhayer, In "Solution Chemistry of Surfactants", K.L. Mittal, Editor, Vol. 1, p.3, Plenum Press, New York, 1979.
7. E. Tomlinson and S.S. Davis, J. Colloid Interface Sci., 76 563 (1980).
8. O. Sinanoglu and S. Obduhur, Fed. Proc. 24, 12 (1965).
9. C.G. Wilson, E. Tomlinson, S.S. Davis and O. Olejnik, J. Pharm. Pharmac. 31, 49 (1981).
10. E. Suzuki, M. Tsukigi, S. Muranushi, H. Sezaki and K. Kakemi, J. Pharm. Pharmac. 24, 138 (1972).
11. S. Gravsholt, J. Colloid Interface Sci., 57, 575 (1976).
12. J. Ulmius, H. Wennerstrom, L.B. Johansson, G. Lindblom and S. Gravsholt, J. Phys. Chem., 82, 2232 (1979).
13. J.W. Larson, L.J. Magid and V. Payton, Tetrahedron Lett., 29, 2663 (1973).
14. T. Nash, J. Colloid Sci., 13, 134 (1958).
15. W. Bruning and A. Holtzer, J. Am. Chem. Soc. 83, 4865 (1961).
16. B.W. Barry, J.C. Morrison and G.F.J. Russell, J. Colloid Interface Sci., 33, 554 (1970).
17. G.I. Mukhayer, S.S. Davis and E. Tomlinson, J. Pharm. Sci., 64, 147 (1975).
18. D.E. Guveli, J.B. Kayes and S.S. Davis, J. Colloid Interface Sci., 72, 130 (1979).
19. B.W. Barry and G.F.J. Russell, J. Pharm. Sci., 61, 502 (1972).
20. C.A. Barker, D. Saul, G.J.T. Tiddy, B.A. Wheeler and E. Willis, Trans. Faraday Soc. 70, 154 (1974).
21. L.S.C. Wan, J. Pharm. Sci. 55, 1395 (1966).
22. E.D. Goddard, O. Harva and T.G. Jones, Trans. Faraday Soc., 49, 980 (1953).
23. J. Kalus, H. Hoffmann, K. Reizlein, W. Ulbricht and K. Ibel, Ber. Bunsen Phys. Chem. 86, 37 (1982).

THERMODYNAMIC STUDIES ON THE INTERACTION BETWEEN LYSOZYME

AND SODIUM n–DODECYL SULPHATE IN AQUEOUS SOLUTIONS

M.N. Jones and P. Manley

Department of Biochemistry
University of Manchester
Manchester, M13 9PL, U.K.

The enthalpy of interaction between sodium n-dodecyl sulphate and lysozyme has been measured in aqueous solutions at pH 3.2 over the ionic strength range 0.01 to 0.21 at $25^{o}C$. It has been found that there is an initial exothermic interaction arising from the binding of n–dodecyl sulphate anions to the cationic residues in lysozyme. This process occurs without a major change in the conformation of lysozyme. At higher binding levels there is an endothermic contribution from protein unfolding characterised by minima in the enthalpy curves.

When mixing protein with micellar solutions of surfactant it is necessary to take account of the thermal effects arising from micelle dissociation (demicellization) in order to obtain the true enthalpies of interaction between protein and 'monomeric' surfactant. These corrections are discussed and have been applied to the observed enthalpy curves at the higher surfactant concentrations. The results suggest that there is a substantial exothermic enthalpy contribution arising from the interaction between lysozyme-surfactant complexes and surfactant micelles.

INTRODUCTION

The measurement of the enthalpies of interaction between surfactants and globular proteins by microcalorimetric techniques have proved useful in interpreting the mode of binding of the surfactant and subsequent unfolding of protein tertiary structure[1-5].

In the particular case of lysozyme the initial binding of sodium n-dodecyl sulphate (SDS) in acid solutions occurs specifically to approximately 18 high energy sites[6]. Lysozyme has 18 cationic amino acid residues (6 lysyl, 11 arginyl and 1 histidyl) and by selective chemical modification of these residues it has been established[5] that they are the sites to which SDS initially binds and that the enthalpy of the interaction is exothermic. Furthermore, this specific binding does not give rise to any large conformational change in the protein. Difference spectroscopy and viscometry measurements indicate that the protein unfolds on further hydrophobic binding when the total number of SDS molecules bound reaches approximately 35[5].

The studies described above were all carried out at low ionic strength (I < 0.01). Under these conditions the free surfactant concentration in equilibrium with the protein-surfactant complexes is below the critical micelle concentration (cmc) of the surfactant. We now consider the important question of the effects of ionic strength on these interactions. However, the decrease in critical micelle concentration (cmc) with increasing ionic strength necessitates making measurements at post-cmc surfactant concentrations. Thus the enthalpies of mixing of protein solution with a surfactant solution above the cmc involve a contribution from the enthalpy of micelle dissociation (called hereafter demicellization). In order to take rigorous account of the enthalpic effect from demicellization to give the true enthalpy of interaction between monomeric surfactant and protein the enthalpies of micellization of surfactant have also been measured over a range of surfactant concentration and ionic strength.

EXPERIMENTAL

Materials. Lysozyme (Grade I 3X crystallized) was used as supplied by Sigma Chemical Co. Sodium n-dodecyl sulphate (especially pure grade) was from British Drug Houses. Spectrapor membrane dialysis tubing (molecular weight cut off 6-8000) was from MSE (Fisons) Crawley, U.K.

Four buffer solutions were used, each contained 50mM glycine plus hydrochloric acid to give a pH of 3.2 and 0.02% (w/v) sodium azide giving a total ionic strength of 0.0119. Buffers of ionic strengths 0.0519, 0.1119 and 0.2119 were prepared from the above mixture by addition of sodium chloride. All the salts used in the preparation of the buffers were of analytical grade and they were made up in doubly distilled water.

Methods. The critical micelle concentrations of SDS in the buffer solutions were measured at 25°C by conductivity using a glass conductivity cell with platinum electrodes in conjunction with a Wayne-Kerr Universal Bridge (B224), or from surface tension measurements with a Du Noüy tensiometer.

Enthalpy measurements were made with an LKB 10700 batch microcalorimeter which utilizes the twin-vessel principle, each vessel being divided into two compartments[7]. In the measurements of the overall enthalpies of interaction between lysozyme and SDS the reaction vessel was charged with 2.0 (+ 0.1)g of protein solution (concentration 0.25%) and 2.0 (+ 0.1)g of surfactant solution of the required concentration in the range (0-30mM). The reference vessel was charged with 2.0 (+ 0.1)g of buffer solution and 2.0 (+ 0.1)g of surfactant solution identical with that in the reaction vessel. On mixing, the enthalpies of dilution of the surfactant solutions cancel. The enthalpies of dilution of the protein solution was zero since the protein solutions were dialysed against buffer and the dialysate buffer then used as a solvent for all the surfactants. The protein concentration after dialysis was checked by measuring the extinction at 281.5nm and adjusted to the required concentration of 0.25% w/v according to the extinction $E_{281.5}^{1\%} = 26.4$[8].

Enthalpies of demicellization were measured as a function of surfactant concentration according to the following scheme where X refers to the concentration of surfactant in multiples of critical micelle concentration. One vessel of the calorimeter was charged with 2.0 (+ 0.1)g of surfactant solution of concentration X cmc and 2.0 (+ 0.1)g of buffer. The other vessel was charged with 2.0 (+ 0.1)g of surfactant solution of concentration (X-1) cmc and 2.0 (+ 0.1)g of surfactant of concentration 1 cmc. On mixing the measured enthalpy corresponds to the process.

$$2g(X)cmc + 2g(0)cmc \rightarrow 2g(X-1)cmc + 2g(1)cmc \qquad (1)$$

i.e. the demicellization of 1cmc of surfactant from a concentration X cmc. This procedure has been discussed in detail previously[9].

The binding of SDS to lysozyme was measured by equilibrium dialysis at 25°C. Aliquots (2 cm^3) of lysozyme solution (0.125% w/v) were equilibrated against aliquots (2 cm^3) of surfactant

solution in the concentration range upto 30mM over a period in
excess of 96 hrs. The free surfactant concentrations in
equilibrium with the complexes was assayed by the Rosanaline
hydrochloride method[10]. The lysozyme concentration of 0.125% w/v
was chosen so as to equal the final lysozyme concentration after
mixing in the microcalorimetric experiments.

THEORY

Before describing the results it is appropriate to discuss the
method of correcting the overall enthalpies of mixing protein and
surfactant for the thermal contribution arising from the demicellization effect. On mixing equal volumes of a protein solution and
a surfactant solution (molarity m) in the reaction vessel of the
calorimeter and equal volumes of surfactant solution (molarity m)
and buffer solution in the reference vessel three cases can be
recognised.

(1) If $m/2 <$ cmc the reaction and reference vessels will be thermally balanced since the extent of demicellization will be the same
in both vessels and no correction is required.

(2) If $m/2 >$ cmc but the free (unbound) surfactant concentration
in the reaction vessel after mixing is less than the cmc then there
will be an imbalance between the reaction and the reference vessels
arising from the enthalpy of demicellization of the amount of
surfactant equal to that bound to the protein plus the difference
between the free surfactant concentrations in the two vessels.

(3) If $m/2 >$ cmc but the free (unbound) surfactant concentration in
the reaction vessel after mixing exceeds the cmc. In this case both
reaction and reference vessels will contain micellar surfactant
after mixing and the thermal imbalance arises simply from the
enthalpy of demicellization of the amount of surfactant bound to
the protein.

Considering case (2) if x represents the cmc in molarity and
the asterisk is used to denote surfactant (S) in micellar form,
in the reference vessel we have;

$$2 cm^3 [(m-x) S*/xS] + 2 cm^3 [Buffer] \rightarrow 4 cm^3 [(\frac{m}{2}-x) S*/xS] \qquad (2)$$

$$\therefore \text{ amount of S demicellized} = \frac{4}{10^3} x - \frac{2}{10^3} x = \frac{2}{10^3} x \text{ moles} \qquad (3)$$

In the reaction vessel, if a protein surfactant complex of
composition $PS_{\overline{\nu}}$ is formed where $\overline{\nu}$ is the number of moles of
surfactant bound per mole of protein and y (which is $< x$) represents
the free surfactant concentration in equilibrium with it then

$$2cm^3[P/Buffer] + 2cm^3[(m-x)S*/xS] \rightarrow 4cm^3[PS_{\overline{\nu}}/yS] \qquad (4)$$

$$\therefore \text{ amount of S demicellized} = \frac{4}{10}3[PS_{\overline{\nu}}] \, \overline{\nu} + \frac{4}{10}3 \, y - \frac{2}{10}3 \, x \qquad (5)$$

where the complex concentration is represented by $[PS_{\overline{\nu}}]$.

The difference between (5) and (3) gives the net amount of surfactant demicellized on mixing.

$$\text{net moles of S demicellized} = \frac{4}{10}3[[PS_{\overline{\nu}}]\overline{\nu} + y - x] \qquad (6)$$

The overall enthalpy of interaction is expressed in $J \, g^{-1}$ of protein and the enthalpy of demicellization (ΔH_{demic}) in $J \, mol^{-1}$ of surfactant thus the enthalpy correction due to demicellization in $J \, g^{-1}$ of protein is

$$\Delta H_{corr} = \frac{4}{10}3 \, [[PS_{\overline{\nu}}]\overline{\nu} + y - x]/\frac{2}{10}3 \, [P]M) \, \Delta H_{demic}$$

$$= \frac{2[[PS_{\overline{\nu}}] \, \overline{\nu} + y - x] \, \Delta H_{demic}}{[P]M} \qquad (7)$$

where [P] is the initial protein concentration and M is the molecular weight of the protein (14,306).

Case (3) above follows immediately from equation (7) when y = x. This third case has been previously considered in more formal terms for the interaction between ribonuclease and SDS[11].

RESULTS

Figures 1 and 2 show the binding isotherms at pH 3.2 for SDS with lysozyme at four ionic strengths. The insert in Figure 1 shows the binding of SDS above $\overline{\nu} = 50$ for I = 0.0119 where a maximum is observed as previously discussed[12]. Binding data (not shown) was also obtained above $\overline{\nu} = 50$ for the other systems and used in the calculations of ΔH. The cmc's of SDS in these media at $25^{\circ}C$ are as follows 5.40mM (I = 0.0119), 1.82mM (I = 0.0519), 0.980mM (I = 0.1119) and 0.720mM (I = 0.2119). The overall shapes of the isotherms are very similar and are characterised by a steep rise at low free SDS concentration which corresponds to the binding to specific, probably largely independent, high energy binding sites. At much higher free surfactant concentrations a further steep rise characteristic of cooperative hydrophobic binding occurs. Figure 3 shows the relevant parts of the Scatchard plots of the data at extremes of ionic strength which are based on the equation

$$\frac{\overline{\nu}}{[SDS]_{free}} = K \, (n-\overline{\nu}) \qquad (8)$$

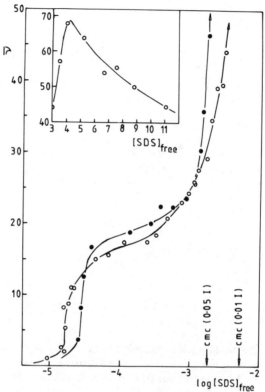

Figure 1. Binding isotherms for sodium n-dodecyl sulphate on interaction with lysozyme at pH 3.2, 25°C. O, I 0.0119; ●, I 0.0519. Lysozyme concentration 0.125% (w/v). The insert shows the maximum in $\bar{\nu}$ plotted as a function of [SDS]$_{free}$ at pH 3.2, I 0.0119.

where K is the intrinsic binding constant and n is the number of independent binding sites. Linear extrapolation of the data gives values of n ∿ 18-20 close to the number of cationic residues in lysozyme.

Figure 4 shows the enthalpies of demicellization of SDS at 25°C as a function of the initial concentration in cmc units. Because increasing ionic strength lowers the cmc and **one** cmc of surfactant is demicellized the errors of the measurements are larger at higher ionic strength.

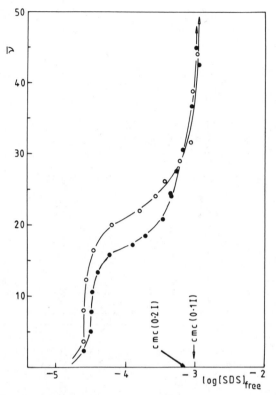

Figure 2. Binding isotherms for sodium n-dodecyl sulphate on interaction with lysozyme at pH 3.2, 25°C. O, I 0.1119, ●, I 0.2119. Lysozyme concentration 0.125% (w/v).

Figures 5-8 show the enthalpies of interaction between SDS and lysozyme before and after correction for the demicellization effect according to equation (7) as a function of the total SDS concentration after mixing. At low SDS concentrations the complexes are insoluble. Both the solubility limits and the SDS concentration at which complexes and micelles coexist are indicated on the figures. In all cases the enthalpy curves change continuously through the solubility limits indicating that any thermal effects associated with the onset of turbidity and precipitation are negligible compared with the enthalpies of surfactant interaction.

The SDS binding data have been used to plot the upper scales in these figures showing the number of SDS molecules bound per lysozyme molecule. In Figure 5 the decreasing values of $\bar{\nu}$ with increasing total SDS concentration reflect the maximum in the binding isotherm. It should be pointed out that the microcalori-

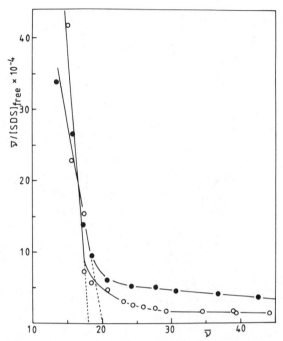

Figure 3. Scatchard plot for the binding of sodium n-dodecyl sulphate to lysozyme at pH 3.2, 25°C. O, I 0.0119; ●, I 0.2119.

metric data are collected within a time period of 30 minutes after mixing protein with surfactant whereas dialysis requires > 96 hours to come to equilibrium, this time scale being determined by the rate of diffusion of the surfactant across the dialysis membrane. We have no evidence to suggest that the binding of SDS to lysozyme is not complete within the time scale of the calorimetric measurements so that the composition of the complexes formed in the calorimeter correspond to those found by equilibrium dialysis.

DISCUSSION

Inspection of Figures 1 and 2 shows that ionic strength has a different effect on specific binding as compared with its effect on cooperative hydrophobic binding. Increasing ionic strength shifts the initial steep rise in $\bar{v}$ to higher free SDS concentrations, suggesting that specific binding is weakened by the presence of increasing salt, in contrast cooperative hydrophobic binding is strengthened by salt and the isotherms are shifted to lower free

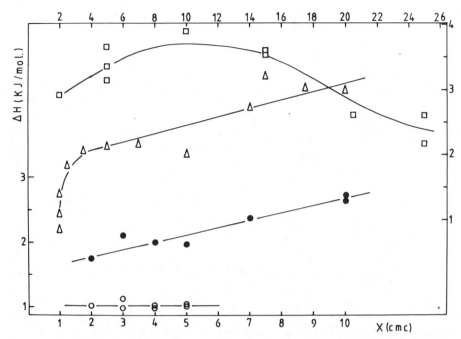

Figure 4. Enthalpy of micelle dissociation (demicellization) as a function of sodium n-dodecyl sulphate concentration in units of cmc according to equation (1) O, I 0.0119; ●, I 0.0519; (lower and left hand axes) Δ, I 0.1119; □, I 0.2119 .(upper and right hand axes).

SDS concentration. Thus the curves in Figure 1 show a clear cross-over point. Although salt weakens specific binding there are no marked changes in the number of specific binding sites within the experimental error. It is significant that the binding isotherms continue to rise above the cmc which is a clear demonstration of the increasing activity of SDS which has been considered by Elworthy and Mysels[13].

The enthalpies of interaction of SDS with lysozyme (Figures 5-8) all show an initial steep rise which corresponds to the specific binding of ∿ 18 SDS ions. This interaction becomes more exothermic with increasing ionic strength and is not accompanied by any major conformational change in protein structure[5]. The change in ΔH from -12 J g^{-1} (I = 0.01) to -15 J g^{-1} (I = 0.21) must be considered together with the shift of the binding isotherm to higher free SDS concentrations. Thus ΔG for the specific inter-action is becoming less negative as I increases. At the lowest I

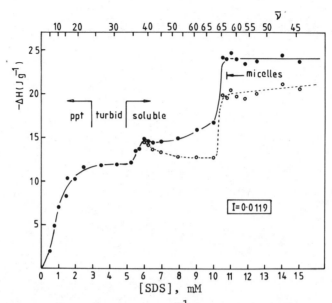

Figure 5. Enthalpy change ($\Delta H/J\ g^{-1}$) on interaction of sodium n-dodecyl sulphate with lysozyme at 25°C, pH 3.2, I 0.0119. The dashed curve represents the data before correction for the demicellization effect. The lower axis gives the total SDS concentration and the upper axis ($\bar{\nu}$) the number of SDS molecules bound per molecule of lysozyme.

$\Delta G = -25$ kJ mol^{-1} (-1.75 J g^{-1})[6] and hence $\Delta S = -493$ J mol^{-1} K^{-1} (0.035 J g^{-1} K^{-1}). It follows that ΔS must become more negative with increasing I.

All the enthalpy curves show a transition in the region of $\bar{\nu}$ ∿ 20–35 and at the higher ionic strengths we observe distinct minima, indicating that there is an endothermic contribution to the overall enthalpy at this level of binding which can be attributed to the unfolding of the native structure. The enthalpy of unfolding of lysozyme has been studied in detail by Pfeil and Privalov and amounts to ∿ 220 kJ mol^{-1} (15 J g^{-1}) at 25°C[14]. It is clear that a significant fraction of this endotherm will be compensated by the exothermic binding of SDS which arises from the exposure of binding sites on unfolding.

Comparison of the enthalpy curves with those before correction for the demicellization of surfactant on mixing (dotted curves) shows that the corrections are very significant. At the lowest ionic strength (I = 0.01) where the cmc is high, the corrections

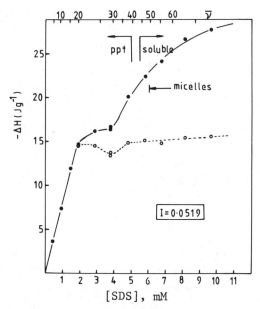

Figure 6. Enthalpy change ($\Delta H/J \ g^{-1}$) on interaction of sodium n-dodecyl sulphate with lysozyme at $25°C$, pH 3.2, I 0.0519. Axes as for figure 5.

are less than 30% in the range upto $\bar{\nu} \sim 65$ and there is a steep rise in ΔH which coincides with the total SDS concentration at which SDS micelles are present in the solution together with lysozyme-SDS complexes. In this particular system a maximum in the binding isotherm is observed at $\bar{\nu} \sim 70$ corresponding to a total SDS concentration of $\sim 10mM$[12]. Thus it would appear that binding is decreasing above $\sim 10mM$ SDS but this decrease is not explicitly reflected in a decrease in the enthalpy of interaction. This observation taken together with the rise in ΔH at $\bar{\nu} \sim 65$ signifies a thermal contribution arising from a source other than SDS binding and most probably from interaction between the lysozyme-SDS complexes and SDS micelles. The relative constancy of ΔH above a total SDS concentration of 11mM may reflect a decreasing contribution from surfactant binding concomitant with an increasing contribution from complex-micelle interaction as the micelle concentration increases. For the systems of higher ionic strength binding is increasing together with the concentration of micelles so that in these systems ΔH continues to rise.

A negative enthalpy for complex-micelle interaction is consistent with the observation that dilution of SDS micelles is

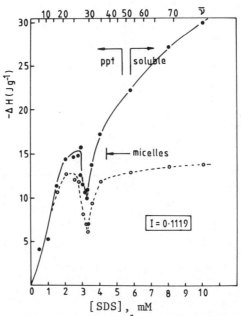

Figure 7. Enthalpy change ($\Delta H/J$ g^{-1}) on interaction of sodium n-dodecyl sulphate with lysozyme at 25°C, pH 3.2, I 0.1119 Axes as for Figure 5.

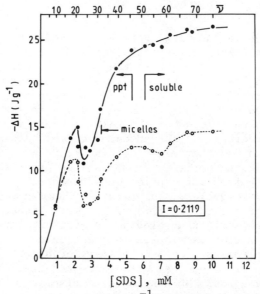

Figure 8. Enthalpy change ($\Delta H/J$ g^{-1}) on interaction of sodium n-dodecyl sulphate with lysozyme at 25°C, pH 3.2, I 0.2119. Axes as for Figure 5.

an endothermic process[9] (hence concentration is exothermic). Since both complexes and micelles are negatively charged, ΔG for interaction will be positive and since ΔH is negative it follows that $T\Delta S$ must also be negative and that $|T\Delta S|>|\Delta H|$. The magnitude of the enthalpy of interaction between complexes and micelles is comparable to that for ionic interactions (e.g. $\sim$ 5 J g^{-1}) and hence might arise from counter-ion binding to the complex and/or micelle surface which would be consistent with the decrease in entropy.

REFERENCES

1. E. Tipping, M.N. Jones and H.A. Skinner, J. Chem. Soc. Faraday Trans I, 70, 1306 (1974).
2. M.N. Jones, H.A. Skinner, E. Tipping and A.E. Wilkinson, Biochem. J.,135, 231 (1973).
3. M.N. Jones and A.E. Wilkinson, Biochem. J.,153, 713 (1976).
4. M.N. Jones, Biochim. Biophys. Acta.,491, 121 (1977).
5. M.N. Jones and P. Manley, J. Chem. Soc. Faraday Trans I, 76, 654 (1980).
6. M.N. Jones and P. Manley, J. Chem. Soc. Faraday Trans I, 75, 1736 (1979).
7. I. Wadso, Acta Chem. Scand.,22, 927 (1968).
8. K.C. Aune and C. Tanford, Biochem.,8, 4579 (1969).
9. G. Pilcher, M.N. Jones, L. Espada and H.A. Skinner, J. Chem. Thermodynamics,1, 381 (1969).
10. F. Karush and M. Sonnenberg, Analyt. Chem.,22, 175 (1930).
11. M.I. Paz Andrade, E. Boitard, M.A. Saghal, P. Manley, M.N. Jones and H.A. Skinner, J. Chem. Soc. Faraday Trans I, 77, 2939 (1981).
12. M. N. Jones, P. Manley and P. J. W. Midgley, J. Colloid Interface Sci., 82, 257 (1981).
13. P. H. Elworthy and K. J. Mysels, J. Colloid Interface Sci., 21, 331 (1966).
14. N. Pfeil and P.L. Privalov, in 'Biochemical Thermodynamics' M.N. Jones, Editor, Chapter 3, p. 95, Elsevier, Amsterdam, 1979.

ABOUT THE CONTRIBUTORS

Here are included biodata of only those authors who have
contributed to this volume. Biodata of contributors to Volumes
1 and 3 are included in those volumes.

Jacqueline Appell is "Chargee de Recherche" of the CNRS in
the Laboratoire de Spectrometrie Rayleigh-Brillouin Montpellier,
France. She received her "Doctorat d'Etat" (Physical Science) in
1972 from the University Paris-Sud (Orsay). Her current interest
is in the ionic and nonionic aqueous micellar solutions.

T. Arnebrant is a Ph.D. student in Food Technology, University of Lund, Lund, Sweden.

J. R. Aston is with the Department of Physical Chemistry,
University of Melbourne, Australia.

Sune Backlund is Lecturer in Physical Chemistry at Abo Akademi,
Abo, Finland. He received his Ph.D. in 1979 and his fields of interest
are electrochemistry, thermodynamics and association colloids.

Arieh Y. Ben-Naim is Professor of Physical Chemistry,
Hebrew University of Jerusalem. He received his Ph.D. degree in
1964 followed by a number of teaching and research appointments.
He has been a Visiting Scientist/Professor at various institutions, the most recent being at Bell Laboratories. His research
interests include theoretical and experimental aspects of the
structure of water, aqueous solutions and the problem of hydrophobic interaction. He has authored two monographs Hydrophobic
Interactions (1980) and Water and Aqueous Solutions (1974).

R. Bennes is with the Laboratoire de Physico-Chimie des
Systemes Polyphases, Montpellier, France.

K. S. Birdi is Assistant Professor ("lektor" in Danish) at
the Technical University of Denmark, Lyngby, Denmark. His research interests are monolayers, biological macromolecules, and
micelles (solubilization).

1417

João B. S. Bonilha is Assistant Professor at the University of Sao Paulo at Riberao Preto, Brazil. He received his Ph.D. degree in Chemistry and did one year postdoctoral work with Prof. D. Whitten. He has ten publications in organic and photochemistry, and is currently interested in photochemistry in organized systems.

Enrico Borgarello has been "assistant de Chimie" of Prof. M. Gratzel at the Swiss Federal Institute in Lausanne, Switzerland since October 1980. His research is directed toward the development of new devices for H_2O splitting and H_2 production by means of solar energy, and in 1981 he won the A. M. Lorgna prize for young scientists devoted to the research on new energy sources. He carried out his thesis research and received degree at the University of Torino, Italy in 1980.

P. E. Bruce is a research student in the Department of Pharmacy, University of Nottingham, U.K.

C. A. Bunton is Professor of Chemistry at the University of California, Santa Barbara. His B.Sc. and Ph.D. degreess are from the University college, London. He has had visiting appointments at universities in Europe and North and South America. His research interests include micellar catalysis and inhibition.

Ludmila Čermáková is Lecturer in the Department of Analytical Chemistry, Charles University, Prague, Czechoslovakia which she joined in 1959 upon graduation from the same university. In 1966 she obtained the M.Sc. and Ph.D. degrees on the basis of the work dealing with the determination of sulfur-containing organic compounds. She is the coauthor of several university textbooks and of a textbook on analytical chemistry for specialized high schools. Since 1970 she has intensely studied the spectrophotometry of the platinum metals and the use of tensides in photometry and has published a number of research papers.

Hernan Chaimovich is Associate Professor in the Department of Biochemistry, Institute of Chemistry, University of Sao Paulo, Sao Paulo, Brazil. He received his Ph.D. degree in Biochemistry and has about 40 publications dealing with enzymology, membrane models and physical organic chemistry. His current research interests include the study of chemical reactions in interphases and models of biological membranes.

Beng Tatt Ch'ng is at present completing his M.Sc. in Chemistry at the University of Auckland, New Zealand where he received his B.Sc. in 1980. He has studied the reactivity of esters and amides in bile salt solutions.

G. Charbit is Maitre Assistant at the Ecole Nationale Superieure de Chimie de Paris. He received his D.Sci. in 1980 from the Universite Pierre et Marie Curie.

Dipti K. Chattoraj is Professor and Head of the Department of Food Technology and Biochemical Engineering, Jadavpur University, Calcutta, India. He obtained his Ph.D. and D.Sc. degrees in Physical Chemistry in 1957 and 1972 respectively from the University of Calcutta. He has published more than 80 original research papers in the fields of adsorption, surface thermodynamics, colloid stability and biophysical chemistry.

Emil Chifu is Professor of Physical Chemistry at "Babes-Bolyai" University in Cluj-Napoca, Romania. He holds a Ph.D. degree in Chemistry and has had visiting appointments at universities in Italy, the U.S.S.R. and West Germany. His research interests include the physical chemistry of interfacial films from thermodynamic and hydrodynamic viewpoint.

L. J. Cline Love is Professor of Chemistry, Seton Hall University, S. Orange, NJ. She obtained her Ph.D. degree in Chemistry from the University of Illinois in 1969. She has published about 50 scientific articles and is the Editor of Advances in Luminescence Spectroscopy (to be published by ASTM). She currently serves as Associate Editor on the journal – Applied Spectroscopy. Her research interests include micelle-enhanced analytical chemistry, fluorescence and phosphorescence, photophysics of licit and illicit drugs, micellar chromatography, automation/control of experimentation, and computer intelligence schemes.

Ron Cohen is currently a Ph.D. student. He received his M.Sc. from the School of Pharmacy, Hebrew University, Jerusalem in 1981.

J. W. Corsel is a physical engineer in the Department of Biophysics, University of Limburg, Maastricht, The Netherlands and is working in the field of blood coagulation.

Iolanda Midea Cuccovia is Assistant Professor in the Department of Biochemistry, Institute of Chemistry, University of Sao Paulo, Sao Paulo, Brazil. She received her Ph.D. degree in Biochemistry. She has 12 publications in the field of micellar modified reactions and is curreintly interested in studying the effect of amphiphile aggregate structure on chemical reactivity.

P. A. Cuypers is in the Department of Biophysics, Rijkuniversiteit Limburg, Maastricht, The Netherlands. Has a Ph.D. in Biophysics and has published many papers in ellipsometry. He is working in the field of blood coagulation.

P. Daniels is a research student in the Department of Pharmacy, University of Nottingham, U.K.

S. S. Davis has been Lord Trent Professor of Pharmacy, University of Nottingham, Nottingham, England, since 1975. Before his current position, he was in the Pharmacy Department, University of Aston in Birmingham. He received his Ph.D. from the Faculty of Medicine, University of London in 1967 and received his D.Sc. from the Faculty of Science of the same university in 1981. He is a Fellow of the Royal Institute of Chemistry and has 190 publications to his credit. His research interests are colloid and surface science, solution thermodynamics, physical pharmaceutics and drug dosage design.

S. Diekmann is currently with the Biochemical Laboratories, Harvard University, Cambridge, MA. He is a physicist and is interested in the properties of polyelectrolytes in solution.

F. Dorion is Maitre-Assistante at the Ecole Nationale Superieure de Chimie de Paris. She received her D.Sci. in 1982 from the Universite Pierre et Marie Curie.

Gerard Douheret is presently a Rechercher of the CNRS at the University of Claremont-Ferrand, Aubiere, France where he received his Ph.D. degree in Physical Chemistry. During 1973-1974 he was a Research Associate at the Royal Institute of Technology, Stockholm. He has published over 40 papers in the areas of electrochemistry and chemical thermodynamics, and his current research is in the thermodynamic properties of micellar solutions and of mixtures of nonelectrolytes.

Dotchi Exerowa has since 1972 been Senior Research Scientist in the Institute of Physical Chemistry of the Bulgarian Academy of Sciences, Sofia, Bulgaria, where she obtained her Ph.D. degree. At the same time she has been Associate Professor at the University of Sofia. Her research activities have concentrated on the physical chemistry of thin liquid films, liquid surfaces, foams, biomembranes, etc.

L. Feely is a research student in the Department of Pharmacy, University of Nottingham, U.K.

M. Fleschar is with the Technisch-Chemisches Laboratorium, ETH-Zentrum, Zurich, Switzerland.

Roberto Fornasier is a Collaboratore Tecnico Professionale of the Italian National Research Council (CNR) at the University of Padova. His research interest concerns organic stereochemistry and the study of micellar catalysts.

J. P. Foussier is Professor at the Ecole Nationale Superieure de Chimie, Mulhouse, France and is also Head of the Photochemistry Department. He received his doctorate from the Universite de Haute - Alsace, Mulhouse in 1975. He has developed an original research program on photochemistry and photophysics in molecules and macromolecules.

Jens Frahm is with the Max-Planck-Institut fur biophysikalische Chemie in Gottingen, West Germany. Is a physicist and is interested in NMR.

J. Francois is Chargee de Recherches (CNRS) at the Centre de Recherches sur les Macromolecules, Strasbourg, France. Holds Sc. Doctor degree and has 50 publications. Research interests include properties of mesophases of soap systems, physical chemistry of macromolecules, polyelectrolytes - hydrosoluble polymers, and interaction between hydrosoluble polymers and small molecules.

D. N. Furlong is with the Department of Physical Chemistry, University of Melbourne, Australia.

R. Gaboriaud is Professor of Chemistry at the Universite Pierre et Marie Curie in Paris, where he received his D.Sc. in 1967. His current research interests are thermodynamics and kinetic studies in nonaqueous or micellar solutions.

America Gilbon has been carrying out Doctor Thesis in Biochemistry and Biotechnology since 1981 about production of microbial surfactants at the Institute of Biochemistry and Biotechnology, Technical University of Braunschweig, West Germany.

Marina Gobbo received her doctoral degree in Chemistry in 1982 from the University of Padova, Italy.

Franz Grieser is currently a Queen Elizabeth II Research Fellow in the Department of Physical Chemistry at the University of Melbourne, Australia where Ph.D. degree was received in 1977. Research interests are in radiation chemistry, photophysical chemistry and surfactant chemistry.

Erdogan Gulari is Associate Professor of Chemical Engineering at the University of Michigan. He received his Ph.D. in Chemical Engineering from California Institute of Technology in 1973. He is the author of 22 papers in the area of liquid solutions. He is currently engaged in research in heterogeneous catalysis, surfactant solutions, non-ideal simple solutions and solution dynamics of polymers.

Thomas W. Healy is Professor of Physical Chemistry and Director of Colloid and Surface Science at the University of Melbourne, Australia. His research interests range from mineral processing, adsorption of hydrolyzable metal ions, hemi-micelle formation, wettability of surfaces, heterocoagulation of metal oxides to surfactant control of interfacial tensions.

P.-O. Hegg is affiliated with the Department of Food Technology, University of Lund, Sweden. Has a Ph.D. in Biochemistry and has published about 20 papers concerning protein gels and protein-detergent association.

H. C. Hemker is Professor of Biochemistry in the Department of Biophysics, University of Limburg, Maastricht, The Netherlands. Holds M.D. and Ph.D. degrees and is working in the field of blood coagulation.

W. Th. Hermens is a theorectical physicist in the Department of Biophysics, University of Limburg, Maastricht, The Netherlands and is working in the field of blood coagulation.

Joanne R. Hicks is a Research Assistant in the Department of Chemistry, Mount Allison University, Sackville, Canada.

Harald Høiland is Professor of Physical Chemistry at the University of Bergen, Norway. He received his Ph.D. in 1977 and his main interest is thermodynamics of aqueous solutions.

Shoichi Ikeda is Professor of Physical Chemistry at Nagoya University, Nagoya, Japan. He received his Dr.Sci. degree from Osaka University. His research interests include adsorption at interfaces, micellization, and conformation of biopolymers in solution.

E. Imre is with the Technisch-Chemisches Laboratorium, ETH-Zentrum, Zurich, Switzerland.

Lavinel G. Ionescu is Professor of Physical Chemistry and Chairman of the Division of Physical Chemsitry at the Universidade Federal de Santa Catarina, Florianopolis, Brazil. He received his Ph.D. in Physical Chemistry from New Mexico State University, Las Cruces in 1970. Before his present position, he had faculty appointments at several universities in the United States. He is a Senior Researcher (Cathegory 1A) of the Brazilian Research Council (CNPq). His current research interests deal with physical and chemical properties of surfactants and micelles, respiratory pigments, chemistry of clathrates and noble gases and history of science. He has published over 80 scientific articles.

P. Jacques is a Maitre Assistant at the Universite de
Haute-Alsace, Mulhouse, France where he received his doctorate in
1977. His research interests are semi-empirical calculations,
photochromism and laser spectroscopy in microheterogeneous
systems.

M. P. Janssen is a technician in the Department of Biophysics, University of Limburg, Maastricht, The Netherlands and
is working in the field of blood coagulation.

M. N. Jones is a Reader in Physical Biochemistry at the
University of Manchester, U.K. where   Ph.D. in Physical Chemistry was received in 1962. Dr. Jones has been in the Department
of Biochemistry since 1970 and the main interests have been in
the thermodynamics of protein-ligand interaction, interfacial
phenomena in biochemical systems and membrane studies. During the
1960's, Dr. Jones worked in the area of surface and colloid
chemistry mainly on thin liquid films and association colloids.
Dr. Jones has published two books and about 80 research papers.

Werner Kanzig has since 1961 been Porfessor of Experimental
Physics of Swiss Federal Institute of Technology (ETH), Zurich,
Switzerland. He studied at ETH and was awarded Dr.'s degree in
1951. He has held a number of positions at ETH including chair-
manship of the Laboratory of Solid State Physics. He has authored
the book Ferroelectrics and Antiferroelectrics and tutorial books
on general physics, and has numerous research publications on
x-ray diffraction, ferroelectricity, defects in solids, magnetic
resonance, molecular motion in solids, and phase transitions. He
was the recipient of the Silver Medal of ETH in 1951 and is a
Fellow of the American Physical Society.

Yuri L. Khmelnitsky is a Research Chemist at the Bakh
Institute of Biochemistry, Moscow, USSR. He received his Ph.D. in
1982 from the Lomonosov State University, Moscow.

Nathalia L. Klyachko is associated with the Lomonosov State
University, Moscow, USSR.

J. Kop is an analytical chemical engineer in the Department
of Biophysics, University of Limburg, Maastricht, The Netherlands
and is working in the field of blood coagulation.

Ove J. Kvammen is a Research Assistant at the University of
Bergen, Norway. He received his M.Sc. in 1980 and he is interested in the thermodynamics of surfactant solutions.

Siegmund Lang has been Research Associate at the Institute
of Biochemistry and Biotechnology, Technical University of Braunschweig, West Germany since 1975 after receiving his Ph.D. degree

from the same university. He has published in the areas of microbial steroid conversion, amino acid production with immobilized cells, and formation of microbial surfactants.

K. Larsson is Professor of Food Technology, University of Lund, Sweden and has published about 100 papers on lipid structure.

J. LeMoigne is Charge de Recherches (CNRS) at the Centre de Recherches sur les Macromolecules, Strasbourg, France. Holds Sc. Doctor degree and has 20 publications. Research interests include micelles, membranes and semiconductors.

Andrey V. Levashov is Associate Professor, Lemonosov State University, Moscow, USSR where he received his Ph.D. in 1970.

Dov Lichtenberg is currently Senior Lecturer in the Department of Physiology and Pharmacology at Tel Aviv University, Israel where he is engaged in studies of cholesterol solubility by surfactants. He received his Ph.D. in 1972 from the Hebrew University in Jerusalem followed by postdoctoral research at Caltech where he was involved in NMR studies of model membranes. Since 1974, he has been a Lecturer at the Hebrew University - Hadassah Medical School and has been active in Model-Membrane Research. His research has included studies of the interaction of various drugs with liposomes, solubilization of these model membranes by detergents, and the use of liposomes for drug delivery.

D. J. Lougnot is a research physical chemist at the Universite de Haute - Alsace, Mulhouse, France where he received his doctorate in 1977. His research interests are laser dyes and fast spectroscopy instrumentation.

Pier Luigi Luisi has been since 1972 Professor at the ETH - Zurich in Macromolecular Chemistry (Section Biopolymers) in the Department of Chemistry (1972 Assistent Professor, and 1980 extraordinarius). Before joining ETH in 1969, he was a Research Fellow at the Institute of Molecular Biology, Eugene, Oregon which was preceded by postdoctoral period at Uppsala. He received "Dottre in Chimica" (cum laude) in 1963 at the University of Pisa, Italy; the Doctor Thesis was carried out partly in Pisa and partly at the Institute of High Molecular Weight Compounds in Leningrad. In 1964 he received "Diploma in Chimica" (cum laude) at the Scuola Normale superiore in Pisa. He has been active in a number of research areas including solubilization in reversed micelles and has published some 90 papers.

P. Manley is a research technician in the Department of Biochemistry, University of Manchester. He is currently registered for an M.Sc. degree at Manchester University and has published eight papers.

Karel Martinek is Professor of Chemistry at Lomonsov State University, Moscow, USSR. He is the author of more than 150 publications in the areas of enzyme and micellar catalysis, mechanisms and kinetics of enzyme action, and enzyme engineering. He is coauthor (with I. V. Berezin) of the book Physico-Chemical Fundamentals of Enzyme Catalysis published in Moscow in 1975 and editor of some other books on Applied Biochemistry and Biotechnology (in Russian).

Ryohei Matuura is Professor of Physical Chemistry, Kyushu University, Fukuoka, Japan. He received his doctor of science degree in 1953 from Kyushu University. He has published many papers concerning colloid and surface chemistry.

N. Mazer has been trained as a medical doctor and physicist at the Harvard Medical School and M.I.T., and is presently engaged in biophysics research at both institutions. During 1979 and 1980 he was a Visiting Scientist in the experimental physics group of Prof. Kanzig of the ETH Zurich and in the physical chemistry group of Prof. Lindman at the University of Lund, Sweden.

Kevin McKeigue is currently a Ph.D. candidate in Chemical Engineering at the University of Michigan. He obtained his B.E. in Chemical Engineering at the Cooper Union in 1979 and an M.S.E. from the University of Michigan in 1980. His doctoral thesis is on the relationship between solution structure and transport properties in hydrogen bonded liquid solutions.

P. Meier is with the Technisch-Chemisches Laboratorium, ETH-Zentrum, Zurich, Switzerland.

Dan Meisel has been a Chemist at the Argonne National Laboratory, Argonne, IL since 1976. He received his Ph.D. degree in 1974 from the Hebrew University, Jerusalem and did postdoctoral research at the Radiation Laboratory of Carnegie-Mellon University. His research interests include photochemistry, radiation chemistry, electron transfer reactions in homogeneous and micro-heterogeneous systems, free radical reactions, and redox catalysis.

Kashmiri Lal Mittal is presently employed at the IBM Corporation in Hopewell Junction, NY. He received his M.Sc. (First Class First) in 1966 from Indian Institute of Technology, New Delhi, and Ph.D. in Colloid Chemistry in 1970 from the University of Southern California. In the last ten years, he has organized and chaired a number of very successful international symposia and in addition to this three-volume set, he has edited 14 more books as follows: Adsorption at Interfaces, and Colloidal Dispersions and Micellar Behavior (1975); Micellization, Solubilization, and Microemulsions, Volumes 1 & 2 (1977); Adhesion Measure-

ment of Thin Films, Thick Films and Bulk Coatings (1978); Surface
Contamination: Genesis, Detection, and Control, Volumes 1 & 2
(1979); Solution Chemistry of Surfactants, Volumes 1 & 2 (1979);
Solutions Behavior of Surfactants - Theoretical and Applied
Aspects, Volumes 1 & 2 (1982); Physicochemical Aspects of Polymer
Surfaces, Volumes 1 & 2 (1983); and Adhesion Aspects of Polymeric
Coatings (1983). In addition to these volumes he has published
about 50 papers in the areas of surface and colloid chemistry,
adhesion, polymers, etc. He has given many invited talks on the
multifarious facets of surface science, particularly adhesion, on
the invitation of various societies and organizations in many
countries all over the world, and is always a sought-after speak-
er. He is a Fellow of the American Institute of Chemists and
Indian Chemical Society, is listed in American Men and Women of
Science, Who's Who in the East, Men of Achievement and other
reference works. He is or has been a member of the Editorial
Boards of a number of scientific and technical journals. Presently, he is Vice President of the India Chemists and Chemical
Engineers Club.

   Yoshikiyo Moroi   is in the Faculty of Kyushu University,
Fukuoka, Japan. He received his doctor of science degree in 1973
from Kyushu University. His research interests are in the area of
micelle formation and micelle-catalyzed reactions.

   R. Nagarajan   is currently an Assistant Professor of Chemical Engineering at the Pennsylvania State University. He received his Ph.D. degree in 1979 from the State University of New
York at Buffalo. His research interests focus on surfactants,
their mechanism of action and their applications. He has published more than 30 papers in the areas of thermodynamics of
micelles, vesicles, solubilization, enhanced oil recovery,
surfactant-polymer interactions, etc.

   Alexander Nikolov   is presently a Research Scientist in the
Department of Physical Chemistry of the University of Sofia,
Sofia, Bulgaria where he has been since 1968, and he graduated in
Chemistry from the same university in 1968. Currently, he is
working in the physical chemistry of liquid surfaces.

   H. Noma   is a student in the Department of Chemistry, Kyushu
University, Fukuoka, Japan.

   Faruk Nome   is presently Professor of Inorganic chemistry at
the Universidade Federal de Santa Catarina, Florianopolis,
Brazil. He received his Ph.D. from Texas A & M University. His
research interests include bioinorganic chemistry and micellar
catalysis and has published over 50 papers.

T. Nylander is a Ph.D. student in Food Technology, University of Lund, Lund, Sweden.

Charmian J. O'Connor is Associate Professor of Chemistry at the University of Auckland, New Zealand, where she received her Ph.D. and D.Sc. degrees. She has had visiting appointments at University College and Imperial College, London, University of California, Santa Barbara, Texas A & M University, and Nagasaki University. Her research interests include micellar catalysis and high pressure kinetics of octahedral substitution reactions and has over 90 publications.

Yochanin Peled is Lecturer at the Sackler School of Medicine, Tel Aviv University, Israel, and is engaged in research on bile and gallstone formation and dissolution. Since 1974, Dr. Peled has been Chief of the Laboratory for Gastroenterology at Ichilov Medical Centre. Received Ph.D. degree in 1974 from Tel Aviv University.

Ezio Pelizzetti is Professor at the University of Torino, Italy. His research interests include properties and application of "functionalized" surfactants, electron transfer reactions in homogeneous and heterogeneous systems (micellar effect, colloidal redox catalysis), and development of devices for photochemical conversion of solar energy.

Gregoire Porte is "Attache de Recherche" of the CNRS in Groupe de Dynamiques des Phases Condensees in Montpellier, France. Current research interest is in the ionic and nonionic aqueous micellar solutions.

Edmondo Pramauro is Associate Professor at the University of Torino, Italy. His research interests include kinetics and equilibria in homogeneous and organized systems, use and application of micelles and microemulsions in analytical chemistry, and the development of colloidal redox catalysts for thermal and photochemical reactions.

Rudolf Preisig has been Professor of Clinical Pharmacology and Medicine, and Chairman, Department of Clinical Pharmacology, University of Berne, Berne, Switzerland since 1967. He had his education at the University of Zurich Medical School. His main research interests are drug metabolism, liver metabolism, hepatic transport, and bile formation, and is a member of the Swiss Drug Control Board.

M. Privat is with the Laboratoire de Physico-Chimie des Systemes Polyphases, Montpellier, France.

Hélène Raous is a student at the Laboratoire de Physico-chimie des Systemes Polyphases, CNRS, Montpellier, France. Received Dr. 3ème cycle in 1982 from the Montpellier University and has 5 publications.

Vincent C. Reinsborough is Associate Professor in the Department of Chemistry, Mount Allison University, Sackville, Canada. He received his Ph.D. degree from Tasmania, and has 25 publications to his credit dealing with molten salts and surfactant chemistry.

L. Stelian Romanesco is a member of the Sarmisegetuza Research Group, Las Cruces and Santa Fe, New Mexico, USA. He received the degree of Licentiate in Chemistry from the University of Cluj, Romania and the Ph.D. in Physical Chemistry from the University of Iasi. He has held various appointments in Eastern and Western Europe. His research interests deal mainly with surface and colloid chemistry.

Laurence S. Romsted is Assistant Professor of Chemistry at Rutgers University, New Brunswick, NJ. He did his graduate work at Indiana University and postdoctoral work at the University of California, Santa Barbara. His research interests include: the theory of micellar catalyzed reactions, ion binding at aqueous interfaces, reaction mechanisms in solution and oscillating chemical reactions.

Jarl B. Rosenholm is Lecturer in Physical Chemistry at the Faculty of Chemical Technology, the University Abo Akademi (Turku), Finland. He received his Ph.D. degree from Abo Akademi in 1978. He has published about 40 papers in the field of association colloids and colloid chemistry.

Eli Ruckenstein has been Distinguished Professor, Department of Chemical Engineering, SUNY Buffalo since 1981, and he was Faculty Professor at the same institution, Faculty of Engineering and Applied Sciences from 1973-1981. He received his Ph.D. degree from the Polytechnic Institute, Bucharest, Romania and was Professor at this institute from 1949-1969. He has been the recipient of the following awards: The George Spacu Award for Research in Surface Phenomena of the Romanian Academy of Sciences (1963); two research (1958 and 1964) and one teaching (1961) awards from the Romanian Department of Education; and the Alpha Chi Sigma Award (1977) of the American Institute of Chemical Engineers. He has published extensively (over 350 papers) dealing with many subjects. His research interests are transport phenomena, catalysis, separation processes, colloids and interfaces, and biophysics. He has been visiting professor at the Catholic University, Leuven, and Technion-Israel, and is listed in World Who's Who in Science.

Kenneth Rundt is presently a Chemist at Wallace oy, LKB Products, Abo, Finland. He holds an M.Sc. degree.

J. Sabbadin is a student at the Centre de Recherches sur les Macromolecules, Strasbourg, France.

Bimal Kr. Sadhukhan is presently a Postdoctoral Research Associate in the Department of Food Technology and Biochemical Engineering, Jadavpur University, Calcutta. He received his Ph.D. degree in Chemistry from the same university in 1982 for his research on the thermodynamics of protein hydration and protein-surfactant binding interaction.

Masahiko Saito is Assistant Professor, Department of Chemical Science, Yamaguchi University, Yamaguchi, Japan. He received his Doctor of Science degree in June 1982 from Kyushu University. His research interests are in the area of micellar solutions and scientific education.

K. Sato is a student in the Department of Chemistry, Kyushu University, Fukuoka, Japan.

Peter Scales is currently working for Applied Chemicals Australia on problems of mineral dust control. He completed a First Class Hons. degree in Physical Chemistry in 1981.

Daniel Schuhmann is Maitre de Recherches at the Laboratoire de Physicochimie des Systemes Polyphases, CNRS, Montpellier, France. He received his Dr. es Sciences in 1964 from Paris University, and has 50 publications in the areas of mass transfer, electrochemical kinetics and impedance, and adsorption. He was organizer of a Summer School (Interfaces Chargees) in 1978 and edited the proceedings.

Peter Schurtenberger is a member of the research group of Prof. W. Kanzig at the Laboratory of Solid State Physics, ETH, Zurich, Switzerland. He received his diploma in natural science in 1980 from the Swiss Federal Institute of Technology (ETH) in Zurich.

Torben Smith Sorensen is Professor at the Institute of Physical Chemistry, Technical University of Denmark, Lyngby, Denmark where he received his Ph.D. in Physical Chemistry in 1973. During 1974-1975 and 1981 he worked in the group of Prof. Ilya Prigogine at UniversiteLibre de Bruxelles. In 1977 he received the "P. Gorm-Petersen Memory Award" for contributions to thermodynamics and physical chemistry. He is the editor of the book Dynamics and Instability of Fluid Interfaces (1979) and has published more than 40 papers in a number of areas including thermodynamics and rational energetics, Marangoni instability of fluid interfaces and fluid films, biophysics of cell division, electrokinetics, and micelle thermodynamics.

E. Stenby is an M.Sc. student at the Fysisk-Kemisk Institut, Technical University of Denmark, Lyngby, Denmark.

Per Stenius is Director of the Institute for Surface Chemistry, Stockholm, Sweden and Associate Professor of Colloid Chemistry at the Royal Institute of Technology, Stockholm. He received his Ph.D. from the University Åbo Akademi (Turku), Finland in 1973. He is a member of the Royal Academy of Engineering Sciences in Sweden. He serves on the editorial boards of Journal of Colloid and Interface Science, and Colloids and Surfaces and is an associate member of the IUPAC commission on Colloid Science. He has published about 90 papers on association colloids, polymer latexes, colloidal stability and surface characterization.

Peter Stilbs is currently Associate Professor of Physical Chemistry at the University of Uppsala, Uppsala, Sweden. He received his Ph.D. from the Lund Institute of Technology in 1974. He has published about 60 papers dealing with the application of magnetic resonance methods for the study of physico-chemical phenomena.

Christoph Syldatk has been carrying out Doctor Thesis in Biochemistry and Biotechnology since 1981 about production of microbial surfactants at the Institute of Biochemistry and Biotechnology, Technical University of Braunschweig, West Germany.

Ilana Tamir is a Senior Research Associate at the School of Pharmacy, Hebrew University in Jerusalem studying the NMR spectra of purines, natural products and model membranes. Received D.Sc. degree in 1966 from the Technion-Israel Institute of Technology followed by postdoctoral research in theoretical chemistry at the University of Wisconsin and University of Notre Dame. Also carried out dynamic NMR studies of organic molecules at Bryn Mawr College, and NMR of enzymatic systems at the Institute for Cancer Research in Philadelphia.

Maria Tomoaia-Cotisel is Assistant Professor in the Department of Physical Chemistry of "Babes-Bolyai" University in Cluj-Napoca in Romania. She holds a Ph.D. degree in Chemistry and has been a Visiting Research Associate at the University of London. Her research interests involve the physical chemistry of surface films with substances of biological importance.

Umberto Tonellato is Professor of Organic Chemistry at the University of Padova, Italy. His research interests are micellar catalysis and physical-organic and bio-organic chemistry.

Emmanuel Tronel-Peyroz is Charge de Recherces at the Laboratoire de Physicochimie des Systemes Polyphases, CNRS, Montpellier, France. Received Dr. es Sciences in 1979 from the Reims University and has 15 publications in the areas of magneto-electrochemistry and adsorption of surfactants.

Pierre Vanel is Charge de Recherches at the Laboratoire de Physicochimie des Systemes Polyphases, CNRS, Montpellier, France. He received Dr. es Sciences in 1975 from the Montpellier University and has 20 publications dealing with a number of topics including adsorption of ions and surfactants.

Andre Viallard is Professor of Physical Chemistry at the University of Clermont-Ferrand, Aubiere, France which he joined in 1964. He obtained his Ph.D. degree from the University of Nancy, and during 1960-1964, he was Associate Professor at the Institute of High Studies, Tunis, Tunisia. He has been interested in chemical kinetics and electrochemistry, and his current research interest is in solution thermodynamics, especially micellar systems and microemulsions.

F. Wagner is Professor of Biochemistry and Biotechnology at the Technical University Braunschweig, West Germany. He obtained his Ph.D. degree in 1961 from the Technical University Stuttgart. He has published about 120 research papers dealing with the biosynthesis, regulation and production of microbial metabolites.

Robert G. Wallace is a graduate student working towards his Ph.D. at the University of Auckland, New Zealand. He received his M.Sc. in Chemistry in 1978. At present, he is studying the decomposition of esters in solutions of bile salts and the effect of the addition of bile salt stimulated lipase on these reactions.

Greg G. Warr is a Ph.D. student at the University of Melbourne, Australia where he is using light scattering and ultrafiltration techniques to study the solution chemistry and aqueous properties of polydisperse nonionic surfactants.

Robert Weinberger is a graduate student working towards his Ph.D. degree in Analytical Chemistry at Seton Hall University, S. Orange, NJ. He is the reciepient of the 1982 Society for Applied Spectroscopy National Student Award for work, in part, that is included in these proceedings.

Paul Yarmchuk is a graduate student studying towards his Ph.D. degree at Seton Hall University, S. Orange, NJ. His research is concerned with the use of micellar mobile phases in liquid chromatography.

Dino Zanette is Assistant Professor at the Federal University of Santa Catarina in Florianopolis, Brazil. He received his Ph.D. degree in Chemistry and has published in the area of organic chemistry. His current research interests include the mechanisms of pesticide degradation and micellar effects on chemical reactions.

INDEX

Pages 1-706 appear in Volume 1
Pages 707-1432 appear in Volume 2
Pages 1433-2174 appear in Volume 3

Acid Base Equilibria
  micellar effects on, 1048-1052
Acrylamide
  polymerization of, in water-
    swollen micelles of AOT,
    1897-1909
Activity Coefficients
  of sodium carboxylates in
    aqueous solution 857-871
Adsorption
  at charged interfaces, 1233-
    1245
  from amphiphilic solutions,
    1291-1298
  isotherms (various types),
    1339-1347
  of AOT on solid surfaces,
    1271-1288
  of carotenoid at air/water
    interface, 1344-1363
  of fibrinogen at chromium,
    1306-1307
  of glycolipids (microbial),
    1365-1375
  of nonionics on quartz, 1337
  of proteins at solid/liquid
    interface, 1301-1311
  of prothrombin at chromium,
    1308
  of SDS on chromium, 1294-1297
  of sodium caseinate on
    chromium, 1298
  on mercury, 1235-1237

Aggregate Geometries
  dependence of emission aniso-
    tropy on, 234
  schematic picture of, 232
Aggregates (amphiphilic)
  interfacial order in, 314
  molecular organization in,
    307-317
  water distribution in,
    310-314
Aggregation of Bile Salt Amphiphiles, 879-884
Alcohol-Hydrocarbon Solution
  thermodynamic properties of,
    1867-1875
Alcohol Metabolism
  effect of ginseng saponin on,
    2107-2110
Alcohols
  molar volumes of, 955-956
  solubilization of, 949-961
Alkanoate-Alcohol-Water, 15-19
Alkanoate-Water, 11-15
Alkyl Chain Mobility in
  Micelles and Microemulsions,
  107
Alkyl Chain Order in Micelles
  and Microemulsions, 107
Alkylsulfonic Acids (Long
  Chain)
  cmc of, 783
  Krafft point, 773-775
  micellization of, 771-787

Alkyltrimethylammonium Bromides
  interaction with sodium
    saccharin, 1391-1400
Ammoniumheptafluorononanoate/
  Water, 76
Analytical Chemistry
  use of micelles in, 1139-1157
  use of surfactants in, 1217-
    1228
Anesthetic-Membrane Interaction,
  2145-2155
Anionic Surfactants in Water
  binding of cations to, table,
    592
AOT
  adsorption of, on solid sur-
    faces, 1271-1288
  and water in oil microemul-
    sions, 1534, 1663-1674,
    1745-1757
  micelles
    reactivity studies in, 1471-
      1481
    solubilization of biopolymers
      in, 999-1011
    solubilization of α-chymo-
      trypsin in, 1075
    solubilization of lysozyme
      in, 1075
Apolar Solvents
  solubilization of biopolymers
    in, 999-1011
Apoprotein, 262
Aqueous Micelles (see also
  Micelles of)
  polymerization of surfactants
    in, 1948-1951
Aqueous Surfactants
  FT-IR studies of, 673-689
Artificial Photosynthesis
  role of membrane mimetic
    agents, 1982-1984
Association of Surfactants in
  Dilute Solution
  effect on their surface
    properties, 1313-1326
Axisymmetric Menisci
  shapes of, in determination of
    very low interfacial tension,
    2113-2120

Benzene
  solubilization in CPC, 939
  solubilization in SDS, 930
Benzophenone
  as a photoinitiator of vinyl
    polymerization, 1184-1188
Benzophenone (Triplet ^{3}BP)
  Dynamics
  in SDS micelles, 1179
Bilayer(s)
  packing of amphiphilic chains
    in, 279-304
  schematic drawing of, 283
Bilayer (Black) Lipid Membranes
  polymerization of, 1851-1853
Bile
  cholesterol solubilization
    and supersaturation in, 981
Bile Acids
  structure and nomenclature
    of, 877
Bile Salt(s)
  aggregation of, 879-884
  enterohepatic circulation of,
    878
  formation of, 876-879
  mixed micelles with lecithin,
    841-842
    schematic model for struc-
      ture of, 842
Bile Salt Amphiphiles
  kinetic applications of,
    884-893
Bile Salt-Lecithin Solutions
  hydrodynamic radius of,
    846-855
  polydispersity index of, 846
Bimolecular Reactions
  in functional, mixed and
    comicelles, 1052-1057
  micellar effect on, 1028-1039
Binding of Surfactants to Mixtures of Proteins, 1249-1268
Biochemical Reactions
  effect of ginseng saponins
    on, 2093-2112
Biological Membranes
  modeling of structure and
    function of, 1084-1086

Biopolymers
  phase transport of, 1004-1010
  solubilization in apolar
    solvents, 999-1011
Black Foam Films, 1323-1326
Black Liquor
  liquid crystal formation in,
    171
Birefringence Relaxation Curves
  in Dilute C_{16}PySal, 433-438
Brij 58, 604

Carotenoid Films at Air/Water
  Interface, 1349-1363
Carotenoids, Structural Formulae
  of, 1351
Cell Model for Micelles, 341
Cetylpyridinium Chloride
  solubilization of benzene in,
    930, 939
Cetylpyridinium (CPX) Micelles
  sphere-to-rod transition,
    specificity of counterions,
    805-822
Cetylpyridinium Salicylate
  birefringence relaxation curves
    of, 433-438
  dynamic light scattering of,
    438-440
  Kerr constant of, 436
  study of viscoelastic solutions
    of, 425-452
Cetyltrimethylammonium Bromide
  binding of, to proteins, 1252-
    1268
  diffusion coefficient of, 462
  fluorescence probe study of,
    568-571
  fluorescence study in, 657-659
  micelles of,
    effect of cosolvents, 789-801
    proton spin lattice relaxa-
      tion, 798
  solubilization of water pene-
    tration probe in, 602
  structure of, on basis of NMR
    spin lattice relaxation, 800

Cetyltrimethylammonium Bromide-
  n-Hexanol-Water Reverse Micellar System
  and preparation of colloidal
    iron boride particles,
    1483-1496
Cetyltrimethylammonium Chloride
  fluorescence probe study of,
    568-569
  fluorescence study in,
    657-659, 665, 668-669
  photoelimination quantum
    yields in, 588
Cetyltrimethylammonium (CTAX)
  Micelles
  sphere to rod transition,
    specificity of counterions,
    805-822
Charged Interfaces
  adsorption at, 1233-1245
Chromatography
  micelle-mediated, 1139-1157
Chromium
  adsorption of fibrinogen on,
    1306-1307
  adsorption of prothrombin on,
    1308
  adsorption of surfactants on,
    1291-1298
  critical surface tension of
    wetting of, 1295
Cholesterol Solubilization in
  Bile, 981-995
Colloidal Stability of
  Liposomes, 2039-2056
Complex Viscosity, 428
Conductivity (Electrical) Mea-
  surements in Microemulsions,
  1522-1526, 1597-1619,
  1805-1819
Conformation Changes due to
  Surfactant Association
  Raman scattering and NMR
    studies, 517-525
Contact Angles in Presence of
  Nonionic Surfactants, 1331,
  1337
Core Radius of Sodium Octa-
  noate Micelle, Table, 494
Correlation Length, 54

Cosolvents
   effect of, on the formation of
      micelles of CTAB, 789-901
Cosurfactant
   requirements of, in tertiary
      oil recovery, 2124
   solubilization and disruption
      of micelles, 1655-1656
Cosurfactant Chemical Structure
   influence on the phase diagram
      and electrical conductive
      behavior of microemulsions,
      1583-1621
Counterions, Specificity of
   in sphere-to-rod transition of
      micelles, 805-822
Critical Binary Mixtures
   light scattering from, 473-475
Critical Coagulation Concentration for Vesicles, 2046
Critical Demicellization Concentration, 134
Critical Micelle, 721
Critical Micellization Concentration (c.m.c.)
   and relation to HLB, 1932
   as a critical nucleation
      concentration, 719-722
   IUPAC definition of, 881
Critical Micellization Concentration (c.m.c.) of
   alkylsulfonic acids, table, 783
   alkyltrimethylammonium
      bromides, 1394
   calcium-n-heptyl sulfate, 338
   cetylpyridinium chloride, 937
      effect of solubilizate, 937
   cetylpyridinium salicylate, 451
   cetyltrimethylammonium bromide,
      795, 1251
   dibutylphosphate, 502
   ionic micelles, 722
   manganese (II) dodecyl sulfate,
      970
   mixtures of sodium laurate with
      sodium perfluorooctanoate,
      table, 138
   nonionic micelles, 723
   potassium trans-3-hexenoate,
      517

Critical Micellization Concentration (c.m.c.) of (cont.)
   Septonex, 1224
   sodium abietate, 161
   sodium alkylbenzene
      sulfonates, 406, 784
   sodium decyl sulfate, 902
   sodium deoxycholate, 883
   sodium dodecyl sulfate, 905,
      937, 970, 1251, 1380
      effect of solubilizate, 937
   sodium n-octanoate, 338, 494,
      755
   sodium octyl sulfate, 1209
   sodium oleate, 163
   sodium taurocholate, 987
   zinc (II) dodecyl sulfate,
      970
Critical Micellization Temperature, 676, 687-689
Critical Nucleation Concentration
   c.m.c. as a, 719-722
Critical Point in Amphiphile
   Solutions, 472
Critical Points in Micro-
   emulsions, 1737-1743
Critical Surface Tension of
   Wetting, 1295
Critical Temperature, 472
β-Cryptoxanthin and Egg Lecithin
   mixed monolayer, 1360
Cytochrome C
   photoreduction by N-methyl-
      phenothiazine, 1479-1480

Debye-Hückel Ionic Micelle
   aggregation number of, by ZST
      principle, 719
Decorated Lattice Gas Models,
   48-56
Diacetylene, Self Polymeriza-
   tion of, 1963
Dialkylphosphatidylchlorine-
   Water, 7-9
Dibutylphosphate Micellar
   Aggregates
   conformations of, 508
   NMR and ESR studies of,
      501-515

Dibutylphosphate Micellar
  Aggregates
  size distribution of, 510
  thermodynamic parameters of,
    table, 508
Dielectric Properties of W/O
  Microemulsions, 1709-1727
Differential Scanning Calori-
  metry of 75 wt% SHBS, 180
Diffusion Coefficient
  for sodium n-octanoate micelle,
    table, 494
  of amphiphiles, for some cubic
    liquid crystalline phases,
    table, 224-225
Dioctadecyl-dimethylammonium
  Chloride
  photoelimination quantum yield
    in, 588
Dioleoylphosphatidylcholine,
  144
Dipalmitoyllechithin
  photoelimination quantum yield
    in solution of, 588
Dissolution of Alkylsulfonic
  Acids and Their Sodium Salts,
  771-787
Divalent Ions (Mn^{+2}, Ni^{+2})
  paramagnetic relaxation induced
    in octanoate micelles by,
    527-539
DNA
  circular dichroism spectra of,
    1003
  solubilization of, 999
Dodecylammonium Chloride/D$_2$O
  Lyotropic Mesophases, 79-91
Dodecylammonium Propionate
  aggregate of, in apolar media,
    669-670
  aggregation number of, 670
N-Dodecylbetaine
  association of, 369-370
Dodecyldimethylammonium Bromide
  Micelles
  aggregation number of, 835
Dodecyldimethylammonium Chloride
  Micelles
  aggregation number of, 835
  light scattering results of,
    826-830

N-Dodecylhexaoxyethylene-Water
  Solution
  dynamic light scattering
    data, 478
  mass diffusion coefficient,
    478
  micellar weight, 478
  osmotic isothermal compres-
    sibility, 477
  scattered light intensity,
    477
Donnan Effect in Charged
  Micelles, 748
Double-tailed Surfactants
  scattering investigations of,
    405-422
Dyes (Organic)
  interaction with cationic
    surfactants, 1217
Dynamic Light Scattering, 157,
  417-421, 438, 455-468, 1663,
  1690
  and microemulsion study,
    1663, 1690
  micellar diffusivities by,
    455-468

Electrochemical Model
  of micellar ionic environ-
    ment, 815-818
Electrophoretic Laser Light
  Scattering
  and zeta potential and charge
    density of microemulsion
    droplets, 1693-1706
  experimental set up for, 1697
Electrostatics of Micellar
  Systems, 897-913
Ellipsometry
  and adsorption of proteins,
    1301-1311
  and adsorption of surfac-
    tants, 1291-1298
Emission Anisotropy of Lipid
  Systems, 229-235
Emulsions
  cohesive energies in, 1933-
    1934
  phase inversion of, relation
    to HLB, 1938-1939

Enzyme Catalyzed Reactions,
  1069-1087, 2102-2105
  effect of ginseng saponin on,
    2102-2105
Enzyme Kinetics in Microemulsions, 1754-1755
EPR Study of Interface Dynamics,
  1485
ESR Study of
  dibutylphosphate micellar
    aggregates, 501-515
  spin labels in surfactant
    micelles, 541-555
Ethyleneglycol Dodecyl Ethers
  interaction between water and
    ethylene oxide groups in,
    93-103
Excluded Volume Effect in
  Micelles, 748

Fibrinogen
  adsorption on chromium,
    1306-1307
Fluorescence Probe Study of Oil-
  in-Water Microemulsions, 1627-
  1646
Fluorescence Quenching
  equilibria studies in ionic
    micelles, 645-660
  in micellar systems, 663-670
  study of aggregation in non-
    ionic micelles, 637-642
  study of size of SDS micelles,
    613-624
Fluorinated Fatty Acids
  liquid crystalline structures
    in, 69-77
Foam Films (Black), 1323-1326
Freeze-Etching Electron Microscopy
  use of, in lipid systems, 237-
    255
Freeze-Fracture
  use of, in lipid systems, 237-
    255
Fromherz Micelle (see also Surfactant-Block Model of
  Micelle), 351
FT-IR Studies of Aqueous Surfactants, 673-689

Functionalized Counterion Surfactants, 1169
Functionalized Micelles
  ester hydrolysis by, 1169-
    1175
  rate enhancement by, 1096
Functional, Mixed and Co-
  micelles
  bimolecular reactions in,
    1052-1059
Functionalized Quaternary
  Ammonium Ions, 1096-1103

Gel-Liquid Crystal Transition
  Temperature, 130
Germanium Surface
  adsorptioin of AOT on, 1278-
    1283
Ginseng Saponins
  effect of, on biochemical
    reactions, 2093-2110
Glycolipids (Microbial)
  interfacial active properties
    of, 1365-1376

Hard Sphere Concept, 1534
Hard Sphere Radius for Sodium
  Octanoate Micelles, (table),
  494
Hartley Micelle, 347
Heptadecafluorononanoic Acid,
  70
  partial specific volume of,
    71
Hexadecyltrimethylammonium Bromide
  solubilization of alcohol in,
    953
Hexaethyleneglycol mono n-Do-
  decylether
  pyrene fluorescence in, 635-
    642
Hexapus System, 356
High Density Lipoproteins, 264-
  268
High Performance Liquid Chromatography, 1141

^{2}H NMR
  in dodecylammonium chloride/
    D_2O lyotropic mesophases,
    79-91
  in liquid crystalline phases,
    93-103
  in monoolein-dioleoylphospha-
    tidylcholine-water, 143-151
  in surfactant systems with di-
    valent counterions, 193
Hydrodynamic Radius, 466
  for bile salt-lecithin solu-
    tions, 846-855
  versus temperature for CTAX
    micelles, 808-809
Hydrophile-Lipophile Balance
  (HLB)
  an overview, 1925-1943
  and emulsion stability, 1941
  and equivalent alkane carbon
    number (EACN), 1942
  and phase inversion temperature
    of emulsions, 1938-1940
  and relation to c.m.c., 1932
  and relation to Winsor's R,
    1935
  group numbers, table, 1929
  methods of determination of,
    1927
  various approaches to, 1928-
    1943

Indicator Equilibria, 1053
Intermediatory Metabolism
  effect of ginseng saponin on,
    2105-2107
Intermicellar Interactions
  effect of, on micellar
    diffusivities, 455-468
Intermicellar Potential, 1857-
  1862
Inverse Micelles (see Reverse
  Micelles)
Internal Reflection Infrared
  Spectroscopy
  and adsorption of AOT on solid
    surfaces, 1271-1288
Ion Exchange Model
  application to O/W microemul-
    sions, 1911-1921

Ionic Micelles
  second virial coefficient of,
    751-752
Iron Boride Colloidal Particle
  Formation in CTAB-n-Hexanol-
  Water Reversed Micellar Sys-
  tem, 1483-1496
Isothermal Osmotic Compress-
  ibility, 54

Kerr Constant, 433
Kinetic Applications of Bile
  Salt Amphiphiles, 884-893
Kosower "Z" Values, 594
Krafft Point, 676
  of alkyl sulfonic acids and
    sodium salts, 773-775

Langmuir's Principle, 308
Laser Light Scattering
  study of nonionic micelles in
    aqueous solution, 471-484
Lateral Diffusion Coefficients
  (table), 222-223
Latices Formed by Polymerization of Acrylamide in Water-
  Swollen Micelles, 1897
Lecithin
  mixed micelles with bile
    salts, 841-842
Light Scattering (see also
  Quasielastic Light Scattering and Dynamic Light
  Scattering)
  and mutual and self diffusion
    coefficients of microemul-
    sions, 1729-1735
  by liquid surfaces, 1991-2012
  from critical binary mix-
    tures, 473-475
  from micellar solutions, 826-
    830
  from microemulsions, 1851-1857
  from solutions of sodium-
    octanoate micelles, 487-499
  study of vesicles coagulation
    by, 2081-2092
Lipid Monolayers, 2015-2021
Lipid Systems
  emission anisotropy of, 229-
    235

Lipid Systems (cont.)
  phase analysis and structure
    determination of, 237-255
Lipoprotein(s)
  classification of, 261
  low density, 268-273
  structure of, 259-273
Liposomes (see also Vesicles),
    183, 1980, 2039-2056
  colloidal stability of, 2039-
    2056
Liquid Crystals
  polymerization of, 1951-1953
Liquid-Liquid Interfacial Tension
    (Very Low)
  determination of, from shapes
    of axisymmetric menisci,
    2113-2120
Liquid Surfaces
  light scattering by, 1991-2012
Loss Modules, 429
Luminescence (Micelle Mediated),
    1139-1157
Lyotropic Nematic Phase
  neutron diffraction, 65-66
  x-ray diffraction, 62-65
Lysozyme
  interaction with SDS, 1403-1415

Macroions
  interaction with small ions,
    749
Manganese (II) Dodecyl Sulfate
  solubilization of phenothiazine
    in, 974
Membrane-Anesthetic Interaction,
    2145-2155
Membrane Mimetic Agents, 1947
Membranes (Model)
  and lamellar phases, 220-221
  surface charge density evalua-
    tion in, 2015-2021
Menger Micelle, 347, 348-350
Metastable Dispersions, 981
Micellar
  catalysis (see Micelle Cata-
    lyzed Reactions)
    near the c.m.c., 1059-1061
    validity of the pseudophase
      model for, 1107-1118

Micellar (cont.)
  chromatography, 1139-1157
    efficiency of, 1148-1150
  chromatograms, 1145, 1150,
    1153
  diffusivities, 455-468
  effects on reaction rates and
    equilibria, an overview,
    1015-1062
  enzymology, 1069-1087
  hydrodynamic radius, 466
  ionic environment
    electrochemical model of,
      815-818
  models
    mass action law model, 862-
      865, 870-871
    pseudophase model, 859-862,
      866-869
  shape, changes in
    from viscosity and ultra-
      sonic studies, 956
  solutions, osmotic pressure
    in, 2034
  solutions, structure in
    a Monte Carlo study, 337-
      345
  structure, studied by NMR
    and optical spectroscopy,
    565-582
  surface dissociation con-
    stants, table, 908
  systems, electrostatics of,
    897-913
Micelle(s)
  demixing, 132
  characterization of, by
    multimethod, 347-357
  charge density, 1240
  fluorescence quenching in,
    663-670
  ion binding to, 591-593
  mediated luminescence and
    chromatography, 1139-1157
  mixed, 130, 165, 841-842
    sodiuim oleate/sodium
      abietate, 165
    bile salts and lecithin,
      841-842
      schematic model for
        structure of, 842

Micelle(s) (cont.)

molecular organization in, 307–317

neutron scattering on, 373–402

NMR studies of probe location in, 569–572

packing of amphiphile chains in, 279–304

schematic drawings of, 283

shapes and structure of spherical and rod like, 836

size distribution
  computation of, from experimental measurements, 731–742
  moments of, 734–736

solubilization of $Py(CH_2)_n COOH$ in, 575–579

stabilized room temperature phosphorescence, 1150

structure
  bilayer model, 322–324
  droplet model, 322–324
  Fromherz model, 321–334, 350–351
  Hartley model, 347–348
  Menger model, 348–350
  surfactant-block model, 321–334, 350–351

surface tension of, 712–714

to vesicle transition, in solutions of bile salts and lecithin, 841–855

tracer self-diffusion studies of, 359–371

water penetration probes of, table, 601

water penetration study of, 565–582, 585–596, 599–609

Micelle-Catalyzed Reactions; or Micellar Effects on Chemical Reactions (same as Micellar Catalysis), 1015–1062, 1069–1087, 1093–1103, 1107–1118, 1121–1136, 1139–1157, 1159–1167, 1169–1175, 1177–1189, 1191–1205, 1207–1214, 1217–1228

acid-base equilibria, 1048–1052

bimolecular reactions with electrophiles, 1038

Micelle-Catalyzed Reactions (cont.)

bimolecular reactions with hydrophilic nucleophiles, table, 1033–1035

bimolecular reactions with organonucleophiles, table, 1036

by cationic micelles, 1101–1103

by CTAB micelles, 1107–1118

by functional, mixed and co-micelles, 1052–1059

by functionalized micelles, 1096–1100

by reverse micelles
  enzyme catalyzed reactions, 1069–1087

dehydrochlorination of DDT, 1112

determination of metal ions, 1217

electron transfer reactions, 908–912, 1159–1167

equilibria shifts
  quantitative treatment for, 1191–1205

ester hydrolysis of functionalized counterion surfactants, 1169–1175

esterolysis of p-nitrophenyl-acetate and hexanoate, 1173

hydrolysis of lithium ethyl-p-nitrophenyl phosphate, 1115

hydrolysis of lithium ethyl p-nitrophenyl ethyl phosphate, 1116

hydrolysis of lithium p-nitrophenyl ethyl phosphate, 1114

ion exchange reactions, 1019–1024

metal (Ni^{+2}) complex formation, 900–906

Ni(II)-PADA reaction, 1207–1214

photoelectron transfer from magnesium tetraphenylporphyrin to viologen, 1471–1479

Micelle-Catalyzed Reactions
  (cont.)
  photoreduction of cytochrome C
    by N-methylphenothiazine,
    1479-1480
  reaction of OH$^-$ and F$^-$ with
    p-nitrophenyldiphenyl-
    phosphate, 1913
  reactivity of nucleophiles
    derived from the dissociation
    of weak acids, 1121-1136
  temporal behavior of triplet
    benzophenone, 1177-1189
  thiolysis reactions, 1126-
    1134
Micelles of
  alkylsulfonic acids and their
    sodium salts, 771-787
  AOT
    polymerization of acryl-
      imide in 1897
    reactivity studies in, 1471-
      1481
    solubilization in, 999-1011
  calcium n-heptylsulfate, 338
  cetylpyridinium (CPX)
    sphere-to-rod transition,
      specificity of counterions,
      805-822
  cetylpyridinium chloride, 607
  cetylpyridinium salicylate, 425
  cetyltrimethylammonium (CTAX)
    hydrodynamic radius vs. tem-
      perature, 808
    sphere to rod transition,
      specificity of counterions,
      805-822
  cetyltrimethylammonium bromide,
    568-571
    effect of cosolvents on, 789-
      801
    micellar catalysis in, 1107,
      1121
  cetyltrimethylammonium chlo-
    ride, 568-569
  dibutylphosphate
    NMR and ESR studies of, 501-
      515
  dodecylammonium propionate,
    669-670

Micelles of (cont.)
  dodecyldimethylammonium
    bromide,
    aggregation number of, 835
  dodecyldimethylammonium
    chloride, 827-835
    aggregation number of, 835
  n-dodecylhexaoxyethylene
    glycol monoether, 478
  dodecyltrimethylammonium bro-
    mide and chromatography,
    1146
  ionic surfactants
    fluorescence quenching
      equilibria studies in,
      645-660
    sphere-to-rod transition
      of, 825-838
  nonionic surfactants
    laser light scattering of,
      471-484
  octylglucoside, 602
  n-octyltetraoxyethylene
    glycol monoether, 471
  potassium trans-3-hexenoate,
    525
  sodium alkylbenzene sulfon-
    ates, 406-422
  sodium alkyl sulfates,
    897-913
    electrostatic properties
      of, 897, 906-908
    metal complex formation in,
      900-906
  sodium decanoate, 956
  sodium dodecyl sulfate, 373,
    456, 565, 572, 581, 587,
    602, 607, 613-624, 635,
    646, 677, 751-753, 834,
    905, 1634
  sodium hexadecyl sulfate, 677
  sodium n-octanoate, 338, 487-
    500, 527-539, 755
    light scattering from, 487-
      500
    NMR study of, 527-539
  sodium octanesulfonate
    Ni(II)-PADA reaction in,
      1210

Micelles of (cont.)
  sodium p-octylbenzene sulfon-
    ate, 359, 362-368
  sodium octylsulfate
    Ni(II)-PADA reaction in,
    1207-1214
  surface active components of
    wood, 153-173
  tetracaine, 2154
Micellization [see Micelles of,
  Micellar, and Micelle(s)]
Micellization as a Nucleation
  Phenomenon, 709-728
Microemulsion(s), 32, 107, 1501-
  1529, 1533-1548, 1551-1580,
  1583-1621, 1627-1646, 1651-
  1661, 1663-1674, 1675-1691,
  1693-1706, 1709-1727, 1729-
  1736, 1737-1743, 1745-1757,
  1759-1778, 1781-1787, 1789-
  1802, 1805-1819, 1821-1827,
  1829-1841, 1843-1864, 1867-
  1879, 1881-1896, 1897-1909,
  1911-1921
  an overview, 1501-1529
  application of ion exchange
    model to O/W, 1911-1921
  conductivity and dielectric
    measurement of, 1522-1526
  cosurfactant type, 1657-1658
  definition of, 1503, 1653
  electrical conductive behavior
    of
    influence of the cosurfactant
      chemical structure, 1597-
      1619
  enzyme kinetics in, 1754-1755
  fluctuation and stability of,
    1781-1787
  interfacial tension with excess
    disperse phase, 1558- 1559,
    1572-1576
  inversion conditions for, 1829
  ionic, properties of, 1867-1879
  light scattering and visco-
    metric investigations in,
    1897-1909
  light scattering by, 1851-1857
  micellar interactions in, 1843-
    1864

Microemulsion(s) (cont.)
  mutual and self diffusion
    coefficients of, 1729-1736
  nonionic, structure of, 1759-
    1768
  oil in water
    fluorescence probe study
      of, 1627-1646
  phase behavior of, 1551-1580
  phase continuity and drop
    size in, 1829-1841
  phase diagram features
    influence of the cosurfac-
      tant chemical structure,
      1583-1597
  phase inversion, 1576-1578
  phase studies and conduc-
    tivity measurements in,
    1805-1819
  percolation and critical
    points in, 1737-1743
  polymerization in, 1898
  polymerization of surfactants
    in, 1951-1953
  pressures in a, 1557
  pseudophases, 1789-1802
  reactions in, 1911-1921
    exchange of $Fe(CN)_6^{-4}$, 1753
    hydrolysis of AOT, 1751-1752
    metal-ligand complexation
      in, 1756-1757
  scatterring methods for char-
    acterization of, 1513-1522
    neutron scattering, 1517-
      1520
    quasielastic light scatter-
      ing, 1520-1522
  structure and dynamics of,
    1527-1529, 1745-1757
  structure, comments on, 1659-
    1661
  surface tension measurements
    in, 2009-2012
  theories of formation and
    stability of, 1504-1513
    mixed film theories, 1504-
      1506
    solubilization theories,
      1506-1508
    thermodynamic theories,
      1508-1513

Microemulsion(s) (cont.)
  thermodynamic stability of,
    origin of, 1555-1559
  transport of solubilized
    substances by, 1881-1896
  use of multicomponent self-
    diffusion data in character-
    ization of, 1651-1661
  virial coefficients of, 1855-
    1856
  viscosity measurements of,
    1526-1527
  water-in-oil, 1533-1548, 1663-
    1674, 1676-1691, 1709-1727,
    1848
    dielectric properties of,
      1709-1727
    photocorrelation technique in
      the investigation of, 1663-
      1674
    position annihilation tech-
      nique in the study of,
      1675-1691
    schematic representation of,
      1848
  zeta potential and charge den-
    sity of drops of, 1693-1706
Microlithography
  use of polymeric multilayers
    in, 1981
Microphases, 1535
Mixed Micelles, 130, 165, 841-842
  bile salts and lecithin, 841-
    842
    schematic model for structure
      of, 842
  sodium oleate/sodium abietate,
    165
Molecular Aggregates
  spin label study of, 559-563
Molecular Pitch of Rodlike
  Micelles, 835
Monolayers and Multilayers
  polymerization of surfactants
    in, 1953-1963
Monolayers on Water, 2007-2009
Monoolein, 144
Monte Carlo Study of Structure in
  Micellar Solutions, 337-345

Monte Carlo Simulation Technique
  investigation of P-B approx-
    imation, 2023-2037

Neutron Diffraction Pattern
  of nematic phase, 65-66
Neutron Small Angle Scattering
  and evaluation of structure
    of plasma lipoproteins,
    259-273
  and investigation of double
    tailed surfactants, 405-422
  and structure of nonionic
    microemulsions, 1759-1778
  in microemulsions, 1517-1520,
    1746-1750
  on sodium octanoate micelle,
    492
  on surfactant micelles in
    water, 373-402
NMR
  ^{13}C studies and alkyl chain
    mobility and order, 107
  ^{13}C study of conformational
    change of surfactants due
    to association, 517-525
  ^{19}F study of size of
    micelles, 1485
  ^{17}O in dodecylammonium/D_2O
    lytropic mesophases, 79-91
  ^{31}P, ^{13}C, ^{23}Na study of di-
    butylphosphate micellar
    aggregates, 501-515
  study of micellar structure
    and water penetration,
    565-582
  study of structure and dyna-
    mics of microemulsions,
    1527-1529
  study of water penetration
    probes, 599-609
Nickel (II)-PADA Reaction as a
  Solubilization Probe, 1207-
  1214
Nitrogen Hyperfine Coupling
  Constant, 545-546
Nitroxide Radicals, table, 542
  ESR line shape of, 543-545

Nonionic Micelle(s)
  aggregation number of, from ZST
    principle, 714-717
  fluorescence quenching aggrega-
    tion number in, 627-642
  laser light scattering study
    of, 471-484
  model of, 6
  second virial coefficient of,
    750-751
Nonionic Microemulsions
  structure of, by small angle
    neutron scattering, 1759-1768
Nonionic Surfactant(s)
  and effect on the wettability
    of solid surfaces, 1329-1337
  as additives for SDS, 624
  average molecular cross section
    area of, 1333
  contact angles in presence of,
    1331, 1337
  phase studies and conductivity
    measurements in systems
    containing, 1805-1819
  polydispersity of, 1330
  surface tension of, 1332
  ultrafiltration of, 1331
  water mixtures
    lattice models for, 35
    phase equilibria in, 35
p-Nonylphenolpolyoxyethylene-
  glycol-Water System
  reverse structures in, 1463-
    1468
Nucleation Phenomenon
  micellization as a, 709-727
Nucleic Acids
  solubilization in reverse (AOT)
    micelles, 999-1011

Octaethyleneglycol Dodecylether
  association of, 368-369
n-Octyltetraoxyethyleneglycol
  Monoether
  phase diagram of, 476
Optical Spectroscopy
  micellar structure and water
    penetration studied by,
    565-582
Organized Surfactant Assemblies
  polymerization of, 1947-1984

Osmotic Pressure
  data to compute micelle size
    distribution, 731-734
  in micellar solutions, 2034
  theory of, 746-747
Oxyethylated Anionic Surfactants
  and electrolyte tolerance of
    petroleum sulfonate, 2121-
    2142

Palmitic Acid Methyl Ester,
  Spin Labelled, 560
Pentanol
  solubilization of, in water +
    sodium n-octanoate+pen-
    tanol, 755-768
Percolation in Microemulsions,
  1737-1743
Petroleum Sulfonates
  electrolyte tolerance of,
    2121- 2142
Phase Diagram of
  AOT-water-decane, 1822
  AOT-water-toluene, 1901
  aqueous solutions of dimyr-
    istoylphosphatidylcholine,
    2061
  C_8E_4-water, 476
  $C_{10}E_4$-water, 46, 52, 95
  $C_{10}E_4$-water-$C_{16}H_{34}$, 39, 41-92
  $C_{10}E_5$-water, 47
  $C_{10}E_5$-water-$C_{16}H_{34}$, 43
  $C_{12}E_4$-water, 1809
  $C_{12}E_4$-water-heptane, 1813
  $C_{12}E_6$-water, 95
  calcium octyl sulfate-decan-
    ol-heavy water, 198
  decanol-water-sodium capry-
    late, 36
  dipalmitoyl PC-water, 9, 11
  di-2-ethylhexyl sulfosuc-
    cinate-heavy water, 199
  dodecylammonium chloride-D_2O,
    84
  glycocholate-lecithin systems
    in buffer, 850
  MO/DOPC/D_2O, 149
  monoglyceride-water, 45
  NaAb/NaOl/aqueous NaCl, 160

Phase Diagram of (cont.)
  potassium caprate-octanol-
    water, 15, 17
  potassium myristate-water, 13
  SHBS/brine/alcohol/hydrocarbon,
    187
  SHBS/brine/hydrocarbon, 187
  soap-water-salt, 37
  sodium abietate/abietic acid/
    NaCl solution, 168
  sodium dodecylsulfate/SHBS, 139
  sodium octanoate-n-decanol-
    water, 209, 216
  sodium octanoate-octanoic
    acid-water, 120
  sodium octylbenzene sulfon-
    ate-pentanol-water-decane,
    1826
  sodium octylsulfate-decanol-
    water, 198
  sodium oleate/abietic acid/
    NaCl solution, 169
  sodium oleate/oleic acid/NaCl
    solution, 167
  taurocholate-lecithin systems
    in NaCl. 843
  Tritan X-100 with phospha-
    tidylcholine, 140
  Tritan X-100 with spingomyelin,
    140
  water-AOT-n-heptane, 1665
  water-AOT-isooctane, 1534
  water-C_6H_6-C_6E_0, 33
  water-decane-$C_{12}(EO)_4$, 1761
  water-n-decane-C_4E_1, 29
  water-n-decane-C_8E_3, 32
  water-dimyristoylphospha-
    tidylcholine/CTAB, 2045
  water/SDS/alkanols/benzene,
    1590-1591
  water/SDS/isomeric pentanols/
    benzene, 1595-1596
  water/SDS/2 methyl-2-butanol/
    benzene, 1588
  water-sodium octanoate (cap-
    rylate)-pentanol, 758
  water-sodium oleate-alcohol,
    1868
  water-toluene-SDS-n-butanol,
    1793

Phase Equilibria in Surfactant-
  Water Systems, 3-21, 23-34,
  35-36, 59-67, 69-77, 79-91,
  93-103, 129-141, 143-150,
  153-173, 175-188, 193-202,
  205-216
  principles of, 3-21
Phase Inversion Temperature of
  Emulsions
  relation to HLB, 1938-1940
Phase Separation at the Inter-
  face, 1339
Phase Transport of Biopolymers
  using Reverse Micelles, 1004-
  1011
  Phenothiazine
    solubilization in anionic
    surfactant micelles,
    963-977
Phosphatidylcholine
  phase diagram with Triton
    X-100, 140
Phospholipids
  chemical structure of, 2062
Phosphorescence Detection of
  HPLC, 1150-1157
Photochemical and Photophysical
  Probes
  and solubilization and water
    penetration, 585-596
Photoconductivity
  and polymeric multilayers,
    1981
Photocorrelation Techniques
  and water-in-oil microemul-
    sions, 1663-1674
Photopolymerization, 1184-1188
  in SDS micelles, 1187
Plasma Lipoproteins
  structure of, 259-273
^{31}P NMR in Monoolein-Dioleoyl-
  phosphatidylcholine-Water,
  143-151
Poisson-Boltzmann Approximation
  breakdown of, in polyelectro-
    lyte systems, 2023-2037
Polarized Emission Studies of
  Cubic Phases and Model Membranes, 219-235

Polyacrylamide
  interaction with surfactants,
    1377-1389
Polyelectrolyte Systems
  breakdown of the Poisson-
    Boltzmann approximation in,
    2023-2037
Polymeric Organized Assemblies,
  Utilization and Potential in
  artificial photosynthesis,
    1982-1984
  membrane modeling, 1980
  photoconductivity-microlitho-
    graphy, 1981-1982
  reactivity control, 1984
  target directed drug carrier-
    molecular recognition,
    1980-1981
Polymerization of Organized
  Surfactant Assemblies, 1947-
  1984
  of surfactants in aqueous
    micelles, 1948-1951
  of surfactants in microemul-
    sions, liquid crystals and
    BLM's, 1951-1953
  of surfactants in monolayers
    and multilayers, 1953-1963
  of surfactants in vesicles,
    1963-1979
Positron Annihilation Techni-
  ques in the Study of W/O
  Microemulsions, 1675-1691
Potassium Picrate
  transport by W/O microemul-
    sions, 1889-1890
Potassium trans-2-hexenoate
  (PT2H)
  ^{13}C chemical shifts of, 523
Potassium trans-3-hexenoate
  (PT3H)
  ^{13}C chemical shifts of, 523
  molecular conformationss of,
    519
  Raman spectra of, 520
Premicellization, 1314
"Primitive" Ionic Micelles
  aggregation number of, 717- 719
Proteins
  adsorption of, at solid/ liquid
    interfaces, 1301- 1311

Proteins (cont.)
  binding of surfactants to,
    1249-1268
Prothrombin
  adsorption on chromium, 1308
Proton-Spin Lattice Relaxation
  Study of CTAB-Water, 798
Pseudophase Ion Exchange Model,
  1017-1018
Pseudophase Model
  an adaptation of, 1191-1205
  validity of, for micellar
    catalysis, 1107-1118
Pyrene
  fluorescence, 635-642
  transport by microemulsions,
    1887-1889
Pyrenyl Probes, 567

Quartz
  adsorption of nonionics on,
    1337
Quasielastic Light Scattering
  in micelles of CPX and CTAX,
    807-811
  in microemulsions, 1520-1522,
    1897-1904
  study of micelle to vesicle
    transition, 841-855

Raman Scattering Study of Conformational Change of Surfactants due to Association,
  517-525
Reactive Counterion
  Surfactants,
  1039-1045
Reverse Micelles, 1435-1449,
  1453-1460, 1463-1468, 1471-
  1481, 1483-1496
  and preparation of colloidal
    iron boride particles,
    1483- 1486
  dynamics of, 1453-1460
  enzyme catalyzed reactions
    in, 1069-1087
  ^{19}F chemical shifts in, 1491
  fluorescence of added probe
    in, 663

Reverse Micelles (cont.)
  kinetic consequences of the self-association model in, 1435-1449
  kinetics of formation of, 1458-1459
  kinetics of solubilization by, 1459-1460
  molecular mechanism of entrapment of protein into, 1073-1074
  phase transport of biopolymers using, 1004-1011
  reactivity studies in
    decomposition of p-nitrophenylacetate in benzene solutions of alkyldiamine bis(dodecanoates),1440-1449
    photoelectron transfer from magnesium tetraphenylporphyrin to viologen, 1471-1479
    photoreduction of cytochrome C by n-methylphenothiazine, 1479-1480
  solubilization of biopolymers in, 999-1011
  substrate specificity of enzymes trapped into, 1083-1084
  technique of entrapment of protein (enzyme) into, 1071-1073
RNA
  solubilization in reverse (AOT) micelles, 1002

Saccharin (sodium)
  interaction with alkyltrimethylammonium bromides, 1391-1400
Saponins
  effect on biochemical reactions, 2093-2110
Scattering Methods for Characterization of Microemulsions, 1513-1522
Second Virial Coefficient (B_2), 747-749
  of ionic micelles, 751-753
  of nonionic micellar solutions, 750-751

Self-Association Model in Reverse Micelles, 1435
Self-Diffusion Coefficient, 360
  of microemulsions, 1729-1737
Self-Diffusion Studies
  and microemulsion structure, 1651-1661
  of solubilization equilibria, 917-922
  of surfactant association, 359-370
Small Angle Neutron Scattering
  of CPySal, 439
  of double tailed surfactants, 405-422
Sodium Abietate
  micelles, hydrodynamic radii of, 162
  solubility in NaCl solutions, 158
  surface tension of, 161
Sodium Alkylbenzene Sulfonates, Homologous Series of, 406
Sodium Alkyl Sulfonates
  c.m.c. of, 774, 784
  Krafft point of, 773-775
  solubility of, 774, 779-780
  thermal transitions of, 773, 776-778
  x-ray diffraction of, 773-774, 778-779
Sodium Borohydride
  and reduction of Fe(III) ions, 1483-1496
Sodium Carboxylates
  activity coefficients of, 857- 871
Sodium Caseinate
  adsorption on chromium, 1298
Sodium Decanoate
  solubilization of alcohol in, 953
Sodium Decyl Sulfate, 61
Sodium Deoxycholate
  solubilization in, 927
Sodium Dodecyl Sulfate
  adsorption on chromium, 1294-1297
  binding to proteins, 1253-1268

Sodium Dodecyl Sulfate (cont.)
  interaction with lysozyme,
    1403-1415
  interaction with polyacryl-
    amide, 1377-1389
  surface tension of, 1315
Sodium Dodecyl Sulfate Micelles
  aggregation number of, 615,
    667, 1634-1642
  and chromatography, 1144-1146
  and counterion association
    effect of alcohols, 959-960
  and nonionic surfactant
    additives, 624
  and reaction rate of Ni^{+2}- PADA
    complexation, 905
  and solubilities of alcohols,
    954
  diffusion coefficient of, 465
  dynamics of ^{3}BP in, 1179
  electron transfer reactions in,
    908-912, 1159-1167
  fluorescence quenching of
    pyrenyl probe molecules,
    table, 581
  fluorescence study in, 648-
    659, 665-668
  FT-IR study of, 678
  ^{1}H NMR spectra of, 572
  hydrophobic radius of, 620
  internal structure of, 399-401
  light scattering for, 838
  neutron scattering on, 373-402
  photoelimination quantum yields
    in, 588
  photopolymerization in, 1187
  pyrene fluorescence in, 635
  radius of, 617
  second virial coefficient of,
    752
  shape of, 395-399
  size of, with various addi-
    tives, 613-624
  solubilization in
    of alcohols, 953
    of benzene, 930
    of phenothiazine, 967-977
    of polar molecules in, 577
    of water penetration probes,
      602

Sodium Dodecyl Sulfate Micelles
  (cont.)
  spin probe study of, 542
  surface dissociation con-
    stant of, 908
  temporal behavior of triplet
    benzophenone in, 1177-1189
  thiolysis reaction in, 1130-
    1136
  water-hydrocarbon interface
    of, 393-395
Sodium 4-(1'-Heptylnonyl)
  Benzene Sulfonate (SHBS), 137
  differential scanning calori-
    metry of, 180
  fluid microstructures of mix-
    tures of, 175-189
  surface tension against
    decane, 177
Sodium Hexadecyl Sulfate
  FT-IR study of, 677
  spin probe study of, 542
Sodium Octanoate-Decanol-Water
  System, 205-216
Sodium Octanoate in Water
  ^{13}C NMR relaxation studies,
    107
Sodium Octanoate Micelles
  light scattering from,
    487-500
  NMR study of paramagnetic
    relaxation induced in,
    527-539
Sodium n-Octanoate-Water-
  Pentanol System, 755-768
Sodium p-Octyl Benzene
  Sulfonate
  amphiphile self-diffusion,
    363-365
  counterion self-diffusion,
    365-366
  micelle self-diffusion,
    362-363
  water self-diffusion, 366-368
Sodium Octylbenzene Sulfonate
  Solution
  interfacial and surface
    tension of, 1237

Sodium Octyl Sulfate
  spin probe study of, 542
  surface tension of, 1316
Sodium Oleate
  micelles, hydrodynamic radii
    of, 163
  solubility in NaCl solutions,
    159
Sodium p-(1-Pentylheptyl)
  Benzene Sulfonate
  micelles of, 405-421
  Zimm plot for, 416
Sodium Saccharin
  interaction with alkyltri-
    methylammonium bromides,
    1391-1400
Solubility of Sodium Alkyl Sulfonates, 774, 779-780
Solubilization, 170-171, 917-
  922, 923-944, 949-961, 963-
  977, 981-995, 999-1011, 1074-
  1083, 1207-1214, 1459-1460
  and microemulsions, 1506-1508
  by reversed micelles, 1459-
    1460
  energetics of micellar, 951-955
  in apolar solvents, 999-1011
  mechanisms of, 949
  nickel (II)-PADA reaction as a
    probe for, 1207-1214
  of alcohols in aqueous ionic
    surfactant systems, 949-961
  of benzene in cetylpyridinium
    chloride, 930
  of benzene in sodium dodecyl-
    sulfate, 930
  of benzene in sodium deoxy-
    cholate, 929
  of biopolymers in apolar
    solvents, 999-1011
  of cholesterol, 981-995
  of Fe(III) ions in reverse
    micelles, 1483
  of phenothiazine and its n-
    alkyl derivatives, 963-977
  of picric acid by Aerosol OT,
    1459
  of probes, 651-652
  of protein (enzyme) in reverse
    micelles, 1074-1083
Solubilization (cont.)
  of pentanol in the system
    water-sodium n-octanoate-
    pentanol, 755-768
  of TCNQ by dodecylpyridinium
    iodide, 1460
  of water penetration probes,
    599-609
  studied by photochemical and
    photophysical probes,
    585-596
  thermodynamics of, 929-935,
    949
  types of, 925
Solubilized Substances
  transport of, by
    microemulsion droplets,
    1881-1896
Solutes
  transport of, by
    microemulsion droplets,
    1881-1896
Sphere-to-Rod Transition
  factors determining the, 833-
    835
  of CPX and CTAX micelles,
    805- 822
  of ionic micelles, 825-838
Sphingomyelin
  phase diagram with Triton
    X-100, 140
Spin Label Study of Molecular
  Aggregates, 559-563
Spin Labels in Surfactant
  Micelles
  ESR study of, 541-555
Storage Modulus, 429
Surface Tension of Micelles,
  712-714
Surfactant(s)
  alkyl chain mobility, 107
  alkyl chain order, 107
  application of, in the deter-
    mination of metal ions,
    1217-1228
  association
    conformation changes, Raman
      and ^{13}C NMR studies, 517-
      525

Surfactant(s) (cont.)
    in dilute solutions, effect
        on their surface charact-
        eristics, 1313-1326
    tracer self diffusion
        studies of, 359-370
    binding of, to proteins,
        1249-1268
    interaction with polycaryl-
        amide, 1377-1389
Surfactant Assemblies
    polymerization of, 1947-1984
Surfactant-Block Model of
    Micelle Structure, 321-333
Surfactant-Water Systems, 3-21,
    23-34, 35-56, 59-67, 69-77,
    79-91, 93-103, 107-125, 129-
    141, 143-150, 153-173, 175-
    188, 193-202, 205-216

Tall Oil
    phase equilibria for, 172
Temperature Induced Micelle
    Formation, 673
Tertiary Oil Recovery
    requirement of a cosurfactant
    in, 2124
Tetracaine
    interaction with membranes,
    2145-2155
Thalliumbromoiodide Surface
    adsorption of AOT on, 1287-
    1288
Thermal Transitions of Sodium
    Alkyl Sulfonates, 773, 776-
    778
Thermally Stimulated Depolari-
    zation,
    and microemulsions, 1715
Thermodynamics of
    liquid crystals, 129
    partially miscible micelles,
    129
Thermodynamics of Micelle Form-
    ation in Aqueous Media, 745-
    753, 755-768, 771-787
Three-Phase-Triangle, 25
Tracer Self-Diffusion Studies
    of Surfactant Association,
    359-371

Trans-3-hexene
    Raman spectra of, 520
Transport of Solubilized Sub-
    stances
    by microemulsion droplets,
    1881-1896
Triton X-100
    phase diagram with phospha-
    tidylcholine, 140
    phase diagram with sphingo-
    myelin, 140

Ultrafiltration of Nonionic
    Surfactant Solutions, 1331
Ultrasonic Studies and Micellar
    Shape, 956-959
Unimolecular Reactions
    micellar effects on,
    1025-1028

Vesicle(s)
    coagulation, study by dynamic
        light scattering, 2081-2092
    coagulation kinetics of
        negatively charged,
        2047-2050
    coagulation, rate constants
        of, 2053
    colloidal stability of, 2039-
        2056
    effect of, on the rate of
        hydrolysis of NPO by HCyS,
        1135
    effect of, on the reactivity
        of nucleophiles, 1121-1136
    electrostatically stabilized,
        2042-2043
    fast dynamic phenomena in
        phospholipid, 2059-2077
    ion binding to, 591-593
    of lecithin-cholesterol, 984
    polymerization of surfactants
        in, 1963-1979
    sodium 4-(1'-heptylnonyl)
        benzenesulfonate, 181-183
    stability and phase
        equilibria of, 2055-2056
    sterically stabilized, 2043-
        2044

Vesicle(s) (cont.)
  transition to, in aqueous
    solutions of bile salt and
    lecithin 841–855
Vinyl Polymerization
  benzophenone as a photo-
    initiator of, 1184
Viologen
  photoelectron transfer from
    magnesium tetraphenyl-
    porphyrin to, 1471–1479
Virial Coefficients of Microemulsions, 1855–1856
Viscoelastic Detergent Solutions, 425–452
Viscosity Measurement of Microemulsions, 1526–1527
Vitamin A and E
  adsorption of, in Wistar rats,
    effect of ginseng saponin,
    2100

Water Distribution
  in amphiphilic aggregates,
    310–314
Water-Oil-Nonionic Surfactant-
  Electrolyte Systems, 23–24
Water Penetration
  photochemical studies and,
    585–596
  probes of micelles, table, 601
  studied by NMR and optical
    spectroscopy, 565–582
  studies, critique of, 599– 609
Wettability of Solid Surfaces
  and the effect of nonionic
    surfactants, 1329–1337

Winsor's R
  and relation to HLB, 1935
Wistar Rats
  adsorption of Vitamin A and E,
    effect of ginseng saponin,
    2100
Wood
  surface active components of,
    micelle formation, 153–173

X-ray Diffraction Study
  of lipid phases, 240–255
  of nematic phase, 62–65
X-ray Diffraction Study (cont.)
  of sodium alkyl sulfonates,
    773–774, 778–779
X-ray Small Angle Scattering
  and structure of plasma lipo-
    proteins, 259–273

Zero Surface Tension (ZST)
  Principle, 709
  aggregation number of non-
    ionic micelles from, 714–
    717
Zeta Potential of Microemulsion
  Droplets
  by electrophoretic laser
    light scattering, 1693–1706
Zinc (II) Dodecyl Sulfate
  solubilization of phenothia-
    zine in, 974
Zisman Critical Surface Tension
  of Wetting, 1295